Mark Canepa

LARGE and DANGEROUS
ROCKET SHIPS

The History of High-Power Rocketry's Ascent to the Edges of Outer Space

On the Front Cover

Adriane L. Carbine's *Triple Threat* is prepared for flight at the Black Rock Desert in Northern Nevada. The three-stage scratch-built rocket was launched several times on multiple N and M motors that propelled the vehicle up to 75,000 feet in altitude, followed by successful recovery. Here, Carbine (center, top) descends the launch tower after arming his onboard electronics. Also pictured are Scott Bowden (far right) and Alex Woolner (center, right). *Rockets Magazine* photographer Neil McGilvray is also shown (left). The high-power launch pad was owned by John Lyngdal. (Photo by the author.)

On the Back Cover

The view from near space: a snapshot from onboard video in a two-stage rocket launched by Kip Daugirdas at the Black Rock Desert on July 4, 2018. His rocket, *Workbench 2.0*, was powered by a Cesaroni N2500 in the booster and an AeroTech M685 in the sustainer. It reached an altitude of 154,068 feet and was returned to the ground unscathed. (Photo courtesy Kip Daugirdas.)

Order this book online at www.trafford.com
or email orders@trafford.com

Most Trafford titles are also available at major online book retailers.

© Copyright 2019 Mark Canepa.

All rights reserved. No part of this publication may be reproduced, stored in a retrieval system, or transmitted, in any form or by any means, electronic, mechanical, photocopying, recording, or otherwise, without the written prior permission of the author.

Print information available on the last page.

ISBN: 978-1-4907-9655-0 (sc)
ISBN: 978-1-4907-9654-3 (hc)
ISBN: 978-1-4907-9653-6 (e)

Library of Congress Control Number: 2019910503

Because of the dynamic nature of the Internet, any web addresses or links contained in this book may have changed since publication and may no longer be valid. The views expressed in this work are solely those of the author and do not necessarily reflect the views of the publisher, and the publisher hereby disclaims any responsibility for them.

Trafford rev. 10/03/2019

 www.trafford.com

North America & international
toll-free: 1 888 232 4444 (USA & Canada)
fax: 812 355 4082

Contents

Book I
Large and Dangerous Rocket Ships
(1981–1991)

Prologue ix

Chapter 1 1981 1
Chapter 2 The Basement Bomber 17
Chapter 3 "Large and Dangerous Rocket Ships" 28
Chapter 4 The Aftermath 45
Chapter 5 Enter the NAR 55
Chapter 6 The Tripoli Rocket Society 77
Chapter 7 The Rocket Builder 92
Chapter 8 The End of the Beginning 108
Chapter 9 The Move West 117
Chapter 10 A New President 136
Chapter 11 The Rocket Scientist 150
Chapter 12 Black Rock 171
Chapter 13 NAR Wars 191
Chapter 14 LDRS X and BALLS 200

Book II
The Reloadable Revolution
(1990–2001)

Chapter 15 The Reloadable Revolution 229
Chapter 16 Showtime 242
Chapter 17 Black Rock, Argonia, and *Down Right Ignorant* 250
Chapter 18 The Visionary 269
Chapter 19 A Code for High Power Rocketry 285
Chapter 20 Theodolite's Demise 302

Chapter 21	"My Own Private Hindenburg"	314
Chapter 22	1996: Orangeburg and the OuR Project	323
Chapter 23	The Rise of the Basement Bomber Part 1	332
Chapter 24	The Rise of the Basement Bomber Part 2	344
Chapter 25	Return to Colorado	356
Chapter 26	The Showman	372
Chapter 27	The Biggest LDRS Ever	386
Chapter 28	Return to Lucerne	392
Chapter 29	"The Fuse Is Lit!"	402

Book III
Here Is Your Temple
(2001–2018)

Chapter 30	Texas	429
Chapter 31	Passing the Baton	446
Chapter 32	The Rocket Challenge	453
Chapter 33	A Space Shot	467
Chapter 34	Rocketry vs. the U.S. Government Part 1	485
Chapter 35	Rocketry vs. the U.S. Government Part 2	499
Chapter 36	Geneseo	509
Chapter 37	The International Launch	525
Chapter 38	Rocketry vs. the U.S. Government Part 3	544
Chapter 39	The March of Time	557
Chapter 40	Rocketry vs. the U.S. Government Part 4	571
Chapter 41	The Saturn V	585
Chapter 42	LDRS Turns Thirty	606
Chapter 43	"Let's Punch a Hole in the Sky"	625

Epilogue 673
Acknowledgments 679
Appendix 1: LDRS Statistics 683
Appendix 2: Tripoli Rocketry Officers/Board Members
  (1985–2018) 685
Endnotes 693
Sources 735
Illustration Credits 755
Index 757

BOOK I

Large and Dangerous Rocket Ships

(1981–1991)

Prologue

On September 20, 2013, Jim Jarvis of Texas propelled a high-power rocket to an altitude of 118,632 feet above the Black Rock Desert in Northern Nevada.

Jarvis, who is neither an aerospace engineer nor an employee of NASA, built the two-stage rocket from scratch in his garage in Texas. The all-black carbon-fiber missile was powered by ammonium perchlorate composite propellant, also called APCP, the same fuel used in the outboard boosters of the space shuttle. Together, the two motors in the rocket—one in the booster and the other in the second-stage sustainer—generated nearly 3,000 pounds of thrust.

Within seconds after the 90-pound rocket cleared its homemade launch pad, which was made from sections of metal available at Lowe's or Home Depot, the missile was traveling more than 750 miles per hour. That's faster than the speed of sound. Moments later, as the rocket streaked into the stratosphere above 30,000 feet, it was traveling at Mach 2—twice the speed of sound or nearly 2,000 feet per second. That's faster than a rifle bullet.

In less than 60 seconds, Jarvis's rocket was nearly halfway to space, more than 100,000 feet high. At apogee, the highest point in the flight of a rocket, the rocket was turning over now, slowly, as the grip of the earth's gravity overcame the vehicle's forward momentum. Now if everything went as planned, onboard electronics would activate a recovery system designed by Jarvis to return his rocket safely to the ground so that it could be flown again.

His rocket vanished high in the sky above, Jarvis had nothing more to do. Now all he could do was wait. Soon, he would find out whether he still owned a rocket or just another hole in the desert.

Forty years ago, a hobby rocket flight to more than 100,000 feet was unthinkable. Higher-power rocketry was not much different from model rocketry. Rockets were small and lightweight and propelled by motors designed to lift a model rocket to an altitude of up to 1,000 feet or maybe just a little bit higher. There were no reloadable high-power motors, no onboard altimeters or sophisticated electronics, no reliable recovery equipment. There were also no advanced materials from which to build lightweight, durable airframes capable of withstanding the forces generated in modern high-power rocketry. All that was still science fiction. There were just small hobby rockets with colorful plastic parachutes.

Yet beginning in the mid-to-late 1970s, hobby rocketry began to evolve from just another branch of the toy industry into a modern tour de force of Space Age fuels, advanced electronics, and high-performance recovery gear. In the ensuing three decades, people from every walk of life, most of them far removed from the professional sciences, created an international organization devoted to the limitless pursuit of citizen-based rocketry. These flyers destroyed thousands of rockets along the way to the upper atmosphere, grappling with the immutable laws of chemistry, engineering, and physics in attempts to send their homemade projects tens of thousands of feet into the air.

To protect their freedom to fly such rockets, flyers sometimes did battle with local authorities, state legislatures, and even the federal government. Sometimes they even battled one another.

This purely civilian journey to the edge of the upper atmosphere began at a time when America's space program was in retreat. The Space Race with the Soviet Union, which ended with the landing on the moon in 1969, captivated young people all over the world, perhaps nowhere more so than in the United States. In the 1960s, hundreds of thousands of young Americans took to their local playgrounds and parks to launch model rockets. Using small black-powder motors, young people fired off their little missiles to several hundred feet or so and then watched them drift back to Earth under parachute—imaginary astronauts returning from their landing on the moon.

After the last *Apollo* mission to the moon in 1972, there was a decreasing political interest in manned space travel. Yet at almost the same time the Space Race became yesterday's news, hobbyists discovered a new and more powerful means by which they could send their homemade rockets higher than ever before. A new propulsion source known as the APCP motor found its way into hobby rocketry. By the late 1970s and beginning in the remote deserts of Southern California, fledgling commercial motor makers working out of ordinary garages began manufacturing and distributing this modern propellant to enthusiastic hobbyists all over America. Starting with a small and very modest selection of rocket motors, these manufacturers grew their product line to include motors of incredible size and power, creating in their wake an entirely new rocketry hobby, one that went far above and beyond model rocketry. This new hobby was called high-power rocketry.

In the 1990s, high-power rocketry spread throughout the United States and then the world. Today high-power vehicles can carry onboard altimeters, GPS tracking units, and enough propellant to break Mach 1 in seconds. They can be flown in configurations that include small handheld rockets that may reach 5,000 feet at a cost of $50, sophisticated midrange missiles that weigh up to 100 pounds and can clear 20,000 feet, or modern-day technological works of art that cost only a few thousands of dollars yet travel higher and faster than any aircraft in any air force in the world. This hobby is now poised to send vehicles to the very edge of space and perhaps even beyond.

This book is the story of how and why high-power rocketry came to be and where it may be headed in the near future.

1

1981

The Smoke Creek Desert in Northern Nevada is an unlikely destination for people. Yet in the late 1970s, Smoke Creek was one of the homes of amateur rocketry in the United States. The temperate climate and wide-open space for flight made Smoke Creek an ideal site for launching rockets. That most of the desert was open for public use was of no small benefit either.

At Smoke Creek, cardboard model rockets propelled by black-powder motors flew alongside small steel missiles carrying solid or even liquid fuel. Launches were organized by small rocketry clubs made up of amateur rocket scientists, students, and weekend hobbyists. Some of the people who launched rockets here were animated by secret visions of reaching space one day. Others were just thrilled by the smoke and the fire and the noise. There were few rules at these events other than the unspoken covenant to always use one's common sense.

In May 1981, more than a hundred people trekked to Smoke Creek for a two-day rocket fest over the Memorial Day weekend. Among them were high school students from Oregon conducting a science project with metal rockets flown on micrograin-type solid fuel. There were also amateur flyers from Washington, Nevada, and California, including a young man from Los Angeles who was building homemade ammonium perchlorate (or AP)–based composite rocket motors. The rocketry pioneers at Smoke Creek that weekend flew rockets of various sizes, some only a foot in length and some larger than a man. A few of these rockets appeared to explode on their launch pads, but most did not.

The typical flight at Smoke Creek that weekend was successful. However, success in those days had a different meaning than it does today. Most rockets were not meant to be recovered after flight. They were designed to arc over at apogee and plummet to the ground to impact the desert floor and be destroyed. Recovery devices, as we now know them, were not the rule at Smoke Creek in 1981. They were the exception in anything but the most elementary model rocket. Consequently, rockets were launched at an 85-degree angle from vertical position, almost like an artillery round, to ensure when they returned to Earth, the impact was safely away from flyers and spectators.

"A 5-foot rocket impacting in hard dirt sticks in 3 to 5 feet down!" exclaimed one observer.

Some rockets vanished entirely into the desert floor on their return, moving through several feet of desert soil like a bullet through water.

Among the onlookers that weekend was a twenty-seven-year-old television engineer from Ohio by the name of Chris Pearson, who was there just to watch. It was the first time he had seen one of the vast deserts of the American West, and the plane ticket from Cleveland to Reno cost him several hundred dollars. But it was worth it, he thought, as he packed for home on Sunday afternoon. He liked what he saw at Smoke Creek, especially the larger rockets with the bigger motors.

The year 1981 was pivotal for high-power rocketry. But strictly speaking, high-power rocketry did not yet exist. There were no national high-power organizations in 1981. There was no event known as LDRS, and there was no Code for High Power Rocketry. No one had heard of Airfest, the Freedom Launch, Midwest Power, Red Glare, or ROCstock; these gatherings had yet to be born. "Dual deployment" was not a part of rocketry's vocabulary. There were no motor certification programs and few, if any, commercial high-power motors. The reloadable motor and the hybrid were unknown. Hobby rocket motors were sold according to their impulse level, measured in Newton seconds, which was typically stamped on the side of a motor. Each discrete impulse level was segregated alphabetically, beginning with the letter A. As one went up the alphabet, motors grew more powerful.

In 1981, the typical hobby rocket motor was in the A through F impulse range, as it had been for many years. A G-impulse rocket motor was rare, and hobby motors larger than a G were rarer still. In the United States, there were a few private amateur rocketry clubs launching rockets with large homemade motors in places like Smoke Creek or the Mojave

Desert in California. But most weekend hobbyists were content with A through F power.

Only one truly national hobby rocketry organization existed in America, the National Association of Rocketry (NAR), and for more than two decades the NAR presided over what was known as model or nonprofessional rocketry, carefully crafting the rules and regulations governing the use of hobby rockets in the United States. In 1981, those rules forbade NAR members from launching rockets weighing more than 1 pound or from using anything larger than an F motor in their rockets. Things were about to change.

In April, a newsletter called *California Rocketry Quarterly* was created by Southern California enthusiast Jerry Irvine. Irvine's first issue, which he dedicated "to the advancement of model rocket technology," contained grainy black-and-white photographs of rockets powered by not only traditional model rocket motors but also the relatively new composite G motor. His twelve-page newssheet included crudely designed advertisements, a few model rocket launch announcements, and the first installment of a three-part series about a new kind of rocket motor technology. The series was written by a young acquaintance of Irvine who had first begun making homemade rocket motors in the mid-1970s. Entitled "Power Freaks Arise: Trends in High-Power Model Rocket Motor Design for the 1980s," the feature suggested a new type of hobby rocket motor, a composite motor, developed in the 1960s, was now poised to lead hobby rocketry into a new frontier in the 1980s. The author of the "Power Freaks" article began his report with a brief recitation of the history of the hobby: "Model rocketry had its beginnings with the development of a pre-loaded rocket motor that required no handling of chemicals by the user. Almost without exception, these motors were fueled by black-powder propellant that was hydraulically rammed into a convolutedly-wound paper casing," he wrote. These would be the common A through D range rocket motors still used today, made by companies such as Estes.

The story then moved to the 1960s and the development of a hobby motor made from "high-energy propellant with a plastic binder" and the subsequent manufacture several years later of phenolic-paper casings and machined graphite nozzles for such motors. These were among the first *composite* hobby motors, he said.

Now according to the "Power Freaks" author, composite motor makers could build rocket motors in traditional sizes of, say, A through

F power but pack more thrust into every motor size with this new type of fuel. In a bold prediction, the writer announced that in the decade ahead, the preferred propellant for rocket motors would not be the traditional black-powder motor, which had fueled model rocketry for nearly twenty-five years, but this alternative form of fuel called APCP. The future of higher-powered model rocketry would be guided by this new fuel, he said, and the scope of the hobby would be limited only by the imagination of a new breed of motor designers to come. His prediction was exactly right.

In fact, beginning in the early 1970s and continuing to the present, APCP has come to dominate hobby rocketry and give birth to not only the launch event known as LDRS but also an international high-power rocketry organization made up of thousands of members whose hobby rocket motors are now lifting rockets to the edges of outer space.

The creator of the "Power Freaks" series was an introspective twenty-six-year-old Southern California man who had recently completed his four-year enlistment with the United States Air Force. Now a civilian again, he was working as a junior engineer for a defense contractor in Los Angeles. In his spare time, he launched model rockets in the desert and dabbled in the construction of larger composite rocket motors he sold or gave away to friends and fellow hobbyists. He didn't know it yet, but he was about to become a leading voice in the embryonic world of high-power rocketry. His name was Gary Rosenfield.

William Rosenfield was the son of a successful entrepreneur and pharmacist in early twentieth-century Minnesota. William's parents tried to steer him toward pharmacy school but his interests were in art and design. He declined to follow his father's footsteps. Instead, he enrolled at the prestigious Art Center College in Los Angeles. His subsequent career as an industrial designer and later as a human factors specialist would carry William to jobs all over the country, including stints with RCA and North American Rockwell and as a contractor with the National Aeronautics and Space Administration, also known as NASA.

Not long after he had moved to Los Angeles, William was introduced to a young woman by the name of Edith Brown. Edith was born in New York City and had recently arrived in Southern California to live with her aunt and uncle. William and Edith dated, fell in love, and married. In December 1955, they had their first child, a son, who was born in Los Angeles. They named him Gary.

In the early 1960s, NASA had the attention of the entire country. Beginning with the *Mercury* program and then to *Gemini* and on through the *Apollo* missions of the early 1970s, a modern manifest destiny was unfolding in space. And with every new step made by man into the near universe, Americans momentarily stopped what they were doing to watch. In primary and secondary schools around the nation, classes were interrupted for every manned space flight. For schools that could afford such luxuries, black-and-white television sets were wheeled into classrooms so students could watch historic launches and recoveries as they happened. At some schools, student representatives were excused from class, sent home to watch a rocket launch, and then required to report on the event to the class. Nearly every manned launch in the 1960s was a national unifying event, connecting Americans in ways typically reserved for national tragedies. The space program was a source of pride for not only the nation as a whole but also its youth. If you were a child in America in the 1960s, you assumed it would never end.

Gary Rosenfield's father was a follower of anything related to the aerospace industry, and his paternal interest in science left a deep impression on Gary. The Space Race was a common theme around the dinner tables of America in the 1960s. The Rosenfield home was no exception. William Rosenfield's design work led to a job working in the *Apollo* program. Sometimes the elder Rosenfield would bring home drawings, photographs, or similar images of projects he was working on with NASA. On other occasions, he brought home strange and unusual materials like aluminized Mylar or honeycombed structures made from exotic substances. It was the high-tech stuff of its day, and Gary reveled in it. It was thrilling to imagine that these things were actually being used in the space program.

Rosenfield took to science as a child. He was a ferocious reader, devouring whatever he could find on anything that caught his attention. He spent hours at the local library after exhausting the books his parents had at home. For Gary, reading was more than just an academic exercise. Books frequently moved him to action. He read somewhere that it was possible to make cornstarch burn by blowing it into a candle. He constructed a makeshift rocket out of an ordinary paper towel tube, some cornstarch, and a shiny bottle cap. On another occasion, he came across a magazine story on lasers. He went to the library and found more articles on lasers, reading each and every one. Soon, he was working on building a laser himself. He was thirteen.

"My first rocket was an Estes *Alpha*, which I flew with an MPC A motor," he recalled. "I went down to the local elementary school to fly the *Alpha*, and it worked perfectly. The parachute came out, and it drifted away somewhere. I never saw it again. But even though I lost the rocket, I was hooked."

Rosenfield's first job was delivering the local Fullerton newspaper. Later, he landed a job with a dry-cleaning store, earning $1.65 per hour—the prevailing minimum wage. He was happy to earn that kind of money. It helped pay for his hobbies.

A "motor" is often defined simply as a device that produces *motion*. In rocketry, propellant-actuated devices that produce motion are called rocket motors. (In contrast, an "engine" is defined as a device that produces motion through the use of a combination of moving parts.)

The late 1960s were an innovative time for hobby rocket motors. For most of the hobby's then brief history, commercial motors were made primarily from a mixture of potassium nitrate, sulfur, and charcoal. This propellant is commonly referred to as "black powder." Black-powder motors are typically around a half inch in diameter and up to a few inches long. The propellant is enclosed in a wound paper-like casing that holds a clay nozzle, a delay grain/smoke charge, and an ejection charge. Black-powder motors are "end burners." They burn from one end to the other, much like a cigarette. They are single use only. They are flown once, removed from the rocket, and discarded.

In 1968, a company called Rocket Development Corporation (RDC) began marketing a new kind of motor fuel using an alternative composite propellant. RDC was founded in Utah in the early 1960s by Irving S. Wait, a rocket engineer employed by the Thiokol Chemical Corporation. Wait had worked on military projects that included the *Minuteman* and *Polaris* missiles. He subsequently moved to Indiana, where he began developing composite-based hobby motors in 1963.

Unlike a jet engine, a rocket motor does not need oxygen from the atmosphere to function. This is because a rocket motor carries along its own oxygen supply. The chemical propellants in the motor always have at least two components; one is the propellant, and the other is the *oxidizer*. The oxidizer provides the atmosphere for the fuel to burn in. The composite propellant designed for hobby rocketry in the 1960s by people like Irv Wait contained both fuel and oxidizer intermixed with a rubbery binding agent that held everything together.

The most common fuel used in high-power rocketry today is APCP. The chemical oxidizer in APCP is AP. ("AP" had been used by industrial giants such as JPL and Thiokol since the late 1940s in full-scale missiles.) The average high-power motor consists of approximately 70 percent AP. The rest of the motor is composed of supplemental metals such as aluminum for fuel, other chemicals serving as burn rate catalysts and antioxidants, and a synthetic rubber binder, typically hydroxyl terminated polybutadiene (HTPB), holding it all together. This combination of agents and binders are molded together to fit into a cylindrical motor case that is typically sealed at one end with a nozzle at the opposite end. Composite motors are not end burners. They have a hollow core that runs the length of the motor. APCP motors are harder to ignite than black powder or liquid-based fuel. This makes APCP safer to handle than traditional rocket fuel. APCP is also a more efficient propellant than black powder. That makes it an ideal fuel for amateur rocketry.

Modern commercial rocketry motors are labeled in a letter/number format, such as the H180. The *letter* in the label rates the total impulse of the motor. Generally speaking, each step up the alphabet ladder denotes a motor twice as powerful as the preceding letter. For example, an H-impulse motor is twice as powerful as a G motor, an I motor is twice as powerful as the H motor, and so on. The *number* in the motor label describes the average thrust in Newtons. A Newton is the force needed to accelerate a mass of 1 kilogram to 1 meter per second squared.

Commercial high-power rocket motors run from the letters H through O. An H180 motor can easily lift a lightweight 2-inch-diameter rocket a few thousand feet high. An N motor, on the other hand, can propel a 100-pound rocket several thousand feet high, depending on the size, weight, and aerodynamic drag of the rocket. In a lightweight minimum-diameter vehicle designed specifically for altitude, an N motor can propel a rocket 50,000 feet or even higher. That's 4 miles higher than the peak of Mount Everest.

Commercial high-power motors today come in six basic diameter sizes: 29, 38, 54, 75, 98, and 152 mm. At the low end of the spectrum, an H motor is typically 29 mm in diameter or a little more than an inch wide. An O motor, at the other end of the scale, can be as large as 152 mm or approximately 6 inches in diameter. The length of a modern high-power motor can range anywhere from several inches to 3 feet or more.

Irv Wait's efforts in creating a hobby motor based on composite fuel culminated with what is purportedly the first hobby rocket ever flown on APCP. This occurred on a freshly plowed cornfield in Southern Indiana on September 12, 1966. Wait eventually marketed his composite motors under the commercial name Enerjet. The Enerjet motor was a big step forward for model rocketry, delivering more thrust and power than almost any black-powder motor then on the market. In 1969, RDC was acquired by Centuri Engineering, owned by Lee and Betty Piester. Centuri expanded the Enerjet line of hobby motors nationwide, marketing E- and F-size composite motors while still using the Enerjet trade name.

Gary Rosenfield was fifteen when he first read about Enerjet motors in a hobby magazine. He couldn't wait to try one. But Enerjets were not legally sold in California at the time.

"Fortunately, my parents drove to Las Vegas a couple of times a year to visit relatives, and during one of those trips, we stopped by a hobby shop that was stocked with Enerjets," he said. "We purchased a few, and we flew them as soon as possible."

When they got back to California, Rosenfield and his younger brother, Bobby, loaded their contraband Enerjet E motor into a model rocket Bobby had modified from an Estes *Goblin* kit. They named their rocket *E-legal*.

"The rocket took off like something out of a cartoon," recalled Rosenfield. "Just a puff of smoke, and it was gone."

The boys were awestruck. Shortly thereafter, the Enerjet motor line was available for sale in California too. Rosenfield started launching bigger model rockets carrying E and F motors. He began to acquire a keen interest in the propellant side of the hobby.

"Enerjet gave us a glimpse of what was possible in rocket propulsion and opened up a realm of performance far beyond what was available from the black-powder technology," said Rosenfield, who joined the NAR in 1971.

The Enerjet line of model rocket motors came to a sudden end in 1974. The Centuri Company was bought by a larger corporation, and the powers that be decided to end the product. It was a disappointment for Rosenfield and other flyers like him who, as they got older, stayed in hobby rocketry because of the increased power potential of composite motors. Bigger motors meant bigger rockets and higher altitudes for those rockets. The possibility of sending their rockets higher than ever before was exciting for hobbyists. Altitude was adventure.

Unfortunately, Enerjet was now history. Without more powerful motors, hobby rockets were not going to get any bigger, and they were not going to go any higher. So Rosenfield decided he would learn how to build his own rocket motors. He started in the library.

"Back then, there were no experimental composite propellant books," he recalled. "All you could do was read about the things that Thiokol was doing and some of the exotic chemistry that they were involved with—the big stuff. You could read these rocketry books, but of course, they never revealed any trade secret information."

The Southern California teen decided to fill in the essential chemistry gaps still missing from his research with personal guesswork and estimates. Not surprisingly, his early rocket motors were dismal failures. American rocketry pioneer Robert Goddard knew all about failure. In the 1920s, Goddard referred to his rocketry shortcomings as "valuable negative information." His long string of mistakes eventually led to success. In the case of Rosenfield's early experiments with solid-fuel composite motors—said experiments often being carried out in his parents' one-car garage—the budding rocket scientist became an expert in valuable negative information.

"You get a very rudimentary understanding of what to do through books," said Rosenfield, "and then it is all trial and error."

A modern high-power rocket motor is made up of a cylindrical outer casing that holds preformed solid-fuel grains. A hollow core in each fuel grain serves as the motor's combustion chamber. An ignition source placed in the open core of the grains, usually through an electrical charge device known as an igniter, ignites the propellant. As the solid fuel in the open core begins to burn, propellant is consumed from the inside outward. The core grows larger as it burns; the resulting gases are expelled out the aft end of the motor through the nozzle. The result is an illustration of Newton's third law of motion: "For every action, there is an opposite and equal reaction." If it all works right, the thrust generated by the burning gases rushing out of the nozzle will propel a rocket into the air—or, if you have enough fuel and the right rocket, into outer space.

It took a tremendous amount of time and patience for the young Gary Rosenfield to create even a simple formula of AP, binder, and other trace materials that might burn effectively as rocket fuel. And finding a chemical formula was only the beginning. While mixing the ingredients for his solid-fuel motors, Rosenfield could not get the propellant to cure properly. He also had trouble creating a practical motor nozzle that would allow for the efficient exit of exhaust gases from the motor's burning core. Ignition of the teenager's early chemical experiments often resulted in either the sudden destruction of the motor—more valuable negative information—or, as he recalled it, a "stinky, smoky mess."

But his persistence paid off. In the fall of 1973, Rosenfield successfully fired his first homemade composite motor. It was a 29 mm phenolic-cased F using polyester-AP propellant and a graphite nozzle and bulkhead that were burned out of an expended Enerjet motor, he recalled. To obtain suitable nozzles and bulkheads, he tossed old Enerjet motors into the family barbecue until the heat generated by the fire allowed the fiberglass casings to disintegrate, disgorging their inner parts. He then reused the nozzles and bulkheads in new motors. The polyester binder for his motor, which held the internal propellant ingredients together, was made from boat resin he had purchased at a local hardware store. He found the phenolic outer motor case in an odds-and-ends bin at a wholesale plastics supply store.

"It took about three tries to get it lit," recalled Rosenfield of his first successful APCP motor. "But when it finally worked, it was one of the most exciting moments of my life."

Rosenfield graduated from high school in 1973 and enrolled at Fullerton Junior College in the broadcasting department.

"If they had a major in rocketry, I probably would have done that. But of course, they didn't," he said.

He soon discovered that broadcasting was not a good fit for him. There were too many people in the program, and he was just another face in the crowd. If he could not excel at something, he wanted out. For the next two years, he looked around for something else to do with his life.

Meanwhile, he filled his spare time with building rockets and motors, steadily gaining more experience in the design and manufacture of his own propellant. By 1975, he was constructing H-impulse composite motors from scratch. This was a powerful hobby rocket motor, capable of propelling a model rocket more than 3,000 feet into the air. As Rosenfield's passion for motor making increased, his parents were supportive yet cautious.

"I was bringing chemicals home for my experiments—one of the things I was using was a Thiokol liquid polymer, really sticky and smelly stuff—and I showed it to my dad," said Rosenfield.

"He thinks he's going to be a rocket scientist," his father told his mother.

One afternoon Rosenfield invited family, friends, and neighbors to a local playground to witness one of his haphazard fuel experiments. "I was making H and I motors, and I tested them in the public playground that adjoined our townhome in Fullerton. I attached an I-sized motor to a piece of playground equipment, and we fired that thing off. Everyone loved it. But we were standing only 15 to 30 feet away from the motor, and it was loud, and it burned for a couple of seconds."

The test was a success. But as Rosenfield packed up to return home, his father came up to him.

"I don't want you doing that here anymore," he told his son quietly.

In 1975, Rosenfield enlisted in the United States Air Force. At the time, he was working at the dry cleaners and attending college. He thought the air force would be a good way to not only serve his country but also help finance his education through the GI Bill. He signed up for four years with the military, with a six-month delayed enlistment.

Just prior to the start of his military tour, Rosenfield and a friend, John Davis, teamed up to form a small motor-making business they called Composite Dynamics. Rosenfield met Davis at a local rocket launch a couple of years before. They shared information on motor-making and composite fuels, and Davis was instrumental in continuing Rosenfield's informal education in rocketry.

"John was about sixteen years older than me and was a senior staff engineer at Hughes Aircraft in El Segundo," recalled Rosenfield. "He designed the transmitters for one of the first Venus probes, among other things. I learned a lot of good things from him, both what to do and what not to do. I also learned a wide variety of methodologies for making motors and how to source out certain materials and chemicals."

Soon, Davis and Rosenfield were manufacturing motors and selling them under the Composite Dynamics label to friends in Southern California's hobby rocketry community.

After his military basic training, Rosenfield was stationed at Nellis Air Force Base in Las Vegas, where he specialized in imagery interpretation. Las Vegas was only a few hours' drive from Los Angeles. So once a month for the next three and a half years, he traveled from Nellis to Southern California to visit his family and work with Davis on the research and development of their composite rocket motors. In late 1978, they designed a line of motors that were eventually flown at a remote rocketry hot spot in the desert near Victorville, California, not far from the intersection of State Highways 18 and 247. This was the Lucerne Dry Lake Bed. The surrounding community was called Lucerne Valley.

"We came up with a design that worked, and I talked to John, and I said, 'You know, if we make some of these motors, I can probably sell them out at Lucerne,'" said Rosenfield. "We made up about thirty or so motors—I may have put in the delays, but John did most of the work—and then I went out to Lucerne and sold them all out."

Rosenfield's service with the air force ended in 1980. He returned to Southern California, and he and Davis tried to shape Composite Dynamics into a business that might pay them a salary. During the next year or so, they sold composite E and F motors and, on occasion, G motors to whoever was interested. However, there wasn't enough money to go around, and disagreements on how to spend the money eventually led to the demise of the business.

Now out of the military, Rosenfield needed a job to support not only himself but also his hobby. He took a position as a junior engineer at the Bermite division of the Whittaker Corporation, a defense contractor with a plant in Southern California. Bermite made initiators, powders, and related military products. They also developed the *Sidewinder* missile for the United States Navy. Rosenfield was not an engineer per se; he did not have an engineering degree. Nor did he have any formal chemistry training beyond high school. Yet what he lacked in a college degree, he made up for in real-world experience and common sense, most of which he learned as a result of his rocketry hobby.

"Bermite hired me as a 'junior engineer' because of my knowledge of HTPB propellant, and nobody else there had any knowledge on it," he said. "I had acquired my knowledge working on my own and with John Davis. I was really excited."

While at Bermite, Rosenfield was introduced to Dr. Claude Merrill, the former head of the solid propellant plant at the rocket propulsion lab at Edwards Air Force Base in the Mojave Desert. Merrill had recently transitioned to the private sector, and he and the young Rosenfield hit it off immediately. Merrill was an endless fountain of knowledge regarding rocketry and chemistry, and Rosenfield soaked up the elder scientist's insights, knowledge, and practical experience like a sponge.

"We talked and carpooled together every day," recalled Rosenfield. "He loved talking about rocketry and chemistry. I obtained information on optimizing propellants, processing, everything. It couldn't have been better if I had gone to college."

Outside of his work at Bermite, Rosenfield, now twenty-six, was living in a cramped apartment in the suburbs of Los Angeles while continuing to periodically build, test, and sell composite rocket motors, flying them at Lucerne in Southern California or the Smoke Creek Desert in Nevada.

"I started going to Smoke Creek in the mid-1970s," said Rosenfield, who was present for the Memorial Day launch in 1981.

At the time, groups such as the Rocket Research Institute, with people like Chuck Piper and Ray Goodson, were launching large metal rockets there.

"They were flying big potassium perchlorate and asphalt rockets," recalled Rosenfield. "It sounded like fun to me, and I went back several times."

The journey to Smoke Creek in the 1970s was not easy. The isolated Washoe County desert location is tucked between the Buffalo and Fox Mountain Ranges, well north of Reno. Paved road stopped many miles before one reached the desert floor.

"Chuck created a how-to brochure on reaching the remote desert launch site safely," said Rosenfield. "He said, 'You better bring two spare tires, not just one, because it is like going to the moon.' He also said, 'You better bring water and be self-contained and have everything you need.' I heard about this, and now I really wanted to go."

Another problem with a Memorial Day launch at Smoke Creek was that in the late spring, there could be water present; early in the season, it can be very muddy, and small lakes can form.

"I even have a movie of a rocket out there coming in ballistic and splashing down in a pond in the middle of the dry lake bed. In the *desert*," said Rosenfield.

In early 1981, Jerry Irvine asked Rosenfield if he would be interested in writing an article on composite motor technology for the inaugural issue of Irvine's fledgling publication, *California Rocketry Quarterly*. Rosenfield and Irvine had met a few years earlier and socialized at model rocket meets in Southern California.

For Rosenfield, writing something for Irvine's newsletter was a way to memorialize the motor designs he had been studying for the past few years. Without revealing any of his early trade secrets, the article was also an opportunity for the young motor maker to reach a wider audience beyond the hobby environment of Southern California. When Irvine extended the offer to write a story, Rosenfield said he would do it, and the "Power Freaks" articles were born.

The three articles written by Gary Rosenfield that appeared in *California Rocketry* in 1981 were carefully couched in terms of model rocketry. But to even a casual observer, Rosenfield was pushing the envelope of model rocketry. Composite motor technology made it possible to create motors with vastly increased performance yet remain within the standard model rocket engine sizes. For example, wrote Rosenfield, one could develop a C-sized motor with D engine performance or a D-sized motor with E or even F-style performance and so on.

Rosenfield briefly mentioned motors in the G through I range in his articles, but he described them as "professional-type rocket motors." The implication for hobby rocketry was clear. With composites, flyers could have higher-performance motors for their model rockets well beyond the power ranges traditionally allowed by the NAR (i.e., beyond an F motor).

Rosenfield also wrote about the performance-enhancing impact of the geometry design of the core of composite motor grains, including center burners, slot burners, star grains, and the recently introduced "mystery jet" grain, which Rosenfield dubbed "moon burner." He described burn times and thrust curves of composite model rocket motors. Several times, he used the adjective "high power" to describe the model rocket technology he was talking about. The final article in the series appeared in the October 1981 issue of *California Rocketry*. It was entitled "Trends in High-Power Model Rocket Motor Design for the 1980s." This was one of the last times someone would use the term "high power" and "model rocket" in the same sentence.

On November 14, 1981, the space shuttle *Columbia* touched down on Runway 23 at Edwards Air Force Base in Southern California. This was the shuttle's second flight, the maiden flight being in April that year, and it marked the first time in the history of the space program that a manned vehicle was reused for a return flight into space. *Columbia* orbited the earth thirty-seven times during its two-day mission, and newly elected president Ronald Reagan visited mission control in Houston to wish the astronauts well during their time in flight.

More than two hundred thousand people turned out to welcome *Columbia* home when it landed at Edwards. The United States had not sent men into space since 1975, and nearly a decade had passed since *Apollo 17* had last landed a man on the moon. America had returned to space, and people were excited about the resumption of manned exploration of the universe.

"Through you, today we all feel as giants once again," Reagan said of the shuttle astronauts that year. "Once again, we feel the surge of pride that comes from knowing we are the best, and we are so because we are free."

Several miles east of the space shuttle's unpaved landing site, on a series of short granite hills known as the Luehman Ridge lay the static motor stands that were used in the 1960s to test-fire the F1 rocket engines of the vehicle that had taken America to the moon: the *Saturn V*. In its heyday during the *Apollo* program and through the early 1970s, every F1 engine was tested here, each motor consuming more than 3 tons of kerosene and liquid oxygen per second. An hour or so east of Luehman Ridge and south of Barstow was the hobby rocketry range near Lucerne.

The space shuttle's second mission was covered in the January 1982 edition of *California Rocketry*. Now nearly a year old, Irvine's primitive newsletter had grown into nearly thirty pages of model rocketry news, local launch reports, and technical articles. It also carried advertisements for Irvine's own rocketry company, U.S. Rockets, selling entry-level and mid-power model rockets and parts. A few other vendors, primarily in Southern California, ran ads for similar rocket products, including rocketry-related T-shirts and launch control systems.

The materials published in the first four issues of *California Rocketry* were focused primarily on the model rocketry community. But higher-powered rocket motors were also discussed by both Gary Rosenfield and others, and there were pictures in Irvine's newsletter of hobby rocket flights that were clearly propelled by G motors or their equivalent. Around the country at the few hobby stores that happened to get a copy of *California Rocketry*, people were starting to pay attention to these new developments in motor technology. The Roto Rocket hobby store in Pittsburgh, Pennsylvania, received copies of Irvine's publication. One of them fell into the hands of a forty-year-old ex–air force technician who was losing interest in model rockets. In a few years, he would become one of the first presidents of the yet-to-be-born Tripoli Rocketry Association.

It wasn't long before *California Rocketry* also caught the attention of the largest hobby rocket organization in the world: the NAR.

2

The Basement Bomber

By the end of the 1970s, there were two groups of organized nonprofessional rocketry flyers in America.

The first group consisted of a very small number of loosely affiliated people who belonged to private semiprofessional clubs like the Reaction Research Society (RRS) in Southern California. Members of the RRS had been making their own propellant—solid and sometimes even liquid fuel—and flying rockets at a specially constructed facility in the Mojave Desert since the late 1940s. On occasion, they would launch at other remote desert locations, like Smoke Creek in Nevada. There were no national rocketry safety codes in those days. For the most part, these flyers were governed by state and local rules regarding the use of "pyrotechnics," a fancy word for fireworks. They obtained flight waivers from the Federal Aviation Administration or the FAA when necessary and, except for a rare article or two in the press, remained off the radar of the public and government regulators. Their hobby was tiny in terms of the numbers of participants; it was sometimes referred to as "amateur rocketry."

The other group of nonprofessional rocketry enthusiasts in America was the NAR. Their realm was model rocketry.

The NAR was twenty-four ears old in 1981. Founded in the wake of *Sputnik* in 1957 by then scientist G. Harry Stine and propelled forward by the subsequent development of mass-produced black-powder rocket motors by Vern Estes, the NAR was omnipotent in hobby rocketry. By the early 1980s, there were NAR chapters in all fifty states and in dozens of countries overseas.

In the beginning, the NAR used its rocketry expertise to influence the rules and regulations governing the hobby in America. In fact, the NAR may have rescued hobby rocketry from an early demise. At or about the time of *Sputnik*, a groundswell of interest in hobby rocketry swept America. This interest was manifested in local unaffiliated rocketry clubs that sprung up all over the country. Such clubs were filled with teenagers building rockets both in and out of their high school science classrooms. This surge in rocketry-related activities was widely covered in the national press by magazines such as *Popular Mechanics*, *American Modeler*, *Mechanix Illustrated*, and *Scientific American*. In one such article, young rocket scientists were portrayed as dreamers with a predilection for smoke and fire and noise.

"If, during a weekend drive in the country, you should happen to see a thin trail of white smoke shoot 100 feet in the air, you will find an unusual group of scientific amateurs near the bottom of it," read the article. "They will be equally proficient in handling explosive chemicals, differential equations, machine tools, and irate policemen."

In 1957, *Life Magazine* ran the first of several rocketry-related stories appearing in its pages over the next few years. Entitled "From Coast to Coast, U.S. Youth Gets Its Rockets Up in the Air," *Life* wrote that "U.S. youngsters seemed to be firing off rockets all over the place." Hobby rocketry was a boon to educators interested in teaching science to young people.

"Backyard rocketry has bloomed everywhere in the United States," echoed *Popular Mechanics* in 1958. "Garages have become rocket-building shops, science classrooms have taken on a space laboratory air, and homemade missiles are shooting skyward from vacant lots."

Amateur rocketry has a variety of attractions beyond the spectacular but short-lived flight of the rocket, observed a professional rocketry society at the time, including not only mathematics, physics, and chemistry but also amateur photography, expertise in miniaturization, and the teamwork and camaraderie necessary to build and launch a rocket into the air.

Most of the press coverage of rocketry in the 1950s was favorable. But there was a darker side to the new hobby too. Young people were not only making their own rockets. They were also making their own rocket fuel. And without any type of motor-making training, safety controls, or national organization to help lead them, they were being injured and sometimes killed in fuel-related accidents.

In Texas, a homemade rocket built out of water pipe and stuffed with match heads for fuel blew up, seriously injuring its fourteen-year-old builder. In Colorado, a teenage boy mixing potassium chloride and other materials for a rocket motor lost part of his hand when his concoction exploded. In another mishap, a science teacher was killed and several students were injured when test-firing a rocket on a school sidewalk. In New England, two teens suffered nearly fatal burns when their "dead rocket" suddenly reignited. And in California, a boy was killed and a friend seriously injured in a rocketry accident. Similar stories were reported in Utah, Rhode Island, Minnesota, and other states.

In a policy study on the hazards of amateur rocketry published in 1959, the American Rocket Society, a private organization of more than twenty thousand engineers and scientists, outlined their view of the hazards of nonprofessional rocket and motor-making activity. The society estimated that there were "at least ten thousand youngsters in the United States within the age group from ten to twenty actively engaged in mixing propellants and fabricating and flight-testing small rockets." Another 250,000 people participated in rocketry as "passive participants," they reported. The society claimed that in less than two months in the spring of 1958, there were more than eighty significant injuries associated with amateur rockets, including four fatalities. The cause of these accidents varied, reported the society, but they were often the result of foolish experiments, reckless behavior, or simple carelessness.

"The rocket didn't go off, so he hit it with a hatchet. Then it did!"
"He was just packing the fuel down with his hammer."
"After loading the rocket, he soldered on the fuse."
"He held it while a friend lit it."
"But I saw another guy do it."

As stories of serious rocketry accidents around the country multiplied, a derogatory term emerged in the press and some legislative circles. The term was later used as part of a lobbying effort to ban hobby rocketry altogether. The term was "basement bomber," and it was used to describe someone who not only built their own rockets but also manufactured their own propellant. The moniker was accepted by early proponents of model rocketry too. One influential model rocketry leader described the basement bomber as "a person who follows no safety rules, who knows very little about what he is doing (but thinks that he does), and who possesses a one-way ticket to suicide by working with highly dangerous chemicals."

According to contemporary estimates, during the early days of model rocketry, fireworks were illegal in at least thirty-four states. As an increasing number of teens began making their own rocket fuel, some states began to limit any type of firework or explosive device, including rockets. By the early 1960s, a few legislative bodies were considering a ban of hobby rocketry. At least one state, Connecticut, purportedly outlawed the hobby altogether.

"The status of model rocketry under the law has often been a questionable one," wrote NAR leader G. Harry Stine in the early days of the hobby. "Our hobby has been variously labeled as fireworks, handling and discharging explosives, public nuisance, which covers a multitude of sins, disturbance of the peace, a hazard to aircraft in flight, dangerous to persons and property, and 'dangerous killer.'"

Following their review of injuries associated with the increased popularity of rocketry, the American Rocket Society agreed with the hard-liners, recommending the elimination of hobby rocketry altogether.

"All practical means must be taken to prevent the manufacture of propellants or rockets by amateurs," urged the society, which called for the prohibition of the launching of all amateur rockets.

The society weighed the risks and benefits of amateur rocketry and decided that one outweighed the other. To help curb the manufacture of homemade motors, the society suggested, among other things, restrictions on the sale of fuel-making ingredients to the public. (Ironically, one of the members of the society was famed rocketry pioneer Werner von Braun, who may have recalled the policy in Germany of the early 1930s to stamp out amateur work in rocketry altogether, giving the military a monopoly on the matter.)

At this delicate juncture in the history of hobby rocketry, the NAR stepped in to separate amateur rocketry from its own *model* rocketry. The NAR lobbied for a set of uniform standards that would not only bring safety to the hobby but also cool the heels of those who wanted to eliminate amateur rocketry altogether.

"As model rocketry grew during its first five years," noted Stine, "the expanding availability of model rocket engines, kits, parts, and accessories by mail order and through an increasing number of hobby shops practically made the 'load your own' amateur rocketeers obsolete and effectively eliminated the basement bomber from the American scene."

"In 1962, General Counsel Richard Shaffer of the NAR drew up the basis for a permissive, educational state statute on model rocketry which could be used by NAR members and model rocketeers in various states who wished to take some sort of action through state representatives to get a good law on the books where needed," continued Stine, who reported that on June 19, 1963, Connecticut became the first state to adopt permissive and educational legislation recognizing model rocketry "as an aeronautical sport, as a hobby, and as an educational tool for the Space Age."

The NAR's influence was also visible in the eventual development of a safety code adopted by the National Fire Protection Association, also known as the NFPA. The NFPA is the primary standards-making organization for public safety in the United States. It is a nonprofit, private organization made up of numerous boards and committees that develop public safety standards dealing primarily with fire safety and related public hazards.

The NFPA codifies standards for scores of activities, ranging from the maintenance of fire sprinkler systems to the installation of carbon monoxide detectors and firefighting services at airports and other public buildings. NFPA standards are created by professionals in a given field who are invited to sit on specialized subcommittees within the organization. These subcommittees then study a particular issue—the use of fire sprinklers in buildings, for example—and ultimately decide what method is safest to carry out particular acts. After meetings and studies and questions and analysis, the subcommittees make recommendations to the NFPA body as a whole. The NFPA then votes to accept or reject the committee's recommendations. If a proposal is accepted, a formal NFPA rule is created.

The NFPA is an unelected body. It is neither a legislature nor appointed by a legislature. The rules and codes adopted by the NFPA have no force of law. Yet because of the stature of its members and the scientific approach of its analysis, the NFPA has become the standards-making organization in America. When its rules or codes are adopted through the local legislative process by states or cities, as they often are, the NFPA regulations take on the force and effect of law. This method of nationwide incorporation of uniform rules eliminates the need for organizations such as the NAR from having to lobby state by state to establish recognized safety codes for their activities.

Throughout the 1960s, NAR leaders worked with the NFPA's Committee on Pyrotechnics to develop a safety code for model rocketry. Influential NAR members obtained seats on the committee and were able to directly influence the committee's recommendations and rules for the hobby. Some of these lobbying efforts required the NAR to convince the NFPA that model rocketry was a safe hobby for adults and children alike and not the province of the risky basement bombers of the 1950s. As a result of the NAR's work, a rocketry safety code was formally announced by the NFPA in 1968. It was known as NFPA Code 41L or the Code for Model Rocketry. It was renumbered in 1976 to NFPA Code 1122, the same number used today. This is the safety code used by the NAR and by which model rocketry has now been governed, subject to periodic amendments, for more than forty years.

NFPA Code 1122 was adopted in some form by most but not all of the fifty states. Once adopted, it guaranteed hobbyists a right, of sorts, to pursue model rocketry. It specifically exempted the hobby from other potentially more restrictive safety code sections dealing with pyrotechnics, fireworks, or explosives. The code's express purpose was set forth in the introduction to Section 1122, drafted by the NFPA. This purpose included, among other things, the prohibition of the "making and launching of homemade rocket bombs" to eliminate the injury and death that resulted when ordinary citizens experimented with homemade rocket fuel.

The Code for Model Rocketry was a significant achievement for the NAR, with obvious benefits for tens of thousands of its members. The code legitimized model rocketry throughout the United States, allowing the hobby to grow within and not outside of the law. As a result of the NAR's efforts and lobbying success, legislative attempts to eliminate hobby rocketry ended. Equally important, serious injuries in hobby rocketry virtually disappeared.

Yet there was a trade-off that accompanied the NAR's important success. The code discouraged the construction of homemade rocket motors. In so doing, it stigmatized the homemade manufacture of rocket fuel. Private motor makers remained basement bombers. They would find no sanctuary in hobby rocketry.

Of course, at the time, a negative view of homemade motors was probably not a bad idea. For the first thirty years of model rocketry, there were few safe and reliable means of making one's own rocket motors. These were the days before modern motor-building seminars, well-written how-to manuals, and the exchange of reliable motor formulas over the Internet or elsewhere. Until the 1990s, formulas for hobby rocketry propellant were closely kept secrets. There were no down-to-earth books on the safe and sane formulation of homemade AP composite motors. There was also no collective experience in research or experimental motors. People without any real training or guidance often learned the hard way, sometimes with disastrous results.

The Code for Model Rocketry also imposed limits on motor size and rocket weight. As a practical matter, the code allowed for the use of only commercially made model rocket motors in impulse sizes A through F. This limitation was practical as there were few, if any, commercially made motors above this impulse range. Model rockets were also limited to no more than 1 pound in weight. If a person flew outside of these power and weight limits, they were beyond the blanket legal protections of NFPA 1122 and the NAR. And in states where NFPA 1122 was adopted as law, such acts might be illegal.

Still, the NAR did not govern all rocketry activity in America. The code applied only to rockets and rocket motors "produced commercially for sale" to the public. Thus, although the code discouraged homemade motors, the language did not make homemade motors illegal per se. There was nothing in the code that suddenly rendered amateur rocketry of the type practiced by groups like the RRS and others illegal. Such clubs continued to function on their own, making their own fuel for private use, launching their rockets in remote locations, and generally being held to the state pyrotechnic laws of their jurisdiction (if any) and, when applicable, the federal airspace rules regarding altitude waivers set by the FAA.

In many ways, the NAR's Code for Model Rocketry boiled down to a simple concept: "This is what we define as model rocketry, and we deem it to be safe when you follow these rules. If you join us, obey these rules. Outside of these rules, you are on your own." The NAR adopted the code as the guiding principle for its members.

The Code for Model Rocketry was the centerpiece of several successful legislative efforts undertaken by the NAR over a period of many years that helped legitimize and protect model rocketry. Other efforts by the NAR included lobbying work with the FAA, the United States Postal Service, the Department of Transportation (DOT), and the Consumer Product Safety Commission (CPSC).

Between 1960 and 1980, tens of thousands of adults and children joined the NAR. Local chapters hosted launches around the nation and disseminated rocketry-related scientific information via magazines, newsletters, and books. The NAR procured liability insurance for its members engaged in model rocketry-related events in a policy amount up to $1,000,000. Insurance was vital to the NAR's growth and the growth of model rocketry. It allowed enthusiasts the ability to provide private property owners of prospective launch sites with a measure of liability protection for NAR-sponsored events. Over time, the NAR earned a reputation as a safe and responsible hobby organization, far removed from the rocketry-related accidents of the late 1950s.

But by the early 1980s and like any large well-established institution, the NAR had grown cautious and resistant to change. In August 1981, barely two months after the Memorial Day weekend launch at Smoke Creek, the NAR hosted its twenty-third annual national gathering, the NAR Annual Meet, called NARAM, near Allentown, Pennsylvania. This multiday event was filled with manufacturer's sales promotions, informational seminars, and launches dedicated to model rocketry. The NAR was and still is a contest-oriented organization. NARAM was host to a series of such contests but only for rockets carrying motors A through F power. Contests that year awarded prizes for the person whose B- or D-powered model achieved the highest altitudes or for those who could keep their rocket hanging in the air longer on a parachute than anyone else. By most accounts, NARAM-23 was a successful event for the NAR.

However, the NAR as a national organization was struggling. Despite more than two decades of success, membership had dropped precipitously low. Although tens of thousands of people enjoyed model rocketry, only a fraction of them joined the NAR. Renewal rates were also low, especially for young people, many of whom left the hobby as they became adults. NAR membership had peaked in the late 1960s as America's race for the moon sparked interest in rocketry of all kinds. In 1970, there were more than 5,500 members. America's subsequent abandonment of the *Apollo* program, among other things, seemed to lead to a commensurate decline in interest for hobby rocketry. Membership dropped to 1,400 by the end of the decade. And in 1979, the NAR was in full crisis mode, almost running out of funds entirely.

In the May 1981 issue of the NAR's monthly journal, *Model Rocketeer*, NAR president J. Patrick Miller announced that the NAR's board of trustees would pursue a new course of action designed to not only attract new members but also help retain current ones. At the time, the NAR's membership had climbed a bit to 2,300 paid members. Miller reported that the age of the typical NAR member was slipping, from an average age of eighteen in 1978 to almost fourteen by 1981. Less than one-third of the membership was electing to renew their membership, said Miller, although he believed that members over the age of twenty-one, also known as senior members, were renewing at a rate closer to 70 percent.

Miller outlined a series of steps to be undertaken over the next year to counter these trends and to increase the NAR's membership. These actions included reduced membership fees, incentives for membership renewal, and increased marketing of the NAR, with the distribution of the *Model Rocketeer* to hobby shops around the nation through which, it was believed, more adult members could be recruited to model rocketry. These and other changes, predicted Miller, could increase the NAR's membership to 2,700 by 1982 and up to 3,800 not long thereafter.

One area of potential expansion for the NAR in the early 1980s was the emerging higher-power rocketry market, comprising rockets that were launched with commercial motors larger than an F engine. However, the NAR was not interested in high power. Although some NAR members had been experimenting for several years with G motors (and higher) at places like Smoke Creek or Lucerne, the NAR as an organization had little interest in taking any action that might require the alteration or amendment of its safety code. One reason for this resistance was the fear that any modification of the rules might lead to a return of the basement bomber days of the 1950s. The NAR's leadership was determined not to allow any changes that might increase the chance of an accident or mishap that could smear the NAR's well-established safety record. And some NAR leaders viewed emerging trends in new motor technology as a potential threat to that safety record. Others thought that the newer large composite motor was a fad that would soon pass.

Ultimately, it would fall upon President Miller and the NAR's board of trustees to ensure that its members refrained from using any motor, composite or otherwise, larger than an F motor or face the consequences from the NAR.

Jerry Irvine sent the NAR president a copy of his first issue of *California Rocketry* in April 1981. Pat Miller replied personally in writing to Irvine, who was a NAR member, with words of encouragement.

"Thank you for the complimentary copy of *California Rocketry*," wrote Miller. "The newsletter is well written, and the graphics are attractively done. I wish you success in this publication." Miller's praise was short-lived. Near the end of the year, Irvine asked the NAR to carry a paid advertisement for *California Rocketry* in the *Model Rocketeer*. The NAR refused, and in a letter written by Miller to Irvine dated December 30, 1981, the NAR president scolded Irvine for using a supposed model rocketry publication, *California Rocketry*, to depict flyers using G motors or their equivalent. Miller observed that recent issues of Irvine's publication contained advertisements for model-type rockets that could be flown on motors that exceeded the limits imposed by the NAR Safety Code. Until these types of ads or stories were removed, Miller told Irvine, *California Rocketry* would not be allowed to advertise in the *Model Rocketeer*.

Whether President Miller's views regarding G motors reflected the opinions of the majority of NAR's active membership at that time is unknown. However, Miller certainly had the backing of the NAR's leadership. Six weeks after the letter to Irvine and at their winter meeting in Houston, the NAR board of trustees issued a new policy statement, leaving no doubt as to the board's view of higher-powered rocket motors. They were banned. The NAR refused to certify any motor with more than 80 Newton seconds of total impulse—in other words, G motors or higher.

As a partial basis for the decision, the board echoed Miller's statement to Irvine that motors in this power class were illegal for model rockets in most states because of the provisions of NFPA 1122. And in an announcement of the new policy, the *Model Rocketeer* observed that "[w]hile the NFPA code is ultimately driven by safety recommendations of the NAR, the board has a low level of confidence that most modelers had the experience, equipment, and launch facilities to utilize G engines safely." In other words, the NAR could try to modify or amend NFPA 1122, but it was not interested.

"Self-limitation for safety considerations is the foundation of this association," said Miller after the board meeting in Houston.

"[T]he technology has outstripped the ability of modelers to handle these engines safely," said another board member in the *Model Rocketeer*, who cited no examples of unsafe operation but revealed an obvious lack of confidence in the abilities of the membership as a whole.

Commercial model rocketry manufacturers were, not surprisingly, in apparent agreement with the NAR's limitations on motor size. In a letter written in 1982, an Estes product control manager repeated the NAR's emerging views on G motors.

"The expansion of this hobby through the legalization of more powerful motors may well account for the first fatal accident in the hobby." When that happens, wrote the author, "government agencies will descend on this hobby like a house on fire."

It was the basement bomber threat again, returned from its long slumber. And by 1982, Estes, like other model rocket manufacturers, had a successful multimillion-dollar industry to protect. At the time, there was little, if any, direct inquiry or polling by the NAR board of its membership regarding the possibility of future expansion into G-impulse motors. As far as the NAR was concerned, the issue regarding higher-powered model rocket motors was seemingly closed.

In early 1982, the NAR added some teeth to its policy. The NAR's safety code was revised to include the following oath: "I will only use preloaded, factory-made, NAR safety–certified model rocket engines in the manner recommended by the manufacturer."

According to the *Model Rocketeer*, this new affirmation "not only serves to publish current board policy in a format highly visible to all modelers but [also] allows for the suspension of NAR membership of violators of this new provision in the safety code."

In other words, fly a G motor, and you may lose your membership in the NAR. The NAR board believed that by clarifying its position on G motors and basing that decision on safety, a legitimate NAR goal, they were not only keeping model rocketry safe but also getting the NAR's house in order for their anticipated future expansion.

Unfortunately, many in the ranks were already heading out the door.

3

"Large and Dangerous Rocket Ships"

In 1982, the economic malaise that had burdened the United States for five years began to dissipate. That fall, the Dow Jones Industrial Average climbed to more than 1,065 points—the highest yet in its eighty-six-year history. Interest rates for home loans, which entered the 1980s as high as 16 percent, were finally dropping, a signal that the country might be heading in the right direction again. And after an absence of more than seventy years, Halley's Comet was spotted by telescope at the outer edge of the solar system on its way to another near rendezvous with the earth.

Retired baseball superstar Willie Mays had a personal telephone in the console of his sports car in 1982. For most Americans, the only phone they owned was still at home. The cellular telephone, the automatic teller machine, and the weather channel had yet to appear. The personal computer was still a toy, and it would be another year before the compact disk player entered the commercial market in the United States. Music stores still sold records in 1982. Few people had heard of debit cards. Unlike today, children did not receive sports trophies for showing up; grown men carried cash in their wallets.

In the small world of hobby rocketry, model rockets propelled by Estes motors were still king. High-power rocketry, if it existed at all, involved single-use G motors or perhaps an H or I motor if you had the right connections. There were no reloadable motors, and viable commercial altimeters were still nearly a decade away. Flyers gauged altitude by optical tracking. It was an established science and was better than nothing. But if a rocket climbed above a few thousand feet high, tracking was subject to the personal skills of the observer and varying interpretation. A partly cloudy day might render tracking useless.

In 1982, Argonia, Orangeburg, and Hartsel were just names on a map unassociated with rocketry. There were no high-power rockets at the Black Rock Desert, which was being used by the military as a gunnery range where fighter and bomber pilots tested their low-altitude flying and shooting skills. There was no Burning Man festival at Black Rock either.

In Ohio that spring, a young woman named Deborah Schultz needed a business license to purchase needlepoint supplies at wholesale prices. Trying to come up with a name for her business, she thought the title "Lots of Crafts" sounded good. She placed that moniker on her license. One day the company bearing that name would dominate the commercial high-power rocket–making industry.

In Colorado, a thirty-seven-year-old electronics wiz by the name of Tommy Billings founded a company that would transform him into a millionaire. At the time, he was not involved in rocketry at all. But his compact electronic products would one day revolutionize high-power rocketry.

In Northern California that fall, Gary Rosenfield, who had left Bermite and moved to Sacramento to take a job with another defense contractor, started a part-time business in his garage making composite rocket motors. He also rented out a small blockhouse-type building on the industrial side of town. Trying to come up with a name for his side business, Rosenfield decided to combine the words "aerospace" and "technology."

He called his business AeroTech.

In April 1982, an announcement appeared in the calendar section of the NAR's official monthly publication, the *Model Rocketeer*:

> **LDRS-1** Sport Launch, 24–25 July 1982, Medina OH (SNOAR). Three unofficial "events," prizes to be awarded (no national contest points). Contact: Chris Johnson, 26481 Shirley, Euclid, OH 44132; (216) 731-3898.

The notice was squeezed in with twenty-two other launch ads and was easy to miss. The print was tiny, and it was third in line from the bottom of the list. It was carried on the very last page of the magazine. The promotion was created by twenty-eight-year-old Chris Pearson of Ohio, who had been a member of the NAR for fifteen years. Pearson was one of a small but growing number of model rocketry enthusiasts who were moving away from model rocketry and into the heady world of G- and H-sized high-power rocket motors. His local rocketry club was the Suburban Northern Ohio Association of Rocketry, also known as SNOAR.

Over the Memorial Day weekend in 1981, Pearson traveled from the Buckeye state to Smoke Creek, where he witnessed his first amateur rocket launch. Some of these rockets were powered by exotic motors containing zinc-sulfur, APCP, or similar compounds. Pearson was impressed by what he saw. It wasn't organized high-power rocketry yet, but it was exciting.

Out on the rocketry range at Smoke Creek that weekend, Pearson overheard one of the local flyers quip, "We're now going to fly some large and dangerous rocket ships!"

"Large and dangerous rocket ships." Pearson liked the sound of that phrase. He used that phrase and its acronym, LDRS, when he placed the announcement for his Ohio rocket launch the following year. Pearson and fellow SNOAR member Chris Johnson ran similar ads in their local club's newsletter too. There was no way Pearson could know at the time, but the launch he had dubbed LDRS would one day become the largest annual amateur rocketry event in the world. And LDRS would pave the way for not only high-power rocketry but also the modern Tripoli Rocketry Association.

Chris Pearson was born an only child to Thomas and Florence Pearson in Cleveland, Ohio, in 1953. His father was a Swedish merchant marine who later became a painter and sheet rock finisher. His mother was a homemaker. Thomas Pearson was from a long line of seafaring mariners, and he immigrated to the United States from Sweden just prior to the Second World War. He met Florence, who was originally from Erie, Pennsylvania, at a dance in Detroit. Florence was the second youngest of eleven children. She and Thomas were married in 1942, and Florence lived briefly in Harlem in New York City, while Thomas served with the United States Army Air Corps overseas.

After the war, the family remained on the Eastern Seaboard. But a few weeks after Chris was born, they relocated to Northern California, where Thomas took a job in the construction industry in the San Francisco Bay Area. Florence obtained a certificate to work as a dietician, and she found a job with a local school district. In 1959, when Pearson was six years old, his father died from cancer. Florence and her son returned to Cleveland in 1961. His mother went to work as a cashier at a local grocery store and raised Chris on her own.

Like most teenage boys of his day, Pearson's role models were men like Alan Shepard, the first American in space, or John Glenn, the first American in orbit. Aviation pioneers later described by author Tom Wolfe as having the "right stuff."

As a child, Pearson recalled his teachers wheeling television sets into class so the students could watch the *Mercury*, *Gemini*, and *Apollo* launches live. One day Pearson came across a model rocketry catalog at a local hobby store. He was fascinated by what he had found inside. Soon afterward, he joined not only his junior high school rocketry club but also the NAR. In 1969, he became president of his local rocket club. That year, another one of Pearson's heroes, Neil Armstrong, became the first human being to walk on the surface of the moon.

Pearson was an active member of the NAR for the next decade, with some interruptions as he completed high school and college. He attended local and regional model rocketry events, participated in NAR-sponsored contests and competitions, and, in 1978, became editor of one of the more influential NAR club newsletters in the country: the *SNOAR News*.

Pearson's experience with higher-powered rockets began as early as 1976, when he started launching vehicles with clusters of motors that exceeded the total Newton seconds allowed for model rockets by the NAR Safety Code—in other words, more power or impulse than an F motor. He flew clusters of D motors and then moved to clusters of E and F engines too. In 1978, he began flying homemade, uncertified composite motors. At NARAM 22 in 1980, Pearson and a friend surreptitiously purchased several "illegal" H and I composite motors from an early commercial manufacturer. They eventually launched the motors elsewhere, a clear violation of the NAR Safety Code, and were subsequently chastised for the flights by the NAR leadership.

In the spring of 1981, Pearson made plans to attend a rocket launch at the Smoke Creek Desert. Pearson had not been west of the Rocky Mountains since he was six, and he had never been to an

American desert. At the time, he was working as a television engineer at a community college, and he could not afford much time off work. Driving to Northern Nevada, 1,900 miles away, was out of the question. So he bought an airline ticket to Reno that cost almost $600. When his plane landed in Nevada, Pearson was greeted by two acquaintances whom he had corresponded with through rocketry—Gary Rosenfield and his wife, Melodi. The trio then picked up another flyer, Matt Otta, and headed off to the high desert with a van full of small rockets.

Pearson was impressed by the vastness of the Nevada desert, although he quickly discovered the launch facilities at Smoke Creek were quite primitive.

"You walked out away from the cars, stuck a rod in the desert floor, and ignited the motors with fuse and a match," he recalls.

Another flyer described Smoke Creek as just "the rattiest place—the roads were so rough that you would just smash your car to pieces." In those days, the pavement ended just a few miles from Gerlach, the closest settlement resembling a town. And the launch site was so far away from Gerlach that everyone just camped at Smoke Creek. It was impractical to commute.

Yet in many ways, Smoke Creek was perfect for rockets. There was nothing else there. That weekend, camped out beneath the stars in the high desert air, Pearson experienced the true breadth of the Milky Way Galaxy for the first time in his life.

"The launch was sponsored by the Reaction Research Institute and was primarily for the zinc-sulfur crowd," Pearson said later. "But they allowed the launching of model rockets along with a lot of professional fireworks. While I was there, I heard flyer Roger Johnson talking about the 'large and dangerous rocket ships' that they were going to fly."

Pearson saw an occasional H or I motor, but most of the rockets were 4-inch-diameter vehicles or smaller and were powered by clusters of either F or G motors. He was disappointed he had not seen more exotic projects. Still, Pearson had a good time. When he returned to Cleveland, he began organizing a high-power rocket launch in Ohio for the following summer. He wanted a catchy name for his event. He liked the phrase he had heard at Smoke Creek—"large and dangerous rocket ships"—so he decided to call his endeavor LDRS.

Pearson's LDRS-1 launch ad appeared in the April and May 1982 issues of the *Model Rocketeer*. In May, the announcement was not only on the last page but also the very last ad. Coincidentally, this was the same issue of the *Model Rocketeer* that had announced the NAR's official decision to ban G motors. Pearson wondered if anyone would see the tiny ad at the back of the *Model Rocketeer*. At least one person did.

"Late one night, a couple of months before the launch, I received a nasty phone call from the editor of the magazine questioning me about the true nature of the launch," said Pearson. "I told him that there had been an interest in flying rockets bigger than the NAR Safety Code allowed, so we went ahead and got an FAA waiver, which allowed us to fly amateur rockets. That was actually stretching the truth a lot. We had always intended to fly large rockets, but I was trying to spin it a bit."

The editor was not pleased.

"He was very upset. I don't recall what he said exactly," said Pearson.

But he did say that the NAR was going to run a counter ad in the next issue.

"Fine," replied Pearson, "go ahead."

The following month, the tiny bulletin for LDRS-1, an unknown and wholly inconsequential event, was removed from the *Model Rocketeer*'s calendar page. In its place, there was a public renouncement:

> **LDRS-1,** previously appearing in this space, has been determined to include intentional amateur activity not announced in the original notice sent to the *Model Rocketeer*. **NAR members are urged not to participate in LDRS-1.** (Emphasis and bold in the original.)

It was a rocky beginning, and it might have been a quick end to what is now the largest annual amateur rocketry event in the world.

The first LDRS took place in midsummer in 1982. At the time, the science fiction films *Tron* and *Blade Runner* were in their first run at theatres around the country. *Star Trek II: The Wrath of Khan* was newly released; its young director and seasoned cast hoped that the sequel might surpass the performance of the first *Star Trek* movie in 1979. In June, the space shuttle *Columbia* launched from Cape Kennedy on its fourth flight. Ronald Reagan, in an address to the British House of

Commons, predicted that the Soviet Union's system of government was destined for "the ash heap of history." On July 24, the song "Eye of the Tiger" by Survivor reached number one on *Billboard Magazine*'s hit list.

That summer, the NAR continued its attempts to dissuade members from attending LDRS. The June notice urging members to stay away from the Ohio launch was repeated again in the July issue of the *Model Rocketeer*. In his opinion column entitled "The President's Corner," Pat Miller did not mention LDRS. But he devoted his message to the NAR Safety Code and the new NAR oath established in Houston that year. He also stressed the importance of avoiding any motor greater than 80 Newton seconds of total impulse.

"Should you feel that you are unable to comply with the NAR Safety Code in all your nonprofessional rocketry activity, the board asks that you resign your NAR membership," he said. If you did not resign, you could be suspended, he added.

There is little written evidence the NAR leadership was actively searching for members who were violating the safety code. But for those who did so and flaunted it, the NAR ultimately acted. Indeed, in mid-1982, a couple of NAR members were disciplined for using illegal motors (including gasoline and fireworks) at a Southern California launch the previous December. One of them was suspended. Other NAR members interested in higher-powered motors were either resigning or, more commonly, simply walking away from the NAR without renewing their membership. One of the members who walked away was Chris Pearson, after having been a member of the NAR for fourteen years.

"When other NAR members asked me the reason, I explained that it was because I wanted to fly rockets that would exceed the NAR's limits, and I didn't want to cause problems by doing so," said Pearson. "I was later told by a NAR official that this was probably the best way to have done it rather than openly flying high power and daring the NAR to do something about it, as some people did."

Mike Wagner was a member of SNOAR. His family owned a farm on Spieth Road on the outskirts of Medina. As the crow flies, the shores of Lake Erie and the city of Cleveland are barely 20 miles north. The entire area is dotted by small family farms. Today visitors to Medina can find the abandoned Lester train depot, another relic of America's past, about a mile away from the Wagner property.

Prior to 1982, SNOAR launched rockets on the Wagner farm on a regular basis. It was a wide-open space, ideal for model rocketry motors, A through F power. At the eastern edge of the farm, there were several small storage buildings and the main house. The home faced west, across more than 20 acres of flat and nearly unobstructed land. In the middle of the flatness in 1982 stood—and, to this day, still stands—a lone tree about 35 feet tall.

Chris Pearson thought that the Wagner farm would be a good place for LDRS-1. In late 1981, he approached Wagner and asked if SNOAR could use his family's property for a high-power event the following summer. Wagner said yes.

The Wagner parcel still includes a grass airstrip that runs parallel to Spieth Road, not uncommon for rural farms in Ohio. Unlike the remote launch locations of the desert west, such as the Lucerne Valley in Southern California or Smoke Creek in Nevada, the land in Medina is lush, green, and civilized. There are houses nearby and plenty of trees only a short distance from the launch site.

Now that he had his rocketry range, Pearson needed to gather launch pads, equipment, and a source of power on the field. For launch pads, he turned to his club. He procured two model rocket racks, with six pads each, which were used by SNOAR. To these twelve pads, he added one more pad: a single high-power pad with interchangeable steel rods of a quarter inch, three eighths of an inch, and a half inch in diameter.

"The lone pad was for the big stuff," said Pearson, "anything that couldn't fly off the model rocket racks."

In 1982, "big stuff" meant G-, H-, and I-powered rockets or clusters of multiple D through F motors. A 12-volt car battery provided all the electrical power that was needed for LDRS-1.

In his years with the NAR, Pearson knew model rocketry enjoyed a blanket exemption from the FAA for launching model rockets. Generally, FAA-related regulations at that time provided a waiver into controlled airspace was not necessary for model rockets as they weighed, by definition, less than 1 pound and carried less than 4 ounces of propellant (i.e., were smaller than a G).

Larger rockets and bigger motors, however, required special permission from the FAA. They required a written waiver. The waiver would result in the FAA issuing a NOTAM, an acronym for "notice to airmen," which makes aircraft in the vicinity aware of an event that might result in objects being hurled into controlled airspace. FAA waivers, when granted, are typically granted to a certain altitude. This means that rocket launches can go up to that altitude but no higher, with the added proviso that if any aircraft accidentally stray into the launch area, all rocketry activity must stop until the craft passes. Out west, professional and semiprofessional rocketry organizations like the RRS obtained FAA waivers to conduct flights over the desert in Southern California and in Nevada.

To ensure that he complied with FAA rules for LDRS, Pearson contacted the local FAA office in Oberlin, Ohio. He asked them what he needed to do to secure a waiver for a rocket launch. The local FAA officials did not have a clue. The question had never come up before, said Pearson.

"They looked it up and got back to me and sent me a waiver form."

Pearson filled out the form, turned it in, and obtained his FAA exemption. The farm at Medina was not far from the flight approach taken by commercial planes to nearby Cleveland Hopkins Airport. So the waiver for LDRS was only 2,000 feet. Ironically, the NOTAM had the opposite effect on the pilots it was supposed to keep away.

"During the launch, it was noticed that there were a large number of light planes flying overhead," said Pearson later. "It turned out that the FAA transponders were telling them we were out there, and they were flying by checking us out."

To demonstrate to the FAA that LDRS-1 was being organized by responsible adults, Pearson also asked those who had planned on attending the rocket launch for their flying plans.

"I asked flyers to send me a list of what they were going to fly because I didn't know what the FAA wanted. So what I did was when I sent my waiver in, I actually sent a typed list of the rockets that were going to fly—not that the FAA probably cared—with what the rockets weighed and the motors they were going to have," he said. This was the extent of preregistration at LDRS-1. "We were very specific in saying, 'No fireworks.' That was part of the respectability thing, and we didn't want anyone to think we were crazies. There was no smoking allowed at the pads or along the flight line either."

To drum up publicity for LDRS, Pearson sent out written fliers in early 1982 to friends and acquaintances in rocketry. He also posted ads in *SNOAR News* and, ultimately, the two ads that appeared in the *Model Rocketeer* in April and May. The NAR printed its anti-LDRS announcements in the *Model Rocketeer* in June and again in July. The antilaunch ads generated some interest too, supporting the truism that sometimes even bad publicity is better than no publicity at all.

The launch was scheduled for Saturday, July 24, and Sunday, July 25. This was just before the NAR's annual premiere launch, NARAM.

"I think I may have looked at the NARAM date and chose a weekend that was not the same so we could get the people who were going to NARAM and who also wanted to come here," said Pearson later.

A number of people expressed interest in attending both events, so Pearson made sure there would be no conflict with the dates. This informal tradition would continue, with a couple of interruptions, for the next thirty-five years.

Despite the NAR's minimal efforts—or perhaps because of them—word of LDRS-1 spread, and flyers began arriving in Medina on Thursday night, July 22. Up the road at the NASA Lewis Research Center near Cleveland, the visitor information center was celebrating National Space Week, with special programming on the first landing on the moon. It was there at Lewis Field that some of the research and development had been done on the first liquid hydrogen rocket motors, which ultimately led to the motors that propelled the *Saturn V*. Late on Thursday night at Cape Kennedy, a new *Pershing* missile exploded 17 seconds after liftoff, showering debris along the coast of Florida and in the Atlantic.

By Friday, people arrived from several other mid-Atlantic states. A few flyers, including *California Rocketry*'s Jerry Irvine, made the trek from as far away as Southern California. Irvine was a guest in Pearson's apartment for the weekend. Others stayed at a local lodge in Medina called the L&K Motel. On Friday evening, the rooms at the L&K were filled with people preparing rockets for the next day. It was the beginning of an LDRS tradition, wherein a local hotel is temporarily transformed into a mini-rocketry university; there are impromptu engineering seminars, show-and-tell presentations, and rocket-building shops. Sometimes rocket construction is finished in guest rooms, followed by a paint job in a parking lot dumpster. At night during launches, hotel revelries go one for hours. No small amount of alcohol is consumed.

The forecast for Medina that weekend was for high temperatures between 80 and 85 degrees, and it was sunny and bright Saturday morning. It was also hot, sticky, and humid. Similar conditions prevailed on Sunday too. Temperatures would reach almost 90 degrees that weekend, and most people in attendance dressed in shorts and T-shirts or no shirts at all.

There were no launch fees at LDRS-1. There were no flight cards. There was no liability insurance. And there were few rules either. But there was a gathering of like-minded people interested in moving above and beyond the weight and motor impulse restrictions imposed by the Code for Model Rocketry. Flying started on Saturday and Sunday at 10:00 a.m. Most of the flights at LDRS would have pleased the NAR as they were conducted with NAR-certified model rocket motors, A through F power.

"There were a handful of launch pads and no PA system," recalls flyer Ken Good years later. "It was pretty loose. You just took your rocket up to the pad, hooked up the leads, and then moved back to launch your rocket . . . Nobody was trying to do anything stupid. They were mostly trying to see what kinds of interesting combinations of rockets they could fly."

Much of the NAR Safety Code was also followed, whether intentionally or not.

"Most of the rockets still had plastic parachutes," observed flyer Matt Steele.

LDRS-1 was not a revolution in high-power rocketry—at least not yet. But there were motors larger than an F, and it was impressive stuff for many hobbyists, most of whom had never seen anything even as large as a G motor in a model rocket. And there were not just G motors. There were a few H motors and at least two I motors, some of which may have been created by Gary Rosenfield and John Davis of Composite Dynamics, neither of whom were in attendance.

Flyer Curt Hughes of Pennsylvania came to LDRS-1 with a rocket kit that was very popular at the time, the *Ace Mongrel*. The *Mongrel* was a bigger-than-life rocket in 1982. It was 4 inches in diameter and stood about 5 feet tall. The four-finned kit came with five motor tubes, an 18-inch-long balsa nose cone, and one-eighth-inch-thick precut plywood fins. The rocket was distributed by Ace owner and rocketry designer

Korey Kline of Southern California, and it retailed for $65—a lot of money for a hobby rocket at the time. Although the rocket weighed only about 2 pounds, a product review in *California Rocketry* called the *Mongrel* "a monster of a model rocket."

Hughes was an active NAR member in 1982. He learned about LDRS-1 while attending a science festival at Kent State University that spring called KentCon.

"I brought a big Ziploc bag of old E and F Centuri Mini Max black-powder motors with me to see if I could sell them," recalls Hughes, who met Chris Pearson at the festival. Pearson collected old motors, and an immediate bond was formed between the two flyers. "We quickly became friends, and he told me about his organization [SNOAR] and also about LDRS, a high-power sport launch that he was organizing for some time in the summer," Hughes went on.

While at Kent State that weekend, Hughes also purchased his first composite high-power rocket motor. The motor was 38 mm in diameter, and both ends were plugged with epoxy. In the aft end, there was a steel washer with a hole drilled though. This hole served as a crude self-ablative nozzle.

For his big flight on Saturday at LDRS, Hughes loaded three motors into his rocket: two Estes D11 model rocket motors flanking one of the composite motors he had bought at KentCon, a central H160 motor. At ignition, the rocket canted somewhat to one side and then roared off into the sky for not only a perfect flight up but also a nice deployment of the parachute. The *Mongrel* then drifted off the Wagner property and into a cornfield and was almost lost forever. Hughes and Ken Good, who would one day become president of the yet-to-be-created Tripoli Rocketry Association, searched for 2 hours before finding the rocket in perfect condition.

"The field wasn't anywhere near what you would think of as a modern LDRS field," recalls Good. "There was big tall corn all around us. It was up over our heads. We saw it come down, and we knew it was in that field somewhere, so we basically just used a grid, and we walked up and down until we found it."

For the flight and recovery of his *Mongrel,* Hughes was later bestowed with the honor of the "most spectacular flight" at LDRS-1. Another well-appreciated launch at LDRS-1 was Mark Weber's I-powered rocket that also carried two F motors. In 1982, the I-impulse motor was king, its ultimate power beyond dispute. The flight earned Weber the "most powerful rocket" award for the weekend. (A few years later, Weber would become one of the founding members of the modern-day Tripoli Rocketry Association.) Jerry Irvine also flew a rocket powered by an I200 motor. Unfortunately, Irvine's rocket came apart in flight. The motor was simply too powerful for the rocket's airframe and fins. This was a "shred."

Clustered motors were popular at LDRS-1 and would continue to be popular for the next several LDRS events. Flyers launched rockets propelled by multiple D, E, or F motors. Scott Pearce outdid everyone by cramming 168 Estes A3 model rocket engines and a lonely C engine into his airframe and then lighting them all "flash in the pan" style. This method of ignition simultaneously lights all motors by igniting a large charge of black powder held in an open pan positioned several inches directly below the rocket. When the black powder ignites via a remotely activated electrical circuit, the resulting flash of fire reaches up to touch the aft end of every motor in the cluster. When it works, it is a sight to behold; it roars away like a nineteenth-century photographer's handheld camera flash. There is a boom and then a large puffy cloud of white smoke. Pearce's clustered flight was perfect, generating large amounts of billowing white smoke, delighting the crowd. The rocket shot out of the smoke and raced into the sky. By Sunday afternoon, the launch was over.

"At the end of the second day of flying, everyone was complaining about their sunburn, not having enough to drink, and sore feet," said Pearson, who also spent time fishing one of his own rockets, another *Ace Mongrel* powered by three F engines, out of a nearby apple tree.

According to reports at the time, there were forty-seven people in attendance over the two-day LDRS weekend—not forty-seven flyers but forty-seven people, total. (In comparison, a modern LDRS might see well over a thousand rocket fights and many thousands of spectators.)

There were few photographs at LDRS-1, and without flight cards, there is no detailed launch record that survived the event. It was a small launch by today's standards, and there was nothing "large or dangerous" about any of the rockets that flew. But it left its mark in amateur rocketry as it brought together in one place flyers interested in higher-powered rocketry from not only the east but the west as well.

"LDRS-1 was nothing short of the most intense, advanced, and fun rocketing event I have ever attended," proclaimed Irvine in *California Rocketry* after he returned home. "Chris Pearson did a damn good job of running LDRS-1, and I look forward to seeing LDRS-2 and any clones that may be started in the near future."

It would be another three years before the Tripoli Rocketry Association would be born and many more years before any LDRS would take on a look that would be recognized as LDRS today. But it was a start.

"The first LDRS was a building block for the modern Tripoli," said Good years later, noting that he and many other future Tripoli members had met for the first time at Medina in 1982.

"The rocketry activity was substantial," wrote Irvine in *California Rocketry*. "But it was a social experience as much as a rocketry event... People were sharing secrets, tactics, methods, and sources of supply. It was a synergistic atmosphere at all times."

NAR member and LDRS participant John Kopplin put it another way: "LDRS-1 was kickass!"

As people packed up to leave Medina on Sunday afternoon, there were casual discussions of holding another LDRS next year. Yet there were no guarantees. Many rocketry weekends have taken place on remote fields and were later forgotten. Only their names remain in discarded back issues of now defunct rocketry magazines: Hayburner, Roar at the Shore, Medicine Wheel, Sci-Power, Summerfest, Danville Dare, Cuban Missile Crisis, etc. These launches came and went, some for a few years, others only once. At the end of the day on Sunday, July 25, 1982, there was no reason to think that LDRS-1 would be any less transitory. To think that it would become an annual tradition and continue for another thirty-eight years was a little more than far-fetched. It was ridiculous.

"We really did LDRS just on a lark," Pearson would say years later. "We never thought it would continue. And then after the launch, someone asked me, 'Well, are you going to do it again next year?' And I said, 'Yeah, why not?'"

By Monday, July 26, 1982, all traces of LDRS-1 in Medina had vanished. There was no mention of the launch in the local press. That morning, the Medina newspaper carried a wire story interview of actor William Shatner, who was basking in the unexpected success of *Star Trek*

II, now a bona fide hit. The movie injected new life into Shatner's acting career, box office receipts eclipsed the original film, and there was talk of another sequel in the series. Another movie was viewed by most people as an unlikely event. Still, Shatner was optimistic.

"I think *Star Trek II* is a wonderful movie," he said in the interview, "which will probably engender another *Star Trek* movie, and that will not make me unhappy."

In retrospect, the significance of the first LDRS does not rest upon the uniqueness of the rockets that were launched there or on the fact that a few high-powered motors were used in those rockets. There was nothing terribly special about the motors used. G, H, and even I motors were rare, but they were not new to hobby rocketry. These impulse-size motors had been flown off and on at launches in the west for several years. Moreover, small composite motors, although exciting and new to model rocket flyers, were nothing compared to the rockets and motors flown by amateur rocketry organizations in the American deserts for decades. Groups such as the RRS had been launching large metal rockets many tens of thousands of feet into the air for years.

Nor was LDRS-1 significant for its size. There were only a handful of people at the Wagner farm that weekend, and the two-day event was smaller than the majority of the monthly single-day regional launches held nationwide today. There were no vendors who pulled into the event in semitrucks, their huge trailers filled with motors, rockets, and rocketry-related parts of every conceivable nature. There was no media coverage. This first LDRS was really just a modified model rocket launch. And throughout the 1980s—and certainly for the next several LDRSs at Medina—it would remain just that: plenty of model rockets and model rocket motors, with a few high-powered flights thrown into the mix.

Yet LDRS-1 was significant for several reasons. It would be repeated annually for the next thirty-eight years, becoming the second longest running annual event in hobby rocketry history, surpassed in age only by NARAM. In high-power rocketry, it is the oldest annual launch.

LDRS also brought together in one location many of the leading voices of the emerging high-power rocket scene of the day. This occurred partly because of the advance adverse publicity and the fact that prominent flyers from the west, such as Jerry Irvine, decided to make the journey to Ohio for this launch. Over the next few years, these contacts would help spread the news of high power across America.

LDRS-1 was also significant because it was more than just a rocket launch. It was a *movement*. Pearson's event demonstrated that current and former NAR members wanted the hobby to change and that the members themselves and not the NAR's national leadership would lead the way. This "in your face" attitude was highlighted by the very name of the launch. Safety was the NAR's central tenet, and for more than twenty years, NAR leaders worked hard to establish its launch events as the exact opposite of anything that could be viewed as "large or dangerous." In the years leading up to 1982, there were periodic G and H motor flights at NAR-sponsored events. But these flights were usually under the radar so as to not encourage confrontation with the national leadership. Occasionally, NAR leaders would get wind of such flights and try to do something about them, but that was rare. "Who flew the G?" was a well-known bumper sticker that mocked the NAR's attempts to rein in such flights.

The LDRS-1 launch threw down the gauntlet: LDRS was an open challenge to the NAR's core principle of safety above all else. Although there was very little that was "large or dangerous" about the rockets at LDRS-1, the name declared that the future of hobby rocketry might be different. Pearson advertised this pursuit of higher-powered rocketry openly. The NAR, in turn, issued a stern warning to stop that pursuit. But instead of backing down, members who were planning to attend LDRS-1 saw the NAR's warning as further encouragement to show up. It was a badge of honor of sorts.

The open nature of LDRS also encouraged the emerging commercial market for high-power rockets and motors. Without commercial manufacturers, there would be no model or high-power rocketry. Following LDRS-1, the number of high-power motor manufacturers and makers of rocket parts would steadily increase, eventually bringing competition and variety in motors, parts, and new technologies. In the future, LDRS would become the annual kickoff point for new products being introduced to the rocketry public. The event would become both launch and show.

The results of LDRS-1 were not mentioned by any NAR publication. But the event was covered in Irvine's *California Rocketry Quarterly* and other contemporary rocketry magazines and newsletters. News of the launch reached flyers all over the country. Some people were excited by the idea of larger motors and bigger rockets; they dreamed of building craft that would go higher than ever before. The altitudes achievable in the early 1980s were still modest, perhaps a few thousand feet or so. But LDRS-1 was the beginning of a marked shift in attitude that would one day lead high-power users, including the NAR, into the future of hobby rocketry.

4

The Aftermath

In the fall of 1982, NAR member Curt Hughes received a letter from J. Pat Miller, the president of the NAR. Miller had just picked up a copy of the most recent issue of Jerry Irvine's *California Rocketry*, which contained detailed coverage of the LDRS-1 launch held in July at Medina. In that article, a rocket flown by Hughes was described as powered by an H160 motor and two D motors. The power plant was clearly in excess of the motor limitations allowed by the NAR.

"All members of the association are expected to follow the NAR Safety Code," Miller told Hughes in his letter. "The safety code specifically requires NAR certification prior to the use of a rocket motor. No motor exceeding 80 Newton seconds total impulse is certified by the association." Miller added that the H motor flown by Hughes at LDRS had a total impulse of 320 Newton seconds. "No such motor is recognized by the association as a model rocket motor," continued Miller. "It is classified by the federal government and the NAR as an amateur rocket motor."

In his note to Hughes, President Miller acknowledged that there were exceptions to the NAR Safety Code that, if applicable, allowed NAR members to launch rockets with power plants that exceeded 80 Newton seconds. These exceptions applied to NAR members conducting research while serving as employees of colleges, government agencies, aerospace manufacturers, or model rocket manufacturers. Miller then put two pointed questions to Hughes about the story that had appeared in *California Rocketry*. He wanted to know first if the story in the rocketry newsletter was accurate and second, if the story was accurate, whether or not Hughes was an employee of one of the exempted services described above.

"Certainly, I would appreciate your response at the earliest possible date," wrote Miller before giving Hughes until November 12, 1982, to reply.

"As a child, I was interested in anything that flew," recalled the then thirty-one-year-old Hughes, an elementary school teacher from North Huntingdon, Pennsylvania. "I remember going outside to look for *Sputnik* in October 1957. I was only six then, but I was already extremely interested in space flight. I saw an ad for an Estes model rocket catalog in *Popular Science Magazine* in 1962, and I saved all my lunch money to buy kits and motors. I was a really skinny kid as a result."

In 1965, Hughes joined a high school science club cofounded by classmate Francis Graham. The club was called the Tripoli Federation. Graham, Hughes, and others conducted science experiments and occasionally launched model and zinc-sulfur high-powered rockets in the late 1960s. Later, Hughes joined the NAR. He used NAR educational materials as part of the science curriculum for his students. Then in 1982, Hughes stumbled across a copy of *California Rocketry*, and he ordered a higher-powered rocket kit from Jerry Irvine's U.S. Rockets. Hughes wanted to travel to Smoke Creek, but it was too far away. When he learned of Chris Pearson's LDRS-1 in nearby Ohio, he knew he had to attend. Shortly before LDRS, Hughes purchased his first H motor.

"I believe it was a homemade composite motor made from green filament-wound G-10 fiberglass tubing with a simple washer as the nozzle," recalled Hughes. "The exhaust gases and pressure formed the correct exit diameter to the core. It was inefficient, but it worked."

In his letter to the Pennsylvania school teacher, President Miller required Hughes to provide facts bringing the flight of his H-powered *Mongrel* at LDRS into one of the NAR's four recognized safety code exceptions. Of course, Hughes did not fall into any of the exceptions. His flight was not related to any college or university, and Hughes was not part of a government agency or an aerospace manufacturer. He was also not a model rocket manufacturer. Hughes was just another flyer who wanted something more from hobby rocketry. He wanted to fly larger rockets and send them higher than ever before, which at the time meant only a few thousand feet above the ground.

Without any exemption to save him, Hughes was purportedly expelled from the NAR for launching an H motor at LDRS-1. For Hughes, the expulsion was not unexpected. Indeed, it was almost an honor. The young flyer was a rebel. He took his chances, ignored the NAR code, and accepted the NAR's decision.

"It didn't really surprise me," he said later of the expulsion, "as it was the end of one thing and the beginning of something else." That something else would be high-power rocketry, which Hughes now enthusiastically jumped into with both feet.

The NAR was serious about its safety code and the rule that its members refrain from participating in high power. The NAR purportedly contacted Chris Pearson prior to LDRS-1, offering to send Pearson a list of all NAR members so he could compare names to see if anyone attending the event belonged to the association. If so, he was asked to forbid them to fly. Pearson declined.

"NAR officials tried to coerce certain members that they knew would be attending, asking them to write down names, take photographs, and generally 'rat' on everyone that was there," recalled Pearson. "To the best of my knowledge, no one volunteered to fink either before or after the launch. After LDRS-1, the nasty grams started to come from the NAR. From what I remember, the NAR was going nuts because they had no control over what was going on. They thought they controlled all of [hobby] rocketry, yet here were some guys going off and doing what they wanted to do and basically thumbing their noses at them."

Pearson believed the NAR leadership was well aware that higher-power motors were already being flown by some of its members in remote areas of California and Nevada.

"But they thought that that was just an isolated phenomenon," he said. "These guys would go out to the Lucerne Valley and do crazy stuff, and the NAR ignored them and figured they would eventually go away." But that was not happening, he added.

The NAR had a different view of things. According to President Miller, the issue was not that the NAR wanted to control all of rocketry. Rather, the NAR board was concerned model rocketry might be threatened if NAR members were involved in a serious accident with uncertified (i.e., high-power) motors. NAR members needed to follow the rules, said Miller, and failure to do so invited action from the board.

In retrospect, this was not an unreasonable stand for the NAR in 1982. After all, the safety code clearly stated the impulse range was limited to A through F motors. If a member could not abide by the code, they were free to leave the NAR. Yet proponents of higher-powered motors didn't all see it that way. Some believed the NAR needed to change its rules immediately to accommodate them. Some NAR members were also unwilling to use the association's internal legislative process, which could and eventually did take years, to implement such changes. Other critics believed the NAR was overstepping its bounds in trying to restrict members from participating in amateur rocketry when they were at non-NAR events on their own time.

"It is the operating philosophy of the NAR that its members and sections agree to abide by the safety code and to restrict their rocketry activity to model rocketry," wrote the editor of the *Model Rocketeer* in late 1982. "This does not affect your personal right to participate in amateur rocketry . . . or anything else. We frankly are indifferent to whether you choose to cluster Is and boost satellites (or bodily parts) into orbit. But you may not do so as a NAR member or a chartered section."

The NAR was not going to get mired down in the minutia of whether a NAR member was on "NAR time" or his or her own "non-NAR time" in the event of a mishap. They wanted a bright line, and the line was clear. If you are a member of the association, participation in amateur rocketry was out of the question.

Curiously, at the same time the NAR was chastising members who were experimenting with G, H, and I motors, the editor of the *Model Rocketeer* asked its readers to submit articles on the subject of high-power rocketry. In September 1982, the *Model Rocketeer* announced it was "actively soliciting articles and plans on the subject of high-powered model rocketry, with a view to producing a future theme issue on this subject." This solicitation from the NAR's official publication came barely nine months after Jerry Irvine was refused advertising space for *California Rocketry* for showing rockets that were launched with G motors and just weeks after President's Miller's warning to NAR members in the July issue of the *Model Rocketeer* to steer clear of higher-powered rocket motors or else. So there was some movement within the NAR on the subject of higher-powered rocket flights. But it was hard to discern in all the noise.

The NAR was moving slowly toward the future on other fronts too. Late in 1982, the NAR announced a committee formed to study the safety of G motors with an eye toward one day certifying such motors. This was a big step for model rocketry. One of the people appointed to the committee was longtime NAR member and motor maker Gary Rosenfield. The NAR knew that Rosenfield was active in higher-powered motors. He was chosen for the committee anyway, a clear signal to those who were listening that the NAR was not as inflexible as some believed.

But as official policy, higher-powered rocketry was forbidden for NAR members. There was not a single story in the NAR's magazine related to true high-power rocketry in 1982 or 1983. The launch at Medina was never mentioned, and insofar as the NAR membership was concerned, other than the expulsion of Curt Hughes, it never happened.

In an unsigned article entitled "NAR Politics" that appeared in the January 1983 issue of *California Rocketry*, the magazine encouraged NAR members to take action: "Given that some will continue to engage in non-NAR rocketry, the only alternative to the rocketeer who wishes to avoid being chastised is to *quit the NAR*."

The NAR would ultimately change its views on high-power rocketry, but it would take time. It would be four more years before model rocketry's most powerful organization would allow the use of the G-impulse motor at its official events. And by that time, the G motor would be yesterday's news.

In his president's column in April 1983, Miller reported that instead of attracting new members, the NAR membership had actually declined in the last two months of 1982.

"It is too early to predict (at the time of this writing) if this is a downward trend over an extended period," said Miller, "but it is quite likely that it is."

Miller explained the decline in membership as a product of the overall national economy, an explanation that undoubtedly had merit. The economy in the early 1980s was sluggish, and the downturn probably impacted the NAR too. But the NAR failed to recognize the possibility that some of its members wanted something more from the organization, and they were taking their interests elsewhere.

The NAR would eventually have to get involved in the emerging high-power market. In the meantime, high power moved forward on its own.

The April 1983 issue of *California Rocketry Quarterly* carried a small announcement for LDRS-2, scheduled for July 30 and 31, 1983.

Once again, Chris Pearson scheduled the launch a week before the NAR's big annual show, NARAM, to allow flyers to attend both events.

The *California Rocketry* advertisement provided little information on the launch except for the date, location, and a phone number to call. It also announced that flyers had to file an FAA clearance ninety days in advance of the launch. This was really an instruction for those interested in launching a rocket to provide Pearson with the information he wanted to pass on to the local FAA authorities prior to the launch. There was no word out of the NAR about LDRS-2 in 1983. And this time, there were no positive advertisements or negative ads about LDRS in the *Model Rocketeer.*

To prepare for LDRS-2, Pearson and his crew pretty much repeated what they had done in 1982. They added another high-power pad, so now there were two high-power launchers and twelve model rocket pads. Otherwise, there were few significant changes in 1983. The launch saw an increase in the size of the motors used by flyers. J-impulse motors would fly for the first time at any LDRS. There was also the appearance of one of the first multi-event deployment recovery systems in LDRS history.

There were still no launch fees. There were still no flight cards. And there was still no insurance. LDRS was a fringe event, and people liked it that way.

NAR member Bob Koenn of Florida could not afford to attend both LDRS and NARAM. He and his friends decided to take a chance on LDRS-2.

"By taking four people in one car and driving up and back in one day, the trip could be done relatively inexpensively," said Koenn.

Around midnight on the last Thursday in July, Pat McCarthy, Paul Rahnefeld, Mike Myrick, and Koenn headed for Ohio.

"About 21 hours later, we arrived, and believe me, we were ready to stay out of the car for as long as possible," said Koenn. At the time, Koenn was a twenty-nine-year-old NASA engineer working in the Space Shuttle Program. He discovered model rocketry in the mid-1960s at a summer camp in Clearwater, Florida. Koenn became infatuated with America's space program and rocketry in general, and he helped start several rocketry clubs in Florida.

"I was the NAR national champ in the C Division in 1980," said Koenn, "and I remember a young Gary Rosenfield flying some of his first batches of motors at NARAM-16 in Virginia in 1974."

Koenn and his friends squeezed a few small rockets into their car for the 1,100-mile trip to LDRS, and they arrived in Medina on Friday evening, July 29. The prelaunch festivities at the L&K Motel were in full swing. Flyers had rockets hanging out of their cars and out of motel windows, in parts and pieces on the beds in their rooms, and fully assembled and standing upright outside their motel doors. People walked through the dark halls of the motel and went door to door to look at everything on display: exotic motors, exceedingly tall rockets, and also the routine items found at any NAR-style rocketry event even today.

"The birds we saw were definitely much different [from] those at any contest or demo we have ever been to," said Koenn. "There were models with clusters of F or larger engines, 15-foot-tall birds, an *Orbital Transport* that was 8 feet tall and 4 inches in diameter, a 469-cluster bird, and other rockets too numerous to mention."

Koenn and his friends were impressed by the quality of construction of the rockets. "Most of the models were very well finished, and all were very well made," he observed.

Koenn noted that higher-powered rockets represent a much bigger investment by the flyer. Hence, there was more incentive to launch the rocket safely so that it could be successfully retrieved, he said. "Suffice to say, I saw no birds that I would not have let fly on the basis of construction," he wrote. "This is more than can be said about many small birds at many contests I have been to, even NARAM."

Flying started around 10:00 a.m. on Saturday, July 30. The field was still a little wet from rain that had fallen the night before.

"There were about a hundred people present and quite a display of big rockets," said Koenn. "You could easily spend a couple of hours walking around, looking at the rockets, and talking about big rockets and large engines. The weather looked like it would be great too, with only a slight wind and sunny skies."

There were flights on motors A through J power at LDRS-2, with the majority of rockets still being propelled by model rocket motors. Scott Pearce, another flyer who had reportedly been expelled from the NAR for his flights at LDRS-1, launched his *Slobodian Avenger*, an eight-finned rocket squeezed tight with nearly sixty motors in the aft end, including at least a dozen E and more than forty D engines. The rocket was more than 12 feet tall and carried multiple model rocketry (Cineroc) cameras.

"It lifted off perfectly, climbed to about 300 feet on top of a huge column of smoke, and popped its chutes," wrote Koenn. "It was a beautiful flight and perfectly safe. Definitely impressive, and only about five engines failed to ignite."

David Flink used even more motors in his cluster. He crammed 469 tiny motors into one of his rockets, a cluster filled with several F motors at its center surrounded by hundreds of mini A motors. Flink's rocket probably held more motors than any other vehicle in LDRS history. The scratch-built three-finned design was on a 4-inch main airframe that contained an inverted shroud at the base into which all the motors were crowded. The 6-foot-long rocket was lit flash-in-the-pan style. It generated huge amounts of white smoke on ignition.

"At liftoff, at least one F100 motor blew and tore the model up as well as tumbling it across the sky," recalled Koenn. "It fell to the ground, where mini As continued to fire at various intervals. There was never any real danger, although quite a bit of work went down the drain."

There were at least two J motors at LDRS-2. Commercial motor maker Scott Dixon of Colorado launched his own Vulcan Systems J-class motor. Dixon's fiberglass airframe and Kevlar fins were design features years ahead of their time. Most "higher-powered" rockets in 1983 were built with unfiberglassed cardboard.

Dixon's fiberglass J-powered vehicle "lifted off with the loudest roar I have ever heard from a model rocket!" exclaimed one spectator.

It quickly disappeared, undoubtedly shattered the waiver, and was thought lost. It was soon sighted drifting back to the ground, unscathed. Chris Pearson also flew a Vulcan J motor, but his rocket was not so fortunate. It disintegrated in flight.

California rocket maker Korey Kline, one of the more creative minds in the hobby at the time, launched a G-powered rocket that carried two black-powder ejection charges. One charge was built into the forward end of the motor, a common design carried over from model rocket engines. That charge would activate near apogee, pressurizing the inside of the airframe and thereby popping off the nose cone to deploy the parachute held inside the airframe. Yet Kline added a second ejection charge too. The second charge deployed a larger main parachute at a lower altitude. Kline activated this second charge by remote radio control. Kline's system was a precursor to the modern multideployment recovery system. It was also years ahead of its time as such a system would not be routinely seen at LDRS or any other high-power launch until well into the 1990s. In some ways, it was probably the first "dual deployment" rocket to ever fly at LDRS, although that phrase was yet to be coined.

One of the most popular flights of the weekend was Jim Dunlap's scratch-built *Orbital Transport*. The aircraft-like rocket had forward canards and twin vertical stabilizers. It was powered by a cluster of three F20 and three F100 motors. At ignition, the transport tore off the pad. The rocket carried a separate vehicle on its back—an aircraft-like vehicle that separated at apogee and glided back to the ground on its own. It was the best flight of LDRS-2, proclaimed *California Rocketry*.

There were many newcomers at LDRS-2. Among them was Gary Rosenfield, who had made the trip from California. Rosenfield would attend every successive LDRS for the next thirty-eight years, an attendance record unparalleled in the hobby. Other first timers included a thirty-six-year-old Ohio machinist named Ron Schultz, and his wife, Debbie, the part-time seamstress with her small business, now called Lots of Crafts. Ron had heard about the launch by happenstance, and he and his wife took their family to Medina to see what LDRS was all about. Ron had dabbled in model rockets for a few years, flying them here and there in Ohio, and with his machining skills, he had built some nice rockets for himself and his family. But he had never been to an event like LDRS. As he walked around the flight line that weekend, he looked at some of the kits being sold by dealers. The quality of the kits was okay, he thought. Yet with his construction and machining skills, he thought he might be able to do better. When he returned home, Ron decided to see if he could build some rockets that he might be able to sell next year at LDRS-3.

Rain cut short the flying activities on Sunday, July 31. This thwarted the plans of several people with larger rockets that were prepared on Saturday night to be flown on Sunday morning. Still, by the end of the weekend, LDRS-2 was deemed another success.

"LDRS-2 was certainly the event of the year," wrote Irvine after the launch. This was Irvine's second trek to Medina from California. "There were minor organizational shortfalls, such as confusion about which pad was being used at any given time, but the activities ran fairly smoothly." Irvine summed up his story in *California Rocketry* about LDRS-2 by stating it was "a fine meet, and the organizers plan on doing it again next year."

"When I meet with friends from those early years, we remember how we were drinking gin and tonic the whole time at LDRS-1 and 2," said Curt Hughes years later. "It was a very loose, hodge-podge type of thing... [T]hat was what LDRS was about. In the early days, LDRS was about trying new things. If a rocket crashed, it was no big deal. People laughed about it."

The size of the rockets had increased slightly between LDRS-1 and LDRS-2, and the crowd had more than doubled, from forty-seven people in 1982 to more than a hundred a year later. By today's measure, these are still small numbers. Yet word was getting out, and in the small but growing high-power community, people were already looking forward to LDRS-3 the following year.

5

Enter the NAR

NAR president J. Patrick Miller was not too impressed with the rise of higher-powered rocket motors in the mid-to late 1970s.

In 1978, Miller was at a gathering of hobbyists conducting burn tests of composite motors in the desert outside Las Vegas, Nevada. Among those present was Gary Rosenfield, who was still in the United States Air Force. At the time, Rosenfield was also making motors with John Davis.

"After we completed the burn tests that day, we went back to a café near Las Vegas to have some coffee and get something to eat," recalled Miller. "During that meeting, Gary was talking about his large composite motors, G impulse and above, and I listened to what he had to say, and then I asked him if he thought these motors had any hobby potential. He said that he thought they had as much hobby potential as any other motors. But I didn't really believe it."

In retrospect, said Miller years later, his early view on high-power rocket motors was a mistake. "My intuition just went against it," he explained. "We had a system in place for twenty years, and it worked, and what Gary was talking about was really outside that system. I just didn't believe at the time that these bigger rocket motors were practical to have around... It was just a gut reaction."

Over the next few years, Miller kept a watchful eye on large motor activities he had read about in model rocket magazines and local newsletters. As president of hobby rocketry's largest organization, Miller was obligated to ensure that the NAR's insurance policy remained in effect for all launches. And the insurance policy required compliance with the NAR Safety Code. Miller's official policy never wavered. The safety code was to be followed at all times. Although it was known that a few NAR members were going outside the bounds of the safety code, these exceptions stayed pretty much off the radar, taking place in remote locations like Lucerne. For the most part, Miller and the NAR looked the other way, confident in their belief that the composite motor fad would pass.

The NAR's laissez-faire approach to higher-powered motors changed in late 1980, when a few NAR members were publicly chastised for surreptitiously flying rockets with G motors at a NAR-sanctioned event. Then following the NAR winter nationals in 1981, also known as NARWIN 3, the NAR board of trustees was forced to take real action against a few members. NARWIN 3 was a model rocketry launch at Lucerne in December 1981. The launch was hosted by the Polaris section of the NAR and was a regional competition event. There was no question that the NAR's safety code would be in full effect at all times.

From firsthand reports written after the launch, it was undisputed that amateur rockets and fireworks were flown at NARWIN 3. One rocket carried not only an illegal G motor but also a quart of gasoline, which was directed into the exhaust stream of the vehicle for a special effects display. Several members present at NARWIN 3 reported the safety code violations to the NAR board afterward. Apparently, there were photographs and even an 8 mm movie film taken at NARWIN 3 that showed the launches in question. This was not a close call for the NAR board of trustees or for President Miller.

"The real problem we had was they were launching rockets that were in flagrant violation of the safety code at a NAR-sanctioned event," recalled Miller. "And so I had to do something in that case."

Miller headed a panel of distinguished NAR board members who would hold a hearing on the matter. The panel included Verne Estes, G. Harry Stine, and Jay Apt, who later became an astronaut who would fly in four space shuttle missions in the 1990s. The panel traveled at their own expense to Southern California, where an informal hearing was held in a church hall in mid-1982, not long before Chris Pearson's first LDRS.

There were three NAR members who had allegedly violated the safety code at NARWIN 3. They were invited to attend the hearing to explain their actions. Two of them showed up. The third, who was the member who had purportedly launched the gasoline rocket, did not attend. At the hearing, the trio of wayward flyers received their "due process" as NAR members. In other words, they were given an opportunity to be heard and explain their actions. There was no denial of what had occurred at NARWIN 3. The result of the hearing was easy to predict. Two of the three flyers were suspended or expelled from the NAR. For the third flyer, the issue was moot. He had already let his NAR membership expire.

"We had to send a message. There was no question about that," recalled Miller, who described the hearing as a test case. "There were other things happening at the time involving alleged code violations, but there was no question about this one. The main problem, in addition to the G motors, was the gasoline-loaded rocket as well as illegal fireworks at the launch. We wanted to send a message out to the rocketry community that if you wanted to engage in this type of activity, that is fine, but don't engage the NAR."

Miller and the board were also concerned about liability at NAR-sanctioned events where the safety code was being flagrantly disregarded. "If there was an accident out on the range and the safety code was not being followed, the insurance would not cover it," said Miller. This meant that liability for any accident would fall directly upon the landowner and the NAR (and perhaps the local NAR section and local NAR members).

Miller's understanding of the insurance policy was accurate. Liability insurance coverage would likely be denied in the event of an accident if the code was not being followed. Yet the need to maintain insurance and follow agreed-upon safety rules was lost on some flyers intent on doing whatever they pleased at whatever launch they attended. At the time, many high-power flyers had little interest in the responsibility of carrying insurance for their activities, which, in many states, might be defined as "inherently dangerous." It would be years before any formal high-power rocketry event would be covered by insurance. Miller and the NAR board did not have the luxury of disregarding the safety code. They were responsible for not only a few flyers who wanted to push the envelope but also the NAR membership as a whole.

There was reaction and a great deal of overreaction to the results of the NAR hearing on NARWIN 3 and the expulsions of the two NAR members. For example, an article in the *SNOAR News* described the hearing as a "lynch mob," with President Miller "out for blood" and acting as "chief prosecutor." The result of the hearing was predetermined, it was alleged, as "the board members had their minds made up long before the hearing." This last criticism, even if true, should have surprised no one. Members of the NAR committed flagrant violations of the organization's safety code at a NAR event, including the launching of a rocket carrying gasoline. These members refused to follow the code, were unapologetic about it, and were disciplined. Go figure.

In retrospect, it is hard to see what other choice such flyers were giving the NAR. As an organization, the NAR had the right to decide what its own rules were and how those rules were to be enforced. The total impulse limit was stated in the safety code, and members who wanted to exceed that limit were free to leave the NAR. Indeed, some members did exactly that. But to remain a member of the NAR and to openly reject the code at NAR-sponsored and NAR-insured events was not reasonable. And to fly a rocket carrying gasoline at a launch was simply ridiculous. It would be equally ridiculous at any NAR or Tripoli rocket launch today. (In the distant future, the yet-to-be-created Tripoli Rocketry Association would likewise exercise its right to suspend and/or expel members for safety code and other Tripoli rule violations.)

Regrettably, a die was cast by the NARWIN 3 incident and the rhetoric that followed. A heightened sense of persecution by the NAR, some of it imagined, was later adopted by some high-power flyers. The perception of the NAR's antagonism to high power was given additional fuel by the NAR's initial response to LDRS in 1982. Miller and the NAR board were painted as antagonists in the way of progress. An "us versus them" mentality emerged, with Miller at the "evil center" of the NAR. It was melodramatic and made for interesting reading in regional rocketry newsletters.

"Articles like these never really bothered me," recalls Miller. "I actually found them very flattering."

Miller firmly believed that he was right to insist that NAR members abide by the safety code, and he stood by his decision that if a flyer could not follow the rules, then he or she should voluntarily withdraw from the NAR.

However, Miller was not altogether inflexible. As president of the NAR, he believed he had a duty to not only protect the organization's present interests but also see over the event horizon into the future of hobby rocketry. Although Miller's early view on higher-power rockets was skeptical, the NAR president eventually got a glimpse of the future that changed his outlook and high-power rocketry forever. One day, in the not-too-distant future, Miller and others would lead the NAR into high power. And the NAR on Miller's watch would greatly influence not only the continued development of LDRS but also the yet-to-be-created

Tripoli Rocketry Association. Eventually, both the NAR and high-power rocketry would join together to fight against larger and much more formidable enemies than each other. And Miller would be an indispensable part of that fight.

James Patrick "Pat" Miller was born in the small town of Portales, New Mexico, in February 1949. His parents, Jamie and Rosemary, met during World War II. Rosemary was an Irish American from New York whose family had immigrated to America in the 1840s during Ireland's potato blight. In 1943, Rosemary worked as a clerk for the United States Navy. While stationed in San Francisco, she met Jamie Miller, who was also working as a navy clerk. They married in 1948 in New York and moved to New Mexico, where Jamie's family had lived for many years.

Jamie Miller's family was originally from Texas and West Virginia; they moved to the Southwest in the early 1900s to set up a homestead near Portales, which was then part of the New Mexico Territory. New Mexico was granted statehood in 1912, and the family's square mile of land became theirs, free and clear. Jamie's grandfather opened a laundry in Portales, and when Jamie and Rosemary were married, they moved to the Southwestern town to help run the family business. Their son, James Patrick, was born the following year, in 1949.

Pat Miller's interest in outer space started early and was initially triggered by the type of personal experience unavailable to most city dwellers.

"I got interested in astronomy when I was about five," he recalled. "You could see the Milky Way Galaxy by just sitting on the front porch."

Miller's fascination for space was further engaged in 1957, when the Russians launched *Sputnik*, and like thousands of other young Americans, he followed the American space program with great interest.

"I was only eight years old when *Sputnik* was launched, and it made a great impression on me," recalled Miller. "But I also couldn't wait until we launched our own."

America was caught by surprise by the Soviet Union's bold satellite launch in October 1957 and tried to catch up as quickly as possible. On December 6, 1957, the United States Navy launched a *Vanguard* rocket from Cape Canaveral. The rocket carried a small satellite that weighed about 3 pounds. The launch was touted as America's response to the Soviet's presence in space. Unfortunately, it didn't go well for the Americans. The rocket exploded on liftoff. Its tiny satellite was propelled by the explosion into the nearby weeds. To add insult to injury, the lost satellite's signal beeped away over Cape Canaveral's radios.

A reporter covering the embarrassing disaster quipped, "Why doesn't somebody go out there, find it, and kill it?"

"I went to school that day and got home thinking for sure that it was in orbit," said Miller. "But I learned later that it blew up on the pad. I was disappointed, but just a few months later, they launched *Explorer*, and that was successful."

As the space program picked up pace and during the subsequent *Mercury* and *Gemini* era, Miller was often absent from school on launch days, staying home with his parents' permission to watch the America's manned rocket ships take off from Cape Canaveral.

When Millar was in junior high school, his mother bought him a telescope for Christmas. Soon, he was reading whatever he could find on astronomy. He observed the rings of Saturn and the moons of Jupiter from his home in Portales. In high school, he constructed his own observatory out of a refurbished building on a friend's farm outside of town.

"We built a flat platform, and then with a crane, we lifted the telescope onto the roof," recalled Miller. "I had my books and my telescope there, and I would go there on the weekends and stay overnight and look at the stars. But anytime you walked on the roof, it jarred the telescope. This was before I knew about the importance of having independent piers for telescopes."

Miller's interest in model rocketry began in 1967, when he and some friends bought an Estes *Astron Scout*. He read about the rocket in the back pages of a science magazine.

"You got a launcher, a rocket, and three engines for less than $10," he said. "I always wanted to send something up, and it was pretty exciting when the rocket launched. We went out immediately after that and ordered some more rockets through the mail."

Miller helped organized his first rocketry club, which he and his friends called ARC-Polaris. "We organized four or five national conferences as ARC-Polaris," said Miller. "We had read about an MIT model rocketry convention, and we said, 'We could do that too.'"

Miller and his friends called their gatherings the Southwestern Model Rocketry Conferences. "We organized our first conference in

1969, and they continued through at least 1973," he said. "We invited guest speakers from Los Alamos, the White Sands Missile Range, and the University of New Mexico. And even NASA sent us a one-third-scale model of the *Apollo, Gemini,* and *Mercury* capsules."

Miller even talked his way into a rocketry conference hosted at the United States Air Force Academy in Colorado Springs. "I drove up there uninvited and just walked in and said that I would like to have a model rocketry conference there," he said. "They said sure. That was all there was to it!"

Miller graduated from Portales High School in 1967 and enrolled in Eastern New Mexico State University. "I still wanted to be an astronomer, but that got sidetracked as they did not have an astronomy program there," he said.

He was encouraged by instructors to take some math courses, and it quickly became apparent that he had a unique talent for mathematics. In 1969, he transferred to the University of New Mexico in Albuquerque, where there was an astronomy program. He graduated in 1972 with a degree in astronomy and mathematics and then earned a master's degree in mathematics in 1974, also from UNM.

Miller joined the NAR in 1973 after he and another student published several articles in the NAR's model rocketry magazine. "The articles had to do with solving certain mathematical equations having to do with the motions of rockets," he said. "It put my name out there, and the NAR invited me to join for free after publishing the articles."

His subsequent rise through the ranks of the NAR hierarchy was meteoric. Between 1975 and 1979 and while earning his doctorate at UCLA, Miller began working for the NAR on various special projects. In 1976, he was elected to the board of trustees.

Miller met NAR founder G. Harry Stine at a NAR function in California. The two became friends for life. Stine also played a role in nominating the young mathematician to be president of the NAR in 1978.

"I was just in the right place at the right time," recalled Miller. "At the time, the president of the NAR, Manning Butterworth, was living in Santa Barbara, and there was another board member in San Diego. I did a lot of volunteer work for the organization, and I was very active as a trustee. Manning was an astrophysicist, and he was appointed to a position at Cardiff, Wales, and so he left in the middle of his term as president."

Miller was nominated to replace Butterworth at the NAR convention in Chicago in February 1978. At the age of twenty-nine, he was president of the NAR. Although this was only an interim appointment, Miller was to serve out Manning's term and then successfully attain reelection on his own. He would remain at the helm of America's largest rocketry organization for nearly two decades.

When Miller took over the reins in 1978, the NAR was a fairly small organization. NAR membership reached a peak of around five thousand in the early 1970s, almost in parallel with the rise of the *Apollo* program. But as the decade progressed, membership steadily declined.

"We had 1,700 members, and we had a very, very tight budget," recalled Miller when he took over.

By 1979, the NAR was almost closed because of a lack of revenue.

"When the board met in Chicago [in 1979], we didn't know whether we had sufficient funds to keep the organization going for another thirty days," said Miller. "To be honest with you, we just sat back, crossed our fingers, and hoped for the best."

It was not the best of times for the NAR. But the board worked well together with Miller at the helm, and they were determined to expand the membership.

"We were all longtime members of the NAR, and we were concerned over the relative inactivity of the organization in the 1970s after the euphoric heyday of the NASA *Apollo* lunar landing program," said Stine later. "Membership was dropping. Services were declining. Estes was the only surviving model rocket company doing much of anything in the marketplace. We could see our beloved organization slowly slipping into obscurity, and we decided to do something about it."

In his new role as president, Miller was tested soon, and he was tested often.

"He first analyzed and reanalyzed the functions and purposes of the NAR in model rocketry and how he might organize a group of wild individuals to give much of their spare time to the NAR as working officers, trustees, and committee chairmen," said Stine.

Miller brought much-needed order to the organization, and he led the NAR through many important changes in his first years as president. He also helped the NAR take on government regulators.

LARGE AND DANGEROUS ROCKET SHIPS

In 1978, the DOT decided to place more restrictions on the shipment of certain model rocket motors in interstate commerce. The restrictions were purportedly created because someone in the government deemed model rocket motors dangerous to ship. This was despite the fact that there had been no shipping incidents during the entire history of model rocketry.

"It would have increased the costs of motors or otherwise prevented manufacturers from making full use of the U.S. Mail to ship motors," said Miller.

As the new NAR president, Miller took quick action to protect the hobby. With assistance from Stine, Miller guided the NAR through a series of burn tests on the El Dorado Dry Lake Bed near Las Vegas, Nevada. The tests were conducted to simulate the worst possible conditions to which large model rocket motors, especially composite motors, might be subjected to in a serious transportation accident. Among those who had participated in the tests were representatives from Estes and Centuri as well as Gary Rosenfield. It was during these tests that Miller and Rosenfield had their discussions regarding the future of composite motors.

"We dug this big hole in the ground, and we tossed cartons of engines into the hole where we had a fire burning, and we watched it all go up in flames," recalled Miller, who said that they would measure how far these motors would fly out of the pit.

The results of the tests provided evidence that the DOT's proposed rules were unnecessary.

"We filmed it all and sent the film and our information to the department, and they issued us an exemption for certain quantities of model rocket motors under DOT Section 7887."

It was an important victory for the NAR, and it helped confirm that the storage and transportation of composite model rocket motors was no more hazardous than storage and shipping methods applicable to traditional black-powder motors.

The NAR began to move toward its own study of higher-powered rocketry within months after Chris Pearson's LDRS-1. In January 1983, representatives of the NAR and the NFPA Committee on Pyrotechnics, the NFPA subcommittee responsible for rocketry-related regulations, convened in Anaheim, California, to discuss the definitions of model rockets and model rocket motors. The meeting was arranged by Harry Stine, who had been on the committee for many years. Manufacturers of commercial rocket motors that exceeded the NAR's motor limits (A to F) were invited to attend and provide their input during the meeting. When the discussions were over, the NAR asked these commercial motor makers to submit proposals to both the NAR and the NFPA recommending changes in the model rocketry safety code. The changes would deal directly with the definitions of both model rockets and model rocket motors. These proposals would then be reviewed by the NAR and the NFPA, among others. The purpose of this work was to explore the possibility of adding G motors to the Code for Model Rocketry.

Following LDRS-2 and in late 1983, the NAR formed a blue-ribbon commission to evaluate proposals for larger motors in model rocketry. The commission would be chaired by Lt. Arthur "Trip" Barber, a respected model rocket flyer and U.S. naval officer and former chairman of the NAR Standards and Testing Committee. Among other things, announced Miller, the commission would address consumer safety and whether or not G motors could be incorporated into model rocketry without unreasonably increasing risks to participants and spectators.

Miller had been president of the NAR for five years when he oversaw the creation of this first blue-ribbon commission in 1983. He saw higher-power rocketry slowly grow in the late 1970s, and he was aware of the success of LDRS over the previous two years. Still, as the NAR official ultimately charged with protecting the integrity of model rocketry, he could not charge forward with abandon, as individual flyers were free to do. He moved forward with the NAR's study of a revised safety code, but he cautioned members to be patient.

"Suppose the definitions are changed to include larger rocket motors. Will there be an increase in the serious accident rate?" asked Miller. "Will there be a hobby-related death? Is there increased risk of property damage? The blue-ribbon commission will study these issues of consumer safety, along with other issues related to this proposed change in the hobby definition . . . Should the definitions be changed and injuries increase or property damage become frequent, it is completely possible that model rocketeers will find themselves back in 1958. Those same state and local laws will be rewritten to prohibit model rocket use, and twenty-five years of a safe hobby will be just so much history."

Actual work by the commission would not begin until early 1984, Miller told the membership. And the review process could take up to five years, he added. In the meantime, the model rocketry safety code must still be followed, he said.

To some outsiders interested in high-power rocketry, the NAR's recent actions expressing some interest in amending the safety code were too little, too late. To others, the organization's movement toward larger motors was simply not fast enough. As president of the NAR, Miller charted his course carefully. He was not in rocketry for the short term. Indeed, he would preside over the NAR for another decade. And under his leadership, the NAR would ultimately embrace not only revisions to the model rocketry safety code but all of higher-powered rocketry as well. But that would come later—much later.

In January 1984, the NAR blue-ribbon commission began exploring the possibility of expanding the definition of model rockets and model rocket motors to include G motors and slightly larger rockets. Commission members were directed to "investigate the safety implications of a proposal to expand the allowed size of model rockets and model rocket engines." They were experienced in model rocketry and included engineers, physicists, and even a physician. In addition to Chairman Barber, the commission included rocket scientist James Barrowman, who was a project manager for certain space shuttle payloads, and model rocket engine maker Vern Estes, founder of Estes Industries, the largest model rocket engine manufacturer in the world. NAR cofounder Harry Stine was also a member.

Miller charged the commission with coming up with recommendations for the NAR board of trustees in one year, by early 1985. In the meantime, Miller decided that it was time that he attended a high-power rocket launch. So the president of the NAR made plans to travel to LDRS-3 in Medina in 1984.

Pat Miller's first visit to LDRS came in an era when higher-power rocketry was advancing quickly.

"It was now apparent to me that high power was an activity that was not going away," he said later. "I wanted to get a sense of who the people were, and when I saw what was there, I decided in my mind that the NAR would have to get involved in high-power rocketry."

Miller had plenty to see at LDRS-3. For the third year in a row, the Wagner farm hosted LDRS, this time on August 11 and 12, 1984. There were few, if any, changes on the range from the prior two events. The waiver remained a paltry, by today's standards, 2,000 feet. And by now, that waiver was routinely violated. The high-power rocket motors at LDRS were getting bigger every year. And they were propelling rockets higher than ever before.

On Saturday morning, August 11, Scott Dixon of Vulcan launched an L-powered rocket at LDRS-3. This was the first L motor to fly at an LDRS. Dixon's rocket was a "minimum diameter" vehicle. This meant the airframe of the rocket was only a fraction larger than the diameter of the L motor. As a general principle, motor thrust, rocket weight, and aerodynamic drag are the three primary factors that determine maximum altitude of a high-power rocket. Minimum-diameter rockets are typically lighter and have less drag than larger vehicles, and with the right motor, they can achieve fantastic altitudes. Dixon's rocket had fins epoxied directly to the motor case, a Vulcan Systems L1780. He and Gary Rosenfield loaded the 3-foot-long rocket on the pad, hooked up the igniter, and retreated for cover. The crowd was on its feet in anticipation of another altitude barrier about to be broken. When the countdown hit zero, the slender missile streaked away in a hail of fire and smoke. The altitude reached was anyone's guess. Dixon's flight inspired all who were in attendance. In just two years, high power at Medina had progressed from G, H, and I motors to an L motor.

"The rocket was truly awesome as it streaked with a deafening roar into the sky," wrote one observer. "There was no mistaking this rocket as a model."

Recovery systems being what they were in 1984, Dixon's 2-inch-diameter missile was never seen again. Presumably, it remains planted to this day, nose first, somewhere in the soft soil in Medina County along Spieth Road.

The crowd at LDRS-3 was larger than the previous year. But there were still no flight cards or formal statistics kept at the two-day event. It was reported that more than 140 people were on hand, leading to the claim that LDRS-3 was the biggest hobby rocketry event in America since the NAR Internationals in 1980. Whether this claim was true or not, LDRS was obviously growing. But it was still small and informal, so informal that an FAA representative hand-carried the altitude waiver to the rocketry field on Saturday. (He must have left soon afterward or been oblivious of what was really happening with the 2,000-foot waiver.)

The L&K Motel continued to serve as the unofficial meeting place for LDRS, with some flyers arriving two days before the launch to socialize. People came from around the nation, and the rocketry activity at the motel lasted well past midnight each day.

LDRS-3 had plenty of model rocket flights, as was the custom for every high-power launch throughout the 1980s. The number of higher-powered rockets at LDRS was growing. There were more G, H, and I motors and several J motors this year. Most of the high-powered motors were built by Vulcan or AeroTech, but rockets were also powered by motors sold by U.S. Rockets or Reaction Labs.

Flyer Doug Forrestor brought two large rockets to LDRS. The first was an 8-foot-tall multiple-motor missile called *Grand Slam*. Forrestor spent Friday night at the motel preparing the rocket, which was propelled by thirteen motors, F through H power. He brought his own launch pad for the flight, an LDRS tradition that would grow in the future for specialized projects. At ignition, the hefty rocket disappeared into cloud cover low in the sky and, like Dixon's L flight, was never seen again. Forrestor returned later in the weekend to fly another big rocket, his four-finned *Blue Jay*, on a J200 for a nice flight and recovery.

There were several flyers from Florida who attended LDRS. One of them was Bill Barber, who launched one of the first high-power two-stage rockets at any LDRS. Barber, who one day would be on the Tripoli board, flew an *Explorer 400* kit that held a Vulcan I500 motor and four Ds in the booster and two AeroTech H120s along with four G motors in the sustainer. Barber's rocket featured a homemade electronic timer used to ignite the motors in the sustainer after takeoff, another rare event for the times. The flight was flawless and undoubtedly broke the waiver. It had a perfect recovery. Barber later flew another rocket on J power that shredded seconds into its ride.

Scale model rockets such as *Pershing* missiles and *Saturn V*s were popular at LDRS-3, as were cluster-powered rockets of all kinds. Many rockets were powered by E, F, and G clusters. The overall number of clusters that worked was increasing, yet some still shredded in flight. The term "CATO" was also working its way into high-power rocketry's evolving vocabulary; a CATO described the catastrophic failure of a motor on takeoff or in flight.

The Ohio machinist and his wife who attended LDRS-2 as observers, Ron and Debbie Schultz, returned in 1984. During the prior year, Ron built many new rockets; most carried custom-sewn bright-colored parachutes made by Debbie. The beautifully painted rockets with their distinctive parachutes delighted the crowd. They brought more than two dozen rockets to sell at LDRS. They sold every one, packed in plastic bags under the name of Debbie's home-based sewing company, Lots of Crafts. In the near future, Ron and Debbie would apply an acronym to shorten the name of their company. It would become known as LOC and would one day become the largest commercial rocket manufacturer in high power.

There was at least one minor controversy at LDRS-3. Upon their arrival at the field, people were told that a "donation fee" was expected from anyone who intended to launch a rocket.

"I went around asking everyone for a donation because up until that time, all the expenses were coming out of my pocket," explained Pearson.

It sounded like a reasonable idea. But some people protested not only the notion of a fee but also the fact that attendees were not informed ahead of time that a fee would be assessed.

"What could possibly cost so much?" wrote one flyer later in a letter to the editor of the *High Power Research* magazine. "The flying field was free, the PA system belonged to someone in attendance at the meet, and there were no port-o-let or restroom facilities."

The fee that Pearson was asking for? One dollar. (LDRS flying fees today can run $50 or higher.)

This minor dispute aside, it was another good year at LDRS. High-power rocketry was still without any formal national organization, and it still lacked a safety code or any type of legal recognition. For the most part, people flew what they wanted to fly and did what they wanted to do. The FAA waiver was all but ignored. There was still no insurance, either; fortunately, there were also no mishaps.

But with a steady influx of new flyers and more powerful motors each year, the day was approaching when launch organizers would have to consider financial responsibility and perhaps even a formal safety code as part of the LDRS plan. And now that the NAR's blue-ribbon commission seemed actively interested in high power, more flyers began to think about the future of the young hobby. Some flyers expressed a desire to see the NAR take control of high-power rocketry, to establish a high-power safety code, and to administer a national high-power association. Clearly, the NAR already had the organizational framework and long-term lobbying experience to embark on such a venture. Other flyers were not so sure.

Pat Miller was invited to speak to all flyers at LDRS-3 on Saturday night. In a bit of historic irony, Miller, the sitting president of the NAR, was the first Saturday night guest speaker in LDRS history. Despite strained relations between the NAR and some members of the high-power community, Miller was warmly received at both the launch and during his presentation on Saturday night.

"We had a town hall meeting that was quintessential Americana in this small town in Ohio with patriotic tapestries and flags," recalled Miller. "I said we need to have a way for the NAR to communicate with high power and talk about the things that we need to talk about."

He enjoyed LDRS and was impressed by what he saw there, and Miller fielded questions at the meeting and talked with flyers on the field throughout the weekend.

Following the launch, Miller informed the high-power community that the NAR wanted to initiate discussions regarding possible joint programs involving both the NAR and high power. To facilitate talks, said Miller, it would be necessary for high-power users to follow NAR bureaucratic procedures. This meant forming a high-power safety committee, developing proposed rules and regulations, and participating in the NAR's annual meeting in February 1985. This would be at or about the same time as the NAR's blue-ribbon commission's study was due.

Three high-power flyers stepped forward to create what became known as the LDRS-HPR Committee: Chuck Mund of New Jersey; Chris Johnson of Ohio; and Bob Geier of Massachusetts. The committee took shape after LDRS-3 in the fall of 1984. One of its first actions was to submit questions to high-power users around the country. The purpose of these questions was to assist the committee in drafting a rudimentary high-power safety code that could be presented to the NAR the following year.

In February 1985, the LDRS-HPR Committee (it was also sometimes called a "commission") formally petitioned the NAR, asking the model rocketry organization to expand its scope to include high power. That proposal, delivered at the NAR's annual meeting in Washington, D.C., was entitled "Petition to the NAR Requesting High-Power Rocketry Service." Among other things, the petition asked the NAR to incorporate high-power rocketry into its organization. The petition affirmed that high-power flyers were interested in several important services through the NAR. These services included flight insurance, motor certification, legal assistance for the transport of high-power motors, and representation on the NAR by means of a high-power rocketry committee. One committee member described the recommendations of the petition as a "merger of high-power rocketry (H motors and above) with the NAR."

As part of this invitation to take command of high power, the committee also presented the NAR board of trustees with a special safety code drafted for high power. The safety code created by the LDRS-HPR Committee was one of the first—if not the first—high-power safety codes created for the hobby. In another bit of historic irony, this particular code was originally designed for the NAR. The code was several pages long and was, in many ways, a limiting document.

According to the proposed code, high-power motors would be defined as "commercial only" and had to be fully assembled and preapproved by the NAR. This meant no research, homemade, or experimental motors would be allowed. The proposed code also set an upper motor limit for high-power rocketry at 5,000 Newton seconds, which is an L class motor. Although this was an enormous motor for its day, it was unlikely that the NAR would consider this proposed expansion in 1985 as the NAR was still studying G motors. Indeed, as L motors had flown at LDRS-3 and Ms were not far off, the acceptance of this proposal might have put the brakes on additional high-power motor development almost immediately. The proposed safety code also stated that a high-power rocket could not exceed 22 pounds, fully loaded. This was a strange number too. Although it was higher than the NAR limit of 1 pound, rockets exceeding 22 pounds had already been launched at LDRS-3 and elsewhere by early 1985.

It is unknown whether the LDRS-HPR Committee's proposed rules had the support of the high-power community, whatever that was, at this time. There was still no national high-power organization, and LDRS was still an unassociated fringe event that was owned lock, stock, and barrel by Chris Pearson. Sure, there were people flying high-power rockets here and there at places like Lucerne, Smoke Creek, or annually at Medina. But it was haphazard. There were very few, if any, rules. Many flyers liked it that way. Even if the NAR had adopted high power at this early date, some flyers would refuse to submit to the NAR's rules and regulations—or the rules and regulations of anyone else, for that matter. And these flyers were not likely to agree to the recommendations of the LDRS-HPR Committee with regard to a new safety code for high power.

The presentation by the LDRS-HPR Committee to the NAR board coincided with the blue-ribbon commission's final report on high-power rocketry, released February 16, 1985. Together, these two studies offered the NAR an opportunity to not only move beyond the model rocketry hobby but also possibly take control of high power on a national level. However, the NAR's interests in higher-power rockets were still fairly limited. Although President Miller and the board were moving toward adopting G motors, the idea that model rocketry rush headlong into H, I, and J motors was not on the table. The NAR would not be ready to take the full leap into high-power rocketry for several more years—and then only after extensive study and research.

The blue-ribbon commission's final report was released in February 1985. Its primary objective was to analyze whether the NAR could modify its weight and power rules without any threat to what the report described as the NAR's "extraordinarily good twenty-seven-year safety record." The report's introduction claimed that since 1957, when hobby rocketry was founded, approximately two hundred million model rockets had been launched. This number seems almost too high to be true. Yet based on this estimate, the commission found that the incident of damages or injuries associated with the hobby was one per million flights. Regardless of the accuracy of these numbers, it was undisputed that model rocketry under the NAR had a good safety record.

Chair Trip Barber's fifty-eight-page written study was impressive. It was presented and then debated at the NAR board's annual meeting, held that year in Washington, D.C. It contained detailed statistical studies regarding, among other things, the potential ballistic hazards of model rockets engines (up through G engines), the analysis of average thrust limits, impact probabilities, and G engine flight tests.

"By far the most interesting (and amusing) of the results were the impact tests," wrote one commentator at the time. "Rockets of varying composition nose sections were fired down 20 meters of guide wires into targets like plate glass, plywood, *and a dead turkey*."

Some of the studies that were not released actually caused concern for the NAR, recalled Miller later. "Initially, Trip Barber ran some tests that kind of scared us in a way, such as a G-powered rocket that penetrated an oak tree! For this reason, we ended up putting some limits on average thrust."

The G engine flight testing was conducted by commission member Harry Stine and others near Phoenix, Arizona, in late 1984. Stine's testing, which involved numerous flights of model rockets on various motors, led him to conclude that an increase in the maximum weight of a model rocket to 2.2 pounds "would not materially decrease the safety of people and property on the ground or in the air."

Stine also concluded that an increase in the maximum total impulse from an F motor to a G motor would not require any material changes in the safety operations then in use under the safety code. What's more, said Stine, increasing the motor size might actually improve the hobby of model rocketry.

"I learned during the flight tests that the general level of craftsmanship in model rocketry has declined in the last ten years," said Stine at the time. "Many of the principles of design and construction that we used for type F model rockets in the 1960s have been forgotten."

Stine observed that model rockets were lighter than they used to be and that many flyers constructed their rockets from kits with thin-walled body tubes and thin plastics. The introduction of G motors would advance the technology of model rockets, including design and construction, observed Stine, and it would encourage craftsmanship "because without craftsmanship, type G altitude models simply will not stay together."

The G motors used by Stine during his tests in the Arizona desert were provided by AeroTech. NAR members were impressed by Gary Rosenfield's propellant, although they noted that one drawback of the composite motors was their virtually smokeless flame, which made them nearly impossible to track.

"Our lightweight rockets left the pad so quickly that they seemed to disappear from the pad," reported one flyer. "By the time I could look up, the model was just an orange dot in the sky."

In its final report, the commission also reported that its members had attended high-power launches during the previous year. They found that generally speaking, high-power rockets had a higher level of craftsmanship when compared to the average model rocket. The commission estimated that at least three hundred people were now involved in high power, "conducting a thousand or so flights each year in six to ten organized launches." Many of these people were former NAR members, they observed, adding that these members may have left the NAR because it forbade its members from flying outside the model rocketry safety code. The commission found that because high-power flyers were mature and exercised good judgment and also because their operations were conducted in remote locations, there had been no major damage or injury in the hobby after several thousand flights. In other words, high-power rocketry was safe—so far.

The commission recommended the NAR raise the maximum take-off weight of a model rocket to 1.5 kilograms, a little more than 3 pounds. It also recommended increasing propellant mass to the equivalent of a G motor. The commission found that generally speaking, heavy but slow rockets were less dangerous than fast lightweight models. This prompted them to devise a system of approved minimum thrust times that would effectively limit average thrust to approximately 80 Newtons. Other changes included but were not limited to greater launch pad safety distance and new launch pad specifications.

It was a big step for the NAR, although it was not enough for flyers who were already using motors above the G-impulse range. The commission's suggestions to increase rocket and motor size required a revision to the Code for Model Rocketry—NFPA 1122. So they presented a draft of a revised model rocket code to be considered by the NAR's board of trustees. Once approved by the board, the draft would go to the NFPA's Committee on Pyrotechnics and then ultimately to the entire NFPA for approval.

The proposed revisions to the Code for Model Rocketry incorporated not only the engine and take-off weight modifications but also several other amendments that commission members believed were necessary to keep the hobby safe as the NAR moved toward high power. The amendments included several common-sense proposals, including the requirement of an audible 5-second countdown for all launches to ensure that everyone in the vicinity was aware of a take-off. It also called for no launches if winds reached 20 miles per hour, no rocket flights into clouds, and that launch rod angles not exceed 30 degrees of vertical. All these suggestions would eventually work their way into the safety code and one day into future codes written exclusively for high-power rocketry.

The NAR board approved the commission's changes at the end of what one writer described as a marathon session of meetings at their February gathering in Washington, D.C., in early 1985. The next step was to present the proposal to the NFPA, which would happen later that year. The process was slow.

"Work has already begun on making the changes with the NFPA and the FAA," Miller told the membership in December 1985, nearly a year after the board approved the action. "It looks like October 1986 is the key target date when most of the changes will be in effect."

Miller also reported that the NAR was working on insurance coverage for higher-powered rockets that he hoped would coincide with the final adoption of the new rules the following year. Until the new rules were adopted, however, NAR members were still limited to F motors.

"I ask that all of you be patient just a while longer as the administrative wheels slowly grind to make the necessary changes," said Miller.

Lost in the shuffle of these discussions and proposed changes between the NAR and the NFPA and others was the LDRS-HPR Committee. Initially, the high-power committee members were pleased with their reception at the Washington, D.C., NAR board of trustees meeting.

"Chuck [Mund] and I presented the LDRS [committee's] proposal to the board, asking for an eventual merger of high-power rocketry (H motors and above) with the NAR," wrote member Bob Geier. "After a short discussion, the NAR board agreed to continue the dialog with high-power flyers by forming a second blue-ribbon commission to discuss our proposal. Since this was exactly what we had hoped for, we were pleased with the board's new openness toward high-power users."

In the short run, the proposals made by the LDRS committee to the NAR board appeared to go nowhere. The big step for the NAR in 1985 was acceptance of the G motor and revising the model rocketry safety code. The idea of the NAR adopting a high-power safety code, flying H through L motors, or taking over high-power rocketry was out of the question—at least at this moment.

President Miller charged the second commission with the responsibility of meeting with the LDRS-HPR Committee and attending high-power launches beginning in the summer of 1985. The goal of this interaction, said Miller, was threefold: (1) establishment of a clear definition of the high-power rocket sport; (2) derivation of safety procedures for high-power consumers to follow; and (3) accommodation on the part of the NAR whereby its members can engage in the sport without jeopardy to their memberships. No guarantees were made that this new commission would accept or adopt high-power recommendations made by the LDRS-HPR Committee, but there would be ongoing dialogue, said Miller.

Among the possible outcomes to be considered regarding this future interaction between the NAR and high power, said Miller, would be to allow NAR members "joint membership" in a yet-to-be-established high-power rocketry organization or even the incorporation of high-power rocketry itself into the NAR. Although it had taken Miller and the NAR a few years, the largest rocketry organization in the world was now clearly moving toward high power. By 1985 and during most of the remainder of Pat Miller's presidency, with one notable exception, the NAR's relations with high-power users would steadily improve.

Not everyone who was interested in promoting high power was enthusiastic about the talks between the NAR and the LDRS-HPR Committee in 1985 or with the NAR's slow movement toward higher-power rocketry in general. Some people remained suspicious regarding the NAR's intentions. Some flyers had trouble believing that the same organization that had limited its members' use of high-power motors a few years before could now be trusted with the future expansion of the hobby. On an even more basic gut level, some high-power users preferred to carry a grudge against the NAR in perpetuity.

Perhaps more importantly, many flyers no longer considered G-class engines to be high-power rocket motors. By early 1985, many flyers were launching rockets beyond the G-impulse range at not only LDRS but also several other locations in America. At the same time that the first blue-ribbon commission approved the use of G motors, many high-power flyers (a growing number of whom had never been members of the NAR) were already launching H, I, and J motors in rockets that weighed up to 20 pounds or more. A few people were even flying L-powered rockets. The blue-ribbon commission's big leap to a G motor for the NAR was viewed by these flyers as no big deal. By 1985, many high-power enthusiasts were simply not interested in G motors anymore. They were already moving further up the motor alphabet.

From this group of independent flyers—those people who had either left the NAR for good, never belonged to the NAR, or just wanted their own independent rocketry organization—a new chapter in hobby rocketry history was about to be written.

6

The Tripoli Rocket Society

In the spring of 1957, a student at Richland Township High School in Gibsonia, Pennsylvania, built a rocket ship in shop class.

He started with a piece of half-inch-diameter gray galvanized pipe that was almost a foot long. The pipe was threaded on the outside edges at either end. He fabricated three fins from sheet metal and welded each fin to one end of his rocket tube. Once the fins were in place, he sealed that end of the airframe with a pipe cap that threaded into place. He filled the opposite open end of the pipe with scores of flammable match heads he had meticulously snipped off tiny wooden matchsticks. He packed the little red match heads into the tube and then sealed the forward end of the pipe with another threaded cap. On top of that cap, he attached a crude nose cone carved out of a hunk of wood. He then drilled a small hole at the bottom of the rocket, and into that hole, he placed a piece of fuse obtained from a farmer friend who would occasionally blow up tree stumps to clear his field.

The student's plan was to light the fuse in the aft end of the rocket, which, in turn, would ignite the match heads jammed together inside the pipe. The ensuing fire would propel the vehicle high into the air—or so he thought.

On Science Day, the would-be engineer carried his small ship to the football field for a demonstration of rocket flight for his class. With his peers and teacher looking on from a safe distance, he lit the fuse and scrambled across the green grass some yards away, where he began a ten count. When the countdown reached seven, the rocket—which was, in reality, a form of pipe bomb—exploded in a bright flash.

There was nothing left other than tiny metal parts, a cloud of smoke, and a deep hole in the gridiron. The "rocket" was no more. Fortunately, no one was hurt. But the student was kept after school and sentenced to cleaning up the mess he had made and refilling the gash in the football field.

The young man learned several lessons that day, not the least of which was the difference between a functional rocket and a real pipe bomb. He also learned of the importance of a working nozzle in a rocket motor and, perhaps most importantly, how close he had come to being maimed or even killed with his dangerous match-head device. It would be nearly twenty years before he would attempt to build any kind of rocket again. Yet one day he would help launch one of the largest amateur rocketry organizations in the world. The student's name was Tom Blazanin.

Tom Blazanin was born in July 1941 in Pittsburgh. His father, also named Tom, was a machinist from West Virginia who worked in steel mills most of his life. His mother, Julia, was a homemaker. The couple met at a Guy Lombardo dance in Pennsylvania in 1939. They had three children: Tom and daughters Cassandra and Christina.

Following high school, Blazanin enlisted in the United States Air Force. After basic training, he was sent to electronics school in Biloxi, Mississippi, where he floundered and ended up graduating "by the skin of my teeth." Blazanin was then asked where he wanted to be assigned for active duty as a military repairman.

"I wanted to go west, where I had never been before," he told his commanders.

In 1960, he was shipped off to Edwards Air Force Base at the edge of the Mojave Desert in Southern California. In the 1950s, Edwards was the center of America's cutting-edge aerospace technologies and was a part of the fledgling space program. Chuck Yeager broke the sound barrier in the *Bell X-1* at Edwards. When Blazanin arrived, the base was home to the nation's most experienced test pilots and future astronauts. Here above the desert, they piloted experimental flying craft ranging from trainers to supersonic fighter jets to the sinister-looking North American *X-15*.

"People said I was lucky to get assigned to Edwards because that is where all the test pilots were," recalled Blazanin. "But I was nervous when I got there because I did not feel that I was the kind of repairman they were looking for. After all, I had trouble getting through training school. And as a young man, I was not sure that I was up to the task."

When he arrived at Edwards, Blazanin was assigned to the flight line, where he worked as a troubleshooter and radio systems repairman.

"The job turned out to be good for me because it forced me to apply myself," said Blazanin. "I paid attention and learned all the stuff they taught me at Edwards, and in a few months, I grasped the concept of troubleshooting."

He also discovered he had a natural aptitude for tracking down the source of an aircraft's problem and figuring out how to correct it. In fact, it was easy for him. He was only nineteen.

Blazanin spent the next four years in the Southern California desert. He saw the origins of some of this country's most sophisticated aircraft. He worked on planes like the Convair *B-58 Hustler* bomber, he saw the futuristic North American *XB-70 Valkyrie* fly, and he watched some of the early flights of the Lockheed *SR-71* spy plane. Once, he hitched a ride aboard a *B-52 Stratofortress* with a window seat, allowing him to watch an *X-15* get dropped from the wing of the bomber into flight.

Blazanin became an expert at troubleshooting, and he was reassigned from the flight line to the repair shop at Edwards, where he remained until his enlistment was up in 1964. After his discharge from the military, he returned to Pittsburgh and went into the car business. Eventually, he started his own business, doing custom body work and painting. But he never lost his love for the American West.

Blazanin had a daughter, Becki, and then a son, Jed. Many years passed, and one day, in 1978, Blazanin told his wife the story of his homemade rocket experiment in high school. Not long afterward, his family presented him with an Estes *Der Red Max* model rocket for his birthday.

"I built the rocket, and my daughter, who was four, was all excited, and she wanted to paint it pink," recalled Blazanin. "So we painted this Estes *Der Red Max* pink, put a couple of stickers on it, and went out to fly it. She really enjoyed it." He enjoyed it too.

In the Estes kit, there was an application to join the NAR. Blazanin mailed it in. Soon, he had an Estes catalog and was ordering more rockets from Estes as well as kits from the other large model maker of the day, Centuri.

"I built bigger and bigger rockets, and my daughter always wanted to come out with me to fly them. We finally bought an Estes *V-2*, and Becki thought that was the greatest thing in the world, flying on a D motor."

Blazanin eventually exhausted his interest in D motors. Then he and Becki discovered E and F engines, and soon, he was modifying model rocket kits to accept larger motors. The higher-powered models would fly higher than ever before, and he and Becki would launch their rockets in abandoned lots around Pittsburgh. However, he soon ran into the NAR's upper limits on rocket size and motor power. On his own and not belonging to any formal club, he began to lose interest in hobby rocketry.

Then in 1981, at a hobby shop in Pittsburgh, Blazanin spotted an early issue of Jerry Irvine's *California Rocketry Quarterly*.

"I thought, 'Holy cow, look at what these guys in California are doing!'" recalled Blazanin. "Korey Kline, in particular, had started a company called Ace Rockets, and I called him and ordered some kits, including a big three-cluster rocket called the *Allegro*."

Kline told him there were some guys flying higher-powered rockets in Ohio, and he suggested Blazanin contact them. He did, and he learned about new developments in higher-powered model rockets. He made more calls to people in California, including Irvine and Gary Rosenfield, looking for information on higher-powered rockets and motors. It wasn't long before Blazanin obtained his first G motor.

"I would fly in the south hills of Pittsburgh in a big dump," he said. "The area is now called Century Three, and they built a mall on it. U.S. Steel used to haul by train the ingot cars with slag from the mill at Homestead, and they would dump these ingots, also called slag, and they would roll down the hill like lava. Well, after the mill shut down, all this area was a barren open rock. It was a big plateau, a beautiful launch site, not far from downtown Pittsburgh. I built a bigger launch pad for these rockets and used an Estes controller, and it was just my daughter and me, and she got pretty excited. This was what I really liked. When I launched the G motor, it was like, 'Wow, this is what I want to do because it was bigger.' I think about this to this very day. When you flew model rockets, it was *pfffst*. But this was *whoosh!*"

Blazanin decided to let his NAR membership lapse as he knew he was going over the F motor-impulse limits on a regular basis. As a flyer of higher-powered motors, he still believed he was alone in the

Pittsburgh area. Shortly after LDRS-1 in 1982, which he missed because he did not hear about it, Blazanin was invited to Ohio to fly with Chris Pearson and his friends. At Medina, he met Curt Hughes, who was about to be expelled from the NAR for flying his high-power rocket at LDRS. Hughes was also from Pittsburgh, and the two became friends.

"Curt was making the motors himself," said Blazanin, who still remembered his own disastrous fuel-making experiment in high school. The idea of making one's own motor reminded him of the basement bomber days of the 1950s. "I said, 'This is dangerous,' and Curt said, 'No, it isn't. It's easier than black powder.'"

Hughes gave Blazanin a demonstration.

"Curt had a band saw, and he said, 'Look how safe this is,' and he went, and he took a G motor, and he cut right through it, and nothing happened," recalled Blazanin, who was shocked the fuel did not explode or even ignite. "I was dumbfounded. And to this day, I still have that motor. It was a Composite Dynamics motor. It was a great magic trick, I thought."

Blazanin attended LDRS-2 in 1983. He was impressed with the rockets, the higher-powered motors, the range setup, and the rudimentary safety precautions in place. Yet what mattered most was he was finally connecting, in person, with other adults from around the country who felt the same way about higher-powered rocketry as he did. In a short period, Blazanin seemed to be everywhere in higher-powered rocketry. He established contacts with several people who were influencing the early growth of the hobby, people like Irvine, Pearson, Kline, and Rosenfield. Blazanin believed there were other people around the country who, like him, would like the opportunity to join a national club dedicated solely to high power. Although he had no experience in creating a rocket club—or any other type of club, for that matter—he decided that he would start a national rocketry organization. He called it the Advanced Rocketry Society.

"I called it a society because to me, a 'club' was local, and I wanted it to be national," he explained. "I got people to join like Korey Kline, Gary Rosenfield, Chuck Mund, and others because they were big people in the hobby."

Blazanin charged members $15 each to join the Advanced Rocketry Society, and he started a newsletter that was two pages long.

"I asked people for articles for the newsletter, and I got a couple," he said, "but it was not enough." He quickly realized that he would be forced to write the articles himself. "I thought, 'Who am I to tell people how to do stuff when I am learning it myself?' It reached a point in time that I was not getting help or input, so I sent everyone their money back."

The Advanced Rocketry Society had about ten members—and then it was gone. Yet Blazanin would be remembered for trying to establish a national rocket society. In the future, he would be asked to try it again.

When Tom Blazanin met Curt Hughes in 1983, Hughes belonged to an obscure science club in Pittsburgh called the Tripoli Science Association. The association, known from 1964 to 1979 as the Tripoli Federation, was created by several high school students in the mid-1960s, including Hughes, Francis Graham, and Art Bower. Between 1964 and the early 1980s, the club's membership ebbed and flowed as its members moved on to other endeavors. But Graham, Hughes, and a few dedicated Tripoli members maintained the organization as a loosely affiliated science club dedicated to all things scientific, including astronomy and rocketry.

Beginning in 1968 and with help from his peers, Graham was one of several editors who helped write and publish a periodic newsletter for Tripoli members. The newsletter was called the *Tripolitan*. When Graham was at the helm, the *Tripolitan* reflected many of his own views of the scientific world, views that had shaped him since he was a boy.

"My grandfather, being an amateur astronomer, would show me craters of the moon, the planet Saturn, and lunar eclipses from our backyard," recalled Graham. "I also remember sitting down and watching the Disney series *Man in Space*. They had spaceships that went to Mars that were umbrella shaped. It was just mind-boggling, what could happen and what would happen."

Graham was captivated by the news on October 4, 1957, when the Soviet Union launched *Sputnik* into outer space. His interest in space travel increased during the *Mercury* and *Gemini* programs of the early 1960s. These programs were fundamentally changing man's relationship with the universe, believed Graham.

"People had different opinions and still do. Some people were totally uninterested, and there were even people who told me this was all nonsense and we should be spending the money on something else," he said. "Knowing what I did, even as a young boy, regarding the size and scope of the universe, I considered it innately absurd that we should be trapped on this little tiny planet."

From the late 1960s through the early 1980s, membership in the Pittsburgh-based science club waxed and waned. During this time, Tripoli members would participate in science-related activities, including model rocketry and, occasionally, forays into launching higher-powered rockets.

"Tripoli in the 1968-to-1970 time frame had a modest and even international expansion of prefectures, and this is the root of our current prefecture system," recalled Ken Good, a Tripoli Federation member and future president of the Tripoli Rocketry Association. "In this period, we had four Pittsburgh-area prefectures plus one in Erie, Pennsylvania, one in New York, one in Texas, and even one in Denmark. But we were too young and inexperienced and ultimately focused on our college educations to hold it together."

One of Tripoli's higher-powered rocketry experiments of the late 1960s was based on a rocket design from Bertrand R. Brinley's classic work *Rocket Manual for Amateurs*, released in the early half of that decade. Graham bestowed a name on the vehicle built by club members from Brinley's plans. He derived the name from a Latin phrase that means "Thus passes the glory of the world." The rocket was called *Gloria Mundi*. *Gloria Mundi* was a four-finned all-metal rocket that Graham, Hughes, Ken Good, and other Tripoli members launched from a quarry on the outskirts of Pittsburgh in 1969.

"We built about fifty prototype rockets after we purchased Brinley's book in preparation for our final flight of the *Gloria Mundi*," recalled Hughes in an interview years later. "We tried different propellant combinations and different tubing. We had lots of failures, with rockets that just melted on the pad."

Most of the early motors for *Gloria Mundi* rockets were based on caramel candy propellant, which was potassium nitrate and sugar, said Hughes. "The challenge with that was you had to melt it before you could cast it into the combustion chamber, and the chamber had to be the same temperature as the propellant. We would take this big giant cast-iron pot out . . . and make an open fire to safely melt this caramel candy propellant and heat the combustion chamber to cast it. The hard part was getting the temperature correct to cast the propellant without cracks forming and not getting small pieces of propellant in the threads of the motor case where you screwed on the nozzle. We had a lot of CATOs when trying to ignite the motors because of cracks in the propellant grains."

"It was the coolest thing we could think about in those days," recalled Ken Good of the rocket. "The airframe was steel tubing that Art Bowers's father obtained for us as he worked for Westinghouse and had connections with machine shops. He also got us the machined nozzle and the nose cone. But we ended up using a wooden nose cone. We thought the metal cone was too dangerous."

Ultimately, the propellant for *Gloria Mundi* was based on a zinc-sulfur mixture created by Hughes.

"We were just not getting anywhere quickly with the caramel candy, and the impulse was too low," recalled Hughes. "Micro-grain [fuel] wasn't much better, but it was easier to work with. You would just mix sulfur flour and zinc dust in the correct proportion."

There were several attempts to launch *Gloria Mundi* using this new fuel in March 1969. When the motors refused to light on the pad, Graham suggested using an inverted Estes model rocket engine to light the propellant. It worked, and the fuel in the 2-inch-diameter rocket ignited instantly.

"There was a big blast, and the rocket took off, and we could hear the whistle as it went through the air," recalled Hughes. "We had a half-inch steel [launch] rod that we had cemented into the ground. After the launch, the rod was bent over, completely horizontal to the ground."

"The darn thing flew perfectly," said Good. "The fins were made out of sheet metal and attached with set screws and a clamp. But we never saw it again. There was a lot of effort to find it, all sorts of expeditions, but it was never found. I think it came down into the nearby Youghiogheny River. We thought it was the glorious crowning of Tripoli as we knew it back then."

Indeed, the *Gloria Mundi* was a true high-power rocket, years ahead of its time.

"My later analysis indicated it was in the L-impulse range," explained Good, who recalled displaying the rocket at a local science fair. "A NAR official at the time described the *Gloria Mundi* as a 'hunk of pipe that will kill you,'" remembers Good. "They clearly thought of us as cowboy rocketeers and basement bombers."

As the years went by, Tripoli members kept in touch with one another via the *Tripolitan*, which was printed on a haphazard basis. At its zenith, the newsletter had a purported circulation of 100 to 125 copies per issue. Issues were created several times a year in the 1960s, dwindling to only a few issues a year—and in some years, none at all—by the early 1980s. The newsletter covered not only science and occasional rocket projects but also politics, economics, popular culture, philosophy, religion, and whatever else came to mind. The *Tripolitan* carried periodic stories about the development of new hobby rocket motors in the late 1970s and in 1981 published an article by Hughes that was similar to Gary Rosenfield's "Power Freaks" series that had appeared in *California Rocketry* earlier that year.

At the suggestion of Hughes, Tom Blazanin joined Tripoli in 1984. By now, the club had only a couple dozen members and held random meetings. A few members, like Hughes, were interested in high-power rocketry. But the club did not have a field to fly on. Blazanin soon found one.

"We would hold regular launches, and I liked it because we were flying bigger stuff. Yeah, we were also flying model rockets, but it wasn't the NAR," said Blazanin. "We had no insurance, and we were just a club, and we couldn't affiliate with the NAR because we were doing experimental stuff."

In early 1985 and over the course of three meetings, the seeds for the modern-day Tripoli Rocketry Association were planted.

On January 24, 1985, in East Pittsburgh, Pennsylvania, members of the Tripoli Science Association began preliminary discussions about splitting the association into two separate but related entities: the Tripoli Rocket Society and the American Lunar Society. Reorganization acknowledged the fact that many Tripoli members were more interested in astronomy than rocketry. Longtime Tripoli member Allen J. "AJ" Reed was named as the first president of the Tripoli Rocket Society, which was rededicated to "advanced rocketry." Reed was a science instructor at a local community college. Hughes was named vice president. Graham became president of the Lunar Society, which would focus on astronomy, a larger part of the Tripoli mission since the late 1970s. (Graham earned a master's degree in astronomy from the University of Pittsburgh in 1982.) He would also remain on as secretary of Tripoli.

The following month and on February 7, 1985, the leadership of the Rocket Society was adjusted again. Bill Kust was named as treasurer of the new club, and Blazanin was named "librarian." Reed and Graham remained president and secretary, respectively. At this meeting, a set of bylaws was adopted by the new rocketry club. These bylaws included several purposes for the new organization, including the goal to "develop and encourage the establishment of a safety program tailored to high-power large-scale advanced rocketry."

That spring and not long after the NAR's first blue-ribbon commission issued its final report, Blazanin was approached by two ambitious high-power flyers from Cincinnati, Mark Weber and J.P. O'Connor. Blazanin had met both men at LDRS-3 in Medina in 1984. According to Blazanin, Weber and O'Connor were trying to organize a national high-power organization.

"They talked with Gary Rosenfield," recalled Blazanin, "and he told them, 'If you want somebody to put together an organization, you should call Tom Blazanin. He tried to start the Advanced Rockery Society, and when he realized it wasn't going to fly, he at least gave everybody their money back.'"

At the request of Weber and O'Connor, Blazanin drove from Pittsburgh to Cincinnati to discuss the creation of a national high-power rocket club. One topic for discussion that evening was the NFPA's Code for Model Rocketry. Although it had been modified slightly since the 1960s, the code still limited *model* rockets to a certain weight, and it limited commercial motors to no higher than an F-impulse range. The NAR was slowly moving toward allowing its members to launch G-impulse motors, but NFPA 1122 was seen as a legal impediment to the future of high-power rocketry.

"At the meeting, Weber and O'Connor outlined their logic for getting around NFPA 1122 based on the belief that if we had a formal organization, we would be exempt by language found in the code," said Blazanin. "Therefore, they thought it was imperative that we start a high-power rocketry organization."

The rationale for an exemption to NFPA 1122 came from the language of the code itself. NFPA 1122 provided, as it still does today, that the code did not apply to certain government entities, colleges, and others. Section 1-1.4, specifically, also exempted "any individual, firm, partnership, joint venture, corporation, or other business entity engaged, as a licensed business, in research, development, production, test, maintenance, or supply of rockets, rocket motors, rocket propellant chemicals, or rocket components or parts."

Weber and O'Connor believed, as Blazanin soon would, that if they were to organize their own rocketry organization, it would be exempt from NFPA 1122 by virtue of subsection (c) of Section 1-1.4. In other words, if they formed an organization that was a "licensed business" and the express purpose of that business was high-power rocketry, then members of that business would be exempt from the Code for Model Rocketry.

The three men did some brainstorming but reached no conclusions at this first meeting. A few weeks later, on April 12, 1985, Blazanin returned to Cincinnati for another meeting with O'Connor and Weber. This time, he brought Francis Graham with him. Also present were high-power flyers Don Carter and Tom Weikamp from Kentucky and Jeff Donatelli and Mike Nelson of Ohio. The meeting was held at Weber's home and started at 8:00 p.m.

"We stayed up until three o'clock in the morning trying to hash out an outline for an organization that could stand up to NFPA 1122," said Blazanin. "But everything had to be their way. They wouldn't take any of our ideas. I was looking for an organization that would benefit the members, but Weber and O'Connor just wanted an organization to get around NFPA 1122 so they could have a market for their rockets and motors."

Weber and O'Connor had a different version of events, and they and others branded Blazanin as the unruly force.

"Basically, Blazanin wanted to run the show from Pittsburgh and tell no one what his intentions were," wrote Don Carter later, who was also present at Tripoli's formation.

For whatever reason, Blazanin won the argument that night. He announced he was happy with the Tripoli Rocket Society in Pittsburgh the way it was, and the meeting wasn't getting anything accomplished, so he suggested they should adjourn. One of the others then asked if the rocketry club in Pittsburgh might be interested in serving as a springboard for a national organization, with the bylaws of the Pittsburgh club being adopted as those of a national rocketry club. Blazanin said he would think about it, and he and Graham returned to Pittsburgh to present the idea to Tripoli's few members, who readily agreed. The Tripoli name and its bylaws would form the basis of a new national club. For Graham, the decision to attach Tripoli's identity to a national rocketry club was an easy one.

"Tripoli was still a very small organization—just a local club, really," said Graham. "If we went national, we could get a lot of members and a lot of energy, and this could enhance rocketry to the point where our hobby might be analogous to aviation and its clubs and societies. Only our organization would not be about flying. It would be about unmanned space access, suborbital, and perhaps even orbital someday."

There was one more minor hurdle: Weber and O'Connor wanted the new club to be based in Cincinnati, and they also wanted control of the organization and Tripoli's modest but established checking account, which held several hundred dollars. Blazanin did not want to turn over all control of the organization to faraway Cincinnati. But he desperately wanted to create a viable national rocketry organization. He struck a deal. O'Connor and Weber would be the first president and vice president, respectively, of the new national Tripoli Rocket Society. Francis Graham would be treasurer and in charge of the money and Blazanin would be the organization's first secretary. Weber and O'Connor agreed.

In the spring of 1985, less than two months after the NAR's blue-ribbon commission issued its final report recommending the adoption of G motors, the Tripoli Rocket Society, the immediate forerunner of the modern Tripoli Rocketry Association, went national. For a time, the club was known as the National Organization of Advanced Nonprofessional Rocketry.

The new high-power club had a name, four officers, and $600 in its bank account. All it needed now were some members.

At Tripoli's reformation as a national entity in 1985, the club had about eighteen members, many of them from the Pittsburgh area. As a Tripoli member, Blazanin had already taken over as editor of the *Tripolitan* newsletter in March 1985. Now, in the second issue with Blazanin at the helm, the *Tripolitan* announced the formation of a national rocketry organization dedicated to high-power rocketry.

"The only solution for the future of advanced rocketry was in the legal formation of a legitimate organization and business dedicated to this area," wrote Blazanin at the time.

Blazanin announced that Tripoli was searching for a lawyer to incorporate the new rocket society and that committees were being formed to establish bylaws, a national system of communication—the

Internet was still years away—and a safety code for high-power rocketry. In addition, a committee was being formed to handle the club's participation at LDRS-4, scheduled for July 1985 in Medina. (At this point, Tripoli had no relationship with LDRS, which was owned and operated by Chris Pearson.)

Tripoli also took out ads in regional rocketry newsletters not only to announce its formation but also to lay down the law with regard to how members of the public would, in the future, obtain commercial high-power motors. In some respects, these ads marked the beginning of a short-lived monopoly that Tripoli had on the distribution of high-power rocket motors. Although the organization was not in the business of direct distribution or manufacture of motors, it had the apparent power to influence distribution and thereby draw high-power enthusiasts directly to the ranks of the organization.

In a nutshell, Tripoli declared that NFPA 1122 prohibited the sale to the public of commercial rocket motors that were not certified by the NAR. This included some model rocket motors and all motors that exceeded the NAR's impulse limits. Tripoli announced that the major high-power motor makers of the day, including AeroTech and Vulcan Systems, had agreed not to sell any of their products in violation of Section 1122. Therefore, proclaimed Tripoli, the average consumer could not buy high-power rocket motors from these companies—unless a flyer decided to join Tripoli.

Tripoli's ad campaign also claimed that since it was exempt from Section 1122, its members were also exempt. Consumers were urged to join Tripoli so they too could *legally* fly high-power rocket motors. Whether Tripoli's claim that membership in the organization conferred some kind of legal standing for flyers is debatable, but the ad campaign worked; people joined Tripoli to fly high-power motors. Tripoli also announced a new partner in its plans for the future of high-power rocketry. Surprisingly for some, that partner was the NAR.

"Tripoli is working with the NAR and the NFPA to possibly create a new version of 1122 to allow *our* type of model rocketry, but this will take time," noted the *Tripolitan* that year. "For any change to occur, the high-power community needs to become organized into one reasonable, responsible voice, working with the government officials who control the 'strings' of our sport. This will not happen though unless we take a more serious approach to organizing ourselves. The NAR and NFPA have agreed to help Tripoli make these changes, but what we really need is you."

The NAR took notice of the Tripoli Rocket Society immediately. On May 7, 1985, NAR president Miller sent a note of best wishes to Blazanin regarding the new club.

"I read with great interest the article about the Tripoli Rocket Society being asked to serve as the nonprofit organization for high-power consumers wanting to fly H+ motors," said Miller. "The choice of your group is a good one as it clearly is concerned about consumer safety and the long-term considerations of the high-power rocket sport."

Miller also told Blazanin that the NAR had "absolutely no interest" in making it difficult for high-power users to enjoy their hobby. It was imperative, he urged, that a motor certification process be put in place to ensure that all high-power rocket motors were thoroughly tested and certified before being released to the public. In his letter to Blazanin, Miller said that by the fall of 1985, he expected the LDRS-HPR Committee to move toward recommending, among other things, a precise definition for high-power rocketry as well as the creation of a new safety code. Miller then suggested three possible scenarios whereby NAR members could one day fly high-power rockets and still remain in good standing with the NAR.

Under the first scenario, wrote Miller, the NAR would absorb high-power rocketry into its association, much like a section of the NAR. A second scenario would be for the NAR to formally recognize a special interest group, such as Tripoli, as an adjunct of the NAR (but still under the auspices of the NAR). Miller's third option was the NAR would recognize joint membership—or the right of its members to belong to both the NAR as a model rocketry organization and also to a separate and independent high-power association. (This third suggestion would ultimately be adopted by both the NAR and Tripoli in the 1990s).

In closing his letter to Blazanin, Miller made it clear that although Tripoli was free to offer input regarding high-power rocketry to the NAR, the NAR did not yet assign this new organization any special significance at the national level.

"Although the Tripoli Rocket Society is to be established as the nonprofit organization to permit the use of H+ motors, the NAR is [still] speaking directly to the HPR Commission. That is, the NAR considers the HPR Commission as the official voice of the high-power consumers," he said.

Tripoli was officially recognized by the NAR—whatever that meant—but it would be a while before the organization would have any national legitimacy from the NAR's point of view.

So as LDRS-4 was about to begin in the summer of 1985, there was (1) an independent LDRS-HPR Committee studying high power, (2) the second NAR blue-ribbon commission studying high power, and (3) a very small new high-power rocket society founded in Cincinnati, Ohio and based in Pittsburgh. There was also a lot of on-the-field speculation as to what would happen next and who would ultimately lead high-power rocketry into the future.

7

The Rocket Builder

When the Tripoli Rocket Society became a national organization in 1985, high-power rocketry was still in its infancy. If one were able to travel back in time, it would be difficult to find a launch that resembles any high-power event today.

The average high-power rocket in 1985 looked like a model rocket, perhaps a little bit larger. The airframe was 4 inches in diameter or less. If you launched a 6-inch-diameter rocket, fellow flyers marveled at the size of the airframe. If you launched an 8-inch-diameter vehicle, you were a superstar, on the cutting edge of high-power rocketry. Airframes in the 1980s were made primarily of cardboard or spiral wound tubing. A few flyers were beginning to add an outer layer of fiberglass to their airframes of cardboard to withstand the higher thrust of H, I, or J motors. Flyers did not dwell on the center of gravity and the center of pressure in their rocket. If a flight went awry, that's the way it goes.

In the summer of 1985, flyers used single-use motors, likely made by AeroTech or Vulcan, although a few other manufacturers were on the field, such as Prodyne and U.S. Rockets. The size of the motor was still in a tight impulse range—nothing bigger than a K at most events, which was an incredibly powerful motor at the time. High-power motors were also expensive; H to J motors cost up to $100 each. There were commercial L motors available for sale, but you had to plunk down several hundred dollars for that flight. With the right connections, an M motor could be purchased at a cost that approached $1,000. This was a lot of money to spend on a flight that more likely than not would end up destroyed because of either an airframe shred on the way up or a recovery disaster on the way down. As a practical matter, L- and M-powered flights were exceedingly rare. A K-impulse launch brought everyone at the range to their feet.

The high-power deployment system in 1985 was nearly identical to that found in the smallest model rocket. There was no dual deployment. At apogee, an ejection charge made of black powder located in the forward end of the motor ignited; the resulting gases overpressurized the inside of the airframe, separating the rocket into two halves. To keep those halves tethered together on the way down, flyers used elastic bungee-type recovery harnesses, which most people called "shock cords." Sometimes shock cords worked; sometimes they did not. As the forces involved in high-power increased, elastic cords snapped during separation and deployment. For many high-power rockets, recovery meant a long tumble to the ground, half of the vehicle smacking the surface without a parachute. If one's parachute did deploy successfully, spectators saw a canopy-style design with many shroud lines. It might be a military surplus parachute, or it might be homemade. There was not much variety in the style of parachutes; there were no X-form designs and few, if any, commercial parachute makers catering to high power.

The method of detecting altitude for rockets in the mid-1980s was crude albeit scientific. Altitude measuring required the presence of several people taking semiprecise measurements from different locations on the field using a device invented in the sixteenth century called a theodolite. This was the same type of instrument used by Goddard and Von Braun to determine altitudes for their early rockets. If high-power flyers had enough volunteers to man multiple theodolites and they recorded the important angular measurements, one could combine these numbers with mathematical equations to obtain an estimated maximum altitude for a rocket's flight. Of course, if a rocket went into the clouds or was simply lost to sight, all bets were off. There were no commercial altimeters yet, no Adept or Missile Works or Perfectflite. Such companies were years away.

In 1985, organized high-power rocket launches were few and far between. NAR-controlled events were still out of the question for a high-power rocket. The NAR had not yet approved the G motor, and Tripoli had no launches of its own. People flew their high-power rockets out west at Lucerne or near the town of Cuba, Ohio, in the east. Or they waited for Chris Pearson's annual LDRS, which was still an independent event located in Medina. At all these high-power launches, most people at the range were still flying model rockets.

When Gary Rosenfield created AeroTech in Sacramento in 1982, his motor-making enterprise was a part-time venture. His primary source of income and the way he supported his young family was his day job. In 1985, that job was at Aerojet, a major U.S. defense contractor. Aerojet built strategic missiles and tactical rockets. They also loaded armaments like the *Polaris, Minuteman,* and *Peacekeeper* missiles.

"I was hired as a process control engineer, which is a fancy way of saying adjusting formulations to match chemical analysis of ingredients," recalled Rosenfield. "It was a production environment, and I worked out of a trailer. Nobody there was a hobbyist. It was more professional than at Bermite. You did your job and then went home. I was one of the younger ones, and sometimes I felt like a fish out of water. There were a lot of long-term Aerojet people there, and I was in my twenties, and most of them were already in their forties."

Rosenfield enjoyed working at his day job with Aerojet. He was promoted to manufacturing engineering, and among other things, he learned how to make six thousand batches of chemicals. At Aerojet, he affirmed some of his hobby-related rocketry knowledge. But much of his work was with larger projects. "Some of the most complicated propellant on Earth," he recalls.

Away from Aerojet, Rosenfield continued to make rocket motors in his spare time. By the mid-1980s, AeroTech was the most recognized name in hobby rocketry. Along with Scott Dixon's Vulcan Systems, AeroTech was a leader in the commercial manufacture of composite propellant for the hobby. These two motor-making companies and a few others like them were creating a viable *national* commercial market for APCP where none had ever existed before. Commercial manufacturers like AeroTech and Vulcan and the dealers who distributed their products were the real engines in the development of modern high-power rocketry. Without them, the spread of the hobby would have been slow, if it spread at all. Companies such as AeroTech and Vulcan were no longer making motors for a few friends or acquaintances. They were building an entire new industry.

Rosenfield joined Tripoli in 1985. He was member number 22.

"I thought it was important to have an organization for high power because the NAR was not likely to embrace it, at least for a while, and I knew that an organization could provide benefits to its members, such as having a voice one day with the NFPA," said Rosenfield. "I knew we could not survive if we were viewed as a bunch of renegades and outlaws forever. Sooner or later, this hobby would have to be legitimized, and a national organization would go a long way toward making that happen."

By 1985, AeroTech made motors for not only model rocketry but also high power. Among its high-power commercial motors were the H120, the J100 (with a 9-second burn time), and the L1507. The L motor, one of the most powerful hobby motors on the market at the time, produced more than 500 pounds of maximum thrust. Rosenfield created his AeroTech motors under primitive conditions that would seem almost quaint by today's standards.

"I was hand-mixing everything," he said. "I had containers strapped to tables, and I developed a motor-making process that was repeatable. You would stir the chemical mix so many times in one direction and then so many times in the other direction. Everything was written down carefully."

AeroTech made motors this way throughout the early to mid-1980s. It was time-consuming and labor-intensive and bears little resemblance to motor manufacturing at AeroTech today.

"I used a thick Delron rod to mix the materials. I also had a vacuum pump and a drill press and not much else. We might have had a radial arm saw to cut tubes. We cured everything at room temperature. We had no employees yet."

Rosenfield still needed his regular job, although AeroTech did take on one employee. The work at Aerojet was still his real security, especially since he now had a wife and child to support. But his successful hobby business was getting bigger.

"I got to the point where it was taking up too much time, and while I was working at Aerojet, I was going home at lunch to make rocket motors, fill orders, and then go back to work. Then I did it all again after work, working until late into the evening. Eventually, I realized that I had to decide on what I was going to do. I couldn't keep working both jobs that way."

Rosenfield then took a leap of faith. He quit his work at Aerojet, gambling that he could make AeroTech successful as a full-time endeavor. The first thing that happened was he took a 50 percent cut in his annual income.

"I was making an annual salary at Aerojet in the thirty thousands, and it was a good rate of pay with benefits," said Rosenfield. "I had one son and another on the way. It sounded crazy that I would quit, and I had no way of knowing where AeroTech was going to go. But people were very supportive, AeroTech was making more money every year, and the trend was good. I had a lot of faith that it was going to work out."

By the end of the year, AeroTech was Rosenfield's full-time job. He believed that to survive, he would have to work harder than anyone else in the motor-making field. At the time, commercial motor makers were, like Rosenfield had been, primarily part-time players whose livelihood did not depend on the quality or quantity of the motors they sold.

"There were a lot of motor makers back in the mid-1980s," recalled Rosenfield. "They would show up at launches, but a lot of them were really small-time, and they were not shipping product like we were. Vulcan was our only real competition."

Rosenfield's big gamble in the mid-1980s to quit his job and risk everything to try to create his own business out of nothing is a familiar American refrain. It would be a familiar story in high-power rocketry too.

Chris Pearson and his supporters at SNOAR made all the necessary arrangements to host LDRS-4 on the Wagner property in Ohio. The launch was held on July 13 and 14, 1985, and was slightly larger than the year before, with more than 125 people in attendance. Pearson's new rocket company, North Coast Rocketry, sponsored several new tripod launch pads, which made range operations smoother. Although model rockets were still common at LDRS, the trend toward larger motors continued.

"The big thrill of flying a G motor only three years ago has been replaced by the big H- and I-motor models," wrote one observer.

On Friday evening, the L&K Motel was again the place to be. Even after midnight, people were milling about and talking rocketry with friends. In and around open motel room doors and between beers, motors were being sold, and high-power rockets were on display. The most impressive display was at the room occupied by Ron and Debbie Schultz of Ohio and their growing collection of rockets under the trade name Lots of Crafts.

This was the third year the Schultz family attended LDRS. In three years, Lots of Crafts had become the most influential commercial high-power rocket builder in America. Much like the early success of AeroTech, LOC's prominence was due to hard work, big gambles, and a singular vision that one company could do things better than anyone else had before. The man with that vision was Ron Schultz, and like Gary Rosenfield, Schultz's efforts would help deliver high power to the future.

Ron Schultz was born in Cleveland, Ohio, the fourth of five children. His father, Frank, was a machinist whose family immigrated to America from Yugoslavia prior to World War I. His mother, Rose, met Frank through mutual friends in the late 1930s. Frank and Rose were married in Ohio in 1939. Ron was born several years later, on July 8, 1947. At the time, Harry Truman was serving out the fourth term of Franklin D. Roosevelt, and the American air force was conducting experiments with the German-captured *V-2* rocket technology in the New Mexico desert.

Ron Schultz grew up surrounded by craftsmen, tools, and machinery. His father worked primarily in the automobile business as a tool maker. The elder Schultz made and repaired dyes and stamping machines in an era before computers and robotics. Ron had his father's interests, talents, and abilities around machinery and tools, and he also had good hands, so he followed the elder Schultz into the machining industry. His mother stayed at home and ran the household.

"Back then, you could have one parent working outside and one parent working at home and have a good life," recalled Schultz.

Schultz graduated from Chanel High School, an all-boys Catholic school, in 1965. He was determined to follow in his father's footsteps.

"I liked working with my hands, and I also liked the appeal of a job that had lots of variety," he said. "You have different products and parts to make every day, and with that came different challenges to work out."

Schultz enrolled in specialized mathematics courses at night that would be helpful to his work as a machinist, and then he took a job at Bowen Machine in Bedford Heights.

"This was 1966, and the Vietnam War was escalating," he said. "Bowen did government work, and there was a lot of work to do. I learned how to use industrial lathes and grinders and milling machines there. While I was there, we were making parts for the *Hawk* missile launchers as well as tank parts."

Over the next few years, Schultz advanced his skills as a machinist. He switched shops a couple of times as better job opportunities arose for him. As a result of his work on government projects, he also obtained a temporary draft deferment. But that deferment ended in 1970, and he became eligible for the draft again.

"There was a lottery at the time, and the 365 days of the year were each marked on a lottery ball that had numbers corresponding to every date on the calendar," he said.

When a lottery ball was picked, all those young men within a certain age range with a corresponding birthday were drafted. But they would be drafted in the order in which the lottery balls were pulled from the bin.

"My December birth date was pulled as number 105," recalled Schultz. "At the time, the war was starting to deescalate. I think they made it up to number 98, and that was it. I was not drafted."

In 1974, Schultz went to work at a large machine shop that was part of Ohio Nuclear.

"We were doing research and development on CAT scans, and later, we did some of the first research and development on MRI machinery," he recalled. "It was my lucky day when I was hired there, and I stayed with Ohio Nuclear, which was eventually owned by Johnson & Johnson, until the mid-1980s."

Schultz was introduced to his future wife by happenstance in 1966.

"I met Debbie at her sixteenth birthday party," he said. "I was hitchhiking down the road with two friends, and some guy, who happened to be a friend of Deb's, offered us a ride. He said that he was going to a big party on Route 21 in Independence, and he asked if we wanted to come along. We said sure. He took us to this house, and there was a big party going on. People were even climbing in and out of windows!"

Four years later, in 1970, Ron and Debbie were married.

Debbie graduated from high school in 1968, and she went to work for the Central National Bank. One of her jobs at the bank was to facilitate the movement of large amounts of old cash, paper money, on its way to be burned by the government. After she married Ron, she went back to school and studied astrophysics and the law, but she eventually focused on what would become a lifelong passion: sewing and tailoring. Between 1971 and 1974, she and Ron had two children, Danielle and Jesse.

In the mid-1970s, Debbie was self-employed, performing freelance sewing, jewelry, and tailor work out of the family home. To help make her business more viable, she needed a vendor's license. This would enable her to buy needlepoint products at wholesale prices. She went to the local county offices, picked up the requisite paperwork, and filled out the forms for her vendor's license. Debbie had to come up with a name for her sewing company. The name she created was Lots of Crafts.

Ron and Debbie Schultz were not involved in rocketry until the early 1980s. As a teenager, Ron had purchased an *Astron Scout* from a mail-order form in *Popular Science*, and he occasionally used his father's drill press to make body tubes and nose cones. Later, he dabbled in Mini-Maxx and Enerjet model rocketry motors in the early 1970s. But he never joined the NAR or any formal rocketry-related clubs.

When their children were old enough, the Schultz family started to fly Estes and Centuri model rockets at local playgrounds and fields near their home in Central Ohio. By the early 1980s, Schultz was making 4-inch-diameter rockets for his own fun and for his children. He preferred Estes tubing for construction material. He eventually learned who the actual manufacturer was of the 4-inch tubing, and he purchased four large cartons of tubes. He used some of them with his family, and he sold some of them to local model rocket flyers in Ohio.

One afternoon, in 1983, Schultz was at a shopping mall in Mentor, Ohio, when he ran into a fellow by the name of Chris Pearson. Pearson and other SNOAR members had set up a rocketry display at the mall. Schultz learned of an upcoming launch called LDRS-2 in Medina.

"At the time, I was living in Macedonia, which was only about 25 miles from Medina," he said. "I decided to go to LDRS, and Debbie and I took our children, and my brother's children also came with us to the launch. It was hot and sticky there, and the kids and I launched some rockets, and we took some video."

While at LDRS-2, Schultz had an opportunity to see the state of commercial high-power rocket kits as they were at the time. He was not terribly impressed.

"I knew we could do better," he recalled, and after he and Debbie returned home and over the next twelve months, they made rockets in their spare time.

The following year, in the spring of 1984, they were test-flying a 40-inch-long rocket they called the *Cruiser*. They were going to sell the rocket. It was 2.6 inches in diameter and carried a single 24 mm motor.

This was the first rocket that would ever fly under the name Lots of Crafts. Ron and Debbie attended LDRS again in 1984. This time, they brought their new rocket products with them.

"We sold dozens of rockets at LDRS-3, all for around $30 or less," said Schultz. "And we also flew twenty-six rockets at the launch. I had them all prepped and ready to go before LDRS, and they all carried handmade parachutes made by Debbie."

The rockets were colorfully painted, and the parachutes were made from a rainbow of bright fabrics sewn together by Debbie. The crowd loved the flights and was equally impressed with how well the kits were built.

"Ron and Debbie Schultz did quite well with their Lots of Crafts kits," said one writer at the time.

After the success at LDRS-3, Schultz began building bigger rockets and more of them. He attended regional events in Ohio, and word of the quality and selection of Lots of Crafts kits began to spread in the rocketry community. He had been building kits in the traditional rocket sizes up to now, but in early 1985, he moved up to the whopping size, for a kit, of 5.38 inches in diameter.

"I wanted something bigger than a 4-inch airframe. Everybody was making those," he said. "I called up a buddy in the tubing business and said I wanted something larger but still less than 6 inches in diameter because that would affect how I would cut my centering rings."

The result was one of the longest running and most successful high-power rocketry kits in the world: the *Magnum*. After obtaining the tubes, said Schultz, the nose cone for the *Magnum* was the hardest work. The first hundred were actually handmade of Aerofoam, he said. After that, a special mandrel was built to machine the cones out of plastic. In the spring of 1985, Schultz received a telephone call from Gary Rosenfield in California.

"I had met Gary at LDRS-3, but we did not have very much contact there," he said. "But he called me to suggest that we plan a joint project for LDRS-4. I would make the rocket, and Gary would supply the motors."

This informal arrangement between Lots of Crafts and AeroTech would become one of the centerpieces of LDRS for the next few years.

"Gary heard about the 5.38-inch tubing I was using, and he suggested that we use that for the joint project," he said. At the time, Schultz was producing the *Magnum* as a cluster with a central 54 mm motor and two outboard 29 mm motors. "Gary said to me, 'Ron, how about we do a central 54 mm motor and [then] surround it with nine 29s?'"

Schultz liked that idea. Construction of the ten-motor *Magnum* began five weeks before LDRS-4, said Schultz, who made centering rings and fins out of one-quarter-inch plywood for the rocket. Arriving at LDRS, Ron and Debbie quickly became the center of attention. The number, size, and assortment of rockets that Lots of Crafts had on display at the L&K Motel Friday night was phenomenal, said flyer Mike Nelson, who also had a chance that night to view AeroTech's new G10 and G15 end-burner motors. But the main attraction that night, said Nelson, was the combined effort of Lots of Crafts and AeroTech: the *Magnum*.

"This gold-and-black monster had been specially prepared for LDRS by Ron, and no detail had been overlooked," said Nelson.

Schultz and Rosenfield rose early Friday morning and prepared all ten motors for flight.

"Friday night at the motel, the rocket stood outside our rooms, majestically, for all to see. They really loved it," recalled Schultz. Privately, however, Ron was a little worried about the upcoming launch. "It was the first time I had designed something like that, and I was worried about the center of pressure and the center of gravity."

As high-power rocket flyers know, the center of gravity in a high-power rocket should be at least one body tube diameter forward of the center of pressure (the point where aerodynamic lift is centered) on the rocket. This helps ensure that the rocket will have a stable flight. The locations of the center of gravity and the center of pressure can be affected by multiple factors. For example, the center of gravity will typically move slightly forward during a rocket motor's thrusting phase as the motor is burned. Flyers concerned that the center of gravity is too far aft often add weight to the nose cone. This brings the center of gravity forward. The center of pressure can be altered by making a rocket shorter, using larger fins, or adding more fins. Schultz had not had the opportunity to fully calculate where any of these markers were located on the ten-motor *Magnum*. He would find out at launch time, when the rocket lifted off.

The weather for LDRS-4 on Saturday morning was perfect for flying, but there were some scattered clouds and a light breeze. The open field at the farm was littered with early-style pop-up tents, and people were busy preparing rockets and visiting with vendors.

"What [was] a pasture the day before was now being transformed into the site for the largest gathering of sport rocketry enthusiasts in the United States," wrote one flyer.

The Lots of Crafts/AeroTech *Magnum* was ready by 11:30 a.m. It was a huge rocket for its day, and there was plenty of excitement on the field as it was raised to vertical position. At ignition, all ten motors in the *Magnum* came to life.

"It lifted off beautifully and soon disappeared into the low cloud cover," said Schultz, who was relieved that his new product was stable in flight. "About a minute into the flight, its 7-foot parachute was spotted, but because of 10-mile-an-hour winds, it drifted downrange."

There were no commercial altimeters in 1985 and no dual deployment. The *Magnum* had reached several thousand feet in altitude and deployed its parachute. Then it simply sailed away with the wind. Schultz and Rosenfield spent more than 2 hours looking for the rocket—to no avail.

"We gave it up for lost," recalled Schultz. "I stopped at one farmhouse, and Debbie asked the farmer to call us at the L&K Motel if he saw the rocket. And she told him that it was armed and should not be opened!"

That evening, the same farmer called the motel to report that he had found a large rocket on his property. He wondered if it belonged to anyone at the L&K. Sure enough, it was the *Magnum*, which was returned in perfect condition.

"We sure were happy to get it back," said an ecstatic Schultz. "We placed it outside our room again that night, but this time, it had a Helix hat on, a pair of eyes, and a button for a mouth that read, 'I'm back by popular demand.'"

Schultz's *Magnum* was the biggest commercial rocket at LDRS-4. But it was not the largest rocket at the launch. That distinction went to a rocket built by Chuck Mund of New Jersey and John Holmboe of New York. They called their rocket the *Big Bertha*.

The *Big Bertha* was based on an upscale version of the Estes model rocket kit bearing the similar name. Mund and Holcomb increased the size of the model to a rocket that was 8 inches in diameter and 10 feet tall. In 1985, this was a monster of a rocket; it may have been the biggest high-power rocket to fly at any LDRS up to that time. The airframe was made from three-sixteenth-inch-diameter spiral-wound paper. The nose cone was scratch-built from Styrofoam covered by epoxy-laden cloth. The rocket carried four film cameras and a parachute ejection system that was operated by radio control. It was powered by a cluster of three AeroTech I140 motors and three H120s. The three-finned vehicle weighed nearly 40 pounds, an astounding weight for a rocket of its day.

The *Big Bertha* was ready for flight on Saturday afternoon. But an errant radio signal activated the remote-control ejection system, resulting in an unexpected bang, followed by the deployment of the parachutes right there at the launch pad. Team members scrambled to reprepare the ejection charges and set everything back in place. Soon, the rocket was ready to go again. At ignition, the rocket roared off the pad, with all six motors firing as planned. The launch, flight, and recovery were spectacular, although the heavy rocket reached only about 1,000 feet. It was recovered intact a couple hundred feet from the pad. Everyone loved the flight, so Mund and Holcomb decided to launch it again on Sunday, only this time, the pair decided to load the rocket with a Vulcan Systems L-class motor.

"This 2-inch-diameter motor was the most powerful rocket motor I had ever observed," said one experienced flyer who had witnessed the flight. "The vibration of the ground and the loudness of the motor were incredible."

Once again, the rocket flew straight and true, and it deployed its multiple 6-foot-diameter parachutes as planned. It was a great weekend for the *Big Bertha*, and after LDRS-4, the rocket would be flown at other launches around the country. It became one of the most recognized rockets of its day, eventually gracing the cover of the *Tripolitan*.

Other notable flights over the weekend, which was slowed by rain showers Sunday morning, included a high-altitude flight by Chris Pearson of an *Enerjet 2560* rocket on three AeroTech G end burners, the first ever launch of the venerable Lots of Crafts *EZI-65* rocket; and Korey Kline's nicely engineered *Monocopter*, powered by two F motors located at opposite ends of a long helicopter-like propeller blade. Reportedly, one rocket cleared 6,000 feet above the Wagner farm.

There were also J, K, and L motors flown. Mike Boesenberg of Delaware launched a rocket with an onboard accelerometer, one of the first such electronic devices to grace the field at any LDRS. Using a Commodore computer on Saturday night, Boesenberg used data from the accelerometer to print out acceleration, velocity, and, reportedly, altitude. Boesenberg and his crew provided flyers at LDRS-4 with a brief look into the electronic future of high-power rocketry.

Tripoli representative Tom Blazanin spent the weekend recruiting as many people as possible to become members of the newly created Tripoli Rocket Society, now a few months old. LDRS was a good place for a recruitment drive for the fledgling high-power club, and among those who signed up were Ron and Debbie Schultz, who became Tripoli members 19 and 38. Throughout the weekend, Debbie Schultz and Melodi Rosenfield spoke to every flyer they could, encouraging them to join the club.

For the second year in a row, the NAR was officially present at LDRS. Dan Meyer was there as a representative of the second blue-ribbon commission. He spent both days taking videotape and talking to high-power flyers about their hobby. On Saturday evening, he replayed some of his video of the days' events for everyone back at the L&K Motel. It was a good experience, and it helped continue the thaw in relations between high-power flyers and the NAR.

"Most of my time at the meet was spent talking to anyone who would listen about what the process was, which the NAR was doing internally to explore the political and legal aspects of high-power rocketry and the NAR," wrote Meyer after the launch.

On Saturday night, Meyer also addressed Tripoli's new members and explained the process of the new commission. Meyer also told Tripoli members that the NAR would deal with any organized group of high-power flyers so long as that group was recognized as the "official voice" of the high-power community. Interestingly, although the LDRS-HPR Committee was, in the NAR's view, the official voice of high power, there was no formal meeting between that committee and Meyer at LDRS-4. Meyer made his pitch directly to the Tripoli crowd.

"I made it clear that some limits would have to be established and defined very clearly if they were ever to have any hope of changing the regulatory environment which high-power rocketry finds itself in," said Meyer. "This would, of necessity, require some changes to the 'anything goes' attitude currently prevalent."

Meyer predicted that changes to NAR and related regulations were likely to take five years, if not longer, and would require a dedicated organization that had the strength and resources to see things through to the end. To help begin the process, Meyer suggested the possibility that the NAR lend a hand in the development of a new high-power safety code.

"I offered the possibility that the NAR [in the person of G. Harry Stine] might be able to assist them in writing a code similar to NFPA 1122 [or possibly an amendment to 1122] whereby high-power rocketry might obtain some legal recognition as a legitimate activity in the codes of the NFPA."

Meyer's suggestion that flyers get the NFPA involved was the right one for high power and the young Tripoli Rocket Society. It was critical that an effective safety code be created soon. Without a safety code, it was only a matter of time before some disaster might spell universal doom for high power everywhere. And to be most effective nationwide, a safety code would benefit from a stamp of approval by the NFPA.

In the 1960s, the NAR had successfully used the NFPA to establish, circulate, and implement its own Code for Model Rocketry nationwide. Now in mid-1985 and at the suggestion of the NAR and others, high-power flyers were sensing the importance of doing the same thing for their hobby. The NAR's experience with the NFPA, if taken advantage of by high-power leaders, could be invaluable in gaining a foothold on the NFPA's committees where safety codes were introduced, debated, and implemented.

The best way to get into the NFPA was for Tripoli to have its own representatives on the committees that might create such a safety code, specifically the Committee on Pyrotechnics. Over the next several years, Tripoli leaders, with the assistance of key members of the NAR such as Pat Miller and Harry Stine, would work toward this goal.

In his report prepared after LDRS-4, Meyer advised the NAR that based on his observations at LDRS, most flyers were willing to recognize the new Tripoli Rocket Society as the official voice of high power.

"There may be some people who object to this, but I never heard anything from them," he wrote.

Although the LDRS-HPR Committee was still viewed by the NAR as the official representative of high power, Meyer's report resulted in Tripoli being informed of all future actions of the second blue-ribbon commission.

Meyer's report also contained a few words of caution. "There are some who may not be willing to accept any type of limits on weight or power, which has been the traditional method of defining where a model rocket ends and an amateur rocket begins," he wrote. "Whether this will develop to be an insurmountable problem or not remains to be seen."

Meyer's observations were accurate. High-power rocketry grew in large part because a few people refused to accept the model rocketry definitions of allowable motors and decided to take matters into their own hands. This sense of independence was invaluable to the growth of the hobby. But it had limits.

By 1985, high-power launches were no longer the tiny, strictly fringe events they had been in the past. More and more people were becoming involved, and the power of the motors being flown was well beyond what had been launched at the first LDRS just three years before. High power was uninsured and still fairly irresponsible. That needed to change too, even if some people didn't like structure or rules. (In an irony yet to be played out in the future, the Tripoli Rocketry Association would one day find itself in the position of the NAR in the early 1980s: a more conservative and established rocketry organization resistant to changes that it viewed as unreasonably dangerous and beyond its own hard-won safety code.)

Overall, Meyer's report to the NAR's leadership was complimentary of LDRS and high power in general.

"One item which was surprising to me was that during the two days of flying that I observed, there were only a handful of failures," he said, adding that two of the failures were from commercial model rocket engines. "The craftsmanship on most of the models was of a high level, and a couple of models were flown which had onboard accelerometers and computers. I saw the data output from one of these flights printed out on a computer in the hotel that evening. It was quite impressive."

Following LDRS-4, Pat Miller wrote a column in which he observed that LDRS was a responsible activity and that high power was, approvingly, growing. This was a far cry from the stance that the NAR president took publicly in 1982. Yet Miller's attitude had changed with the times. He wrote that it is important to portray high-power rocketry as an activity conducted by responsible adults who are concerned about consumer safety issues.

"From what I have seen at the launches I have attended," he said, "this is precisely what the high-power sport is."

LARGE AND DANGEROUS ROCKET SHIPS

Chris Pearson and SNOAR were praised for organizing the fourth LDRS launch; flyers were looking forward to LDRS-5 next year. Pearson had done a good job again, and the annual launch in Medina was still growing.

Privately, however, Pearson needed a break. After running LDRS for four years, Pearson was beginning to feel time pressures because of his other obligations. He had a full-time job outside of rocketry, and in late 1983, he also cofounded a commercial rocketry company, North Coast Rocketry, selling mid- and high-power rocket kits and components. It seemed he had less time each year. He needed to cut back somewhere. He could not run LDRS indefinitely.

There were other pressures too. The Wagner farm was seemingly shrinking, not in its physical size, which remained the same, but by virtue of the new motors that were simply overrunning the dimensions of the 20-acre property. J, K, and L motors were now propelling rockets many thousands of feet into the air along Spieth Road. And more and more flyers were spending their time off the Wagner farm searching for rockets that were landing on private property in every direction. It was time to find another location for high power's national event.

There were informal discussions after LDRS-4 that it might be time to hold the launch out west somewhere. At a dinner on Sunday night, it was suggested that LDRS-5 might be held in Nevada and that LDRS-6 might go forward in Colorado, hosted by Scott and Ramona Dixon of Vulcan. For Pearson, this was all good news. Perhaps someone else would do all the work.

8

The End of the Beginning

On January 28, 1986, the space shuttle *Challenger* broke apart 73 seconds into flight, killing all seven astronauts on board. It was the first disaster in the American space program in nearly two decades, the last deaths having occurred in 1967, when Gus Grissom, Edward White, and Roger Chaffee were killed during testing of the *Apollo 1* command module on the ground.

Since its first flight in 1981, the success of the shuttle became routine, so much so that this launch was covered live by only a few television stations nationwide. The *Challenger* was the twenty-fifth flight in the program. The television cameras at Cape Canaveral that morning recorded the spacecraft during its magnificent takeoff and ascent—and then as it suddenly burst apart over the Atlantic Ocean. It was later determined that the cause of the disaster was an O-ring failure in the shuttle's right solid-fuel rocket booster.

That evening, Pres. Ronald Reagan addressed the nation regarding the tragedy. He finished his remarks with a quoted from the poem "High Flight" by John Gillespie Magee Jr.: "We will never forget them nor the last time we saw them this morning as they prepared for their journey and waved goodbye and 'slipped the surly bounds of Earth' to 'touch the face of God.'"

In early 1986, a full-page ad in the *Tripolitan* announced a four-day launch to be held at the El Dorado Dry Lake Bed near Las

Vegas, Nevada, that summer. The event was called LDRS WEST. The advertisement proclaimed that LDRS WEST would be, "without question, the largest high-power, large-scale, nonprofessional advanced rocketry event ever held in the United States." What's more, promised the ad, the launch would be held at the best location ever and would offer the best competitions, best prizes, best seminars, and best social gatherings in the history of high power. LDRS WEST promised to be the next step in the evolution of high-power rocketry.

After LDRS-4, Pearson was approached by western flyers who wanted to use the LDRS name for a launch in the Nevada desert. Pearson was tired of running the Medina launch every year. If someone else wanted to hold LDRS in 1986 and they wanted to hold it out west for the first time, that was fine by him.

At the time, at least one rocketry publication announced that LDRS WEST would be held in addition to LDRS-5. For the first and only time in LDRS history, some flyers were contemplating two LDRS events in a single year.

Las Vegas could be an ideal location for a high-power rocket launch. But that was dependent on the season. Although there were no obstructions for miles, a rocket launch in the summer would have to withstand almost unbearable heat. Still, it would allow West Coast flyers the opportunity to host LDRS in their own backyard instead of having to make the journey to Ohio again.

LDRS WEST never happened. The organizers could not get it done, the FAA waiver wasn't granted, and not enough volunteers stepped up to the plate to help get the event off the ground. At the last minute, Pearson agreed to host LDRS again in Medina. One day the annual event would be held in the desert outside of Las Vegas. For now, the launch would remain in Ohio. It was a false start. But high-power rocketry's move west was about to begin.

In the first year following the creation of the national Tripoli Rocketry Association, Tripoli leaders asked members for legal assistance to help incorporate their new venture. Tripoli member Dale Gardner of Alaska was a lawyer. Gardner offered to incorporate Tripoli pursuant to the laws of the Alaska Corporations Code. With his assistance, the modern-day Tripoli Rocketry Association Inc., an Alaska nonprofit corporation (entity number 38589D), was legally born on July 18, 1986, two weeks prior to LDRS-5. Tripoli was now a legal entity.

Every corporation has bylaws that define in general terms the purpose of the corporate entity. Among other things, Tripoli's stated purpose was the following:

> [T]o promote, encourage, and advance research, competition, technology, and recreational activities related to advanced nonprofessional rocketry in a safe and legal manner.
>
> [T]o conduct, as a licensed business, research and testing of rockets, rocket motors, and rocket components and parts.
>
> [T]o educate the membership in the proper and safe use and storage of commercially available rocket motors and materials.
>
> [T]o establish regulations and procedures for testing, training, and authorizing members of the corporation to acquire and use commercially available Class "B" solid propellant rocket motors in a safe and legal manner.

The incorporation of Tripoli was another step forward for what was still an obscure association of a few people around the country interested in launching bigger rockets with bigger motors. It was hard to imagine that it would survive for very long. At the time of its incorporation, Tripoli had $324.45 to its name.

On Friday evening, August 1, 1986, the night before LDRS-5, the first corporate meeting of the board of directors of the Tripoli Rocketry Association Inc. was held at the L&K Motel in Medina.

Francis Graham, one of the founders of the science club whose namesake was now poised to become the leading edge of high-power rocketry, recorded the minutes. Among those present at the first corporate meeting were Tom Blazanin, Bill Barber, Curt Hughes, Ed Tindell, Gary Fillible, Chuck Mund, and Graham. Also present were Gary Rosenfield, Ron Schultz, Scott Dixon, and Korey Kline. Jerry Irvine, creator of *California Rocketry*, the newsletter that had played a role in the initial spread of high power, was also present.

Blazanin read a letter to board members from Tripoli's attorney explaining the formalities of corporate existence. The first order of business was the election of the corporate board. On Blazanin's motion, seconded by Bill Barber, the board chosen the previous January was moved in as the new corporate board, with all terms running from January 1, 1986. A similar motion was passed to confirm the corporate officers selected earlier that year.

The new corporation also took up money matters, selecting a bank account in Houston, Texas, where treasurer Ed Tindell lived. Board members debated raising the membership fees for Tripoli to $20, but that was rejected as too expensive. It was decided to keep the fee at $15. Several committees were formed, and funding was established for the continued publication of the organization's official publication, the *Tripolitan*.

On unanimous motion, the board also voted to affirm the use of a "confirmation" card, which would allow cardholders—and theoretically only cardholders—to purchase high-power motors directly from manufacturers. Ironically, this new Tripoli rule amounted to a restriction of a flyer's ability to use high-power motors. Up to this point, almost anyone could buy a high-power motor. This ease of acquisition of motors came with no safeguards to ensure users had at least a minimum level of competency. Tripoli now sought to change all that.

The details of who would actually confirm candidates and how candidates would obtain their confirmation were sketchy and left to future discussions. Eventually, any Tripoli board member and several prominent high-power manufacturers were designated to confirm a flyer for the purchase of high-power motors. A list of people able to confirm flyers was printed in the *Tripolitan*. That list would grow every year.

The experience threshold for confirmation was low. If a person successfully launched and recovered a rocket propelled by a single-use H motor, they were confirmed. With their confirmation card, flyers were then authorized to purchase and launch any high-power motor they could get their hands on. There were no stepped certification levels, no testing or training of any kind. Tripoli's modern certification system was still nearly a decade away.

The safety benefits of the confirmation process were debatable. However, one immediate benefit was Tripoli now had de facto control over the means by which flyers could buy and launch high-power rocket motors. With the apparent acquiescence of the most prominent commercial motors makers, Tripoli announced that the only way one could purchase and fly a high-power rocket motor at a Tripoli event was to be confirmed. Motor manufacturers went along with the policy, which was a boon to the early membership drive of the young corporation. For the time being at least, the confirmation process gave Tripoli a virtual monopoly on high-power rocketry.

For the second straight year, Ron Schultz of Lots of Crafts and Gary Rosenfield of AeroTech teamed up to fly a large demonstration project at LDRS.

"Gary and I wanted a vehicle that would be awesome in both size and boost yet have a good margin of safety built in," said Schultz.

Early that year, Schultz located airframe tubes that were nearly 8 inches in diameter. From that foundation, he created one of the most popular big rocket commercial kits of its day: the LOC *Esoteric*. The *Esoteric* was a four-finned rocket more than 10 feet tall. It weighed 17 pounds empty. From its aft end, the airframe started off 8 inches in diameter, but then transitioned several feet up to a payload diameter of 5.6 inches. The rocket housed five 54 mm motor tubes for clustered flight.

"Motor selection could be varied from a mild 'just a nice flight' to 'all hell breaking loose,'" said Schultz, who, in the spring of 1986, created six *Esoteric* rockets, the first of many more.

The *Esoteric* was on display Friday afternoon at the L&K Motel, where it garnered plenty of attention. In terms of its relative size, the *Esoteric* was to the commercial rockets of its day what a full-size *Nike Smoke* is to a high-power rocket today. It was bigger than life. Following the Tripoli board meeting Friday evening, Schultz and Rosenfield went right to work on their joint project, finishing assembly of the rocket and preparing the motors. At 2:00 a.m., they loaded the *Esoteric* with a J700 and two I140s for flight on Saturday.

The weather in Medina Saturday morning was excellent, and flying started early. The crowd was larger than any LDRS before. At least one estimate reported nearly two hundred people present. There were still no statistics kept and no record regarding the number of flights at LDRS-5. For the third year in a row, NAR representatives were also on hand at LDRS, including Vern Estes, Jim Barrowman, and Chas Russell. All three were members of the second blue-ribbon commission. President Miller was present too. Like everyone else, the NAR representatives were there to observe and enjoy high-power rockets propelled by high-power rocket motors.

At a little after noon, the LOC/AeroTech *Esoteric* was carried out to its pad on the shoulders of four men, three high-power motors loaded and ready in the aft end. At ignition, the rocket came to life instantly, flying straight and true but not very high. The central J700 lit briefly and then shut down unexpectedly. Still, the two remaining I motors helped propel the rocket to about 1,500 feet. The rocket landed safely under parachute, to the applause of the crowd. On inspection, it was discovered that the J motor was gone. This was still the era of primitive motor retention in high power. The single-use AeroTech J was simply blown out the back of the rocket when its ejection charge fired. The recovery crew carried the rocket back to their tents, and plans were made to fly the *Esoteric* again on Sunday.

There were several other interesting flights at LDRS-5. Vulcan premiered its new line of Smokey Sam motors, which emitted a thick cloud of black smoke during flight. Canadian Bill Dennett thrilled the crowd with his rocket holding a cluster of two Fs, eight Gs, and a central J motor. The J motor failed, and the rocket was engulfed in flames. Jerry Irvine launched a two-stage rocket powered by G, J, and K motors; it shredded in flight. Throughout the day, the dry grass at the Medina field caught fire every now and then. It had been a hot summer. The fires were quickly extinguished.

On Sunday, Schultz and Rosenfield launched the *Esoteric* again. This time, the propellant was a central K1500 and two outboard I motors. Rosenfield wrapped extra masking tape around all three motors for a good friction fit in the motor mount tubes to prevent the loss of any motor after the onboard ejection charges fired.

"After a long visual check for aircraft in the vicinity, the countdown was started," recalled Schultz. "Everybody on the flying field had their cameras and eyes affixed on the one lone rocket in anticipation."

The rocket surged from the launch pad, with the big K motor emitting a 6-foot-long bright white flame. All three motors worked perfectly. The 23-pound rocket achieved more than 4,000 feet and then drifted away in the light winds. For more than 2 hours, the team searched for the rocket—without any luck. Later that afternoon, the rocket was located on private property more than a mile from the Wagner farm.

Tripoli held its first ever corporate general assembly meeting on Saturday night. The meeting was open to all Tripoli members and was held outside the L&K Motel. Among other things, board members explained the new confirmation process, and there were questions and answers regarding various new committees being formed. Some members said they wanted Tripoli to host a series of flying contests, like the NAR did, in the future. There were also questions about Tripoli's club sections, called prefectures, and an explanation as to where the name came from and how individual prefectures in Tripoli were to operate. In essence, prefectures were simply a local club of at least three people who agreed to support and uphold the goals of the Tripoli Rocketry Association.

At the suggestion of Gary Rosenfield, Tripoli made Lee and Betty Piester, formerly of Centuri Engineering, honorary lifetime members of Tripoli for their contributions to rocketry. This began a tradition of sorts, and at LDRS events in the future, others would periodically be honored with a lifetime membership.

NAR members with the second blue-ribbon commission observed activities all weekend long. They also had discussions with not only the LDRS-HPR Committee but also Tom Blazanin on behalf of Tripoli. These joint meetings helped establish "a solid groundwork for the peaceful coexistence of the NAR and high-power rocketry," wrote one commentator at the time.

Some of these meetings focused on insurance. At the time, the NAR was insured through the American Modeling Association (AMA). It was reported that the AMA had been reluctant to extend insurance even to model rocketry; high power, they thought, might make matters worse. Other talks involved the direction of high power in the future. According to one report, the NAR was contemplating "absorbing" high power but was reluctant to do so because of potential liability issues.

"If the liability picture improves in the future, the NAR is positioned to take advantage of the improvement by expanding its activities," said one observer.

As part of these discussions at LDRS-5, it was assumed that by 1987, the NAR would formally approve a rule to allow its members to participate in high-power non-NAR rocketry activities without facing any sanctions from the NAR. As a practical matter, this was already a common occurrence; the NAR needed to make things official, and it needed to protect its own liability coverage. The NAR was therefore considering a "no-mixing rule" that meant that high-power and model rocketry would be kept separate and apart so as not to endanger the NAR's insurance coverage. The no-mixing rule was yet to be defined. But among the possibilities being discussed was a proposal that model rocketry activities not be held on the same day and in the same place as high-power events. This would eliminate any confusion as to what was covered by insurance and what was not covered.

LDRS-5 was another successful event. There were now as many as half a dozen high-power launches around the country, some of which had more flights than the annual gathering at Medina. Yet LDRS was clearly the premiere high-power event in America. Much of the credit for this distinction and for seeing to it that LDRS was sustained year after year rested with LDRS founder Chris Pearson.

In five years, Pearson created a tradition that epitomized the high-power movement. From a handful of people flying primarily model rocket motors at LDRS-1 in 1982, LDRS was now the high-power event of the year. It was the Detroit Auto Show of high-power rocketry. Rocket makers, commercial motor builders, and manufacturers of all things rocketry related gathered every year at LDRS to show off the latest in high-power technology. Although the first five LDRS events bear no resemblance to the mega-launch LDRS is today, they were a large part of the foundation upon which high power and the modern Tripoli Rocketry Association were being built. Pearson's trek to Smoke Creek in 1981 and his perseverance organizing the launch year after year made him one of the most recognized flyers in high power.

Still, Pearson had had about enough. LDRS was not a money-making event in the 1980s. Pearson paid for many of the annual expenses out of his own pocket. As the event grew larger, the costs, responsibilities, and headaches increased too. Most rocketry events do not last for more than a few years. Pearson had guided LDRS through its first five years. He was not going to do it forever. He had a life outside of rocketry.

The event had also grown too big for the field in Medina. Flyers were now launching L motors at the farm, and M motors were right around the corner. This could not continue on a small plot of ground in the suburbs of Cleveland. LDRS needed a bigger field—a much bigger field. And the FAA waiver needed to be higher too. The waiver that began in 1982 was routinely violated every year. Some minimum-diameter high-power rockets were now clearing several thousand feet in altitude, and in a few years, high-power flyers would be launching rockets to the edge of the stratosphere. This could not happen in Medina or, for that matter, anywhere else in the state of Ohio.

Shortly after LDRS-5, Pearson made a decision that would have a profound effect on the future of the Tripoli Rocketry Association. Pearson gave the LDRS event to Tripoli:

> It is with regret that I announce that LDRS-5 will be the last national launch held in Medina, Ohio. The reasons for this are many. Mike Wagner, the field owner and longtime SNOAR member, has announced that the field which LDRS has been held on in the past will be rented to an outside interest. The field will be plowed and planted with crops, eliminating the use of it as a flying field. Furthermore, after acting as director of the launch for five years, spending over $1,000 of my own money and hundreds of man hours in the process, I am reluctant to continue . . . I now leave it to other parties to continue the LDRS tradition. I have given the Tripoli Rocketry Association permission to use the copyrighted term "LDRS" for use in conjunction with their national launch in the future. At the present time, LDRS-6 is tentatively planned to be held in Colorado Springs in 1987... I wish them all the best.

No one knew it at the time, but the summer of 1986 would see the last LDRS ever held in Ohio. And it would be another decade before the national launch returned to any location east of the Mississippi River. LDRS was heading west.

9

The Move West

On November 18, 1986, the NFPA amended the Code for Model Rocketry, redefining model rockets and model rocket motors. Beginning on January 1, 1987, NAR members could fly certified G motors in rockets weighing up to 3.3 pounds. It was another step in the NAR's slow yet steady progression toward higher-power rocketry. The first G motor to be certified by the NAR was Gary Rosenfield's 29 mm AeroTech G25.

As another sign of improving relations between the NAR and high-power users, Pat Miller invited Tripoli president Tom Blazanin to the annual NAR board of trustees meeting as an observer. Blazanin accepted the invitation "in hopes of obtaining better insight to the NAR's operation and, if called upon, to aid in the subject of establishing accommodations between the NAR and the Tripoli Rocketry Association." (In the not-too-distant future, NAR leaders would invite Blazanin to join the NFPA on behalf of Tripoli.)

Meanwhile, Tripoli's board was laying the groundwork for what one day could become a nationally recognized safety code for high-power rocketry. In early 1987, the board unveiled its Advanced Rocketry Safety Code. This was the first safety code endorsed by Tripoli as a corporation. It was created by the Motor Listing Committee: Gary Fillible, Edward Tindell, and Tom Blazanin. The safety code was offered to the membership as a *guideline* "meant as a means to allow you to enjoy advanced rocketry, not to stifle creativity but to encourage safe and constructive involvement in rocketry with the maximum of enjoyment and participation"—whatever that meant.

Since the Tripoli guideline was not mandatory, it was really just a collection of suggestions. It was a weak start on the road to a real safety code. Yet it was a step in the right direction. As the NAR had done in the early 1960s, Tripoli, whether it knew it at the time or not, was moving toward a safety code that would ultimately be accepted by the NFPA. Tripoli leaders realized that without a recognized safety code, high power was subject to the whims of local legislators nationwide, much like model rocketry was in the distant past. Perhaps even more importantly, board members understood that without a safety code, high-power rocketry would remain unduly hazardous.

A few of the guidelines set forth in the Advanced Rocketry Safety Code would be recognized today. The code required the use of only certified motors at launches, minimized metal in rockets, and set limits for maximum winds at takeoff. There were rules regarding the mandatory use of blast deflectors and rules regarding ignition systems. There were no safe distance tables yet. But the code provided all Class B motors be launched a minimum of 150 feet from the spectator line. Motors exceeding a J-impulse level had to be fired at least 200 feet away from spectators. The code called for a minimum 5-second countdown prior to the launch of any rocket.

Some of the 1987 code rules would be altered or even abandoned. For example, the code allowed launch rods at an angle up to 30 degrees from vertical. That rule was eventually modified to 20 degrees. The code stated that every launch pad was required to have a 5-gallon container of water within 10 feet of the pad. That rule was simply abandoned. The code even provided a rule for launch event photographers, who were permitted within 75 feet of the launch pad for any motor. This rule too would eventually be cast aside. At the time of Tripoli's first safety code, the association had two hundred members.

California Rocketry was an important publication in the initial spread of higher-powered rocketry in the early 1980s. However, like so many other hobby publications, its marketability in hobby stores never caught on. By 1987, the magazine was closed, and its few remaining subscribers were offered refunds for their subscription money or a choice of other options to compensate them for the magazine's demise. It would become a familiar story in the history of high-power. Flyers love to read stories about new motor or rocket developments, exciting launches, and rocketry personalities. But over the next thirty years, it would remain difficult for any rocketry magazine to survive without some kind of subsidy. A magazine must have advertisers to survive. And advertisers require lots of readers to justify their ads. For high power, the business model to obtain both readers and advertisers in the world of print journalism would remain elusive.

In February 1987, Tripoli announced that LDRS-6 would be held in Colorado. Scott and Ramona Dixon, who supplied high power with Vulcan Systems motors, AeroTech's only real competition, secured a launch site at the United States Army base at Fort Carson, just south of Colorado Springs. The flying field was the base's impressive Howitzer cannon range. As a backup plan, the Dixons also secured the use of a private property parcel of some 80,000 acres high in the Rocky Mountains, not far from the city of Colorado Springs.

LDRS-6 was scheduled for August 7, 8, and 9, 1987. However, there would still be only two days of actual flying. The first day was set aside for socializing. The launch dates were chosen so as to avoid conflict with NARAM.

Tripoli president Blazanin asked for member support early in the year to help run the corporation's first ever LDRS. He called for volunteers to serve as optical altitude trackers, range safety officers (RSOs), and launch control officers (LCOs) for both days of the event. Blazanin said he also needed "paper pushers" to help organize this first ever western LDRS.

"All Tripoli members planning on attending should volunteer for something," said Blazanin, who signed himself up to haul most of the needed launch equipment from his home in Pennsylvania to the distant Colorado launch site.

There would be at least two flying contests at LDRS-6, not unlike the contests that were traditionally held each year by the NAR at NARAM. There was a G-motor water loft duration contest in which flyers would loft 16 ounces of water in the air for as long as possible and an H-motor duration event in which the objective was to keep an H-powered rocket in the air for as long as possible, utilizing only a streamer for recovery.

Tripoli designated board member John O'Brien as event coordinator for LDRS-6. O'Brien would arrange all scheduling for the launch, and more importantly, his word would be the ultimate authority on the field.

This way, there would be one person who could make final decisions at LDRS, much like Pearson did for the first five years in Ohio. Tripoli chose the Howard Johnson Motel in Colorado Springs as official lodging for the launch. The Tripoli board of directors and the second general assembly would be held at the hotel.

As it turned out, LDRS-6 would not take place on a cannon range at Fort Carson, Colorado. The army decided to conduct firing exercises over the same weekend scheduled for the launch in August. That location was scrubbed. Fortunately, Scott and Ramona Dixon still had their backup plan: an 80,000-acre grazing range on private property near Hartsel, Colorado, less than an hour from Colorado Springs. LDRS would proceed on schedule. The Advanced Rocketry Safety Code would govern the event.

In addition to an increase in physical space for the launch, the waiver for LDRS-6 was also a big leap forward for Tripoli. The FAA gave flyers in Colorado 8,500 feet to play with, with brief windows to 20,000 feet on Saturday and Sunday. This was an incredible altitude for high power at the time. Flyers were advised that all persons were welcome to fly at LDRS-6—you did not have to be a member of Tripoli—but only Tripoli members holding a signed confirmation card would be allowed to launch high-power motors. In other words, if you wanted to fly high power at LDRS, you must join the Tripoli Rocketry Association.

Tom Blazanin released another issue of the *Tripolitan*, which was still the official journal of Tripoli, on the eve of LDRS-6. For nearly two years, he had almost single-handedly created issue after issue of the black-and-white *Tripolitan* in a new larger magazine-type format. The journal did not come out on time very often. But it was the undisputed voice of high-power rocketry in America.

The editor of any publication chooses a cover shot for each issue, usually from a selection of many photographs at his or her disposal. It was the same for the *Tripolitan*. For the cover of his August 1987 issue, Blazanin chose a picture of two unknown high-power flyers at the Lucerne range in Southern California. The pair was depicted placing a short three-finned rocket on a tripod pad earlier that year. The man on the right was Gary Price, who was from a small town in Utah. Price would one day sit on the Tripoli board of directors.

The flyer on the left—wearing a baseball cap, blue jeans, and a T-shirt—was also from Utah. He made the drive to Lucerne that weekend to fly rockets with his wife and five young children. The rocket on the pad was his recently finished LOC *Warlock*. He was new to high-power—until now, he had been too busy with his regular job for hobbies of any kind—and he had only recently joined Tripoli. However, in only a few years, this unknown flyer would carry the banner of the Tripoli Rocketry Association into every hobby store in America, changing the course of high-power rocketry forever. He was Tripoli member number 232. His name was Bruce Kelly.

Hartsel is an unincorporated town in Park County, Colorado. It was founded in the 1880s and named for Samuel Hartsel, a local cattle rancher. Even today, the population of the town is less than a thousand. It is located near the geographic center of the state, leading some to refer to Hartsel as the "Heart of Colorado." U.S. Route 24 passes through the center of town, which rests at an elevation of more than 8,800 feet, making Hartsel the highest location of any LDRS ever held.

The open range near Hartsel is ideal for rocketry. There is barely a tree for more than a mile in any direction. The landscape is slightly rolling but appears relatively flat from a distance. The grass in the summer of 1987 was short and dry, and the vista was unobstructed for miles. Flyers could see snow on top of the peaks of the Rocky Mountains in the distance.

Access to this launch site outside of Hartsel was easy once one made it to Colorado. The range was an hour away from the hotels in Colorado Springs, and the state highway brought cars right to the edge of the field. It was a longer drive than the trip from the L&K Motel to the field in Medina had been, yet the trade-off in terms of waiver, space, and freedom from worry about landing in cornfields or on someone else's property was well worth it. The only nuisance besides the rapidly changing summer afternoon weather was an occasional cow wandering onto the flying range.

President Blazanin traveled from Pennsylvania to Colorado for LDRS-6. His trip was plagued with disaster. Early in the week, fellow flyers arrived at Blazanin's Pittsburgh home to pick up Tom and more than 600 pounds of range equipment he was hauling to the launch. These were the days before rocketry clubs banded together to provide their own equipment locally at an LDRS.

"The equipment came from the stuff I had put together for the Pittsburgh launches," recalled Blazanin, who came up with the idea of Tripoli having its own "traveling equipment" for big events like LDRS.

It was a not a bad idea, but it would prove logistically impractical over the years. In 1987, everything was still being tried at least once.

"About 90 percent of it was mine," he said. "But a few people brought their own launch pads to LDRS. We shared everything."

Blazanin filled a huge wooden box with hundreds of pounds of gear and waited for his ride. When his fellow flyers arrived, the big box was lifted on top of a trailer, and Blazanin and his crew tried to fit everything else into their vehicle. But it wouldn't work. There was just too much stuff. The wooden box was taken down, Blazanin's equipment was unloaded, and he sent them on their way to Colorado without him. He would find another way to get there with everything, he told them. So Blazanin and his fourteen-year-old daughter, Becki, who made the long trip to Hartsel with her father, loaded everything he had into his ailing two-door 1976 Mercury sedan.

"The vehicle was completely overloaded," he recalled. "It looked like a low rider as it took to the highway."

In retrospect, packing the car was the easy part of the trip. In Kansas, the overheating radiator of the Mercury blew up. Later, near the Colorado border, the car lost a tire. Finally, near Colorado Springs and almost to its destination, the car's transmission simply quit. Tripoli members lent Blazanin a hand and removed the rocketry equipment from his disabled car and split up the parts to be driven out to Hartsel in separate vehicles. The Mercury, however, went no farther. It was sold for scrap in Colorado. After LDRS, Blazanin and his daughter returned home on a Greyhound bus.

The first Tripoli-run non-Ohio LDRS launch had other growing pains too. Tripoli was now in charge of LDRS, but did anyone in the organization have the experience needed to run the biggest high-power event in America? At first glance, the answer to that question appeared to be no. On Saturday morning, malfunctioning launch equipment frustrated flyers to no end. Dead launch system batteries with frayed wiring and rusty alligator clips wreaked havoc.

"I would think that before a launch of this magnitude, someone would have snipped and soldered new wires and clips at the pad," said one disappointed flyer.

The launch pads themselves were not much better. They were dilapidated and needed to be replaced. There were complaints regarding not only the Tripoli-sponsored contests, which were inexplicably called off on Saturday, but also the confusion created by too many people at the combined RSO/LCO table. LDRS was growing, and with more people in attendance, it was clear that the RSO and LCO functions needed to be isolated from one another at the range head.

Another problem at LDRS in 1987 was member participation in helping run the event—or rather the lack of it. Volunteers who had previously agreed to work failed to show up for their shifts. This resulted in numerous delays as the work of many was carried out by an overburdened and unappreciated few. It was clear LDRS needed tighter controls, more planning, and more experience at the helm.

Still, amid delays and grumblings and calls for a new launch system, the flying at Hartsel went forward. For the first time at LDRS, flight cards were used to log individual flights. And for the first time at any LDRS, M-impulse motors were launched. This increase in motor power thrilled the crowd to no end.

Jim Dunlap built a rocket he called *Athena*, powered by an AeroTech M1000. The single-use motor retailed for $900. People were excited to see it. As a special effect and to aid in tracking his rocket, Dunlap attached four small smoke generators to the aft end of *Athena*. The smoke charges were lit just prior to motor ignition, encircling the rocket in thick smoke on the pad before flight. Unfortunately, at ignition of the M1000, something went terribly wrong. The rocket barely moved as the solid fuel simply burned up on the pad, along with most of the bottom end of Dunlap's rocket, which just sat there ablaze. There was no flight, just lots of water being tossed on the burning rocket to douse the fire. Fortunately for spectators, there was one more M motor at LDRS-6: Chris Pearson's flight of a Vulcan Systems M2000. This was one of the first—if not the first—successful launches of an M-powered rocket at any LDRS.

"It was a scaled-down version of the N motor that Vulcan Systems was building for the NCR/E-Prime Aerospace project we were working on," recalled Pearson. "The full-sized motor was an N5000, so the M motor was about an M2000. It was a 3-inch-diameter motor about 2 feet long. The rocket was just the motor case with fiberglass fins, a small parachute, and a solid epoxy nose cone. It was Smokey Sam propellant."

At ignition, the rocket roared off the pad, leaving a long black smoke cloud in its wake as it disappeared high above the field into wispy cloud cover. When it reappeared in the sky, it was clear something was wrong with the deployment system. The rocket's airframe was fluttering back to the earth like an open tube dropped out of a plane. It was missing two fins and its payload transmitter. What remained of the rocket was pretty much destroyed on impact.

"I flew it never expecting to get it back," said Pearson. "But someone brought it back the next day . . . The nose cone had ablated away almost by one-half. The paint had melted off the body and flowed around the fins."

Despite their respective flight problems, these two single-use M motor flights at LDRS-6 represented the crossing of an important threshold for the national launch. Over the next fifteen years, M-class motors represented the pinnacle of high power. And LDRS would become the best place in the world to see M-powered rocket flights. By the late 1990s, the common use of M motors would be a distinguishing feature separating LDRS from almost every other launch in America. This would not change until well into the first decade of the new century, when a few regional launches would grow to rival LDRS itself, at least in terms of M-motor flights.

Lots of Crafts, recently renamed LOC Precision, continued its reign as high power's premier commercial heavy rocket builder in 1987. The LOC *Esoteric* rocket, introduced at LDRS-5 in 1986, was flown by several flyers at Hartsel in 1987. Tripoli board member John O'Brien modified his *Esoteric* with an extended payload section, resulting in a total rocket length of more than 12 feet. He applied gold leaf to the fins and equipped the rocket with a heavy $2,000 camcorder to record the flight. The rocket, nicknamed *Warp Factor*, was powered by a single L1507 and four J-class motors. The *Warp Factor* was late in being ready, and on Sunday afternoon, the flight crew hurriedly prepared the rocket as thunderstorms advanced toward the range.

"As we pushed to accomplish this last flight of the meet, a small but costly event took place," recalled O'Brien later. "The emergency backup parachute system, which I was trying for the first time, was accidentally triggered. Knowing that there was no time to reprep this system, we proceeded without it. This later proved to be a fatal mistake."

The rocket had a spectacular liftoff and disappeared into cloud cover.

"There I stood with my mouth open, on the verge of wetting my pants as a year's worth of work and $2,600 in costs rapidly became just a smoke trail high in the clouds," said O'Brien.

When *Warp Factor* was sighted on its way back to Earth, the payload section was separated from the booster, and the parachute was long gone. The hapless booster was also on fire, ultimately landing in a small creek. The payload section carrying the expensive camcorder was not found until the next day. It was another big-rocket-deployment lesson for all in attendance. It was also another crowd-pleasing flight.

Ron Schultz premiered another new rocket product at LDRS-6: the *Top Gunn*, a seven-motor rocket that would be a familiar vehicle at high-power launches for the next decade. *Top Gunn* was 7.67 inches in diameter and was a classic "cluster" rocket, meaning it could hold multiple engines in the aft end of the airframe. Cluster rockets require special precautions. Most importantly, the engines in a cluster must ignite in a timely manner. If they do not, the rocket can become unstable and may not fly straight. *Top Gunn* featured seven 54 mm motor tubes. It retailed for $210, an expensive purchase in the days before carbon fiber and fiberglass airframes. For LDRS, Schultz loaded all seven motor tubes with propellant: a J700, two J200s, two I140s, and two I125s. Everything worked as planned, and there was a magnificent launch, reportedly the most powerful at LDRS-6.

Chuck Mund and John Holmboe teamed up again to fly another big project at LDRS. This time, it was their scratch-built *Mercury Redstone*. The stunning rocket stood more than 8 feet tall and launched on an AeroTech I motor. At motor burnout, three more motors on the escape tower were fired, pulling the capsule clear of the rocket body. All the parts were recovered.

LDRS-6 brought flyers from throughout the United States, from Southern California to Massachusetts, Florida, and all places in between. There were more than two hundred people in attendance, with more than sixty flyers. Although a small LDRS event by today's standards, this was truly a national event, and the central location in Colorado brought more flyers to the launch than the Medina events. Chris Pearson, who had graciously turned over the launch to Tripoli in 1986, was present at LDRS-6, but this time, he was free to fly rockets and let others handle the details of hosting the event. Pearson was also honored at the Tripoli general assembly meeting on Saturday night with the presentation of the lifetime membership award for his contributions to high-power rocketry through his perseverance with LDRS.

Tripoli board member Chuck Rogers published a list of optically tracked altitudes at LDRS that included several flights that exceed 6,000 feet above ground level. A LOC *King Viper* owned by Utah flyer Bruce Kelly purportedly carried a barometric pressure sensor that reported an altitude of 5,500 feet. Another rocket flown by Mark Grubelich of Applied Physics Labs also carried a homemade onboard altimeter. Grubelich's 2.6-inch-diameter rocket cleared more than 13,000 feet on a K420 motor. A tally of the flight cards revealed some interesting statistics. There were more than 230 flights spread among 66 flyers at LDRS-6. Of motors burned, more than 115 were H motors or larger. Almost half of all the rockets were built from scratch.

Tripoli's corporate business was conducted at LDRS-6 on Friday and Saturday nights. The membership committee announced that Tripoli now had 324 paid members, with new prefectures in Pittsburgh, Houston, Cleveland, Massachusetts, and Florida. The Safety and Motor-Testing Committee reported that plugged nozzles at launches around the country were causing many catastrophic motor failures. This problem was common with nozzles that rested on flat blast shield plates at the pad, said committee member Ed Tindell of Texas. The plates were closing off the nozzle, he reported. A new safety rule was created that required standoffs be designed to keep rocket motor nozzles elevated off the blast deflector.

President Blazanin reported he had been recently approached by a member who thought that Tripoli should change its name "to better represent our purpose." Some flyers were uncomfortable being associated with an organization that shared the same name with the Libyan city of Tripoli. Some Tripoli members thought a name change would be appropriate because the association's current name had terrorist overtones.

The suggestion to change the organization's name did not gain much support with the Tripoli board, which voted against the proposal six to two. However, at the general assembly meeting the following night, there was additional membership support for the name change. There was also some very vocal opposition. One member suggested that Tripoli hold a contest to pick a new name for the organization. Ultimately, no decision on any name change was made at LDRS-6. But this issue would remain on the table and be debated several times for the next two years.

Apparently, not even the name LDRS was safe in 1987. At LDRS-6, some flyers suggested that the phrase for which LDRS was an acronym be changed too, from Chris Pearson's "Large and Dangerous Rocket Ships" to "Large and *Dynamic* Rocket Ships." A few Tripoli members believed the word "dangerous" cast the launch in an unfavorable light with the public. This begged the question of whether the public at large even knew that high-power rocketry existed. The debate over the name of LDRS lingered for months after the launch. Taking out the adjective "dangerous" would improve the image of the event, claimed one flyer. Another flyer suggested that the acronym be changed to "Let's Do Rockets Safely." That suggestion and others like it brought the following response from one Tripoli member:

> Who are we trying to kid? High power has an element of risk, and that's why we do it. We may as well call it *"Let's Do Rockets Stupidly"* or perhaps *"Lotta Dumb Rocket Stuff"* or *"Love Dem Rocket Ships"* or *"Little Dirty Rocket Ships"* or *"Little Dainty Rocket Ships."* Let's face it. These are large and dangerous rocket ships, so why evade the issue or try to pull a fast one on an insurance company?

No decision on a name change for LDRS was made in Colorado. Yet like the issue of the name change for Tripoli itself, the matter was taken seriously by the board, which agreed to leave the subject open and ask for membership input in the months ahead. Later that year, the membership would vote on it, rejecting any change.

Despite the logistical issues and problems with launch equipment, the Hartsel location was a hit with flyers who attended LDRS-6. Utah NAR member Randall Redd launched rockets at LDRS in Medina. He was one of the many flyers impressed by the new site at Hartsel.

"Medina was a beautiful place for the birthplace of high-power rocketry," he said. "It was a moderate-sized cornfield that was big enough for G, H, and I motors, which were flown all the time, but it was too small for J, K, and L motors, which drifted into surrounding cornfields."

Hartsel, on the other hand, observed Redd, was a wide-open cattle range big enough for the higher-powered motors to come. Redd joined Tripoli while at LDRS-6 in Hartsel. He became member 300.

Before leaving Colorado, the board also discussed where to hold LDRS-7 in 1988. No site was chosen. Alternate locations near Houston, Las Vegas, Salt Lake City, and even Canada were discussed as possible venues for the next big event. At the conclusion of LDRS-6, the board of directors agreed to hold the next general assembly at LDRS-7 "wherever that may be."

There were calls for change in the way high-power events were run after LDRS-6. Some of the more compelling suggestions came from Tripoli board member Ed Tindell of Texas.

In the October 1987 issue of the *Tripolitan,* Tindell authored a long opinion column on the conduct of a high-power launch. In Tindell's view, based in large part on his experiences at LDRS-6, there were several important changes that needed to be made at future LDRS events and perhaps with high-power rocketry in general. These changes were necessary to improve both the safety and the quality of the LDRS experience, wrote Tindell. First, LDRS must be run by experienced people and more of them, he said. And they had to have help.

"The last five LDRSs were held at the same location and run by the same small group of people," he observed. "They had a pretty good system worked out, and things usually went well."

With the Tripoli Rocketry Association now in charge of LDRS and the site location changing, said Tindell, more preparation had to be done far in advance of the event. Next, said Tindell, improvement was needed in the application of rocket-science fundamentals to high power. Rockets were getting bigger, a lot bigger in a short period, and stability issues involving the center of pressure and the center of gravity had to be taught to all Tripoli members to reduce the incidence of unstable rockets.

"[A]s our hobby grows and we try more and more new things for the first time at these launches, stability becomes very critical," said Tindell. "Our rockets must be both statically and dynamically stable."

Tindell also called for better high-power motors with fewer failures and an improved type of igniter. He demanded better launch pads and more of them, each equipped with stainless-steel rods and azimuth and elevation controls.

"The pads should be well spaced out as advanced rockets require more clearance between themselves to avoid jet blast when launching."

The launch control system was a source of serious disappointment at LDRS-6. Tindell argued that heavy-duty components were needed, such as superior electric relays, better-secured electric cables, and a portable generator. These improvements would ensure the LCO at LDRS could consistently and reliably light high-power motors and clusters, he said.

Tindell suggested that the flying field should change too. The days of a launch site with only a few tens of acres for launch and recovery were over, he said. Ranges for LDRS needed to be vast, like the field at Hartsel or Lucerne in California—hundreds, if not thousands, of acres of open space at a location where the local club could secure a suitable FAA waiver far in advance of an event. It was also necessary that there be contacts with the local fire and emergency services to let them know a rocket launch was scheduled, he said. Finally, said Tindell, LDRS needed an experienced launch team comprised of an RSO, an LCO, and pad managers.

"All persons not working at the time or unconnected with rockets being launched should spectate from a safe distance and not bother the launch crew," he said. "Large organized launches such as LDRS are not the same as small sport launches, with just a few people and your own equipment. Some necessary evils such as stringent safety checks are mandatory."

Tindell was not the only person in high power who recognized the need for such improvements to the hobby. Indeed, his column reflected common sense more than anything else. Today most of the suggestions for improvement made by Tindell in his 1987 column are simply taken for granted as a given at not only LDRS but also every other large regional launch in the country. One day soon, Tindell would be given the chance to steer Tripoli himself.

For more than two decades, the NAR restricted its member's participation in "nonprofessional" rocket activities. Although the NAR formally approved the use of G motors in early 1987, the restriction on launch activities beyond the confines of a NAR-sponsored event remained in place. Members of the NAR who also wanted to be part of Tripoli and participate in Tripoli-sanctioned events had no choice but to run afoul of the rules. For flyers who wanted to remain in model rocketry and also participate in high-power, there was still no real alternative. They had to leave the NAR or take their chances. In August 1987, all this changed.

At the association meeting on August 3 held at the University of California in Irvine and under Pat Miller's guidance, the NAR approved a change to its bylaws that would profoundly impact high power, the NAR, and Tripoli. The NAR removed the restriction on member participation in non-NAR rocketry events and, for all intents and purposes, allowed its members to fully engage in high-power rocketry without violating the NAR Safety Code.

There were many factors leading to this change, probably best summarized by one prominent NAR member shortly thereafter: "The high-power crowd made the logical and moral argument that mere membership in the NAR cannot and should not revoke an adult's government-granted privilege to fly legal nonmodel rockets as long as all government regulations are being satisfied. It was a persuasive argument and hard to refute."

The rule change was also the result of the eighteen-month study conducted by the second blue-ribbon commission, chaired by former NAR president Jim Barrowman. Based on the commission's findings and recommendations, the NAR adopted what became known as the "3/48 Rule." To preserve the NAR's ability to insure its model rocketry activities, the NAR would allow its members to participate in non-NAR rocket activity so long as there was a clear line of separation, both in physical distance and in time, between model and high-power events.

Under this rule, a NAR member could enjoy rocketry activity other than model rocketry as long as the member did not fly model rockets and nonmodel rockets (i.e., high power) within 48 hours of each other at the same location. The rule also provided that if the member wanted to fly model and nonmodel rockets within 48 hours of each other, he could do so as long as the launch sites were at least 3 miles apart.

Although subject to some confusion, the 3/48 Rule was another step forward for the NAR. As one NAR leader stated at the time, "it ensures that model rocketry and other types of rocketry will be kept separate and distinct." The physical and temporal separation between model rocketry and high power allowed the NAR to retain its liability insurance footing without having to worry about nonmodel rocketry events undermining its coverage. The 3/48 Rule also decreased the risk that an accident at any NAR launch could be interpreted as a high-power accident, which would cut off insurance coverage for model rocketry.

"After a simple, clear criterion has worked for thirty years, it is never easy to make the mental adjustment necessary to abandon it in favor of a new untried principle," observed NAR trustee Chris Tavares at the time the rule was announced. "Yet the board spent over a year doing so, appointed a commission to double-check all the safety claims, and spent hours in meeting discussing legal ramifications."

The result, said Tavares, was this more "liberal" restriction of the 3/48 Rule.

"You can control your level of risk by personally limiting [how] far outside the bounds of model rocketry you are willing to wander," said Tavares. "But when you do indulge in nonmodel rocketry activity, do it responsibly. Please don't do it at the expense of another's launch coverage or NAR membership."

One immediate result of the rule change was that NAR members would no longer have to leave their organization to explore high power.

"Tripoli has received a large membership increase since word of this ruling has come out," observed the *Tripolitan* in the fall of 1987.

The *Tripolitan* also noted an increase of flights on E through G motors at recent launches, including LDRS. Some of this increase was due to the fact that NAR members were now free to fly model rockets at high-power events in addition to exploring high power, said the *Tripolitan*. "Many hardcore NAR members are reaching out to be legal in their large rocket activities."

The 3/48 Rule was the first step in what became joint membership in the NAR and Tripoli. Although a bit confusing and far from perfect, it was also the first "no-mixing rule" in modern hobby rocketry. This concept would appear again several years later, when Tripoli tried to deal with criticisms that it too was not in step with the technological advances of the times. Only the latter no-mixing rule would be applied to flyers who wanted to fly experimental or homemade motors at Tripoli-insured monthly events.

The NAR's relaxation of the rules with regard to non-NAR activities applied only to flying high-power rockets. The new rule was not applicable to the manufacture of rocket fuel. President Miller reaffirmed the NAR's position that all rocket motors used by NAR members must still be manufactured "by qualified professionals." There was no allowance for experimental or research rocketry in the NAR at any launch at any time. The NAR continued to refer to people who mixed their own fuel as "basement bombers." And for all practical purposes, so did the Tripoli Rocketry Association.

LDRS-6 was the largest LDRS to date. But it was not the biggest high-power event of 1987. Rocket launches in Southern California, Ohio, and Pennsylvania continued to bring out an increasing number of people interested in high power.

One such event was near the town of Zelienople in Pennsylvania that fall. There were more than three hundred flights at an event called Z4, likely making this launch the biggest high-power event of the year. At least one hundred motors at Z4 were H impulse or above. The two-day launch attracted flyers from all over the East Coast. Chris Pearson, Ron Schultz, Tom Blazanin, and other Tripoli leaders were among those in attendance, and there were plenty of spectators too. John Holmboe and Chuck Mund replayed their success at LDRS-6 with the repeat flight of their *Mercury Redstone*. Ron Shultz again flew his seven-motor *Top Gunn*.

The field at Zelienople was really just a back-filled strip mine outside the tiny Western Pennsylvania town. Permission to fly there was somewhat nebulous. Organizers actually spread out their launch and recovery activities over multiple properties, most of them private. The primary landowner would appear from time to time, exhibiting a benevolent malevolence toward the rocketry crowd.

"Do what you want!" he shouted, allowing the flyers to remain on his land with the following warning: "But if someone gets hurt, you're trespassing!" This was the fourth big launch held at Zelienople. Unfortunately, it also was the last. There were a high number of CATOs at the event, and some people later complained that safety practices should have been better. Then on Sunday afternoon, a high-power rocket ended up on private property well off the field. The rocket was owned by none other than Ron Schultz.

"We flew the *Top Gunn* and the *Ultimate Max* in Colorado that year, and whatever we flew out west, we also flew in the east," explained

Schultz. "So I flew them at Z4 but with less power. We put three Vulcan System I motors in the *Top Gunn*, and the rocket was perfect. In fact, it landed only a few hundred feet away."

Then, said Schultz, he prepped his seven-motor *Ultimate Max*, filling the 4-inch rocket with seven H and I motors. The *Ultimate Max* was a much smaller and lighter rocket than the *Top Gunn*. The H motors would be airstarted, meaning they would not be ignited until the rocket left the pad.

"It was gone like a bat!" said Schultz, who believed everything on the *Ultimate Max* worked perfectly on the way up.

The rocket arced over high in the air at apogee, but the 10-second delays for motor ejection on the H45s failed to work.

"We waited and waited, but the parachutes never popped open."

The rocket disappeared in the distance, he said. Schultz and his wife then set out in search of the rocket. They found it a short while later, impaled in green grass on a golf course nearly 2 miles away.

"It must have been screaming when it came in undeployed," said Schultz. "Someone said it landed near a guy who was getting ready to hit a ball."

There was almost nothing left of the rocket. The golf course owners were not sympathetic. They called the authorities, and the police made their way to the launch site. With three rockets sitting on their pads awaiting final countdown, the entire launch was shut down. Afterward, Tripoli made peace with the golf course and nearby property owners. Yet the police refused to budge. According to one report, there could be no more launches anywhere in the township without the flyers first posting proof of financial responsibility.

It was not an unreasonable request. People who send high-powered missiles into the air in populated areas should be required to post proof of some kind of financial responsibility in case of an accident. However, the Tripoli Rocketry Association was not responsible yet. In many ways, high-power rocketry was still a bunch of hobbyists doing whatever they wanted with as few controls as possible. Insurance was for the NAR or for other people who were responsible, not for Tripoli. So the field at Zelienople was lost forever. It would be nearly a year before another high-power event would go forward anywhere on the East Coast.

One evening in 1987 and not long after LDRS-6, Tom Blazanin received an unexpected telephone call at home.

"There is a Mr. Harry Stine on the phone, and he wants to talk to you," said Blazanin's wife as she handed him the receiver.

Blazanin knew who Stine was. Everyone in rocketry did. Stine was model rocketry personified; he was the NAR in the flesh, the father of model rocketry. Blazanin, who had some lingering doubts about the relationship between Tripoli and the NAR, assumed there was trouble brewing.

"But I got on the phone, and Mr. Stine says, 'Tom, we have this NFPA committee, and we are rewriting the model rocketry codes, and I think it would be good if you could have a seat on the committee as a Tripoli representative.'"

Blazanin quickly agreed. Following his call with Stine and in October 1987, the president of the Tripoli Rocketry Association traveled to St. Louis, Missouri, to attend his first NFPA meeting. At that meeting, the final draft of the Code for the Manufacture of Model Rocket Motors (NFPA 1125) was finalized. With a representative from high power informally present, Tripoli was, for the very first time, able to get a firsthand look at the inner workings of the important NFPA Committee on Pyrotechnics. If a permanent seat could be obtained for a Tripoli representative, the organization would be in a position to add its voice to the upcoming rewrite of the Code for Model Rocketry, anticipated to begin in 1989. And with a seat on the committee, Tripoli would also be in a position to perhaps influence the NFPA to consider a code for high-power rocketry too.

It was a great opportunity for Blazanin and the Tripoli Rocketry Association, and it would provide dividends to high power in the future. And the opportunity was handed to Tripoli by G. Harry Stine and the NAR. Soon, additional Tripoli members obtained seats with the NFPA on the Committee on Pyrotechnics, of which Stine was the chair. The Tripoli members chosen were Blazanin as the official member, with Ed Tindell as the alternate. Gary Rosenfield, who represented model rocket motor manufacturers, was already on the committee. Together with Blazanin and Tindell, Rosenfield added his voice to a contingent of high-power users whose presence would one day influence the NFPA and ultimately its full acceptance of high-power rocketry.

LARGE AND DANGEROUS ROCKET SHIPS

Tripoli became an international organization in late 1987, when Jussi Leander of Finland joined the club. Leander became Tripoli member 400. The NAR, meanwhile, was also growing again, leaving behind its membership slump of the early 1980s. Despite the NAR's initial resistance to high power, NAR membership climbed from its low point of 1,870 members in 1983 to 3,500 members by late 1987. NAR membership would continue to outpace that of Tripoli and would reach more than five thousand by the end of the decade. Many of those new NAR members were now looking forward to their own version of high-power rocketry, a version that would soon be upon them.

10

A New President

In late 1987 and with a year remaining on his term, Tom Blazanin stepped down as president of the Tripoli Rocketry Association. Blazanin was indispensable in the reformation of the national organization in 1985. He presided over the creation of the confirmation program and high power's first safety code, and he was Tripoli's first official representative on the NFPA's Committee on Pyrotechnics. Blazanin enjoyed the challenges that went along with the creation of Tripoli. Now he believed it was time for someone else to lead.

Since 1985, Blazanin had also published Tripoli's official publication, the *Tripolitan*. From time to time, he asked the board for help creating the journal of high-power rocketry, but there were no serious takers. Ultimately, serving as both the president of Tripoli and the creator of the *Tripolitan* was too much work for one individual. The *Tripolitan* was usually behind in not only the number of issues that were produced per year but also the timeliness of its content. Blazanin stepped aside as president in large part so he could continue running the magazine.

"I felt that if Tripoli were to survive, it would need to be able to continue under someone other than me," he said. "One person running everything was not for the better of the whole."

In early 1988, the board elected Tripoli treasurer Ed Tindell as the second national president of the Tripoli Rocketry Association. With its new president living in Texas, Tripoli's official business address was changed to Houston, where Tindell resided. He was a married engineer working for Houston Lighting & Power. He was also a model rocket flyer who discovered high power in 1984 and the dynamic leader of Tripoli Houston, one of the first Tripoli prefectures in the United States. Prior to joining Tripoli, Tindell was a NAR member and the editor of a local NAR newsletter. He believed a Tripoli prefecture could bring advanced rocketry to local flyers in Texas.

"I figured the response would be enthusiastic and open-minded because most of the people flew high power anyway, and they all loved the hundred or so photos that I brought back from LDRS-4 in 1985," he said.

Initially, however, setting up a local Tripoli club did not go as smoothly as Tindell had hoped. Although some flyers signed up immediately, Tindell ran into trouble with a local NAR member who decided that anything that was not a model rocket was illegal.

"My home was threatened with visits by the local fire marshal, our [NAR] section charter was threatened with cancellation, and my every move was scrutinized," said Tindell. Ultimately, NAR president J. Pat Miller came to Tindell's aid and ended the interference.

Tindell was a true self-starter. His enthusiasm for rocketry helped establish Tripoli's presence in the Lone Star State.

"I decided to do things by example," he explained in the mid-1980s. "I started flying kits from LOC, [North Coast Rocketry], and Ace at local model rocket launches and public demonstrations. People were impressed and started asking questions and telling me about their experiences with high power. If I gave out the addresses of LOC, NCR, or AeroTech once, I gave them out a hundred times."

Tindell convinced a hobby shop owner in Houston to stock his shelves with high-power supplies. He also helped create the first high-power launch in Texas, called Lone Star 1. As a result of his efforts, Tripoli Houston flourished, and Tindell subsequently earned a reputation as someone who could be counted on to get things done.

Tindell's enthusiasm for rocketry led him to action on a national level too. He volunteered for several Tripoli committees and was a contributor of technical articles and launch stories to the *Tripolitan*. He became Tripoli's treasurer in 1986, and he oversaw a significant increase in revenues as the national organization grew in size. By early 1988, Tripoli had more than five hundred members, and that number climbed to more than six hundred by the middle of the year. Tindell wanted to continue to grow the organization's membership, and he believed that Tripoli could one day "provide the infrastructure to make advanced high-power rocketry the top amateur scientist endeavor in the United States." Now, as Tripoli's new president, he was determined to see this goal accomplished.

On the morning of May 4, 1988, a series of explosions ripped through a PEPCON manufacturing plant located outside Las Vegas in the quiet community of Henderson, Nevada. The strength of the explosions was so powerful, the blasts registered as a small earthquake on seismographs hundreds of miles away in Southern California. The resulting shock waves not only obliterated the PEPCON plant and several buildings nearby but also shattered windows in nearby Las Vegas and buffeted a Boeing 737 passenger jet on its approach to McCarran Airport. Two people were killed in the blast, and more than 350 people were injured. More than seventeen thousand people were evacuated from their homes. It was the biggest news story in America that night, and the lawsuits that followed would last nearly a decade. The result of these claims was legal settlements approaching $100,000,000. Although a definitive cause was never determined for the disaster, one thing was clear. The explosions were the result of a plant fire followed by the accidental detonation of millions of pounds of AP, the oxidizer used in the manufacture of composite propellant for solid-fuel rocket motors.

PEPCON was one of only two manufacturers in America who produced AP. The other producer, Kerr-McGee, was also located in Henderson. The AP from these two plants supplied America's solid fuel for military and nonmilitary rockets, including the fuel used in the outboard boosters of the space shuttle and the composite motors used in high-power rocketry. In the months after the PEPCON disaster, there was a temporary shortage of AP in the United States. Some people even proposed regulations that ranged from reclassifying AP as a class 1.1 explosive to an outright ban on its production.

"Try to imagine the world of high-power rocketry today with large black powder and hybrid motors," wrote Tripoli member Chuck Piper, an engineer who served as one of the many technical advisers in the lawsuit that followed. "[T]hat's kind of what it would be like had 'cooler heads' not prevailed and halted the implementation of certain proposed regulations concerning the manufacture, transportation, and storage of this material."

The PEPCON accident had an immediate impact on rocketry.

"Losing the largest producer of AP in the world created a major shortage of hobby rocketry motors," recalled Bruce Kelly, who believed that much of the remaining reserve of AP was quickly acquired by the Department of Defense. "When news leaked out that there was going to be a motor shortage, a couple of us went from dealer to dealer, buying every rocket motor we could get our hands on. The shortage did not let up for a year."

PEPCON ceased to produce AP in Henderson after the 1988 explosion, ultimately going through corporate restructuring and name changes before relocating its manufacturing facilities to an isolated area near the small town of Cedar City, Utah. One day, in the distant future, AeroTech would make its home there too.

The development of the minimum-diameter rocket continued to move forward in 1988. And slowly but surely, high power was gaining altitude. That spring at Lucerne, a rocket was tracked to 11,460 feet above ground level, purportedly a new record. A few months later, that record climbed to 14,158 feet in a vehicle flown by Bruce Kelly of Utah. Kelly's record did not last long either; altitude records were broken and reset all year long. Of course, tracking high-altitude flights in the late 1980s was by optical means only. Here or there, an experimental payload contained a crude version of an altimeter, but such flights were exceedingly rare.

As high-power rockets continued to evolve, the need for improved safety practices was becoming more apparent too. Tripoli still had no *enforceable* safety code. There were no serious injuries reported in 1987 to 1988. But anecdotal evidence suggests that advancements in rocket and motor size were outpacing safety and sometimes common sense. Yet with no national training or certification programs, most flyers learned the hobby the hard way: via trial and error.

Unlike today, flight anomalies were the rule and not the exception for most large rocket projects. And the bigger the rocket, the more likely the flight would end in disaster. Heavy motors in short, small, and midsize rockets were a particular recipe for trouble; such rockets would frequently take off, fly a few feet straight up, become unstable, reverse direction, do a loop or two, and then roar off on a course parallel to the ground, 100 feet high, for hundreds of yards. This anomaly was called a "cruise missile." If the same rocket careened into the ground and skidded along the earth under power, it was called a "land shark." Sometimes rockets would inadvertently deploy their parachutes midair under thrust while the motor was still burning. Such rockets would twist and turn and gyrate in every possible direction, trailing fire and smoke the whole way. This was a "skywriter."

Recovery problems were another plague on the hobby. Shreds (failure of the airframe during boost), zippers (airframe slices caused by recovery line mishaps), broken shock cords, and separated airframe sections falling from altitude were common. Rockets sometimes came in ballistic—that is, with no deployment at all; they just went up, turned over, and came down fully intact—and impaled the ground at high speed. This was either a "lake stake" or a "lawn dart," depending on whether you were flying above a barren desert lake bed in Nevada or a green sod farm in Georgia. Sometimes a rocket popped off just its nose cone at apogee, but the parachute would not deploy. The open-ended airframe would hurtle to the ground, sans the cone. On impact, the rocket would crush itself against the ground, the open end of the vehicle filling with tightly compressed dirt. This was known as a "core sample."

One flyer sized up the increasing safety problem in high power this way:

> Last fall, I left the Z4 launch in Zelienople before the premature closing of the launch by the local authorities. I came away with the feeling that the launch had more than its share of impacted rockets . . . I had my sister-in-law draw up a cartoon of a gopher wearing a hard hat and a T-shirt that said, "I survived Z4." I never submitted it to the *Tripolitan* for two reasons. First, the subject of free-falling rockets was not one to be taken lightly with levity. Second, the gopher looked more like a beaver.

Another flyer wary of too many unstable rocket flights suggested kit makers publish the center of pressure for their rockets so that flyers could be sure that the center of gravity remained ahead of the center of pressure, thereby ensuring basic in-flight stability. Similar concerns were expressed by manufacturers too. On the eve of LDRS-7, LOC owner Ron Schultz reminded all Tripoli members of the importance of thrust curves, burn times, and careful motor selection in high-power rockets.

"I cannot overstress the importance regarding the recommended motor usages that are made for our kits," said Schultz. "Gain experience in rocket flight analysis. Know your rockets. Learn their limitations and capabilities. If you are unsure of what motor to use in a built kit, ask the manufacturer. After all, he's the guy who has already done the preliminary guesswork for you."

Tripoli continued its confirmation program into 1988. It was a far cry from the certification system used today. A person could still fly an H rocket successfully in a small rocket, and they were good to go with anything else they could get their hands on.

In the summer of 1988, Tripoli obtained insurance for LDRS-7. This was the first time modern high-power rocketry was insured for a launch event. The general liability coverage was $1 million, with a small deductible. There would be no medical payments, and coverage was restricted to spectators only.

"This may seem like a limited deal," wrote the editors of the *Tripolitan* at the time. "But in reality, it is a very great deal and a giant step forward for advanced nonprofessional rocketry."

Insurance coverage for LDRS-7 was not without conditions. Yet the impositions of conditions also helped Tripoli. The insurance company made coverage dependent on rules to minimize the presence of spectators and other nonessential persons in the launch area during the firing of rockets. Further, only commercial motors were allowed and then only from a preprinted list of nineteen manufacturers (a remarkably large number compared to the short list of commercial motor makers today). Finally, coverage was dependent on the *strict* observance of what was now being called the "Tripoli Safety Code." A representative from the insurance company was present at LDRS to observe activities and ensure compliance with these rules.

The Tripoli Safety Code of 1988 was virtually identical to the Advanced Rocketry Safety Code published in early 1987. These codes were offered only as *guidelines* to Tripoli members, like a set of unenforceable suggestions. Thanks to the insurance company requirement that these rules be followed at LDRS-7, the Tripoli Safety Code finally received the teeth that Tripoli, in its reluctance to sound like the NAR (i.e., financially responsible, adult like, and ready to enforce safety rules), lacked at this moment in its history. A formal safety code was finally in play, and it would be followed at LDRS.

Orville H. Carlisle, NAR member number 1, died from natural causes at his home in Nebraska on August 1, 1988, just days before LDRS-7. Carlisle was credited by G. Harry Stine as the inventor of the model rocket. The two men helped launch the NAR in the late 1950s.

Stine met Carlisle, a lifelong resident of Norfolk, Nebraska, in 1957. At the time, Stine was a White Sands engineer living in Las Cruces, New Mexico. In his spare time, he wrote a regular column on space and rockets for a national how-to science magazine called *Mechanix Illustrated*. One day Carlisle read one of Stine's articles, and he decided to write the engineer about his own passion for rocketry. In his letter, Carlisle described the homemade rockets he and his brother had developed. He asked Stine if he would like a few samples to fool around with. Stine said, "Sure." Their subsequent correspondence and Stine's fondness for Carlisle's rocket design led Stine to travel to Nebraska to learn more about Carlisle's work.

"When I visited him in October 1957, he taught me how to make black-powder model rocket motors. In time, I passed this information along to Vern Estes," said Stine.

"Orv created the model rocket by combining his knowledge of pyrotechnics with his brother's model airplane knowledge," Stine went on later. "And Orv was an expert in fireworks. He served as an expert member of the NFPA Committee on Pyrotechnics from 1974 until his death . . . [F]or many years, his original Mark 1 and Mark 2 model rocket designs were on display at the National Air and Space Museum of the Smithsonian in Washington, D.C."

Orville Carlisle was seventy-one.

For the second year in a row, high-power enthusiasts assembled in Central Colorado for LDRS, held on August 5 to 7, 1988. LDRS-7 was still a two-day launch event; only Saturday and Sunday would be flight days. The fee was $10 per flyer, and all flyers were required to sign a liability waiver. LDRS would continue to feature several flying contests too.

There were no M flights at LDRS-7. But rockets continued to get bigger and more powerful, with LOC leading the way. Ron Schultz again used Tripoli's national event to premiere another new rocketry product. Schultz's goal this year was simple: to launch the biggest, most powerful rocket ever flown at any LDRS and to make the same rocket available in kit form for sale to the public. This year's LOC star was 7.67 inches in diameter and more than 12 feet tall. It weighed 25 pounds, unloaded, and housed three 54 mm motor mounts. The rocket was supplied with quarter-inch-thick centering rings, a set of fiberglass fins, and a 10-foot-diameter parachute. It would retail for $295, another pricey creation for its day. Schultz called his rocket the *Mother Lode*.

"For our trip to LDRS-7, the *Mother Lode* was carefully wrapped to avoid any damage," he said. "Upon arriving at the motel, it was then placed outside our room in a specially made display cradle. There, it received double takes from all passersby. Most people asked, 'Will it work?' My answer would be 'On paper, it looked good!'"

During the prior two LDRS events, LOC teamed up with AeroTech to launch its new projects. This year, Schultz asked Scott Dixon at Vulcan for propellant. Dixon recommended three Vulcan L750s. These motors would burn for 5.3 seconds with a combined total impulse exceeding 11,500 Newton seconds or nearly 500 pounds of thrust. It was an ambitious project for 1988. On the pad and loaded with three L motors and several pounds of counterbalancing weight in the nose, the unfiberglassed *Mother Lode* weighed 50 pounds.

With a crowd of more than two hundred on hand, the *Mother Lode* blasted off Saturday afternoon. Two of the three L motors lit on cue, and Mach diamonds—a formation of standing wave patterns that appear like bright jewels in the supersonic exhaust plume of a rocket—were seen in the *Mother Lode*'s flaming wake. The third L motor delayed ignition for a moment and then lit when the big rocket was several hundred feet in the air. The *Mother Lode* climbed high into the overcast sky, reaching an altitude of 15,349 feet above ground level. It was yet another new Tripoli altitude record.

Tripoli board member and altitude tracker Charles "Chuck" E. Rogers said after the flight, "The rocket reached an estimated Mach 1.33, making it undoubtedly the largest diameter paper rocket ever to go supersonic."

The *Mother Lode* separated at apogee via motor ejection. The rocket was then lost in the clouds, and when it was spotted again, the big airframe was whirling around in a flat spin, descending rapidly toward the ground. There was no parachute deployment.

"I felt my knees weak and ready to buckle as we watched [the rocket] plummet to the ground," said Schultz, who hurried to the scene of the crash. "When we reached it, the sight was not pretty."

The rocket landed hard, destroying much of the airframe. The nose cone split open, and the 10-foot-diameter parachute was nowhere to be found. Schultz realized later he should have used additional lines to the parachute for a more secure attachment to the rocket's shock cord. Despite the failed recovery—or perhaps because of it—the big rocket was viewed as a spectacular highlight of LDRS-7.

The *Mother Lode* was the most powerful and highest-flying rocket at LDRS-7. Yet it was not the biggest. Richard Zareki arrived with his scratch-built 22-foot-tall *Aurora IV* that was even bigger. Zareki's four-finned rocket was powered by four Vulcan K420 motors and weighed 62 pounds on the pad. This was probably the heaviest rocket to yet fly at any LDRS. At liftoff, the Smokey Sam motors emitted a huge column of black smoke that trailed the rocket high into the air. Unfortunately, there was no deployment of this rocket's parachute either. The rocket tumbled back to Earth, bounced high off the ground, and was destroyed.

John O'Brien returned to Hartsel with a reconstructed version of his ill-fated flight at LDRS-6. He called his new rocket the *Warp Factor II*, and it would be another highlight of LDRS this year. O'Brien completely revamped his rocket, a modified LOC *Esoteric*, using bungee line instead of elastic strap for the shock cord, and he strengthened all aspects of construction. On Sunday morning, the *Warp Factor II* ripped off the pad under the thrust of five AeroTech J motors. At an apogee of 6,000 feet, the 30-pound rocket separated cleanly, deploying two 5-foot-diameter parachutes.

The rocket was drifting safely under chute, but O'Brien was not through yet. At 600 feet above ground level, O'Brien pushed a handheld remote-control button that fired an onboard backup ejection charge. The charge overpressurized the upper section of the airframe and blew the tethered nose cone free of the rocket. This pulled out a final "emergency parachute" attached to the nose. It was an example of a dual deployment–like system and was years ahead of its time. The rocket landed gently.

There were many other interesting flights at LDRS-7, including SST-replica rockets, UFO-looking rockets, and several camera-equipped vehicles, which were becoming more sophisticated every year. Doug Gerrard's *Eye in the Sky* carried a 35 mm camera for photos at altitude under I and J power. Gerrard's first flight suffered a broken shock cord. The payload bay and its camera ended up soaked in a shallow pond. He dried out the mess, patched things together, and had a fine second flight that yielded photos of the entire rocketry range. Mort Binstock also obtained impressive photos at LDRS-7 with multiple high-power flights of camera-carrying rockets.

Guy Soucy had no luck when his sophisticated two-stage rocket, powered by more than eight total motors, sustained multiple CATOs immediately after takeoff. But Paul Binion had success with his J700 to J125 rocket called *Nightmare*, which purportedly reached an altitude of more than 16,000 feet before succumbing to yet another failed recovery.

Binion's rocket was not the highest altitude reported at LDRS-7. That honor went to Chuck Mund, whose 44-pound *VideoRoc* was optically tracked to more than 20,000 feet. Mund's 6-inch-diameter rocket was powered by a single K and eight I motors. And his altitude—or the reported tracked altitude of 20,539 feet—quickly became suspect. Not that Mund had anything to do with it—he didn't do any of the tracking—but the tracking results of LDRS-7 seemed to some observers to be far too high for way too many rockets.

In a detailed analysis written after LDRS by Chuck Rogers, who participated in the tracking at Hartsel, several altitudes were called into doubt by computer simulations suggesting the numbers were inadvertently inflated through faulty readings by inexperienced trackers and unusual field conditions. The flight simulations cast doubt on all the tracking at LDRS-7, wrote Rogers, including Mund's flight and the flight of Binion's two-stager. However, Rogers let stand the altitude claimed by Ron Schultz's *Mother Lode* to 15,349 feet—nearly 1,000 feet higher than the tallest mountain in the continental United States—as simulations revealed the rocket may have flown that high.

"Of course, the tracking data for the *Mother Lode* could be as invalid as for the other two rockets," said Rogers. "Ron just may have lucked out and got a track that approximated his rocket's altitude capability."

It was clear that a better way had to be found to reliably track a rocket's altitude. Unfortunately, practical commercial altimeters for high-power rockets were still several years away.

Tripoli president Ed Tindell published another massive compilation of statistics for LDRS-7. Tindell's statistics give modern observers a detailed look into the average LDRS event of the late 1980s. Tindell went through the 232 flight cards from LDRS-7, and he listed every possible piece of information, from the total number of flyers (91) to total rocket weight flown (1,045 pounds) and overall flight success rate (86.5 percent). Tindell listed each flyer by name and described their rocket's performance, along with details of the causes of unsuccessful flights. Mishaps at LDRS-7 also provide a snapshot of high-power design issues at the time. Troubles included but were not limited to stripped parachutes, failed ejection charges, instability issues, and structural failure (shreds).

Six different motor manufacturers were represented at LDRS in 1988, from which 422 motors were ignited, wrote Tindell. The most high-power motors were made by AeroTech, followed next by Vulcan Systems. There were only seven reported motor failures. Not surprisingly, LOC-created rockets were the most numerous of any commercial builder at LDRS-7, with twenty-four. Other commercial rocket makers at the time included North Coast Rocketry (seven) and U.S. Rockets (six). At least seventy-two rockets were scratch-built, reported Tindell.

Tindell noted high-power rockets were carrying more payloads than ever. Payloads at LDRS included heavy camcorders, 35 mm cameras, Estes Astrocams, and several Super 8 movie cameras. Rockets carried beepers, smoke cartridges, and experimental timers. Tindell reported that AeroTech launched a rocket at LDRS-7 with a new demonstration motor. Gary Rosenfield had been searching for a name for this special mix of propellant, which emitted a bright long flame and copious amounts of thick white smoke. He ultimately settled on something catchy. He called his new propellant "White Lightning."

Of the 232 rockets flown at LDRS in 1988, 60 were clusters. At least twenty-five new high-power confirmations took place. The total Newton seconds expended at Hartsel was, according to Tindell, 144,656.

The most pressing issue addressed by the Tripoli board at their annual meeting at LDRS-7 was the acquisition of a professional motor test stand. Tripoli needed to start a high-power motor certification program, just like the NAR had been doing for model rocket motors for many years. The testing and certification of commercial motors was one of the stated purposes of Tripoli in its corporate charter. A motor certification process would help ensure commercial motors were safe and reliable and that they performed as advertised; it would also establish Tripoli as the authority for high-power motor certification. Even in 1988, the NAR was slowly moving toward high-power rocketry. If the NAR began certifying high-power motors first, not only might Tripoli lose credibility as the advanced high-power rocketry organization, but also, it could become irrelevant.

The problem for Tripoli was money. A good motor stand would cost approximately $2,000, Tom Blazanin told the board at LDRS-7. And Tripoli was still short on funds. On August 5, the board voted to bring the matter to the general assembly the following night, following the first ever LDRS banquet. More than $200 was collected that evening to start funding the test stand project. This was a small amount given the importance of a test stand. But volunteers came forward to get the project moving in the right direction. Still, another several years would pass before Tripoli achieved its goal of certifying high-power motors.

For the fifth time at LDRS, an honorary lifetime membership was awarded at LDRS-7. This time, the honor went to Blazanin for his efforts at establishing the Tripoli Rocketry Association as a national organization. And for the first time at any general assembly meeting since Tripoli took over the event, a decision was made as to the location of next year's LDRS. A vote taken revealed that a majority of the flyers in attendance wanted to return to Hartsel in 1989. LDRS-8 was scheduled for Colorado on August 4 to 6, 1989.

On September 29, 1988, Space Shuttle *Discovery* was launched from the Kennedy Space Center. This was the twenty-sixth flight in the shuttle program, and *Discovery* was the first launch since the *Challenger* disaster in 1986. The *Discovery* crew spent four days in space and landed safely at Edwards Air Force Base. Within a short period, the shuttle program would be in full operation again, flying up to six times per year over the next decade.

In late 1988, high-power rocketry was still dominated by the single-use motor. From coast to coast at high-power events in California, Texas, Ohio, and Virginia, flyers launched their rockets unaware that new motor and electronics technologies were just around the corner. These technologies would soon change the hobby forever.

In Southern California, three men working independently of one another were conducting secret work on their own visions of high-power rocketry's future. Two of them were experimenting with their designs for a *reloadable* high-power rocket motor. Shortly, they would begin covert testing of their products in distant corners of the Lucerne Valley rocket range. The third man, who would also test his products in secret at Lucerne, was working on a motor that was made of an entirely new type of propellant, made up of a combination of solid fuel and gas. And in Colorado that fall, an electronics expert was about to begin testing of an onboard device that would use air pressure to accurately gauge the altitude of a high-power rocket.

By 1988, Tripoli had more than six hundred members, with flyers from almost every state. California had the most Tripoli members with eighty-three, followed by Ohio and Florida, with forty-two and forty, respectively. There was only one member each in South Carolina, Montana, and Kentucky. There were no Tripoli flyers in Idaho, Arkansas, or New Mexico. To the north, Canada now had nine members. Tripoli had a high-power monopoly in 1988. But the NAR, like a slow-moving ship changing course, was picking up speed as it churned toward a future in high power for its members too. Within four years, the NAR would be fully immersed in high-power rocketry.

In September 1988, Virginia had its first Tripoli-organized high-power event. It was called "Fort AP Hill-1." The launch was held at the United States Army base of the same name and drew flyers from throughout the mid-Atlantic region. This was the first high-power launch on the East Coast since the range at Zelienople was lost in 1987.

The field at AP Hill was used by the army as a parachute drop zone. It was wide open, with only a few scattered trees as obstacles. Light winds blowing in the wrong direction sent several rockets into those trees. Still, the 12,000-foot waiver was outstanding for an East Coast launch in the 1980s.

Tom Blazanin worked as a guest LCO at AP Hill, and he helped debut Tripoli's new electronic launch controller constructed by Tripoli Pittsburgh and tested in Colorado at LDRS-7. This was also another opportunity for Blazanin and Tripoli to recruit flyers into the ranks. To use a high-power rocket motor at AP Hill, flyers were required to join Tripoli and become confirmed. As a result of this rule, new people joined the association so they could fly H motors and above. There were more than 150 flights at the launch, A through J power. Ron Schultz launched the very first rocket of the event, a LOC *Hi-Tech H45*. The largest rocket of the event was a LOC *Top Gunn* that blasted off on multiple I motors. In the distance, flyers could hear machine gun and mortar fire at the army base all weekend long.

High-power action that fall shifted to the West Coast for the Octoberfest event at Lucerne. Blazanin trekked across the country for the launch. He brought the new Tripoli control system along with him. In the 1980s, Lucerne launches were somewhat of a free-for-all, with flyers doing their own thing with regard to launch pads, controllers, range setups, and sometimes even safety.

"There was this one group who came in, and we called them the Hillbillies," recalled one longtime flyer looking back at Lucerne in the 1980s. "They drove this old broken-down station wagon, and they looked like those guys from *Duck Dynasty*. So they would camp out on the lake bed at night, and in the morning, we would drive in from our motel and pull into our spot, and they would be belly-up to the sun, and they were drunk the night before, and wherever they fell, that's where they were all night. [They] would then get up, take a pee, and get a rocket ready for launch. They would put their rocket on the antenna of the station wagon, using the antenna as launch rod. Then they would put the car in drive and pull the station wagon right up to the launch line. They would stick the leads of the igniter into their car's cigarette lighter and launch the rocket. And that's how they flew their rockets, G through H power."

For the first time and with the new Tripoli launch control system in place at Octoberfest, Lucerne saw some order to its weekend. Flyers also filled out Tripoli flight cards, reportedly another first at Lucerne.

Most of the flights at Octoberfest were in the midrange of high power, H through K motors. However, there were at least two L flights, and several rockets cleared 10,000 feet in altitude. One rocket was visually tracked to 15,818 feet. For the first time at any Tripoli launch, an N motor was also flown. The motor was made by Reaction Labs and was carried in a rocket built by Tav Niche. On October 1, 1988, Niche's four-finned rocket thrilled the crowd with a spectacular liftoff under N power. The airframe shredded at 800 feet, showering motor parts, rocket fins, and airframe debris all over the desert floor. The cause of the mishap was reportedly a motor CATO in flight. Despite the failure of the flight, the N motor threshold had been crossed by a high-power flyer at a Tripoli event; others would soon follow.

11

The Rocket Scientist

Ed Tindell was a positive force in high-power rocketry from the moment he became involved with Tripoli in the mid-1980s. As the Houston prefecture founder, as a member of the Tripoli board, and as Tripoli's second president, Tindell played an integral role in Tripoli's growth in the latter part of the decade. However, Tindell's management style rubbed some of his fellow board members the wrong way. By early 1989, some board members were concerned that Tripoli was moving in the wrong direction, particularly with regard to relations between Tripoli and the NAR.

"It seems that the attitude of Tripoli has shifted in the past eighteen months," wrote one commentator at the time. "Rather than working with the NAR, as it has in the past, Tripoli seems to be on a confrontation course with the NAR. This is incredible, considering the NAR changed its bylaws to accommodate its members who want to be active Tripoli members!"

Tindell had disagreement with some Tripoli members over the selection of Colorado as the host site for LDRS-8. Some flyers wanted the launch to be held outside Las Vegas. He also got into a spat with members who were upset that LDRS-6 and LDRS-7 were both scheduled on the same days of NARAM, the NAR's premier annual event. Tindell was not solely responsible for the decision to hold LDRS on the same dates as NARAM. But while he was president, LDRS-8 was scheduled to conflict with NARAM again. This forced flyers and vendors to choose between attending either LDRS or NARAM rather than offering everyone the ability to attend both, as had been done during the first five years of LDRS. Some people believed Tindell was the force behind these decisions.

"I have no knowledge of the dates for NARAM," wrote Tindell at the time, "and quite frankly, I am not concerned with NARAM ... Like most Tripoli members, I do not attend NARAMs, and I cannot see catering to the few that do. That sort of thing would set a precedent to accommodate the NAR. Tripoli is separate from the NAR and will remain so. We have a right to plan our activities when and where we choose. The NAR will not move NARAM to accommodate us because they would then set a precedent for accommodating us. If you decide to attend NARAM over LDRS, well, that is your choice ... Advanced high-power rocketry is no longer the hobby of an elitist few because Tripoli is succeeding beyond all expectations to reach out and let everyone participate."

Tindell's remarks may have added support to the belief by some that Tripoli was heading in the wrong direction. The fact that NARAM had been in place for nearly thirty years and was typically held in the first week of August and that many Tripoli members were also members of the NAR was of no consequence to the Tripoli president.

"My predecessor, Tom Blazanin, was not in contact with the NAR, and I am not in contact with the NAR," declared Tindell.

Tindell may have been unaware that the NAR and Tripoli had grown increasingly cooperative under Blazanin's watch. The two clubs had some serious disputes still ahead of them, but the days of threatened suspensions for participating in high-power activities were long gone. Indeed, by the late 1980s, the NAR was assisting Tripoli in gaining a foothold with the NFPA. The NAR also helped pave the way for even better relations between the clubs with new rules that allowed for joint membership in both organizations.

In an opinion column in the *SNOAR News* in January 1989, the editor called for Tindell's replacement as Tripoli's leader and predicted that with the recent election of three new Tripoli board members, it was likely Tindell would not be reelected as president by a reconfigured Tripoli board. It seemed Tindell's days as president were numbered. But as it turned out, the Tripoli election of 1988–1989 was not as final as some had assumed. In fact, the election was thrown out.

Until 1989, the election of board members was usually held at the end of every calendar year. Once the election was decided, the new board would directly elect its president and officers. During the election held in late 1988/early 1989, three new Tripoli board members were elected by the membership. Afterward, the election was invalidated. And the reasons given for the "election redo" were nebulous. In an unsigned editorial in the *Tripolitan* in mid-1989, an explanation was offered as to why the election had to be repeated. Among other things, the explanation for the "redo" was that (1) some members were allegedly confused by the ballots they had filled out, (2) some members did not understand what the issues were, and (3) more than two hundred Tripoli members apparently did not get their ballots on time. Not surprisingly, the newly elected board members who were now in limbo and those in the association who had voted for them were outraged at the invalidation of the election.

Tindell sent out new ballots with the *Tripolitan*, and in a move that may have alienated him from fellow board members, he included in the new ballot a space for Tripoli members to write in their preferred choice for president of the corporation. This move was contrary to the corporate structure of Tripoli and the precedent of prior elections. At the time, Tripoli presidents did not get elected this way. The president was elected by the board. Tindell was bypassing corporate bylaws and tradition by calling for direct election of the president by the membership. The election debacle was, for many, the last straw. Tripoli board member Chuck Rogers of California was one of Tindell's critics. Rogers was running for president, and he believed the original election results, which were thrown out, were in his favor.

"Do you like the idea that the entire election has to be redone?" he asked the membership as the redo was still pending. "I feel that in the last cycle of the officers' election gyrations, I had the solid support of the majority of the Tripoli board members to be elected president by the board. Yet now the election of the president is being done by the membership when the bylaws explicitly state that *only* the board members can elect the president!"

As LDRS-8 approached, the controversy regarding the election and who would lead Tripoli became more intense. Yet Tindell was not without supporters. Tom Blazanin thought Tindell was a good president, perhaps a better president than he was. And Tripoli cofounder Francis Graham came to Tindell's defense in early 1989 in response to comments by some in the rocketry press.

"Ed Tindell is an honest and straightforward guy," said Graham, "and he served with blemishless distinction as treasurer in the Blazanin administration." He added that Tindell's personal "honorable conduct" was beyond reproach and that ad hominem attacks on Tindell did not ring true to "anyone who has had even a passing acquaintance with Ed."

Although Tindell appeared antagonistic toward the NAR and his management style was not for everyone, it was hard to deny he worked hard to do what he thought was best for Tripoli and high power. Tindell wrote a series of lengthy articles for the *Tripolitan* in 1987 and 1988 calling for volunteers in a wide range of high-power activities. He had also called for the formation of new committees to strengthen Tripoli nationwide.

Tindell improved the logistics of LDRS, and just prior to LDRS-8, he purchased twelve stainless-steel high-power pads at a cost of $1,100 to Tripoli, which is nearly $2,400 in today's dollars, to help improve flight operations to make the launch run smoother. Tripoli flyers admired the pads at LDRS-8 and thought they were a good addition. Indeed, the new launchers were an improvement to the event. Yet Tindell was criticized for buying the new equipment without approval from the board. He replied that he had repeatedly asked for input on the pad purchase from board members but had received no response from anyone.

As LDRS-8 approached, the biggest problem facing Tindell and the board was the election mess. Several board members found the election situation intolerable, and they blamed Tindell. Prior to August, plans were made to try and replace Tripoli's president at LDRS-8 in Colorado. In 1989, there was no requirement that Tripoli board members attend LDRS. It was rare that all the board members would show up in any given year. But the majority of the board stayed away from LDRS-8. Of the eight then current directors—Blazanin had just resigned and had yet to be replaced—only three board members made it to Colorado: Ed Tindell, Chuck Rogers, and Curt Hughes.

A fight for control of Tripoli was brewing, and Rogers knew he needed board member votes. He arrived at LDRS with three notarized proxy votes for a new president and one verbal proxy from a fourth absent board member. Theoretically, he now had support of a majority of the board. If he played his cards right, Rogers might outmaneuver Tindell when the time was right.

Most Tripoli members attending LDRS-8 were unaware of the conflict within the board. For the eighth year in a row, at least out in the field, nothing mattered except for the smoke and the fire and the noise. This was the first LDRS with three full days of flying. It would also be the first LDRS where an N-powered rocket would be flown.

Tripoli members arrived at Colorado Springs on Thursday, August 1, 1989. At the last turnoff from the main highway to the flying range was a large hand-painted sign to greet the rocketry faithful from all over the country:

LDRS-8 Rocket Heaven—Welcome.

The size of the crowd at Hartsel increased from the previous year, with up to three hundred people milling about, preparing rockets and meeting friends. Tripoli's twelve new high-power pads, each carrying an interchangeable stainless-steel rod attached to a stand with four legs painted yellow and black, were an improvement in the event's infrastructure. Each pad was made of stainless steel and had elevation controls. The pads were arranged in two rows of six pads each, with the bigger rockets flying from the back row.

All weekend long, people walked the field lugging heavy shoulder-mounted video cameras recording launch preparation and flights. During heavy rainstorms, which were frequent, people hurriedly retreated to makeshift tarps, lean-tos, cars, trucks, and tents to wait out the soaking. When the skies cleared, rockets were made ready on the ground in the short-cut range grass or under well-secured tarps.

There are no statistics for the event. Videos and other anecdotal evidence suggest there were at least a couple hundred flights at LDRS-8, with most rockets being in the model-to-mid-power range up through H power. There were plenty of I- and J-powered rockets and some K flights too. The primary motor manufacturers were still Vulcan and AeroTech. Other motor makers such as Prodyne and Reaction Labs were also represented. Long-burning I-impulse motors (such as the I65 by AeroTech) remained popular, along with the Smokey Sam motors produced by Vulcan. As had now become common at LDRS, the biggest of the rockets, the ones powered by the largest motors or the heaviest or tallest or fastest rockets, were the stars of the show.

One ambitious flyer thrilled the crowd on Friday with the flight of a large yellow-and-black rocket that carried an L585 motor, a showstopper of a motor in 1989. It also housed a big 1980s-style camcorder. After a quick five count by the LCO, the rocket lifted off rather slowly and began to pick up a modest amount of speed. The entire vehicle then came apart, pouring down debris all along the pads.

"Keep your eye on that rocket!" yelled the LCO into the microphone.

It was hard to do. All the rocket's fins had been stripped off, and many more pieces filled the air. The L motor was not that powerful by today's standards, but the rocket was built to model rocketry specifications. As the fins and debris fluttered to the ground, the parachute drifted away, attached to nothing. The expensive camcorder was reduced to junk on impact.

Young flyer Caz Sienkewicz of New England thrilled everyone Friday with the successful launch and recovery of his one-third-scale *Black Brant*. The rocket took off on a Vulcan K575 and roared into the sky before returning fully deployed under a good parachute. Another crowd-pleaser on Friday was an upscaled *Maxi-Mosquito* by Chuck Sackett and Mike Ward of Florida. The 4-foot-tall *Mosquito* took off on a J250, climbed in a long slow arc, and then hurled itself into the ground under power. In a few years, Sackett and Ward would become well-known for some of the most ambitious flights in the history of high power.

One of the more interesting rocket experiences at LDRS-7 involved a scale version of a *WAC Corporal* built by a team of flyers from Tampa who had hurriedly finished their rocket in Florida just before LDRS. The drive to Colorado would be a long one.

"The group wrapped it in a blanket to protect it during transportation," recalled observer Art Markowitz years later. "However, the paint was not completely dry, and the blanket stuck. [They] had to shave it and repaint it in the parking lot of the motel before they could fly it."

On Saturday and after the rocket had been duly groomed near a hotel dumpster and made ready for its maiden flight, the *WAC Corporal* was transported to the range and was then carried on the shoulders of four men to Pad 12. It was then heaved into vertical position. Powered by four 54 mm motors, the rocket's aft end came to rest a few inches above the round metallic blast shield. An empty can of Mountain Dew served as a standoff between the blast shield and the rocket. The crew members wired up the igniters and then ran off the field as the countdown began from ten.

At ignition, the *WAC* was momentarily motionless as a ball of flame erupted under the airframe and danced around the base of the pad.

Then the vehicle moved quickly up the rod and off the pad. Almost immediately after clearing the launcher, the entire upper end of the rocket separated from the booster and fell backward to the ground. The booster, with some of its multiple motors still spewing fire, continued upward under thrust. It was like some whirling dervish, rolling furiously in the air, moving away from the pad and the crowd, twisting and turning for a few hundred feet before collapsing to the ground a smoking heap.

"At that time, the fraternity [of high-power flyers] was small, and you could keep in contact with almost everybody," remembers Markowitz. "Because the number of flyers was small, after a bad flight, the group would gather around and perform an 'autopsy' to determine what went wrong and how to prevent it in the next flight."

Inspection of the debris of the *WAC* revealed at least one of the motors failed on ignition. Either that failure or the premature deployment of ejection charges—or perhaps both—led to the rocket's quick and fiery destruction.

At just before 1:00 p.m., a thunderstorm raced across the field and shut down the range. People huddled under their tarps, holding firmly as the wind picked up anything not tied down securely. But just as quickly as it arrived, the storm moved on, and flying resumed. A LOC *Warlock* was brought to Pad 10 for a confirmation flight.

"Do we have the confirmers ready?" shouted the LCO as the rocket was made ready.

Moments later, the rocket took off under I power and was recovered under parachute for a successful confirmation and the applause of the crowd.

Next up was one of the biggest flights of the weekend. Kelly Badger's rocket carried the first N motor ever flown at any LDRS: an AeroTech N3050. The rocket weighed more than 60 pounds, loaded and ready to go on the pad. At ignition, it took off slowly and then disappeared into the clouds for most of its flight. The altitude of the rocket was never known as visual tracking could not be accomplished. Unfortunately, the details of any recovery were not recorded either.

As usual, there were some impressive mishaps at LDRS. On Sunday, one rocket equipped with six G100 motors and a single I motor went unstable as soon as it left the pad. It flipped over and impaled itself in the ground. A short time later, another flyer launched a LOC *Esoteric* packed with five motors, including an L750, two J285s, and two J250s.

"This rocket has a total of 8,970 Newton seconds," said the excited LCO as he brought the entire crowd to its feet in anticipation of the launch. "Everyone on your feet for this one."

The rocket lifted off slowly, climbed several hundred feet, and then came apart at the seams, either by motor CATOs or some other major mishap. Rocket parts rained down on the field.

Randall Redd teamed up with friend John Rahkonen to launch a rocket with a special high-performance motor on Sunday. Redd built the rocket; Rahkonen made the fuel. Rahkonen was a retired fuels engineer for propellant giant Morton-Thiokol of Ogden, Utah. He had been designing rocket motors most of his life, and he had worked on the propellant used in the space shuttle solid rocket boosters.

"He and some Thiokol associates formed a small company to build high-power rocket motors," said Redd.

The company was called Propulsion Dynamics, also known as Prodyne. In the early 1960s, Prodyne produced solid fuel in motor sizes D through F power.

"John used to make black-powder motors in the early days of model rocketry, and when he discovered high power, he became very excited because they were using the same materials he had been using at Thiokol," recalled Redd.

Indeed, Rahkonen was a pioneer in higher-powered model rocketry. By the mid-1960s, he was marketing Prodyne motors in national hobby magazine ads. These were black-powder motors with phenolic cases and ceramic nozzles. Harry Stine praised Rahkonen's new motors as early as 1963 in the *American Modeler* magazine. Stine reported that model rockets using Prodyne motors reached altitudes of up to 3,000 feet, if not higher. Prodyne briefly manufactured its own line of model rockets too.

At LDRS-8, Redd's rocket was powered by a K700 built by Rahkonen. On ignition, the small red rocket was instantly destroyed in a massive CATO. Pieces of the vehicle shot off the pad, tumbled, crashed, and then burned nearby. Two years later, a Rahkonen-powered rocket purportedly streaked to over 20,000 feet at LDRS-10, said Redd.

"I never saw a rocket move faster or more vertically straight previous to this one," echoed longtime NAR member Bradley Ream, who was also on hand for the flight.

Rahkonen passed away in January 2010 at the age of seventy-nine. His rocketry life was the subject of a Finnish movie entitled *A Strange Message from Another Star*.

Some of the action at LDRS-8 was back at the hotel in Colorado Springs. At the Tripoli board meeting at the Hilton Hotel on Thursday night, Chuck Rogers announced that a majority of the board wanted the pending elections results published immediately. He also asked for the resignation of Pres. Ed Tindell.

Tindell responded quickly to criticisms aired against him. He defended his purchase of the new LDRS launch pads, and he addressed the problem of board members who did not return his calls as it related to association decision making. He pointed out that his term for president would not expire until Saturday, August 5. In his view, he had at least two more days as president no matter what the board wanted to do. Tindell agreed he would have the election results by Saturday, and seeing the writing on the wall, he then announced he would resign because of demands that his duties as Tripoli president were putting on his personal and professional life.

Two days later, at the general assembly meeting on Saturday, August 5, Tindell formally resigned as the second president of Tripoli. Later, he put his resignation into writing:

> In the past, I had time to devote to run headquarters, arrange Tripoli activities, work on Tripoli projects, and make administrative decisions as president. However, my life has changed . . . both professional and personal, for which I have resigned . . .
>
> I am not burned out on rocketry. No way, no how! I simply cannot continue to support Tripoli at this time. I had every intention of continuing my present support of Tripoli up through the end of 1990, when my current board term would have expired. My only regret is that I am unable to achieve that goal.

Tindell remained prefect of Tripoli Houston for a short period. He would attend at least one more LDRS, and he would continue to fly, write, and remain active with Houston-area rocket flyers. A regional high-power launch was even created by flyers in his honor. However, within a few years, he left high-power rocketry.

In Tindell's place and by a vote of the board of directors, including the four absent proxies, Charles E. "Chuck" Rogers, the rocket scientist, became the third president of the Tripoli Rocketry Association. The membership poll authorized by Tindell was also tallied, and the results were the same. Rogers was president by a vote of 129 to 71.

When Chuck Rogers became Tripoli's third president in 1989, he was an aerospace engineer at the Air Force Flight Test Center at Edwards Air Force Base in California.

Since the 1950s, the Flight Test Center was home to many of America's most ambitious experimental aircraft, including projects that helped put the United States into space, such as the North American X-15. Rogers had worked at the Flight Test Center since graduating from college in 1984. And for this ambitious thirty-year-old originally from Upstate New York, his work in Southern California was a lifelong dream come true.

Rogers was born in Buffalo, New York, in 1959, the son of Jim and Cynthia Rogers. His grandfather, Earnest James, was of Irish-Scottish descent whose family immigrated to North America in the mid-1800s, settling on a small farm in Canada on the outskirts of the city of Perth, in the province of Ontario. Earnest James became a dual citizen of the United States and Canada, and he went to work as an electrician for the Niagara Mohawk Power Company and raised a family in the small town of North Tonawanda, New York. Chuck's parents, Jim and Cynthia, met in North Tonawanda.

"My dad was a guest at a country club that Cynthia's dad, Charlie, was a member of," said Chuck. "My dad was at the bar at the golf course, and he and Charlie met, immediately hit it off, and at one point, Charlie starting talking about his daughter, and he said, 'Gosh, you should meet my daughter.'"

Jim and Cynthia were set up, and eventually, they married, a union that would last for nearly fifty years until Cynthia's death. When Chuck was born, he was named after his father-in-law.

Jim Rogers was a chemical engineer for the Union Carbide Corporation, located in Buffalo, New York. Later in his career, Jim became general manager of the O-Seal division of the Parker Hannifin Corporation. The Parker Hannifin O-Seal division was a supplier of elastomeric seals for the aerospace industry, ultimately manufacturing the O-rings for the space shuttle solid rocket boosters.

Jim's work moved him around a lot, and in the early 1960s, he moved to Wisconsin briefly and then to Southern California. He and his young family—Chuck Rogers was only three, and he had an elder sister and younger brother—settled in Garden Grove, a small community south of Los Angeles. Chuck Rogers was curious about rocket-propelled vehicles even as a child.

"My dad saw me out in the backyard one day with a garden hose with a hose nozzle that I had mounted to a couple of pieces of wood," he recalled. "I took a straw with some makeshift fins, and I attached it to the nozzle. When I turned on the water, the straw would launch into the air. Perhaps I got the idea for this kind of thing from television shows of the time, like *Fireball XL5*."

Rogers' parents were avid readers, and his mother was a librarian at a local elementary school.

"She instilled in us a love of books, and that's why I have tons of books today," said Rogers. "Before she died, they actually named the children's wing of the Garden Grove Central Library after her."

When Rogers was twelve, his parents bought him an encyclopedia set called *Above and Beyond* that was about aircraft, rockets, and space vehicles. Rogers read each and every volume.

"I remember a picture in there of this swoopy-looking hypersonic airplane powered by a rocket engine, and I thought, 'This really looks wild,'" said Rogers. "I wanted to work on planes like that. I was sure even then that I wanted to be an aerospace engineer. "My parents were very supportive of my desire to be an engineer. In seventh grade, I was getting D grades in math, and my Dad told me, 'If you want to become an engineer, you're going to have to get good grades in math.' I had a friend whom I could not hang out with anymore because I had to study instead of play. This made him very angry, and he and a couple of his friends would wait for me along the route I took when I rode my bike to school to chase after me on their bikes and try to catch me to beat me up. Because I wanted to be an engineer and therefore had to spend more time studying to improve my grades in math, I ended up having to take different routes to school, changing which routes I took each day, and riding my bike as fast as I could to get away from them."

Like many young Americans, Rogers was captivated by America's space program. He watched the televised moon landing in 1969. His heroes were men like engineer/test pilot-turned-astronaut Neil Armstrong and the creator of the *Saturn V*, Werner von Braun.

"We went to Florida when I was a teenager to visit our grandparents one year," recalled Rogers. "We only had so much time to be there. My parents said we could go to either Disney World or Cape Canaveral and the Kennedy Space Center. Of course, I wanted to go to the Cape, but everyone else wanted to go to Disney World. So my mom and my sister went to Disney World, and my dad and my brother and I went to the Cape. I remember touring the museums and standing next to rocket engines. It was such an exciting time. It seemed like the space program would keep moving and moving."

Rogers started launching rockets when he was eight years old.

"I got an Estes *Alpha III* as one of my first rockets," he said. "When I painted it, I held the spray can about 1 inch from the surface of the rocket and started painting. Of course, the paint immediately ran, and then I tried to sand it right away before it was dry. It was a really big mess. But I launched it anyway."

Rogers and his friends would fire their model rockets at Lake Elementary School in Garden Grove.

"I always liked the bigger motors, like the Bs and Cs, and I would just go out to the field at the school and launch them," he said. "One day the police drove up and told us it was illegal to launch rockets there. But the officer then told us that we could launch rockets at a place called Mile Square Park in nearby Fountain Valley. When I got my driver's permit, I would drive there and launch rockets."

Mile Square Park was an old military training auxiliary airfield, with part of the airfield still in place and the rest converted to a park and golf course. It was also the home to a local NAR club. The NAR club had insurance, and insurance was required at the park to fly rockets, said Rogers. He joined the NAR when he was fifteen so he could fly at the park. At Mile Square, Rogers also met a man who not only would become a great rocketry friend but also played a key role in Rogers's future. That man was Tom Kolis.

Kolis was in his thirties, and he worked for the NASA Dryden Flight Research Center (now the Armstrong Flight Research Center) at Edwards Air Force Base. Rogers was flying larger motors at Mile Square Park, and he met Kolis during the course of one of the launches.

"We started talking about rocketry, and we became fast friends," recalled Rogers.

NASA Dryden had a technical library, and when Kolis met Rogers, the library was in the process of discarding many technical reports from the 1950s and 1960s, including those containing early information on scramjets and hypersonics. These reports were piled in an area of the library where Dryden employees and visitors could sift through the stacks of reports and take the ones they wanted before the rest were taken out to a trash dumpster to be thrown away. Kolis saved many of these technical reports and gave them to the young Rogers, who read through every page of every document. The idea of traveling at high speeds and the theory behind hypersonic flight intrigued him.

In high school, Rogers built larger rockets with bigger motors, such as clusters powered by F motors built by manufacturers like Flight Systems Incorporated. His rockets soon exceeded the total impulse limits allowed by the NAR, and they were also getting too big for the flying field at Mile Square.

"Finally, I launched a three-stage minimum-diameter rocket—an F100 to an F100 to an F7, big black-powder motors—and something went wrong with the recovery system," said Rogers.

The third stage left the park, crossed over a road bordering the park, and hit the roof of a building over a half mile away.

"The park people came back with the wreckage of the rocket, and they wanted to know who launched it. I decided then and there that I needed to go somewhere else to fly my rockets."

Rogers heard about a desert location northeast of Los Angeles where people launched rockets. There were few limits there, and bigger rockets and larger motors were being used. That location was the Lucerne Dry Lake, and beginning in 1977, Rogers drove to the desert as often as he could to fly rockets there.

"In those days, Lucerne was the Wild West of rocketry," recalled Rogers. "The launches were pretty small, and there were no facilities, nothing at all. We just slept in our cars there. But we had fun. We were learning to do all this new stuff, and we felt like we were renegades because we knew the NAR didn't want us to launch some of these rockets."

Rogers graduated from Rancho Alamitos High School in 1978 and enrolled at UCLA, where he studied engineering. After a couple of years, he transferred to the California State Polytechnic University Pomona

(Cal Poly Pomona), where he completed his aerospace engineering degree in 1984. He continued flying rockets, supporting his hobby by delivering pizzas in Anaheim, California, near Garden Grove. In college, he landed a part-time job as a computer operator with the business services division of Honeywell.

While he was at Cal Poly Pomona studying aerospace engineering, Rogers wrote one of the first altitude and drag prediction computer programs ever created for high-power rocketry. He offered the programs for sale in 1984, taking out advertisements in *California Rocketry*, under the trade name Rogers Aeroscience. A few years later, Rogers teamed up with Fred Brennion to create the Rogers Aeroscience CD2 and ALT4 drag coefficient and altitude prediction computer programs. These programs were widely used by high-power flyers through the 1990s.

(Years later, Rogers and David "Coop" Cooper developed the RASAero aerodynamic prediction and flight simulation software, first released in 2008. These computer simulation and altitude prediction programs are today used to accurately predict flight characteristics for supersonic high-power rockets up to and exceeding 100,000 feet.)

At Lucerne in the late 1970s, Rogers met many of the most influential young people in the emerging high-power hobby scene, people such as Gary Rosenfield, Korey Kline, and Jerry Irvine. When composite motors first became available at Lucerne, Rogers flew them. He knew the NAR was warning its members away from higher-powered motors, and he realized that he could be thrown out of the NAR for violating the safety code. The rules did not deter him.

"I did a few bad things," he recalled.

Fortuitously and like many other NAR members of the time, Rogers decided to let his NAR membership expire. Not long afterward and in 1981, Rogers was one of three flyers taken to task at a Southern California NAR hearing, headed by Pres. Pat Miller, for violating the Code for Model Rocketry.

"The only reason I was not thrown out of the NAR was because I wasn't actually a member when I did these horrible things," said Rogers. "But I put on a tie, and I went to the hearing. We were already to the point where we were done with the NAR."

Rogers received his aerospace engineering degree from Cal Poly Pomona in 1984, and he had several employment prospects in the aerospace industry.

"One that I now laugh about was an offer to design rocket-powered ejection seats at Rockwell," said Rogers. "My rocket experience was going to be very helpful for that work!"

Rogers accepted a verbal offer from Rockwell, but the company was slow in finalizing the deal, he said. He also interviewed for a civilian engineering position at the Air Force Flight Test Center.

"This was still the Cold War era, and things were ramping up at the time," he said. "The Flight Test Center is located at Edwards Air Force Base, and if you were into aerospace, you knew Edwards was where all the hot planes were tested."

Rogers's first set of interviews with the different divisions at the Flight Test Center did not go well.

"I talked to all sorts of people, but I really did not like any of the jobs they were talking about," he recalled. "At the end of the day, I am in a meeting with the head of engineering, and he asked me, 'Well, of all the groups you interviewed with, which would you like to work for?'"

Rogers didn't like any of them.

"I don't think I want to work here," said Rogers.

The head of engineering was taken aback.

"He was mad," said Rogers, "and in his response, he almost spits out at me, 'Well, what do you want to do?'"

Rogers wanted to work on space shuttles and on high-speed aircraft and on lifting reentry vehicles.

"That's what I want to do," he told the bewildered head of engineering. "And he looked at me, and then he said, 'I know just the guy for you to talk to,' and he sent me over to Bob Hoey at the Office of Advanced Manned Vehicles."

The next interview went fine. Hoey was surprised the young Rogers knew so much about scramjets and hypersonic-type vehicles. After all, this was 1984, and scramjets and hypersonics were known in only a relatively few small circles.

"Of course, I had all this knowledge about scramjets and related vehicles because my friend Tom Kolis had saved so much information in those discarded technical reports, which I had read many times, and I had also taken a course on hypersonics at Cal Poly Pomona."

Rogers was hired as an aerospace engineer, and he moved to a tiny apartment in the desert town of Lancaster, not far from Edwards Air Force Base. His salary was $29,000 per year.

"It was a lot of money in those days," he said.

Rogers remained at the Flight Test Center as a civilian engineer for the next twenty-two years. Programs he worked on included the *X-30* National Aerospace Plane and the *X-33*. He also worked at the NASA Dryden (now Armstrong) Flight Research Center as a project manager on the *Orion* Multipurpose Crew Vehicle Abort Flight Test Program, for which he received a NASA Outstanding Leadership Medal in 2011.

One of the later highlights of his career included a technical study on a new lunar landing training vehicle, which culminated in an all-day technical briefing to several astronauts, including none other than Neil Armstrong.

"I got to ask Armstrong detailed technical questions on his moon landing and on his training on the *Apollo* lunar landing training vehicles," recalled Rogers.

Chuck Rogers read about LDRS in *California Rocketry* in the early 1980s. He traveled to his first LDRS in 1986.

"I heard that this might be the last LDRS at Medina, and I was really glad that I got to go there because I was able to experience the atmosphere of the place," he said.

Rogers flew a rocket that was then lost in a cornfield across the street from the Wagner farm. Still, he got the full effect of Medina.

"It was a nice, almost Woodstock-like environment. At that time in the mid-1980s, the two centers of the high-power universe were Lucerne and Medina. I had already flown at the desert in Lucerne, and I wanted to check out Medina. I remember the big green field—not big by West Coast standards, but it sure was green!"

Rogers joined the Tripoli Rocketry Association soon thereafter.

"A couple of people tried to start a national high-power association before, and there were two or three false starts," recalled Rogers. "After the second and third false starts, I decided to wait a while before joining Tripoli to see if it really lasts."

Yet Tripoli did last, said Rogers, primarily because of the work of one key individual: Tom Blazanin.

"He was the guy who really made it happen," said Rogers, who became Tripoli number 137 in 1986. "I joined Tripoli because I really wanted to support high-power rocketry, and I felt that we needed to be organized. It was pretty clear that there was strength in numbers, and when you are organized, you can defend something that you want to do—just our right to be able to fly rockets and to enjoy our hobby without being hassled."

Rogers ran for a Tripoli board seat soon after he joined. He was defeated the first time, but he ran again later.

"This was a time of the transition from the 'Wild West days' to actually becoming respectable as a national organization," said Rogers.

Rogers was appointed to the Tripoli board as a temporary replacement for another board member who resigned in 1987. In time, he would be elected to the board in his own right, and he would remain on the board for nineteen years, becoming one of the longest-serving board members in Tripoli's history.

When Rogers took over as Tripoli president, the association's problems were many. Tripoli had grown to more than a thousand members. Yet infighting was rampant. Among other things, Rogers had to deal with noncommitted board members, a perceived "East Coast versus West Coast" factionalism, an almost defunct *Tripolitan*, insurance problems (or the lack thereof), and the slow but steady movement of the NAR into high power.

"I thought the organization was in really big trouble at the time," recalled Rogers. "We were not working together, we were fighting among ourselves, and we were just falling apart."

Rogers acted quickly, and he started at the top. First, he addressed the issue of lackluster board member participation.

"One of the problems we had was that board members were not showing up at board meetings or for the annual meetings at LDRS," he said. "We even had one board member [whom] we had no idea where he was for eighteen months. We wondered if we would have to have him declared legally dead to remove him from the board."

To remedy the situation, Rogers made a motion that required anyone running for a seat on the Tripoli board to sign a pledge that, if elected, they would attend the annual board of directors meeting at LDRS. The motion passed. Next, Rogers followed up on a suggestion first made by Tindell, taking steps to put into place a telephone conference system—in this age before the Internet—to better connect board members during the year and to otherwise conduct Tripoli business in a modern fashion.

No decision had been made at LDRS-8 as to the location of LDRS-9. Rogers immediately assigned a search committee to continue working on the matter, and a date for LDRS-9 was picked soon thereafter. LDRS would be held on August 16 to 19, 1990, weeks after NARAM. In the future, hobbyists and manufacturers would never again have to choose between conflicting dates for NARAM or LDRS. They could attend both.

The manner in which the site for the next LDRS was selected evolved under Rogers too. This was the third year in a row LDRS was in Colorado, and some people wanted the launch to move elsewhere, but where? At the general assembly on Saturday night, several recommendations were made for a new location. Possible locations included Las Vegas, Oklahoma, South Dakota, and Lucerne.

One person suggested an obscure location in Northern Nevada, not far from Smoke Creek, where, in July 1989, a small high-power launch had recently been held. However, this location had its own problems. The accommodations were primitive, and the site was in an "environmentally sensitive area." It might not be a good place for high power's premier event—or any other rocket launch, for that matter. The location did have an interesting name. It was called the Black Rock Desert.

In late 1989, the LDRS search committee recommended three sites for LDRS-9: Lucerne, Las Vegas, or Hartsel. Each location was put to a vote by the general membership. Once again, Colorado prevailed—quite easily, in fact—indicating that Tindell was right about the membership's preference for that location of LDRS-8. The vote was 195 for Hartsel, 91 for Las Vegas, and 86 for Lucerne. The next LDRS would be held in August 1990 in Colorado. To respond to complaints that Tripoli's premier event was neglecting its eastern-based flyers, Rogers announced that after LDRS-9, the event should alternate between the East and West Coasts.

"LDRS-10 should be held on the East Coast," said Rogers, who added that "a tenth-anniversary LDRS in Ohio would be quite a special event, with immense sentimental value."

In his first twelve months as president of Tripoli, Rogers also tackled another perennial problem for Tripoli: insurance for high-power rocketry. Insurance issues plagued Tripoli in the late 1980s. There was coverage for LDRS-7. But board member squabbling resulted in no coverage for LDRS-8. Rogers knew this had to change. Tripoli needed insurance to protect spectators and landowners, as the NAR had done for years. With Blazanin's help, Rogers pushed through an increase in annual Tripoli dues of $5 per year, from $25 to $30, to help cover the cost of insurance for nationwide launches, excluding LDRS, at "Tripoli-sanctioned events." Coverage for LDRS would be negotiated separately. Rogers believed that since the insurance covered *only* Tripoli members, it would be an added incentive for high-power flyers to join the organization.

The dues increase was effective January 1, 1990. The subsequent insurance coverage was for $1,000,000, with a $500 deductible. The policy did not cover individual flyers per se, but it did cover spectators and others who might be injured at a "sanctioned event," whether that injury was caused by rocket mishap, stepping into a hole, or some other premises liability type of claim. The insurance provided landowners with liability protection, arguably expanding the number of possible sites that might become available to Tripoli flyers around the nation. Insurance coverage for LDRS would be negotiated separately at a later date.

Motor certification was another problem that Chuck Rogers tackled early on. Unfortunately, this issue remained a problem for Tripoli for some time to come. Despite at least three years of discussing a certification process, Tripoli was still unable to certify high-power motors. (The final insurance coverage requirement, which only certified motors be used at launches, was ignored for some time.) To certify motors, Tripoli had to have a good thrust stand. Rogers made the purchase of a thrust stand part of his campaign when he initially ran for president.

"I believe Tripoli should finally step up to its responsibility for Class B motor certification to provide its members with safe, reliable, certified motors," said Rogers during his campaign. "Some motor manufacturers 'claim' total impulse is 20 percent to 30 percent higher than actually delivered. You, the customer, deserve to know the truth. Tripoli should purchase a thrust stand and find a facility to locate it where a group of rocketeers have been active for more than five years to provide continuity in the manning of the facility over the long haul."

But it was still slow in coming. By the end of 1989, Rogers announced that Tripoli had agreed to use a thrust stand owned by Gary Rosenfield and AeroTech to conduct Class B motor certification. Rogers said that the goal was to allow only certified motors at Tripoli launches within two years, no later than the 1991 LDRS. Bill Wood of Southern California would head up the committee that would coordinate and publish motor-testing data.

In May 1989, G. Harry Stine wrote to NAR president J. Pat Miller regarding the expansion of the NAR into the realm of high power.

"You have been asking me for my inputs regarding the future course of the association," said Stine. "I have pondered . . . for several months. Finally, I feel I have given enough rigors to my personal observations to set them forth in a letter to you. First of all, the name of the organization is the National Association of Rocketry. It is not the National Association of Model Rocketry or the National Association of Space Modeling."

In Stine's opinion, the NAR had strayed off course from its original intent. The association had turned its back on modern developments in sport rocketry—his euphemism for high power—and had instead grown into what he called "the spokesman for the 'toy rocket' segment of nonprofessional rocketry." The notion that the NAR still represents the nonprofessional rocketeer is nothing more than a charade, said Stine.

"I don't know how long we can keep this up. Certainly, a strong sport rocketry organization could severely impact this and degrade the position of the association."

Stine believed "sport rocketry" was in turmoil and that a lack of strong safety standards in the newly developed field was threatening the entire hobby.

"Here is the biggest danger," posited Stine, "the sword of Damocles that hangs over the association at this moment. Public safety officials and the news media care not that sport rockets aren't 'model rockets' by the NAR definition. They look the same and fly the same and are only bigger and heavier. When the inevitable 'big accident' occurs in unorganized sport rocketry, model rocketry will be at enormous risk."

Stine called for a renewal of the NAR's original intent, which he suggested was *preeminence* over the field of "nonprofessional rocketry." In essence, Stine was calling for the NAR's full entrance into high power.

"Many NAR members will seriously and with quite firm personal convictions say that the NAR cannot or must not do this and will provide all sorts of reasons why," said Stine. "But if we had taken this approach thirty-two years ago, the NAR and model rocketry would not have become a reality. The association can continue to adequately represent model rocketry while, at the same time, expanding its horizons and programs and services to include the rest of nonprofessional rocketry. We can do it! Furthermore, we must do it! That is what we set out to do in the NAR thirty-two years ago."

In August 1989, the NAR took another big public step toward high-power rocketry. At the board of trustees meeting, the NAR established the Commission on Advanced High-Power Rocketry. The purpose of the commission was to evaluate the possibility of the NAR's full entry into high power, not just G motors this time but everything. The commission's task, suggested by Stine, was to study and help create a step-by-step procedure by which the NAR might move beyond being a strictly "model" rocket organization and into the future, a future where the NAR would be a full competitor in the high-power consumer market.

The Tripoli Rocketry Association's brief monopoly on consumer high-power rocketry was about to come to an end.

12

Black Rock

Legend has it Steve Buck discovered the future of high-power rocketry by happenstance in the late 1980s. Buck was a Tripoli member and the owner of High Sierra Rocketry, which specialized in mid-and high-power-related rocketry products. He launched rockets at the Lucerne range in Southern California and, occasionally, at Smoke Creek in Nevada. To get to the Smoke Creek Desert from his home in Reno, Buck would sometimes drive north on State Highway 447 and then make a left-hand turn at a fork in the road just past the small community of Gerlach.

One day and in lieu of his usual left-hand turn toward Smoke Creek, Buck took the road less traveled; he turned *right* instead. Minutes later, he came upon a vast, perfectly flat expanse of dusty-white nothingness. It was another huge desert, like Smoke Creek. But it was larger, smoother, and emptier than any stretch of desert he had ever seen. He could drive on the desert floor, which was almost as firm as asphalt, as fast as he wanted and in any direction for miles. There were no roads on the desert surface—none were needed on this level plain—only the temporary tracks left by the occasional vehicle that arrived before him. The dusty tracks from such vehicles trailed ahead for miles like imprints on the surface of the moon; then they disappeared over the horizon.

He didn't know it yet, but Steve Buck had discovered the future of high-power rocketry in America. It was called the Black Rock Desert.

The Black Rock Desert takes its name from a black rocky prominence along the northeastern edge of the desert playa. Geologists believe the prominence is made up of volcanic rock and limestone and is a remnant of an ancient island chain that through plate tectonics traveled hundreds of miles from the sea to its present location in Northern Nevada.

Like its neighbor Smoke Creek to the southwest, Black Rock was originally part of a prehistoric body of water called Lake Lahontan. Until 12,000 years ago, the lake covered more than 8,000 square miles of Nevada, Oregon, and California. At Black Rock, the water was more than 500 feet deep. Rock terraces formed by wave action along ancient beaches of Lake Lahontan are still visible along the paved highway northeast of Gerlach, between the 3- and 10-mile playa access points. At Black Rock, signs of climate change are nothing new; it is evidenced here for thousands of years, long before the advent of man, and likely caused the evaporation of Lake Lahontan at or about the same time the last glaciers of the west also disappeared. Artifacts discovered in the region suggest man first appeared here nine thousand years ago.

American frontiersman John C. Fremont was reputedly the first European American to see the Black Rock Desert. Fremont had a reputation of being fearless. In his years of exploring the west, he navigated uncharted rivers, battled fierce Indians, fought in several wars, and was one of the first governors of California.

When he had first encountered the Black Rock Desert, he said, "The appearance of the country was so forbidding that I was afraid to enter it."

In the middle of the nineteenth century, Black Rock was a small portion of the Applegate Trail, named for the Applegate brothers who established an alternative Southern route along the Oregon Trail from St. Louis to the Pacific. The trail avoided the treacherous waters of the mighty Columbia River, where several members of the Applegate family drowned in 1846. There was no threat of any water crossing at Black Rock.

"The earth appeared to be as destitute of moisture, as if a drop of rain or dew had never fallen upon it from the brazen heavens above," wrote one thirsty traveler. "Nothing presented itself to the eye but a broad expanse of a uniform dead-level plain, which conveyed to the mind the idea that it had been a muddy and sandy bottom of a former lake . . . and having its muddy bottom jetted into cones by the force of the fire of perdition."

Today evidence of the pioneers' difficult passage through this desert can be found at several points along the edge of the playa. Remnants include wheel rut trails visible from outer space, discarded globs of axle grease along the ground, and even abandoned wagons half-submerged in the soil, slowly returning to dust.

"It looked like some fossil landscape that had long since been left behind by the rest of terrestrial evolution," wrote Tom Wolfe in *The Right Stuff* when describing the dry lake beds in Southern California where test pilots trained as astronauts in the 1950s. His narrative could certainly describe the Black Rock Desert too. "When the wind blew the few inches of water back and forth across the lake beds, they became absolutely smooth and level. And when the water evaporated in the spring and the sun baked the ground hard, the lake beds became the greatest natural landing fields ever discovered and also the biggest, with miles of room for error."

This flat and wide expanse of nothingness attracted several twentieth-century thrill seekers to Black Rock. On October 4, 1983, Great Britain's Richard Noble set a land world speed record here, averaging more than 633 miles per hour in his jet-powered *Thrust 2* car. Fourteen years later, Noble returned to Black Rock with the jet-powered *Thrust Supersonic Car* or *Thrust SSC*. On October 15, 1997, *Thrust SSC* set the first supersonic world land speed record, averaging 763 miles per hour.

Steve Buck was born in Reno, Nevada, in 1954. He launched model rockets as a young boy, and then in the early 1980s, he rediscovered rocketry when he came across a small advertisement in *Popular Science* that discussed bigger rockets with larger motors. He eventually made his way to the launch site at the Lucerne Valley, nearly 10 hours south of his home in Northern Nevada. Buck launched his first high-power rocket there, and he was hooked. Soon, he was moving on to bigger and more sophisticated projects.

"I bought a kit from Ron Schultz of LOC that was a huge three-motor cluster called the *King Viper III*," recalled Buck. "We flew it on three H motors, which, at the time, was really big. Everyone said, 'Oh my god' when it took off."

At the Lucerne range, Buck met many of high power's early movers and shakers. He joined Tripoli not long after its incorporation, becoming member number 82. One day Gary Rosenfield approached Buck and asked him if he would be interested in selling AeroTech motors as a vendor. Buck said yes, and High Sierra Rocketry was born. Buck eventually became a commercial rocket vendor too, selling LOC kits for Ron and Debbie Schultz. High Sierra Rocketry ads were a staple of the *Tripolitan* in the mid-to late 1980s, and Buck had a high profile in the Southern California rocketry community. He even organized a launch at Smoke Creek in the late 1980s.

Then he found Black Rock.

Buck believed Black Rock was an ideal location for a high-power launch. An early supporter of his vision was future Tripoli president Bruce Kelly.

"[Buck's] description of the Black Rock Desert was beyond belief," said Kelly. "I could not comprehend such a place as he described—a huge, vast, flat wilderness with almost no bounds for as far as the eye could see. It was a place he described as 'flat and large enough that you can launch a rocket to almost any altitude, get in your van, and drive right to the spot of landing.'" Buck organized the first high-power rocket launch ever held at the Black Rock Desert. To attract flyers to the remote location, he ran a full-page ad in the *Tripolitan*. Rather than billing his event as "a launch in the middle of nowhere," he dubbed the gathering as the "advanced rocketry sport launch at the site of the world's land speed record." He called the launch "Black Rock One."

To prepare for Black Rock One, Buck contacted the Bureau of Land Management (BLM), which had authority over the desert, the FAA for possible airspace clearance, the Washoe County sheriff, and the Gerlach Empire Volunteer Ambulance and Fire Department.

"On July 8 and 9, 1989, a group of people will be camping and flying rockets on the dry lake bed of the Black Rock Desert," Buck wrote to authorities in early 1989. He provided officials with a map of the desert showing the location of the proposed launch. "Some of the participants and spectators will be camping. Others will be staying at the motel [in Gerlach]."

He estimated a dozen flyers and perhaps thirty spectators would show up, and he told the government that all flyers would be adhering to the Tripoli Safety Code, although this was not a Tripoli-sanctioned event. Buck spoke with Lynn Clemons of the BLM about his rocketry event. She told him his planned event was so small, it did not warrant any fees, let alone a permit. FAA officials were equally unimpressed. They told Buck that controlled airspace at Black Rock did not begin until 10,500 feet. As long as rockets remained below this altitude, no waiver was necessary, and no waiver was obtained for Black Rock One. However, Buck still asked flyers to fill out forms describing their rockets and anticipated altitudes, and these forms were purportedly turned over to the FAA. Flyers who signed up for Black Rock One received an information packet with a copy of the Tripoli Safety Code and instructions on finding the launch site. To access the desert floor, Buck told flyers to drive 11 miles up the highway from Gerlach and turn right.

"A short gravel access road allows you onto the lake bed," explained Buck. "A Cadillac could handle this maneuver without scraping the bottom. At a distance of 3.4 miles from the highway, you should see two pipes sticking in the ground. Another mile, and you should find two posts with an old wooden sign that faintly reads, 'Bombing Range.'"

The camping site was just beyond, just off the desert floor, with the launch pads 2 miles east of where the camp was located, he said. The flyers packet also contained instructions on camping in the high desert.

"The camping area is a small cove that will offer protection from any high winds that may develop," said Buck. "Numerous anthills can be found in the camping area, so you may desire to bring a can of gas or kerosene to control them."

He promised that event organizers would bring firewood for "one nice fire each evening." Otherwise, people were told to being their own wood. "You won't find any wood lying around," he said.

As for water, Buck was equally clear. "Bring your own—and lots of it," he warned. "And if you are planning on tent camping on the edge of the playa, watch out for scorpions too."

The launch pad equipment and two portable bathrooms for Black Rock One were provided by Buck and High Sierra Rocketry. The pads held rod sizes that ranged from a quarter of an inch in diameter to almost an inch in diameter.

"In accordance with the Tripoli Safety Code, Class B motors must be flown from a pad no less than 150 feet from the spectator line," said Buck. "Large clusters and K motors must be out to 200 feet. If your launch control wires aren't long enough, it's okay. You get to push your own button."

Twenty-two flyers showed up for Black Rock One on July 7-8, 1989, and they logged 108 flights, including perhaps the first ever high-power beer loft competition, in which flyers lofted rockets carrying a payload of one 12-ounce can of cold beer. The goal of the contest was to see what flyer could launch, safely recover, and then completely consume the can of beer in the shortest time possible. The "Brewloft" would become a staple at Black Rock events in the future.

Otherwise, Black Rock One was a typical high-power launch for its day, a mix of model and mid-power sport rockets, with a few high-power flights thrown in to establish its high-power credentials. Many rockets were loaded with D through G motors. There were several Js, a couple of Ks, and even an L flight. The L-powered rocket was owned by Buck. It was an all-fiberglass rocket standing nearly 6 feet tall. It carried an AeroTech L585. The rocket shredded under thrust shortly after takeoff.

Tripoli board member Tom Blazanin heard about Buck's launch, and he made the journey from his home in Pennsylvania to check things out. Like Buck, Blazanin had bad luck flying rockets there. His *Scorpion* rocket, powered by a J700, had a good ride up but came in undeployed, taking a core sample of the desert floor.

"The nose had to be left behind as it was solidly lodged about 2 feet in the ground," said Blazanin. "While I was bummed out some, I made up for it knowing that I now was somehow personally part of Black Rock." Blazanin fell in love with the west when he was stationed at Edwards Air Force Base years earlier. At Black Rock, he was overwhelmed.

"This place can, should, and will be the new capital of advanced rocketry," he predicted a few months after the launch, adding that one day an LDRS might even be held there.

Other Tripoli members who attended Buck's inaugural launch included Chuck Rogers and Kelly. They too were impressed by the freedom to fly Black Rock afforded to high-power rocketry. This first high-power rocket launch at the Black Rock Desert in 1989, like Chris Pearson's LDRS in 1982, was not a big event. But like LDRS-1, the launch left a deep impression on those who made the trip.

"Flying at Black Rock is beautiful," noted one flyer at the time. "Rockets reached some nice altitudes and drifted several miles to be easily retrieved. One had only to sight the rocket against a peak on a mountain range [60 miles away], get in the car, and drive toward that peak, and eventually, you would come to your rocket! All rockets were retrieved, except the ones that the owners didn't want."

Another early Black Rock launch brought this comment: "Believe me, you can easily lose your sense of direction when you are out in the middle of this immense, flat, vegetation-free terrain. If an alien space probe landed at Black Rock, it could give the impression that life is scarce on this planet. On the other hand, the raw and rugged beauty at sunset along with the clear star-studded sky makes the camping at Black Rock a special treat."

Blazanin chose Black Rock One as the cover story for the November 1989 issue of the *Tripolitan*. The story introduced high-power enthusiasts to not only the remote high desert location but also a dusty old motel in Gerlach that had about forty beds, a tiny restaurant, a dimly lit bar, and a cantankerous owner who promoted everything about the tiny desert town. Buck chose this motel as unofficial headquarters for Black Rock One. It was close to the playa and, unlike the desolation of Smoke Creek, only a short drive from the rocketry range. The motel was owned by Bruno Selmi, an Italian immigrant who had come to Gerlach at the age of seventeen from Tuscany, Italy, just after the end of the Second World War. On July 7, 1989, high-power rocketry came to his motel for the very first time. It would remain there for another three decades. The motel was called Bruno's Country Club and Casino.

The modest success of Black Rock One inspired Buck to host the event again the following year. In early 1990, Buck and friends ran ads for the launch they called Black Rock Two. This time, the launch was cosponsored by High Sierra Rocketry and another vendor, West Coast Rocketry. The launch would also be overseen by a new Tripoli prefecture recently formed in Northern California called the Association of Experimental Rocketry of the Pacific—or AERO-PAC for short. It would be held July 21 and 22, 1990, less than a month before LDRS-9 in Colorado.

All history is retrospective. Historians search through past events using the benefits afforded them by the lens of future developments. Demarcation points in history are often not recognized until later. People in the midst of events may not realize where they are going or if they are going anywhere at all. Is this the beginning of something new? Or is this just a one-time deal? Only time will tell. In retrospect, Black Rock Two was the biggest leap forward in motor technology between consecutive launches of any event in high-power rocketry history. This was the dawn of a new era in hobby rocketry.

For the first time at any Tripoli launch, there were high-power motors in nearly every impulse category. At Black Rock One in 1989, there were a few Js, a couple of Ks, and an L motor. These were standard big-ticket motors at the time, seen at launches all over the country, from the large Midwestern high-power events at Danville, Ohio, to Lucerne in California and to LDRS, wherever that was. At Black Rock Two, there were rockets propelled by M, N, and O motors. One rocket even carried a P-impulse motor, the first P motor launched at any Tripoli event anywhere. Here in Northern Nevada, in July 1990, high-power rocketry completely separated itself from model and sport rocketry. As LDRS had cleared the way for high power's rapid advance in the 1980s, Black Rock Two would be the breakout launch of high power for the 1990s. More than any other rocketry event of its time, Black Rock Two was a glimpse into the future of high power. Over the next thirty years, this desert would become the leading edge of amateur rocketry in the world. One day high-power flyers would clear 100,000 feet here, reaching for the boundaries of outer space.

Seventy flyers registered for Black Rock Two, three times the number who had made the trek the year before. They came from all over the country. There were more than 160 flights, including several minimum-diameter and two-stage rockets. Many of these rockets were attempts to set new Tripoli altitude records, the beginning of a Black Rock tradition.

The altitude waiver for Black Rock Two was obtained by William "Bill" Lewis, the commissioner of AERO-PAC. Earlier that year, Lewis obtained the first FAA waiver for high-power rocketry ever granted at Black Rock. That waiver was for AERO-PAC's inaugural launch as a club, called Aeronaut I. For Black Rock Two, Lewis secured a waiver of 20,000 feet above ground level for the entire weekend, with brief windows up to the incredible altitude of 55,000 feet.

The FAA waiver for the Black Rock Two event was issued by the FAA's Western Pacific Region branch office in Los Angeles. It commenced at 4:50 a.m. on July 21, 1990, and ran through July 22 at 7:20 p.m. The waiver came with a number of special provisions, including, among other things, the requirement that AERO-PAC notify the

Oakland Air Traffic Control Center no later than 24 hours prior to the initiation of the launch, the posting of four radio-equipped observers at different points away from the range head to check for approaching aircraft, that operations must cease if surface visibility was below 5 miles, and that the trajectory of all rockets shall be contained within a 5-mile area from their launch point. Tripoli member Tom Binford made the cross-country drive to Black Rock Two from his home in Georgia, more than 2,300 miles away.

"In 1988, I saw an ad in the back of *Popular Science* from High Sierra Rocketry," recalled Binford. "It said 'model rocket motors D-J.' J motors? I have to see about this!"

Binford mailed for a copy of Buck's rocketry catalog and thereby learned about Tripoli, high power, and the confirmation process. He joined Tripoli and was confirmed as a high-power user in March 1989. He traveled to LDRS-8 in Colorado that summer, and when he saw the ads for Black Rock Two, he knew he had to go.

"I was kind of in a daze from having to drive, but my first impression was just how flat it was here," recalled Binford. "Once driving out on the playa, it was amazing to see how empty it was. There were not even any bugs. There hadn't been a trace of rain the previous winter, and when driving on the desert, you didn't leave any tracks, but the truck did raise a dust trail. Being able to drive anywhere at any speed was fun!"

Binford brought two Vulcan Systems O1536 Hellfire motors to Black Rock. These powerful commercial motors cost $1,500 each. On Saturday, July 21, 1990, Binford launched the first O-powered rocket at any Tripoli event anywhere. It was also the first O motor to race into the skies above Black Rock. The rocket for Binford's O-powered flight was his *Cloudbuster 110*, a scratch-built 4.5-inch minimum-diameter vehicle with a nose cone made of solid basswood. The rocket weighed 49 pounds on the pad, almost 30 pounds of which was propellant. There was also 5 pounds of dust chalk on board. The chalk dust would be released at apogee and would help Binford track the rocket at high altitude from the ground.

After a slow ten count and with everyone on their feet, the Hellfire O motor in Binford's *Cloudbuster* roared to life at the away cell. In an era when Tripoli altitude records were still in the 20,000-to-25,000-foot range, Binford was aiming for 40,000 to 50,000 feet. This was not an unrealistic goal given the rocket's size and propellant. Unfortunately, tracking in 1990 was still visual only, and at nearly 20,000, feet the rocket and its smoke trail vanished from sight. Tripoli president Chuck Rogers, one of the altitude trackers that weekend, followed Binford's rocket to 19,832 feet before it was lost. Rogers said the "rocket apogee altitude was considerably higher."

Recovery of the *Cloudbuster* was unsuccessful. Binford's rocket had ejection charge problems. There was no deployment of the parachute, and the rocket was destroyed on its return.

Binford came prepared. On his cross-country trip from Georgia, he had carried another nearly identical *Cloudbuster* rocket for his second Hellfire motor. He hauled it to the launch pads again on Sunday morning. The motor on Saturday had an ejection charge delay advertised as 40 seconds long, said Binford. After Saturday's experience, he thought this delay was too long. The lengthy period before the ejection charge fired might have contributed to the rocket ending up buried in the ground. He reduced the delay time for his flight on Sunday. Perhaps this would help ensure a successful recovery, he thought.

Binford's second *Cloudbuster* flight quickly disappeared but soon met the same fate as his rocket the day before. There was no deployment and no recovery. For some reason, ordinary black-powder ejection charges failed to perform at higher altitudes. This would be a scientific and engineering problem for flyers at Black Rock for at least another decade. High-altitude flights would come in ballistic year after year, their airframes of fiberglass, carbon fiber, or even metal embedded deep in the ground, totally destroyed. One day, in the early 2000s, that problem would be solved by a California general contractor and a friend who taught engineering at Stanford. They would determine that thinning air at high altitudes, especially above 30,000 feet, could render traditional black-powder ejection charges ineffective. Unfortunately, this discovery was years beyond Black Rock Two. Had Binford's rocket safely deployed its parachute and recovered, he would have set a new Tripoli altitude record. But for now, he was heading home with a garbage bag full of expensive debris.

There were several other bigger-than-life flights at Black Rock Two, including three N-powered projects launched on Sunday. Bruce Kelly and his son Brian launched an N2000 in a 5-inch-diameter airframe dubbed *Nancy One*. The all-fiberglass rocket tore out of its tower and left an impressive trail of thick white smoke, only to shred to bits at somewhere between Mach 1 and Mach 2. Tom Blazanin's fiberglass N-powered *High Venture* met a similar fate at altitude.

The third N attempt was an N-to-O two-stage rocket owned by Jerry Irvine. The vehicle suffered a catastrophic motor failure at the pad that partially melted the all-metal launch tower. There were also several M-powered rockets, plenty of L motors, and several two-stage rockets, including a K-to-K missile called *Hawaiian Punch* that was optically tracked to burnout of the upper stage at nearly 17,500 feet and then lost to sight. The rocket probably went higher, but without onboard electronics, there was no way to determine the maximum altitude for the flight. Still, it was recovered intact after a successful deployment.

Perhaps the most daring launch at Black Rock Two was by Chuck Mund of New Jersey and his team of flyers. Mund's crew included John Holmboe, Keith Murdock, Eric Mund, Doug Caldwell, and an older and somewhat eccentric Southern California rocket scientist who liked flashy short beach pants, tie-dye T-shirts, and building his own experimental motors.

The airframe of Mund's four-finned rocket was 6 inches in diameter and stood nearly 10 feet tall. Much of the rocket's construction was completed on the desert playa prior to and during Black Rock Two. Mund and his crew sweated it out in the desert, applying layer after layer of dripping epoxy and fiberglass cloth to the body tube during daylight hours, and they brought a portable generator for light so they could continue work on their rocket at night. When it was finished, the big rocket tipped the scales at nearly 200 pounds. It carried a P-class AP composite motor with a case crafted from a machined steel tube that was threaded at both ends. The exhaust nozzle was made of pure graphite. The four Smokey Sam propellant grains inside the motor case were manufactured by Scott Dixon at Vulcan Systems. The motor weighed almost 150 pounds.

This was high-power rocketry's first Black Rock Desert P project. The rocket was placed 5 miles away from the rest of the event's participants, and Mund's team spent the early part of the day on Sunday setting heavy 4-by-4 wooden planks into 3 feet of concrete hauled out to the desert and poured directly into the playa to help stabilize the pad. It was ready to go Sunday afternoon. Projected altitude was 38,000 feet. Flight simulations suggested the rocket might achieve Mach 1.7.

With a deafening roar, Mund's P-powered rocket streaked away from the steel launch tower, trailing huge columns of smoke. The trajectory was straight and true, and for the first few thousand feet, it was a perfect flight. Then it all came apart like some airborne piñata struck by an invisible bat high in the sky. Witnesses described the rocket as several thousand feet in the air when the vehicle just shredded, raining down motor parts, fins, and thousands of other tiny pieces all over the desert floor. It was an ambitious flight. But it ended as most of the big motor projects ended that weekend. In 1990, motor thrust outpaced airframe design by a mile.

A few weeks after Black Rock Two, LDRS-9 went forward in Colorado on August 16 to 19, 1990. For the first time at any LDRS, there were three days of flying and two evening launches. The Hartsel property was again the site of the launch. But the location of the range was changed slightly, providing members with more unobstructed space than ever before.

Attendance at LDRS was growing steadily each year at Hartsel, attracting flyers from all over the country. Tripoli had more than a thousand members, including NAR members who recently joined the association following the enactment of the 3/48 Rule. Many NAR members attended LDRS. For the most part, they were impressed with what they had seen. By this fourth year at Hartsel, the launch was well organized and ran smoothly. The launch pads purchased by Ed Tindell in 1989 were a good addition, flight cards were universally used, and the national launch had settled comfortably into the Colorado Springs hotel and launch site. It was familiar and fun, like Medina was in the early to mid-1980s. Afternoon thunderstorms at Hartsel were a threat, but the wide-open space and generous FAA waiver made up for any temporary inconvenience posed by the weather.

One longtime New York NAR member observed at the time, "The launch site is huge and seems to go on forever, without a tree in sight. The land we are flying on is part of a large ranch that has been used by Tripoli for several LDRSs. The owner of the land has many animals on his ranch. Some are your garden-variety cows and goats, while some are of a more exotic nature, like llamas and buffaloes. The rancher provides some of the cowboys to help recover stray rockets that land past the fences that surround the launch site. This is very important because one of the rules in using the land is that you never cross the fences. Even if your rocket lands inside the fence, it can be a very long walk to find it."

By today's standards, LDRS-9 was still a modest event. There were only twelve launch pads (in comparison, at LDRS-29 in Lucerne in 2010, there were at least eighty pads). And the pads were simple in design—rods only; there would be no rails for years to come. The rest of the facilities were primitive too. The LCO sat on a small stool facing the Tripoli launch controller, which rested on a large toolbox in front of him. There were few frills or extras. LDRS was still the best venue in America to view the largest selection of every kind of model, sport, and high-power rocket available. It was also the place to see the most exciting and catastrophic flights in hobby rocketry. And people loved catastrophic flights.

Tripoli's annual event was becoming well-known for in-flight rocket disasters of all kinds. Cruise missiles, CATOs, shreds, skywriters, lawn darts, core samples, and land sharks described flights that went terribly awry and, more often than not, resulted in destroyed projects. A full-length commercial video of LDRS-9 began with short clips of more than fifteen rockets coming apart in every fashion, all to the tune of the Surfaris' 1963 hit, "Wipe Out." Rockets of the 1980s could meet destruction at any time, from the moment of ignition to the end of recovery.

"When the RSO calls, 'Heads up!' you'd better do it!" said one NAR member witnessing the carnage at his first LDRS this year.

Motor failures were common, particularly with new manufacturers, of which there were far more than there are today. There was still no certification process for commercial high-power motors. Some poorly designed motors CATO-ed immediately upon ignition; others failed in flight. Motor-based ejection charges oftentimes failed to work on time, or they failed to separate the rocket at apogee, or they failed to work at all.

Problems with rocket design were legion. Even in 1990, many flyers continued to use model rocketry construction techniques for high-power rockets. Lightweight glues and balsa wood and basswood fins were common; rockets shredded during flight, some right at takeoff and others when the rocket reached Max Q, shedding fins and pieces of airframe all over the range.

This is not to say that all high-power flights resulted in failure. There was a steep learning curve in high power, but trends were emerging.

Some flyers were successfully launching sophisticated rockets, and they were getting them back intact too. However, unlike today, these people were far less common.

There were no bigger-than-life projects at LDRS-9. But one of the stars of the launch was Mike Ward's upscale version of the Estes *Deep Space Transporter*. The vehicle was finely detailed and nearly 9 feet long. At just under 6 inches in diameter, the *Transporter* had a wing span of 4 feet, twin booms on either side of the main fuselage, and multiple aft rudders and stabilizers. It carried a 12-foot-diameter custom-made parachute and was launched on an L750 motor. The rocket had a spectacular takeoff and rotated slowly on its ascent, which was perfect.

"It boosted up very high and straight, but there was no deployment of the chute," wrote one observer. "It nosed in at high speed with an eerie whistling sound and impacted downrange from the crowd with a loud thud." Destroyed.

The majority of the flights at LDRS in 1990 continued to be midsize rockets in the G- to K-power range. Clusters remained popular, as they had been since LDRS-1. There were several rockets with L motors but no M flights this year. Long-burning motors were common at LDRS-9, but more often than not, the rockets carrying these motors would disappear high in the sky, only to reappear later without a deployed parachute, tumbling in a flat spin to the ground.

AeroTech and Vulcan remained the most popular high-power commercial rocket motors at Hartsel, with AeroTech leading Vulcan by a substantial margin. A new line of motors sold under Jerry Irvine's U.S. Rockets label was flown in several rockets. These motors contained a unique combination of propellant ingredients that showered bright sparks on ignition and during flight. Irvine called them Firestarters, an apt name for a motor that would set fire to absolutely anything combustible within several yards of the launch pad.

LOC continued to dominate the commercial rocket-maker scene, but another new rocket maker emerged at LDRS-9 too. The business was owned by Tripoli flyers Gerald Kolb and Frank Uroda of Michigan. They called their company Public Missiles Systems. Like LOC, Public Missiles would one day be one of the most recognized brands in all of high power.

Flyers Jim Gentry and Ken Monai made the trip to LDRS from Ohio. They launched a LOC *Ultimate Max* on seven G motors. With help at the pad from Ron Schultz, who jury-rigged a pair of vise grips to use as a standoff for the rocket, the duo watched their rocket scream off the pad for a great ride up. The rocket disappeared high in the sky and had a failed recovery. There was a hard impact downrange. A wispy cloud of dirt and debris rose on the horizon—destroyed.

Altitude tracking by visual sight continued at LDRS-9. According to Chuck Rogers and his team of trackers, the highest flight was a 4-inch-diameter rocket launched by Charles Hurlburt. The rocket was powered by a long-burning AeroTech K250. Hurlburt's rocket weighed less than 9 pounds, and his K motor burned for almost 10 seconds. The rocket reached more than 18,000 feet, said Rogers. Another high-altitude rocket was put up by a team led by high-power flyer Randall Redd of Utah. The *Wasatch Rocketry Ascender* weighed in at 10 pounds and zoomed to nearly 12,000 feet on a Rahkonen Prodyne K700. The motor worked fine. But the rocket was destroyed on its return, another deployment victim.

One rocket at LDRS purportedly carried a homemade electronic payload consisting of an accelerometer and a barometric altimeter. Unfortunately, the data from that flight does not appear to have been published anywhere.

Among those in attendance at Hartsel in 1990 was an emerging commercial high-power motor maker making his first trip to LDRS. He was a participant at Black Rock Two a few weeks earlier. He was part of Chuck Mund's P-motor rocket team. With a graduate degree from MIT and a broad knowledge of the manufacture and use of rocketry propellant, he would soon become a fixture at high-power events throughout the western United States. All weekend long at Hartsel, clad in shorts and a white T-shirt with a DOT explosives marker on the back, this rocket scientist educated younger flyers about fundamental theories, engineering, and mechanics of propellant. One day he would be one of the most well-known figures in high-power rocketry. But at Hartsel in 1990, he was an unknown on the field. His name was Dr. Franklin "Frank" Kosdon.

By now a tradition at LDRS, the Tripoli members' general assembly meeting and annual banquet was held Saturday night in Colorado

Springs. This year's special guest was NAR president Pat Miller. He was well received at LDRS-9.

"His speech expressed his desire for unifying the groups in dealing with issues that would arise in years to come," wrote one observer at the banquet. "I would have to say that he was treated with respect and was given a good round of applause after his talk."

Former Tripoli President Ed Tindell also returned to LDRS in 1990, speaking about his recent involvement as a witness in an FBI case involving some IRA terrorists who were allegedly trying to make weapons with high-power rocketry technology. The foiled plot would eventually result in prison sentences for several alleged terrorists. All in all, it had been another good year for LDRS in Colorado.

The experience of Black Rock Two in July was still fresh in the minds of people like Chuck Rogers, Tom Blazanin, and Bruce Kelly when LDRS-9 went forward in August. All three Tripoli members believed Black Rock might be a good location for a future LDRS. But was it the right time yet? Would people from the eastern United States travel all the way to Northern Nevada, to an absolutely desolate location, for Tripoli's premier event? Some Tripoli members had complained that Hartsel was too far west for LDRS. But Colorado was in the center of the country compared to Black Rock.

At the board of directors meeting on Sunday night, August 19, the subject of next year's LDRS was raised. The discussion began late in the evening as other agenda items kept the board busy most of the night. When the topic of the next LDRS came up, it was announced that Chris Pearson and others were working on securing a field in Ohio for LDRS-10 in 1991. Some believed that a return to Ohio would be fitting for the tenth anniversary launch. Yet not everyone was feeling that nostalgic. By 1990, some Tripoli members had grown too accustomed to the wide-open spaces of Colorado. They would resist the selection of any field with limited space and a low-altitude waiver. Board member discussion regarding LDRS-10 dragged on late into the night. Eventually, three venues emerged as possible locations for LDRS-10: Ohio, Colorado, andthe Black Rock Desert in Nevada. Ohio and Colorado were well-known entities. Black Rock, on the other hand, was obscure. It was known to a few influential Tripoli flyers, but it was not on the radar of the average Tripoli member. It seemed an unlikely venue for LDRS-10.

Tripoli president Chuck Rogers had been to Black Rock Two, and he liked the idea of LDRS being held there. He was also sensitive to the concerns of eastern-based flyers. After four years of cross-country travel to Hartsel, they were anxious to have an LDRS a little closer to home. So Rogers made the following motion. He proposed Tripoli hold LDRS-10 in Ohio, with a backup plan of returning to Colorado if the Ohio supporters failed to find a suitable field.

"It was a peacemaker move," explained Rogers later. "There was some controversy with [the Black Rock] location, and I thought it was best to give the people in Ohio a chance to find a suitable field."

However, Rogers's motion failed. By a vote of five to four, the board rejected Ohio as the first choice for LDRS-10. Board member Rich Zarecki then made an alternate motion. Zarecki moved that Black Rock be chosen for LDRS-10. Gary Rosenfield, who had also been at Black Rock Two, seconded Zarecki's motion. Another vote was tallied, just past 11:00 p.m., and by seven to two, Zarecki's proposal carried. The Black Rock Desert would be the site of the tenth anniversary LDRS in 1991. As an alternate location and if there were unforeseen problems with Black Rock, the board unanimously agreed that Colorado would serve as backup location.

Within days of the board's decision, some Tripoli members expressed disappointment and even outrage that the board had picked a launch site not only unknown to most flyers but also farther west than Colorado. It was time for an eastern LDRS, they said. Other Tripoli members questioned the timing of the board's vote on Sunday night at LDRS. Why was the vote taken so late in the evening? they asked. After all, most of the LDRS-9 participants had already gone home. Why wasn't the vote conducted on Saturday night, when more members were present? A few flyers referred to the selection of Black Rock as a shadowy conspiracy, taking place at "a secret board meeting on Sunday night."

President Rogers responded to such complaints immediately, quelling much of the controversy by announcing that the matter was not set in stone. He announced that despite the board's prior vote for Black Rock, he had also asked Chris Pearson to continue his search for a suitable Ohio location for LDRS-10.

"If one is found in the near future, board members can call for another board vote," said Rogers, noting the narrow margin by which

Ohio was defeated during the first board vote on Sunday night at LDRS. "I am confident that Ohio will win the second time around," he said. "But the key is that a site needs to be found, preferably by December, and one has not [yet] been located."

Rogers dismissed the idea that there had been any secret vote, and he defended the board's decision late on Sunday night after most flyers had gone home.

"Did I have a deeply thought-out Machiavellian plan to do this? No," said Rogers. "We rolled into LDRS-9 in Hartsel, and next year's site was unresolved. We had a chunk of the board who wanted to go to Black Rock, and we did not resolve it Thursday through Saturday. Now it was Sunday, and we didn't want to leave before we had decided on next year's location."

To those who insisted the issue should have been discussed on Saturday night, Rogers reminded them of Saturday's night's busy schedule.

"Pat [Miller] was a historic invited guest with an important message for Tripoli members, and we were all quite interested in Ed [Tindell's] work with the FBI and the Justice Department in recent investigations where he has been of considerable service to both Tripoli and sport rocketry in general."

Rogers later acknowledged that politically, the board may have made a mistake in reaching its decision so late on Sunday night.

"In retrospect, this should have been discussed with the members at the Saturday night members' meeting, even if the meeting had to run until two in the morning," he said.

Rogers knew when to retreat and also when to offer the opportunity—or, in this case, the burden—to the other side. He publicly reversed the board's midnight decision, offering those who opposed the choice of Black Rock the chance to come up with a suitable alternative in the east. Now it was their turn to produce a viable site. And they had until December 1990 to do it.

In the meantime, Rogers acted on his own. Rogers had initially proposed Ohio as the location for LDRS-10. But his motion was defeated. Now that this controversy erupted and the board had extended the chance for an Ohio site to be picked, he decided it was time to make his own personal argument, not for an Ohio LDRS but for an LDRS at the Black Rock Desert in Nevada. In October 1990, he took his case directly to the membership:

Why Black Rock for LDRS-10? The Black Rock Desert in Nevada is the best launch site in the nation for high-power rockets. It is one of the largest dry lakes in the country, being over 50 miles long and 26 miles wide at its widest point, with clear blue skies and FAA clearances up to 100,000 feet. The camping is excellent, with the nearby town of Gerlach providing somewhat limited facilities that would be adequate for the banquet and meetings of an LDRS . . . The site has a well-organized prefecture running launches, AERO-PAC, which recently held the very successful Black Rock Two launch, where there were six N engines, two O engines, and a P engine flown in serious altitude attempts up to 50,000 feet. But best of all, most of the town of Gerlach is owned by a guy called Bruno who has offered Tripoli the opportunity for the first time to basically take over a town!

Rogers acknowledged there were also sentimental reasons for holding a tenth anniversary LDRS in Ohio. But the fact of the matter was there may not be a suitable location found in the Buckeye State, he said. More importantly, Rogers knew Tripoli was at a crossroads, and the membership needed to consider carefully the direction in which the high-power organization was headed. He believed it was time for Tripoli members to not only choose a location for LDRS-10 but also decide what kind of rocketry organization Tripoli was to become in the future. According to Rogers, a decision to pick Black Rock for LDRS would say a lot about Tripoli's future:

> [Are] Tripoli and its national LDRS launch to be paper rockets flown on moderate power from small Midwest launch sites with FAA altitude limits of 4,000–8,000 feet or large high-power rockets flying 10,000–30,000 feet with large motors at excellent western dry lake bed launch sites? Is LDRS a social event for families or a place to fly serious rockets, or what are we hoping to do at Black Rock—both? Are very high-power West Coast rocketeers underrepresented at LDRS because they have no reason to attend because the best launch sites are out west? Or are they overrepresented on the board because they are some of the leading rocketeers of the association?

Today, with the benefit of nearly thirty years of hindsight and at a time when Black Rock is hailed as the premier location in the world for high-power rocketry, it is hard to imagine any controversy in holding a launch there. But this was not just any launch the board was debating. It was LDRS. And as Rogers and others could see, within the Tripoli membership, there were still lingering questions as to what kind of club Tripoli was becoming and what kind of launch LDRS was.

The search for an LDRS location in Ohio went forward in the fall of 1990. It was unsuccessful. On December 2, 1990, the Tripoli board was informed that a proposed Ohio location was turned down by the landowners. LDRS-10 would not be held in Ohio. Indeed, Ohio would likely never be the site of any LDRS again. And it would be several more years before a suitable location would be found for LDRS anywhere east of the Mississippi River. In early 1991 and by a vote of eight to one, the Black Rock Desert was confirmed as the site for the tenth anniversary of LDRS. There were no complaints from the membership. It was the beginning of a new chapter in high-power rocketry history.

In February 1990, the NAR board of trustees extended the timeline for its Commission on Advanced High-Power Rocketry to complete its work. The commission's deadline was moved from August 1990 to February 1991.

"The additional six months are needed by the commission to properly address all the complex issues involved," reported Pat Miller.

The final report from the commission would contain recommendations to the board as to how the NAR might proceed on the advanced high-power rocketry issue, added Miller, who promised to keep NAR members up to speed on the commission's progress.

"It is important for you to know that there is a dialogue growing between Tripoli and the NAR," he added. "The issues relevant to both organizations are being carefully discussed."

The relations between the Tripoli and the NAR were excellent—or so it seemed.

13

NAR Wars

At the NAR's annual association meeting (NARAM-31) in August 1989, Harry Stine took to the floor to read aloud a letter he had written to J. Pat Miller about the future of hobby rocketry. In his letter, Stine expressed disappointment in the NAR's failure to live up to all the goals he had envisioned when he founded the organization in the late 1950s. These goals, he said, consisted of serving all forms of rocketry, not just model rocketry. Stine wanted a return to his "original vision," and his meaning was clear. The NAR founder was calling for the NAR's entry into high-power rocketry. Stine was applauded by those in attendance. A resolution was immediately passed asking the board of trustees to study the NAR's potential role in higher-powered rocketry (also sometimes called nonprofessional consumer rocketry).

Shortly thereafter, President Miller announced the creation of the Commission on Advanced High-Power Rocketry. The commission was chaired by Jim Barrowman of Maryland. Gary Rosenfield, who was making products for both model rocketry and high power, was also a member. Martha Sienkiewicz of Massachusetts, a Tripoli member and well-known figure at high-power launches, was also chosen. Miller and the NAR board also asked for an official representative of Tripoli to sit on the commission. Tripoli board member Chuck Mund of New Jersey was tapped for the job.

"The purpose of the commission is to study how the NAR might implement programs for consumers of advanced high-power rockets," said Miller. "The commission is responsible for developing a step-by-step procedure which the NAR might use to do this." Miller promised that by August 1990, the commission would present its final report to the board of trustees. "At that time, the board will decide if the NAR will, in fact, incorporate [advanced high-power rocketry]."

In October 1990, the NAR board reviewed the recommendations of the commission at a three-day meeting in St. Louis, Missouri.

"I have an important and historic report from the NAR board of trustees which I want to share with you," Miller told the membership after the meeting in December.

He then informed the world that the NAR was now in the business of high-power rocketry. Miller outlined the rough details of the move in the NAR's now official publication, *American Spacemodeling*, in December 1990. Among other things, he said, the NAR was ending the 3/48 Rule. NAR members could now participate in high-power launches without having to worry about a 3-mile or 48-hour separation between model rocketry and high-power events—provided that the range was approved for all types of rockets and all applicable safety codes were being followed. And once a NAR high-power safety code was in place, he added, members would be able to launch high-power rockets even at NAR-only gatherings.

Miller reminded members that the rule change did not yet come with insurance coverage for high-power activities. That would take some time, he said. He also reported the NAR would have to expand its motor certification program to cover high-power motors. The NAR would also create at least two "Tiger Teams" of experts to address, among other things, a high-power safety code for hobby rocketry. Ultimately, Miller and the NAR would move toward a revision of NFPA 1122 for the purpose of accommodating high power.

The NAR's entry into high power was neither sudden nor unexpected. Such a move had been coming for quite a while, literally for years, and Tripoli was well aware of the actions of the NAR commission that led to the board's decision. Yet some people in high-power rocketry were apparently surprised. Worse, as a result of either a serious misunderstanding or perhaps stray remarks by influential NAR members, Tripoli's leadership received a very different back-channel message from the NAR's actions. Tripoli thought the NAR was going to force it out of business and take high power for its own.

As president of Tripoli, Chuck Rogers had kept abreast of the NAR's possible entry into high power since the announcement of the formation of the NAR's commission at NARAM-31 in 1989. In fact, Pat Miller had originally invited Rogers to be the Tripoli representative on the commission.

"I declined," explained Rogers at the time. "[I felt] that having the president of Tripoli serve on the commission would imply a preendorsement of its findings. I felt it was more appropriate to select a designee, Chuck Mund, to serve on the commission in my place."

It was a wise political move by Rogers, who realized that entry by the NAR into high power would not only formalize the end to the NAR's internal resistance to larger motors and bigger rockets but also mean that Tripoli would be faced with real competition for consumers in the high-power market. Tripoli's monopoly on high power would be over. Rogers was not anti-NAR, and he was well aware of the strength of the NAR and what its entry into high-power rocketry might mean for Tripoli. In late 1989, he even raised the possibility that the NAR and Tripoli might one day merge.

"It is pretty obvious why the NAR has a newfound interest in high-power rocketry," wrote Rogers. "Basically, Tripoli has shown that high-power rocketry works. With the confirmation process for Class B motor purchases, FAA clearances, and now insurance for Tripoli events, the Tripoli Rocketry Association has shown that high-power rocketry is a viable hobby that has captured the adults who previously left rocketry because there was nowhere to move up to beyond model rockets and their limited activities. Tripoli now has over a thousand members, whose average age is thirty-three! We can talk numbers! The NAR has taken note of the fact that Tripoli is an organization of adults already one-fifth the size of the NAR. The NAR commission will study many options, including the formation of a NAR high-power division and a possible merger with Tripoli sometime in the future . . . Things have certainly come full circle, and while the NAR has come a long way toward our position, will they ever come far enough to satisfy Tripoli members?" Tripoli would participate in the NAR advanced high-power rocketry commission, said Rogers. "But the members of Tripoli will have the final say on our response to the commission's recommendations."

When the commission released its report in the fall of 1990, Rogers was in the middle of several internal Tripoli struggles. There were money issues, the club's national journal was in disarray, certified motor testing was still delayed, and the fight over the location for LDRS-10 was brewing. The NAR's high-power announcement quickly took center stage. Word reached Rogers and the Tripoli board that the NAR was not only moving into high-power but also planning a rewrite of NFPA 1122. And the claim was made—allegedly supported by either written material authored by someone in the NAR or inflammatory statements made on an early Internet forum—that the NAR was going to ask the NFPA Committee on Pyrotechnics to eliminate the corporation exemption under which the Tripoli Rocketry Association was formed in 1985. In other words, the allegation was made that the NAR was trying to make Tripoli *illegal*.

What followed was a terse exchange of letters between the NAR and Tripoli (by presidents Miller and Rogers, respectively) in late November and early December 1990. Rogers publicly and also privately confronted Miller regarding the rumors of the NAR's alleged activities. Miller responded by vehemently denying them. With the rumor mill in full swing, Rogers decided he needed to take drastic action to protect Tripoli's existence.

In late 1990, the Tripoli president asked all Tripoli members to immediately join the NAR. Tripoli members received a copy of a NAR membership application with every issue of the December 1990 *Tripolitan*. The copies were reproduced from an actual NAR application. In an editorial in the *Tripolitan*, Rogers and newly appointed *Tripolitan* editor Bruce Kelly asked every Tripoli member to fill out the application and mail it to the NAR. The plan was to flood the NAR with new Tripoli members and then use this new political influence to take control of the NAR from within the organization.

"Tripoli will run candidates for the NAR board of trustees this coming year," explained Rogers to the membership. "We ask those of you who are already members of the NAR to remember Tripoli in that election. Tripoli was the organization that allowed you freedom to fly what you wanted all along, and it will be Tripoli that will continue to provide you that freedom."

The NAR had approximately 3,700 voting members in late 1990, and like many other private associations, including Tripoli, only a handful of its members participated in the annual election process. Rogers wrote to Miller and told him that unless the NAR sat down with Tripoli to rewrite NFPA 1122 together, then by sheer numbers alone, Tripoli would seize control of the NAR and take care of the issue itself. The NAR already had many members who held joint membership in Tripoli and the NAR, warned Rogers, and they were "encouraging members of Tripoli who are not already members of the NAR to join the NAR and to vote for Tripoli candidates."

According to Rogers, "only 127 NAR members bothered to vote in the last NAR board of trustees election." Rogers claimed that twice that many Tripoli members had voted in the most recent Tripoli election despite the smaller membership base.

"All your board seats come up for election at one time, and their terms are for three years," he told Miller, promising that Tripoli would run for at least three of those seats in the upcoming balloting.

If Tripoli members really did join the NAR and then cast their votes accordingly, they really could impact the NAR board of trustees in a big way. It was akin to a hostile corporate takeover. Rogers was threatening to go to the mat, if necessary, to ensure that Tripoli's voice would be heard by the NFPA and that the exemption in NFPA 1122 be left alone.

Pat Miller and the NAR board of trustees responded to Tripoli's call to action in several ways. First, Miller denied the NAR was trying to rewrite NFPA 1122 to make Tripoli illegal.

"No one was trying to get rid of Tripoli," said Miller, "and it was nobody's plan to rewrite the safety code to foreclose Tripoli from being a legitimate organization."

Next, he invited Rogers and the entire Tripoli board to the NAR's annual trustees meeting held in February 1991 in Texas.

"I did this because we needed to resolve this issue," said Miller. "We needed to get things settled down, and it would have just escalated. It had to stop."

Finally, and to absolutely ensure that Tripoli's planned corporate election raid was unsuccessful, Miller effectively ordered all the recent applications from Tripoli members to be ignored—at least for the time being.

"There was a flaw in the printing of the NAR application form by Tripoli," said Miller. "And I began getting a flurry of these applications with the flaw in them, and so I had all such applications routed to me in Texas at our headquarters. I then sent out letters to these people and said, 'Thank you for your application form, and now I am putting it before the membership committee for consideration, and we will be back in touch with you soon.'"

But the applications were not acted upon; they were not even processed until after the election was over. It was an astute political move by Miller, and it deflected much of Tripoli's electoral attack. Rogers recalled that the NAR returned some of the applications to Tripoli members.

"We published the NAR application," said Rogers, "but the NAR figured out that certain applications were from us, and so they would return the application and the money, saying that they could not accept it because it was either not the right form or that it could not be a copy."

In the short run, Miller's decisive action helped prevent a possible Tripoli takeover of the NAR. The back-and-forth wrangling between Tripoli and the NAR soon gave way to an uneasy truce. Then Rogers and the Tripoli board decided to accept Miller's invitation to join the entire NAR board for a joint discussion in Dallas in February 1991.

"At least five Tripoli board members will be attending," promised Rogers in response to Miller's suggestion. "We are planning a Tripoli board meeting in Dallas coincident with your board meeting to decide our final course of action."

The Dallas meeting between the board of trustees of the NAR and the Tripoli board of directors on February 16, 1991, was one of the most important meetings in the history of high-power rocketry.

"It was held in a dumpy, unrestored Ramada Inn," recalled one Tripoli board member.

The entire Tripoli board showed up, along with all the members of the NAR board of trustees. Not surprisingly, the meeting got off to a rough start.

"Vern Estes and people from Tripoli were yelling back and forth [at each other] at one point about Tripoli's interference with the NAR's election process," wrote one session participant, who took Estes aside and said that the fighting with Tripoli had to stop.

NAR leaders vehemently denied having any intention to get rid of Tripoli with a rewrite of the safety code, and they claimed it was never an issue at all. But Tripoli board members were not so sure.

"We aired our grievances to the NAR, especially the 1122 issue," remembered one Tripoli board member, and the grievances included the claim that the NAR was trying to eliminate Tripoli.

"Tripoli members were accusing the NAR of doing a lot of things in secret and in closed session," recalled Miller. "And they claimed that all of Tripoli's work was done in the open, in open sessions. We eventually came to one issue, and I don't remember what it was, but the Tripoli board said that they would have to talk about it for a while, and they basically went off in closed session. When they returned, I said to everyone that the NAR board of trustees would like to welcome the Tripoli board of directors back from its *first ever* closed session!"

Everyone on both boards laughed, said Miller, and eventually, the two groups realized that they had more in common than not. Rogers also believed that Tripoli's aggressive action with regard to the NAR's upcoming election process had to be abandoned as the relationship between the two national rocketry clubs was vital to the future of each.

"What began to weigh heavily on my mind and in the minds of the other Tripoli board members was this," said Rogers. "Was Tripoli about to undertake actions that rightly or wrongly were going to poison relations with the NAR for the next ten years? Would NAR members in the year 2000 still be talking about how Tripoli 'ruined' the NAR and/or itself after a successful or attempted takeover?"

Any winner from this war would only have a pyrrhic victory at best, thought Rogers, who also realized, as did Pat Miller, that an open line of communication between the NAR and Tripoli had to be created to prevent similar disputes in the future.

"What Tripoli found out was that the differences between our positions and those forming within the NAR were far closer than we thought possible," said Rogers. "Was external pressure from Tripoli responsible for moderating the NAR's positions, or were the NAR's positions always there to be seen if we had taken the time to look rather than being so quick to rush off and declare war?"

As the Dallas meeting progressed, the leaders of both organizations resolved to embark on a new course of relations, divorced from not only the recent election acrimony but also the previous years of mistrust that periodically surfaced between the two organizations. There were still some NAR and Tripoli members who would hold on to that acrimony, but the two boards made amends in Dallas, and the future consequences for all of rocketry would be profound. "That night, the Tripoli board treated the NAR to dinner at a Chinese restaurant in Dallas," recalled Miller years later. "I had the NAR vice president issue a toast, and then it came time for us to open our fortune cookies, and I turned to him, and

I said that whenever you read your fortune, you need to add the words 'in bed'. Everyone went around the table reading their fortunes aloud and adding the words 'in bed' to the fortune.

"'Imagination is more important than knowledge . . . *in bed.*'

"'Sometimes a stranger can bring great meaning to your life . . . *in bed.*'

"'Your ingenuity will bring fine results . . . *in bed.*'

"'Life will be happy until the end, when you'll pee yourself a lot . . . *in bed.*'

"By the time we got around the table, we were all saying it in unison, and I think the collegiality had been restored between the two groups. We were now just a bunch of guys sitting around a table who all wanted to fly rockets."

Immediately after the Dallas meeting, both organizations published a joint resolution for their respective memberships. Among other things, this declaration of cooperation pledged mutual support and respect to each other and a policy of noninterference in the internal affairs of each other's association. Tripoli would abandon its attempt to take over the NAR in the election process, and NFPA 1122's exemption would remain unchanged. And Tripoli and the NAR agreed to work together on the upcoming rewrite of the NFPA codes applicable to high power, also described by some in the NAR as "sport rocketry." Later that year, Miller invited Rogers to NARAM to speak directly to the NAR members; Rogers reciprocated, inviting the NAR president to speak to Tripoli flyers at LDRS.

The revision of NFPA 1122 was of great importance to the long-term viability of high-power rocketry. The NAR and Tripoli would gain strength from each other's cooperation on the Committee on Pyrotechnics. In the long run, the mutual efforts of both organizations would be indispensable in repelling additional, more formidable attacks on the hobby by outsiders, especially government regulators. Indeed, within only a year, the NAR and Tripoli presidents would be negotiating together with the CPSC, issuing joint statements to their respective memberships regarding rocketry's dealings with the government. After the meeting in Dallas, President Rogers reported to the Tripoli membership that some members of the NFPA were going to be shocked when they found out the extent of high power already in place and that flyers were already launching rockets of significant size.

"They are going to require some convincing to write a code that will allow the activity to grow to its full potential," Rogers wrote. "If the rocket members of the committee can't come out of the upcoming NFPA pyrotechnics committee rocket caucus with a unified position on high-power rocketry, then frankly, we don't stand a chance of getting the kind of code we want passed by the full committee. The NAR also realizes this. We are not only going to be friends with the NAR. We are going to be allies. Strange world, we live in, isn't it?"

14

LDRS X and BALLS

For most people, the final leg in their journey to the Black Rock Desert begins at exit 43 on Interstate Highway 80, about a half hour east of Reno, Nevada. Once off the Interstate, the rest of the trip is due north on State Highway 447. The destination is Gerlach, a windblown collection of old buildings that resemble the set of a Western movie. There are a few bars, a restaurant, a lone gas station, and a couple hundred people living in a few dozen small homes and trailers at the southwest edge of the vast desert.

Gerlach is 75 miles up Highway 447, a finely paved two-lane road passing through the communities of Wadsworth, Nixon, and Empire. None of these towns are more than a mile long, and all three are dotted with abandoned buildings, battered trailers, and beat-up cars. The first two towns are speed traps. There are no motels along 447 until you reach Gerlach. But there is one great oasis: Pyramid Lake, which originates at Lake Tahoe in the Sierra Nevada Mountain range west of Reno. The Truckee River leaves Lake Tahoe high in the mountains and descends eastward into the high desert. The river cuts through the middle of Reno and then parallels Interstate 80 for 20 miles before abruptly turning north near Fernley. In Nixon, just before Pyramid Lake comes into full view, the river makes its final few lazy turns, transforming hundreds of acres of arid high desert into lush grasslands and a few horse farms before emptying its last drop of snowmelt into the big blue lake.

Once past Marble Bluff at the southeastern end of Pyramid Lake, the color green all but disappears from view. This is the most desolate stretch of the trip. The drive is another hour of rocky prominences, gray-white lake beds, scrub brush, and barren terrain. You can drive for another 50 miles without seeing a car. To your right is what the locals call Winnemucca Lake. It's really a dried-up lake bed at the base of the rugged Nightingale Mountains. To your left, there is nothing, occasionally a few coyotes, antelope, or a herd of wild mustangs roaming the land scape of the Lake Mountains. Near the end of the drive, the Granite Range comes into view directly ahead. At the base of that mountain range is Gerlach, population 200, elevation 3,946 feet. This feels like the middle of nowhere. Directly east of town and visible from the center of Gerlach is the Black Rock Desert, one of the flattest spots on the face of the earth. It's most prominent feature is its isolation. In the summer, it is ruled only by the sun.

"The sun doesn't just rise over the Black Rock Desert in Nevada," observed *Popular Science Magazine*. "It ignites."

Up through the mid-2000s, when the phony hippies at Burning Man decided they needed cell towers to stay in touch with their parents, the Black Rock Desert truly was off the grid—no computer access, no cell phones, no nothing. There was a single pay phone outside Bruno's Country Club and Casino in Gerlach. When you left Bruno's and went out into the desert, you were on your own. Most adults preferred it that way.

LDRS-10 began on Friday, August 16, 1991. The weather was terrific, with a bright blue sky and a few clouds lying low along the western horizon. Visibility was unobstructed for miles. The ground was flat, so flat that a person walking away from camp in a straight line literally disappeared over the horizon in a couple of miles. Just prior to the first rocket flight, an eight-engine Boeing *B-52 Stratofortress* passed low over the desert south of the rocketry range head. Black Rock was used by the United States Air Force as a gun and bombing range for many years. Travelers walking the desert in the 1990s would find scores of empty 20 mm cannon shell casings littering the ground. This particular bomber was heading home to Castle Air Force Base in Merced, California, a few hundred miles south. Tom Binford of Georgia returned to Black Rock for the second year in a row. He was at Black Rock Two in 1990. Binford was an early flyer at LDRS-10, launching one of his favorite rockets, a LOC *EZI-65* on an I115 motor.

"This is the thirty-sixth flight of this rocket," said Binford as he carried the dented and well-worn missile out to the pads.

At ignition, the rocket tore off the pad. A poor separation at apogee led to the rocket's demise; there would be no flight 37. Later that weekend, Binford would fly his *Cloudbuster 54* rocket to more than 16,000 feet on a tower-launched Smokey Sam Vulcan L-impulse motor. Binford was one of several flyers who achieved significant altitude success at LDRS in 1991. More than twenty-three flights reached 10,000 feet or more. At least seven rockets cleared 15,000 feet.

John Cato's 5.5-inch *Nike Smoke* was another crowd favorite on Friday. Cato's rocket was powered by four G motors and a central J motor. At 5 feet tall, it was still considered a big rocket for its day. The *Nike* had a slow takeoff, climbed low in the sky, turned over, and then impacted the hard playa. Flyers discovered how hard the ground was at LDRS-10. It took some doing to pry fins, airframes, and nose cones out of the dusty desert floor. But if a rocket came in undeployed from even a few thousand feet, it would slice into the ground with ease. Four fins and a few inches of body tube are rocketry sign posts at Black Rock. Among the debris scattered around Cato's *Nike* was a digital watch with wires hanging out that he had mounted to a stand in the rocket. Crudely designed by today's standards, such homemade electronics were state-of-the-art in 1991.

LDRS was becoming more efficient each year; this year's event was no exception. The pace on Friday was slow but comfortable. The crowd was moderate in size, and there were twelve pads again, as there had been in Hartsel; most of the launchers held steel rods for guidance. There was at least one tower, and it was used frequently. Rails were being introduced into the world of model rocketry in 1991; they were still uncommon in high power.

To recover rockets from almost any altitude, flyers simply jumped in their vehicles and drove to the very spot where their rocket landed, whether it was a mile away or five miles away. The speed limit was easy to remember—there wasn't one. The desert floor was hard enough to drive on—and drive at high speed too. Fast-moving cars and trucks would race through the desert, trailing a rooster tail of fine dust that could be seen for miles. Once disturbed, the lightly colored Black Rock surface was so fine, it penetrated nearly everything, even locked car doors with rolled-up windows. You can drive this desert with your eyes closed. And when you opened them again, you might have a hard time telling if you had actually moved. That is how vast the desert seemed from any location on the playa.

The waiver for LDRS-10 was the highest yet in high power, with windows that reached as high as 60,000 feet. It was clear the Tripoli board had made the right decision to bring LDRS here. The minor conflict over the selection of this location as the tenth anniversary launch quickly vanished from the collective memory of the Tripoli Rocketry Association. This was the new home of high-power rocketry, and everyone knew it.

David Bolduc drove to Black Rock all the way from his home in Springfield, Massachusetts, with a scale model of a *Nike Hercules*, one of the favorite flights of LDRS-10. Bolduc took two years to build his two-stage missile, and it was painted in military colors with a four-finned army-green Nike booster topped off by a sleek white sustainer. The booster held multiple motors, and the big rocket stood several feet tall. It was carried on the shoulders of three men who set the rocket gently on the pad while a group of excited onlookers edged closer for a better look.

Bolduc's rocket carried an impressive amount of firepower for its day: two J700s and two K500 Firestarters in the booster and a Vulcan K-impulse Hellfire in the sustainer. At ignition, the 60-pound *Hercules* tore off the pad and had a good ride up, emitting sparks, flame, and lots of black smoke. The sustainer motor failed to ignite, and at apogee, the upper half of the rocket turned over and plummeted into the ground, undeployed. Most of the rocket was destroyed. But the memory of the Nike survived; a smiling Bolduc with his rocket on the pad was the cover shot chosen for the *High Power Rocketry* magazine's story on LDRS-10.

The size of the crowd increased on Saturday. With only a dozen or so launch pads, there were long lines as flyers waited patiently for their turn on stage. Clustered flights of F and G motors were popular in 1991, and the LCO insisted that all in attendance rise to their feet even for these flights.

"This is a heads-up flight!" barked the LCO before one such launch. "And in case you don't know what that means, it means get up on your feet!"

A two-stage rocket took off, assumed a horizontal flight path high in the sky, and then launched its upper-stage sustainer directly toward the ground—with predictable results.

One of the largest rockets of LDRS-10 took off on Sunday: Rich Zareki's scratch-built *Aurora*. Zarecki prepped the *Aurora* using his car as the rocket stand. The big rocket was long and slender, close to 10 feet in length. With the front windows lowered on either side of the car, he and his crew slipped the rocket's airframe through the passenger's front window, suspended above both seats, and out the driver's window on the other side. The aft end of the rocket poked beyond the passenger's door, and Zarecki worked there, carefully wiring up the four K550 motors that would be used for the *Aurora*'s takeoff. Preparation of the nose and the recovery system went on outside the driver's side of the car. When assembled, the rocket was carried out to the pad in a solemn procession of crew members and anxious onlookers.

At ignition, *Aurora* took off in a rush of billowing smoke trailing a bright AeroTech flame. The rocket had good boost and stayed intact through apogee. Then one of the motors was ejected out the aft end, and the rocket's parachute failed to deploy. The big rocket turned over and came screaming in for a terminal event on the desert floor, not far from its launch pad—another typical big-rocket recovery for its day.

On Sunday, several members of a team of flyers from the University of Central Florida launched the highest flight of LDRS-10. Their *Starshot* rocket was a two-stage vehicle they had attempted to launch at LDRS in Colorado, but they were thwarted by bad weather. The rocket housed two Vulcan L motors, one in the booster and one in the sustainer. It had a good takeoff and flight, trailing black smoke out of its tower and climbing to more than 22,000 feet.

There were no reported M flights at LDRS-10. But at least ten L motors were burned. There was also one N flight. The N was in a rocket built by Chuck Rogers, Larry Liggett, and Moose Lavigne. Their *Astrobee-G* was powered by a long-burning Ace Aeronautics N. The 4-inch-diameter 8-foot-long vehicle weighed more than 40 pounds on the pad and roared skyward on a perfect trajectory. The rocket was in search of a new Tripoli altitude record, and it looked like a contender until its nose cone popped off at motor burnout, deploying the recovery gear too early. Still, the rocket cleared 14,000 feet and was one of the first N-powered vehicles to ever fly at an LDRS.

LDRS-10 ended on Sunday afternoon, August 18, 1991. It was a good weekend, with more than a hundred flyers launching scores of rockets, many of which were A to G power. But the launching of rockets at Black Rock was not over. For the first time in LDRS history, flying would continue for a fourth straight day. There would be another launch on Monday at the same location, only on Monday, the event would no longer be called LDRS. It had a new name. And like LDRS-1 some ten years earlier, this one-day launch was the beginning of another revolution in high-power rocketry.

The fourth day of flying at Black Rock began as a small advertisement that appeared as an ad in the *Tripolitan* in August 1990. The ad announced a new high-power launch that would take place in the summer of 1991 at Black Rock. This event was restricted. All flights were required to have a minimum of 1,300 Newton seconds in motor power, meaning a K motor or higher. There would be no model rockets allowed, no mid-powered rockets either, nothing but high power.

When the ad was first placed, there was still no date picked for the launch. But for the registration price of $30, flyers signed up for the event, date unknown, and received a T-shirt and commemorative patch. The money was sent to a post office box in Reno, Nevada. There was no person's name associated with the address, only the name of the launch. In fact, Steve Buck placed the advertisement. And the event he helped create was called "BALLS."

The origin of BALLS, Tripoli's annual experimental launch, coincides with the rise of one of Tripoli's longest-running and most successful prefectures: AERO-PAC.

AERO-PAC was created in the fall of 1989 by a group of Northern California flyers led by Bill Lewis of San Jose, California. Lewis was a forty-seven-year-old industrial designer who had earned a Purple Heart while serving with the U.S. military in the Vietnam War. He was one of only a handful of people who attended Steve Buck's Black Rock One launch in July 1989. In August of that same year, Lewis traveled to LDRS-9 in Colorado, where Buck addressed the membership about the Black Rock One experience and the benefits of flying in the desert near Gerlach.

"Steve Buck gets up at LDRS and started talking about Black Rock," recalled Lewis. "He told everyone how desolate and unbelievable it was and that we needed to start doing launches out there." Lewis couldn't agree more.

After LDRS-9, Lewis approached Buck with a proposal.

"We've got the population in Northern California, I told Steve, and you've got the site in Nevada. Why don't we start a Tripoli prefecture for Northern California and Nevada together at the Black Rock Desert?"

Buck said yes, and AERO-PAC was born. Lewis came up with the name for the club after talking with Tripoli member Bill Woods, who lived in nearby Santa Clara.

"I was interested in experimental or amateur rocketry, and Bill was talking about forming a rocketry club," recalled Lewis in an interview in 2016. "He wanted to call it Aero, but I wanted to unite flyers from Northern California and Nevada, and I came up with the name Association of Experimental Rocketry of the Pacific. Ultimately, I combined the words 'aero' and 'Pacific' to come up with AERO-PAC."

The first meeting of Tripoli's Black Rock prefecture was held on November 17, 1989, at the Minolta Planetarium at De Anza College in Cupertino, California. Within several months, the young club pulled together equipment, resources, and almost thirty members. The club looked forward to its first high-power launches in 1990 at Black Rock.

Black Rock One was organized by Buck in 1989, which preceded the formation of AERO-PAC and did not have any altitude waiver. This restricted flights to 10,500 feet above ground level. For Lewis, this limit on altitude was unacceptable. He approached the FAA in the spring of 1990 to inquire about securing a real waiver for AERO-PAC's inaugural launch in May (called Aeronaut I) and then later that summer for the Black Rock Two event in July.

"I called up and spoke to an FAA representative on the phone and told him we wanted to launch some rockets at the Black Rock Desert and I would like to send an application for a waiver to do so," said Lewis in an interview in 2016. "I said we were part of a national organization called Tripoli Rocketry and we have liability waivers that would be signed by flyers and we would get permission from the BLM for the launch. He also knew something about rocketry, and he told me that we were the first ones to ask for such a waiver in the region."

On March 1, 1990, Lewis mailed the FAA not only the waiver application form but also the BLM paperwork he had prepared. In his waiver application, Lewis described the proposed rocket launch as an "invitational launch for recreational and research rockets using nonmetallic construction and commercial manufactured motors" 15 miles northeast of Gerlach, Nevada. On April 21 and after reviewing the waiver application submitted by Lewis, FAA manager Robert Brekke issued a certificate of waiver for AERO-PAC's Aeronaut I launch to be held on May 19 to 20, 1990. This was the first high-power rocketry waiver ever granted at the Black Rock Desert. Lewis asked for 25,000 feet. He got 26,000.

Following the Aeronaut I launch, Lewis submitted another application to the FAA for the Black Rock Two event. This time, he attained a waiver that included windows between 30,000 and 55,000 feet. These were the highest FAA waivers ever obtained for high power up to that time. In many respects, these waivers and what they allowed flyers to do at Black Rock quickly made AERO-PAC one of the most important prefectures in all of Tripoli. With the skies above the Black Rock Desert opened up to extraordinary altitudes, flyers began the construction of new rockets to take advantage of every foot of the desert air that was granted to them.

For high-power rocketry, there suddenly seemed to be no limit as to the altitudes that might one day be achieved.

The first BALLS launch was conceived during Black Rock Two in 1990 and likely started as a conversation among several people, including Steve Buck of High Sierra Rocketry, Tim Brown of West Coast Rocketry, and Bill Lewis. Planning for BALLS began even before Tripoli had selected Black Rock as the site for LDRS-10. An announcement for the new launch appeared in AERO-PAC's newsletter in September 1990 with a proposed launch date of June 1991. Later, once it was determined that Tripoli's national launch was coming to Black Rock in 1991, a new date for BALLS was chosen.

"At the Winterfest 1990 launch at Lucerne, Tim, Steve, and I got together briefly to discuss general possibilities, including having AERO-PAC host [BALLS] as a nonsanctioned event in the same manner as Lucerne was conducting its launches," recalled Lewis. "I believe it was at this meeting that Steve [Buck] suggested that the launch be held on the Monday following LDRS." In Buck's view, the BALLS event would be like no other high-power launch ever held. "If you have motors smaller than a K, don't bring them," he said. Buck discouraged spectators too. "It's going to be a dangerous launch," he said, fully expecting that the majority of the rockets launched would be destroyed. "Back then, the bigger rockets were still blowing up half of the time. They would end up thrashing around on the ground on fire. There was still no motor certification for many high-power motors. People would show up with a new motor with little to no testing, and we would say, 'Let's try it!'"

The date for LDRS-10 was August 16 to 18, 1991. The date chosen for BALLS was August 19, 1991. The first national ad for the launch appeared in the *Tripolitan* in the fall of 1990. The ad read, in part, as follows:

Special Announcement

B-A-L-L-S

Big Ass Load Lifting Suckers

001

Black Rock Desert Dry Lake

Gerlach, Nevada

Summer 1991

There is no dispute that the name chosen for this first experimental launch at Black Rock was BALLS. However, when the ad was placed in the December 1990 issue of the *Tripolitan*, only the acronym remained; the longer name for BALLS was dropped. Then the name for the launch was altered yet again on the application forms filed by AERO-PAC with the FAA (waiver) and BLM (land use). The alteration was performed by Lewis, who was tasked with getting the waiver and related permits from the government.

"I had no problem with [the BALLS] name," explained Lewis later. "But I was the guy who had to sell it to the BLM and the FAA. I was the guy out front getting the waiver and permits. And I decided to put a name on the government permits that was in a professional manner. Of course, I wanted to keep the rocketeers happy too. So, I changed the name of the event to Fireballs 001 on all the official papers that went out."

After conferring with the AERO-PAC board of directors and following a short discussion with Chuck Rogers, Lewis sent a letter to the FAA regional office thanking them for their support in 1990 and informing them of the intended AERO-PAC launches to be held in 1991. This letter contained the initial basic plans and information on the Fireballs event. The Fireballs name chosen by Lewis did indeed stick for a while. The experimental launch at Black Rock would be called Fireballs off and on for the next several years until it went back to BALLS in 1997. By the spring of 1991, preparation for Fireballs was well underway. Although the launch was originally to be run by commercial sponsors, AERO-PAC ultimately sponsored and ran the launch. Still, this was not a Tripoli-sanctioned event, and there would be no insurance. All participants were advised they would have to sign a liability waiver and fill out preregistration forms outlining the details of their proposed flight.

"This being the first launch of this nature, how, when, and under what conditions future experimental launches will be conducted by AERO-PAC at Black Rock will be shaped by this launch," Lewis told those interested in attending.

On July 5, 1991, Lewis submitted an application letter for a waiver for Fireballs 001 to the FAA regional office located in Hawthorne, California. In that letter, Lewis described the event as a "custom launch," not to be confused with the waiver he had already obtained for Tripoli and LDRS-10.

"This is a separate request in that the motors used for this launch will be custom, one-of-a-kind, with a minimum thrust of 1,300 Newton seconds in liquid, hybrid, or solid propellant configuration," Lewis told the FAA in his letter. "Typical launch vehicles would/could consist of rockets with motors that were not submitted for Tripoli certification (typical older, no longer in production, or new experimental motors) that are required for the three-day national event. Airframes for this one-day event designated 'Fireballs 001' may also be made of experimental materials such as aluminum alloy, metal/composite, metal honeycomb/epoxy, graphite fiberglass, [and] fiber/thin wall steel that are being explored for high performance evaluation (Tripoli currently sanctions nonmetallic airframes only)."

At the end of the application, Lewis proposed three waivers for Fireballs. He asked for a general waiver to 36,000 feet (AGL) for the entire day starting at 7:00 a.m. and running until 6:00 p.m. Then he asked for two windows up to 60,000 feet, one in the morning and one in the afternoon. Finally, he asked for two more windows to a whopping

100,000 feet, the last of which would run from 4:00 p.m. to 6:00 p.m. The request for 100,000 feet was to accommodate amateur rocketry groups planning on attending Fireballs.

"When I received a phone call from Dean Oberg of Space Delivery Systems (SDS) in Buffalo, New York, I quickly realized that Fireballs was being interpreted beyond the scope of high-impulse commercial rocket motors," said Lewis. "Apparently, the word had spread, and several organizations such as SDS were looking to utilize the Fireballs launch as a site to test their 'custom' motors. In Dean Oberg's case, this was a hybrid propellant motor with the capability to reach 65,000 feet [in] altitude. In my discussions with Dean, I asked him to send me a set of drawings for presentation to the AERO-PAC board. At the next board meeting, the proposal was unanimously approved."

Not long afterward, Lewis received another call, this time from George Morgan of the Pacific Rocket Society. Morgan told Lewis that they would arrive at Fireballs with a "large liquid-fuel rocket that was capable of reaching 100,000 feet." It was these two rockets, said Lewis, that led to the first request for the 100,000-foot waiver from the FAA at Black Rock.

On August 15, 1991, just four days prior to launch, the waiver for Fireballs 001 was granted in its entirety. Lewis had secured high-power Rocketry's first ever waiver to 100,000 feet. This was 40,000 feet higher than the waiver for LDRS-10. At that moment, the future of high power shifted imperceptibly to the Black Rock Desert.

(By 1994, the 100,000-foot waiver applied to multiple AERO-PAC launch events at Black Rock. In April 1994 and on a motion by Chuck Rogers and Dennis Lamothe, the BALLS launch became an official Tripoli event.)

As the summer of 1991 approached, BALLS was shaping up to be an important launch. Not only was it attracting large rockets from around the country, but also, it was connected on the calendar to LDRS-10—stretching the flying at Black Rock to four full days. The BALLS event was also the only place where Tripoli members could fly rockets with uncertified motors that weekend at Black Rock. This is because the first day of LDRS-10 was supposedly the "line on the sand" date by which all motors used at Tripoli-sanctioned events had to be certified by Tripoli. Certification of rocket motors was a continuing problem for Tripoli in the late 1980s, and President Rogers made it clear in 1990 that uncertified high-power rocket motors would not be allowed on the field beginning at LDRS-10.

"We've all enjoyed the freedom during the last six years to fly whichever motors we pleased, but there has been a lot of shoddy product on the market," explained Rogers. "Thrust curves or even the most basic data on the motors have been hard to come by, even by major manufacturers . . . Starting a motor-testing and certification program has been crucial in convincing those involved in the NFPA codes and regulations process that the high-power rocketry industry is ready to police itself and that Tripoli is ready to play a leading role in making that happen . . . I would hope that the members would appreciate that without this hard deadline, we wouldn't have had half the motors submitted from the manufacturers that have already been tested."

Despite plenty of good intentions, the Tripoli board was beset by problems that ultimately led to a last-minute cancellation of the LDRS-10 motor certification deadline. One of the stated reasons for lifting the deadline was that Tripoli's motor test stand was recently damaged. As a result, the testing of some motors could not be completed prior to LDRS. Another reason was that some Tripoli members still possessed older uncertified motors that were never going to be submitted for commercial testing at all. The board decided to give flyers one more chance to ignite their old motors at LDRS. Another reason for the deadline extension was that some manufacturers said that they had not been given enough time to get their new motors certified. By a vote of seven to zero on August 15, 1991, the first day of LDRS-10, the board suspended the certification deadline. No new deadline was set.

Tripoli's decision to lift the motor deadline was criticized as reflecting a continuing weakness in the organization's ability to responsibly conduct its affairs. Privately, many flyers were more than pleased. They were allowed to launch rockets at LDRS-10 using uncertified motors after all. And since uncertified motors could be flown at LDRS, fewer flyers decided to stay over after Sunday to launch rockets on Monday at BALLS.

There were not many flights at Fireballs 001. And for those who did launch their own motors, AeroTech's Gary Rosenfield assisted AERO-PAC in evaluating the safety characteristics of the experimental motor designs.

"Gary would ask the motor makers about their motors to be sure they were okay," said Lewis. "It was just a loose type of approval. We did turn down a few of them. I think one flyer that was turned down did go ballistic."

One of the unique aspects of the first BALLS launch was the pad layout on the desert floor. The information packet mailed out to registered flyers contained a diagram that broke the launch into at least four distinct zones: The first zone was the spectator area, well behind the flight line. Next was the "main prep area," a section just behind the flight line where rockets were prepared for flight. The third zone was forward of the flight line and was called the "core launch area." This was where most of the rockets at the event would be launched from their pads. The core area was broken into three sections, with motors of M power and above slated for pads at least 500 feet from the flight line.

The final zone was designated by Bill Lewis as a "cell" area. The AERO-PAC leader created three separate cells well beyond the area for M-power rockets. In these cells would be rockets that were sufficiently large or complex so as to be placed farther away from other flyers and spectators. People who launched from the cells were expected to have everything they needed to launch their rocket, including pad equipment and electronics. They were also required to be in radio contact with the LCO along the flight line. One day, at this and many other Tripoli launches, these faraway pads would simply be referred to as "away cells."

"The first rocket on the pad was Rick Loehr's two-stage telemetry down-linked vehicle powered by two L motors in the booster and one L motor in the upper stage," recalled Lewis. "Rick's rocket brought everyone to their feet as the twin flames from his custom Space Dynamics L motors lit up the morning sky."

Another flyer at Fireballs was Bill Morrow, whose minimum-diameter rocket, powered by a U.S. Rockets L-class motor, reportedly tracked to more than 16,300 feet—the highest altitude of the launch and one of at least seven rockets to break the 10,000-foot barrier at BALLS. New Tripoli member Frank Kosdon teamed up with friend Bob Baker to launch two rockets propelled by L motors Kosdon created himself. Both rockets cleared 11,000 feet, and their *Starfinder* vehicle tracked to more than 12,000 feet.

Martha Sienkiewicz became the first woman to fly at a BALLS event and also the first woman to break the 2-mile-high mark with her *Mach Buster*, a rocket she had created with Dave Bolduc that launched on a Vulcan L750. Her rocket reached 13,743 feet by visual tracking. Jay Orr launched his *Titan* on a long-burning K250 to 10,826 feet, and Jim Cotriss sent skyward his *Experimental Ionosphere*, a 2.26-inch-diameter rocket, to 10,826 feet, also on a K250. Young flyer Mark Clark of Arizona launched his rocket called *Swift* on an AeroTech K550. Clark's rocket reached more than 11,000 feet. Within six years, Clark and friend Robin Meredith would lead the Arizona High Power Rocketry Association (AHPRA) into the annual sponsorship of the BALLS event. (Their joint oversight of the annual BALLS launch would last until 2015.)

"As it turned out, neither the hybrid nor the liquid-fuel rocket materialized for this event," said Lewis after BALLS. "Although no records were broken and many individuals expressed disappointment at not seeing the hybrid or liquid-fuel rockets fly, most people were content with the results."

Epilogue to Book I

The tenth anniversary LDRS at the Black Rock Desert in 1991 was the antithesis of the first event held in Medina in 1982. In only ten years, the modest, primarily model-rocketry launch created by Chris Pearson had evolved into a high-power event that ranked among the biggest regional shows of the year. By 1990, there were more people, more rockets, and more powerful motors at LDRS than ever before. And in the upcoming decade of the 1990s, LDRS would be reinvented again.

Tripoli's national launch would become unlike any amateur rocketry event in history. Attendance would increase dramatically. A single LDRS would see one thousand flights or even more. The launch would never return to Ohio. Yet it would spread to every other region in the country, from the Black Rock Desert in Northern Nevada to the wheat fields near Wichita, Kansas; from the Bonneville Salt Flats in Utah to the green sod farms in South Carolina. By 2000, LDRS would be the most exciting amateur rocketry event in the world. And as LDRS grew, so too would the fortunes of the Tripoli Rocketry Association.

The technology of high-power rocketry would also be transformed in the 1990s. Only a few rocketry-related manufacturers from the 1980s would survive to see the end of the 1990s. Nevertheless, the total number of commercial vendors in high power would increase dramatically. New rocketry-related companies would bring improvement to every component of the high-power vehicle, from the tip of the nose cone to the aft end of the rocket. Exotic composites such as Kevlar, G-10 fiberglass, and other materials would find their way into the hobby.

The means by which a rocket motor was built, the method by which a rocket was tracked in flight, and the manner in which deployment and recovery were initiated—all these things would change in the coming ten years. The Mercury device, the remote-control timer, and the theodolite would all but disappear from the hobby. In their place emerged a dazzling marketplace of onboard altimeters and other electronics capable of tracking rockets to altitudes thousands of feet high. Rockets would get stronger, and yet they would be lighter. Rockets would also get bigger—much, much bigger—and they would fly higher than ever before. High-power rocketry's reach for the edges of outer space would begin in the 1990s.

Perhaps the most dramatic change in high power's upcoming decade involved the very heart of the hobby itself: the commercial high-power rocket motor. New motor technologies were emerging in the early 1990s that would fundamentally alter the rocketry landscape forever. Commercial solid-fuel motors were going to get bigger and more powerful, yet they would become cheaper to buy. A hybrid motor, based on both a liquid and a solid fuel, would sweep through the hobby in the middle of the decade. Even the mythical basement bomber would emerge from the shadows to become a common figure at some of the most popular rocketry events in America. The experimental or "research" launch would be born.

The 1990s would be a memorable decade for hobby rocketry. And the changes began almost immediately, with the demise of the single-use high-power rocket motor.

Gary Rosenfield with one his rockets at the Lucerne Valley rocketry range in Southern California in the mid-1970s. Rosenfield's talent for the design and manufacture of ammonium perchlorate composite rocket motors led to the founding of high-power rocketry's most iconic brand, AeroTech, in his family garage in Sacramento in 1982. Less than a decade later, Rosenfield would lead the Reloadable Motor Revolution, which dramatically reduced the cost of high-power motors and fueled the continued expansion of the hobby well into the Twenty-First Century. Nearly forty years after the founding of AeroTech, Rosenfield continues to supply hobbyists worldwide with high-power motors of all sizes, including the long-burning M and N motors propelling rocketry to the upper limits of the Earth's atmosphere.

2

Chris Pearson recovers his Level 3 Certification rocket at the Black Rock Desert in 2009. In 1982, Pearson created the launch he called "Large and Dangerous Rocket Ships," or LDRS. Held on a small farm on the outskirts of Cleveland, LDRS grew into an annual showcase for high-power rocketry. Pearson's decision to transfer ownership of LDRS to the newly-formed Tripoli Rocketry Association in 1986 helped ensure the long-term survival of the young rocketry club. Today, LDRS is held in different parts of the country each year and attendance is well into the thousands. The modest event conceived by Pearson forty years ago has resulted in four national television programs and is now the largest annual rocketry show in the world.

Former U.S. Air Force technician Tom Blazanin of Pennsylvania was a founding member and key political figure in the creation of the modern-day Tripoli Rocketry Association in 1985. He was Tripoli's first corporate president in 1986 and publisher and editor of the association's journal, the *Tripolitan*, from 1986-1990. During his tenure as Tripoli president, Blazanin presided over the creation of high-power's first safety code in 1987. He was also the first Tripoli representative to obtain a seat with the NFPA's Committee on Pyrotechnics. He returned to the Tripoli Board of Directors in 2016.

3

Francis Graham of Pennsylvania was a co-founder along with Art Bower and Curt Hughes of the Pennsylvania high school science club called The Tripoli Federation, created in the mid-1960s to explore all things scientific, including rocketry. Twenty years later, Graham served as Secretary on the first corporate board of the modern-day Tripoli Rocketry Association, Inc. Even in the mid-1980s, Graham believed Tripoli members might one day use their rockets to reach for the edges of outer space.

Ron and Debbie Schultz of Ohio at a Southern California rocketry range in the mid-1980s. In 1983, the couple attended LDRS-2 at Medina, Ohio, where Ron discovered high-power rocketry. His experience there inspired the machinist to create a line of mid- and high-power rocketry kits. In 1984, he and Debbie returned to LDRS-3 marketing rockets under the name of a sewing company Debbie had created in the 1970s called Lots of Crafts, or LOC. Today, LOC is one of the premiere rocketry brands in high power.

Jerry Irvine of California was among the early pioneers of the high-power rocketry movement in the West. Irvine's *California Rocketry* newsletter was one of the earliest national publications calling for a new direction in the hobby rocket industry. His company, U.S. Rockets, was a household name in high power through the mid-1990s. He is shown here at the Lucerne Valley range in the mid-1980s.

Charles E. "Chuck" Rogers of California on the rocketry range at Lucerne in the 1970s. Rogers was elected third president of Tripoli in 1989. He was a professional aerospace engineer at the U.S. Air Force Flight Test Center and an early leader in high-power rocketry. During his tenure as president, he advocated for the first Black Rock LDRS and helped guide Tripoli into an amicable relationship with the National Association of Rocketry. Among his many other achievements, Rogers would serve almost 20 years as a Tripoli board member and would be an important part of the CXST Space Shot team in 2004.

PRESENTS

BLACK ROCK DESERT

TRIPOLI SANCTIONED LAUNCHES

AERONAUT I — MAY 19 - 20

BLACK ROCK II — JULY 21 - 22

16,000 AND 20,000 FT BLANKET FAA CLEARANCE
50,000 FT MSL FAA WINDOWS
20 MILE DIA SANITIZED FAA ZONE

SPONSORED BY

HIGH SIERRA ROCKETRY
AND
WEST COAST ROCKETRY

ADVANCE MOTOR RESERVATIONS ARE RECOMMENDED. ALTITUDE TRACKING AVAILABLE SPECIAL EVENTS BREWLOFT[BOTH LAUNCHES] H STREAMER DURATION [AERONAUT I] HOLE IN ONE GOLF BALL DRIVE LAUNCH [BLACK ROCK II] A STEVE BUCK SPECIAL

LAUNCH PACKAGE 5 DOLLARS PER LAUNCH 9 DOLLARS FOR BOTH LAUNCHES CONTACT LAUNCH DIRECTOR PHIL HAYTON AT 3690 PEACOCK CT #4 SANTA CLARA CA 95051 408 247 6723 OR PREFECT BILL LEWIS 408 267 1915

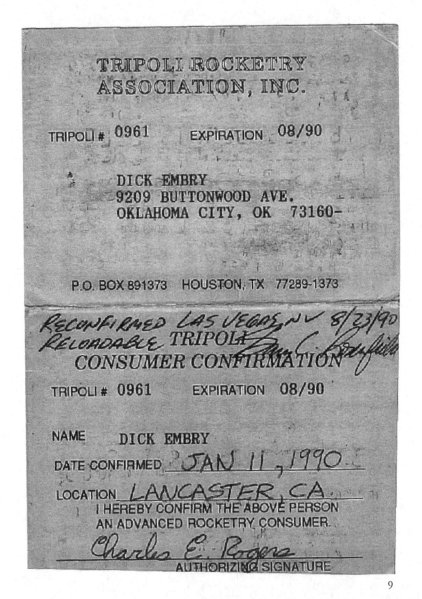

A Tripoli Confirmation Card from 1990. In the days before the modern certification process, the Confirmation Card was needed to purchase and launch high-power rockets at Tripoli events. This particular card belonged to flyer Dick Embry of Oklahoma. The Card is signed by then-Tripoli President Charles E. "Chuck" Rogers in January of 1990. It is re-signed in August of the same year by Gary Rosenfield, authorizing Embry to purchase and launch AeroTech's (ISP) new reloadable motors. In 2002, Embry would be elected President of the Tripoli Rocketry Association.

Steve Buck of Nevada was the owner of High Sierra Rocketry, a mid- and high-power rocketry supplier in the 1980s. In 1989, and years before Burning Man would arrive there, Buck discovered the Black Rock Desert for high-power rocketry. He helped organize the first high-power launch at Black Rock in 1989 and, in 1991, he was instrumental in the creation of what is now the premiere Research launch in the world: BALLS. With FAA clearances to the upper limits of the atmosphere, rockets flown at Black Rock have cleared more than 100,000 feet in altitude. In the near future, it is likely that high-power rocketry will reach the edges of outer space from Black Rock's remote location, scouted by Buck more than thirty years ago.

LAUNCH ANNOUNCEMENT!
BALLS
- 001 -
Black Rock Desert Dry Lake
Gerlach, Nevada
Summer 1991
1300 N-s Per Flight
MINIMUM

For dates and registration package Send $30 (Inc. T-Shirt & Patch) To:
BALLS
P. O. BOX 5127
RENO, NV 89513

12

Four members of the Tripoli Federation of the mid-1960s: (clockwise from lower right) Ken Good, Curt Hughes, Art Bower, and Francis Graham. Graham, Hughes and Bower co-founded Tripoli. Good joined a couple of years later, and he went on to become President of the Tripoli Rocketry Association, Inc., from 2004-2010. This photo was taken at the Black Rock Desert in the mid-2000's at a commemorative launch of the original *Gloria Mundi* rockets flown by the team in the late 1960s near Pittsburgh, Pennsylvania.

13

DEPARTMENT OF TRANSPORTATION
FEDERAL AVIATION ADMINISTRATION

CERTIFICATE OF WAIVER OR AUTHORIZATION

ISSUED TO
Association of Experimental Rocketry of the Pacific (AERO-PAC)

ADDRESS
469 Heatherbray Court
San Jose, California 95136

This certificate is issued for the operations specifically described hereinafter. No person shall conduct any operation pursuant to the authority of this certificate except in accordance with the standard and special provisions contained in this certificate, and such other requirements of the Federal Aviation Regulations not specifically waived by this certificate.

OPERATIONS AUTHORIZED

Launching of unmanned rockets from a point 17 nautical miles northeast of Gerlach, Nevada 40°51'30"N/119°08'00"W (LOL314051) operations not to exceed altitude specified in special provisions and a radius of 5 statute miles of the launch.

LIST OF WAIVED REGULATIONS BY SECTION AND TITLE
FAR 101.1 (ii): (a) and (c)
FAR 101.23 (b)

STANDARD PROVISIONS

1. A copy of the application made for this certificate shall be attached to and become a part hereof.
2. This certificate shall be presented for inspection upon the request of any authorized representative of the Administrator of the Federal Aviation Administration, or of any State or municipal official charged with the duty of enforcing local laws or regulations.
3. The holder of this certificate shall be responsible for the strict observance of the terms and provisions contained herein.
4. This certificate is nontransferable.

NOTE.—This certificate constitutes a waiver of those Federal rules or regulations specifically referred to above. It does not constitute a waiver of any State law or local ordinance.

SPECIAL PROVISIONS

Special Provisions Nos. _one_ to _ten_, inclusive, are set forth on the reverse side hereof.

This certificate is effective from August 19, 1991 sunrise to August 19, 1991 sunset, inclusive, and is subject to cancellation at any time upon notice by the Administrator or his authorized representative.

BY DIRECTION OF THE ADMINISTRATOR

Sabra W. Kaulia
Sabra W. Kaulia
(Signature)

Western-Pacific Region
(Region)

August 13, 1991
(Date)

Manager, System Management Branch
(Title)

FAA Form 7711-1 (7-74)

William "Bill" Lewis of California was one of the founding members of the Tripoli Prefecture known as AERO-PAC. He also obtained the first FAA waivers for high-power rocketry at the Black Rock Desert. Such waivers allowed flyers to aim for altitudes never before attained in hobby rocketry. This waiver was for the first-ever BALLS event held in the summer of 1991. The waiver was for one day only, from sunrise to sunset. The maximum allowed altitude secured by Lewis that day was to an incredible 100,000 feet above ground level. It would be another 25 years before high-power rocketry actually attained such heights at the Black Rock Desert. But this was the beginning.

15

"The appearance of the country was so forbidding that I was afraid to enter it," said pioneer and frontiersman John C. Fremont when he first laid eyes on the Black Rock Desert in the mid-Nineteenth Century. This barren desert in Northern Nevada was originally part of Lake Lahontan, a prehistoric body of water covering more than 8,000 square miles of Nevada, Oregon and California. Artifacts discovered in the region suggest Man first appeared at Black Rock 9,000 years ago. In 1983 and 1997, the world land speed records were set here. High-power rocketry discovered Black Rock in 1989 and has never left. From here, the hobby is now poised to reach for the edges of outer space, and beyond.

BOOK II
The Reloadable Revolution
(1990–2001)

15

The Reloadable Revolution

By 1991, AeroTech had been in business almost ten years. Gary Rosenfield took a big chance leaving his aerospace job in 1984 to work full-time in hobby rocketry. But his gamble paid off. AeroTech was now the largest commercial motor vendor in high power, selling motors in the E- through M-impulse range all over America. AeroTech also manufactured its own line of mid- and high-power rocket kits and was at home in spacious new headquarters just outside Las Vegas, Nevada. AeroTech had competitors in the composite motor industry, most notably Vulcan Systems, run by Scott Dixon of Colorado. Yet Rosenfield's start-up company enjoyed unparalleled name recognition in the hobby. The future looked good for AeroTech, even if the status quo was maintained. But Rosenfield had come up with a new idea. And that idea would fundamentally alter the landscape of high-power rocketry forever. He was going to develop a reusable high-power rocket motor.

Rosenfield's pursuit of reloadable motor technology began quietly in the mid-1980s when he was approached by a contractor who had asked him to develop a rocket-based parachute for ultralight aircraft that ran into trouble during flight. Rosenfield jumped at the chance to expand his business beyond hobby rocketry, and he was confident he could handle the propellant aspects of such a system. All he needed was some help with the electrical engineering portion of the project.

While working at Aerojet in the early 1980s, Rosenfield became friends with another Aerojet employee by the name of Dan Meyer. At the time, Meyer was not only a NAR board member but also a skilled electronics expert. Rosenfield turned to Meyer for help with the electronics portion of the ultralight rocket system he was building. Since the project was not directly related to hobby rocketry, Rosenfield decided to keep AeroTech separate and apart from the ultralight venture. He and Meyer created a new company. They called it Industrial Solid Propulsion or "ISP." ISP began testing its new ultralight product by loading a rocket motor, firing it, cleaning it out, and using it again with new fuel.

"We would make motors out of aluminum for these ultralight planes," said Rosenfield. "We used a little lockwire that fits in a groove that is in the nozzle and in the casing, and you kind of feed it in. It's an old-style fastener that has been used in the military for many years."

Soon, ISP was busy creating reusable rocket motors for the ultralight aircraft industry. Then not long after forming ISP, Rosenfield had an epiphany.

"I thought, 'Hey, I wonder if this could work as a hobby product,'" he said. "But we would be traversing two rocketry taboos. One was that we would be using a metal casing, and the other was that we would be making a *reloadable* motor."

The notion of a user-reloadable rocket motor, let alone a rocket motor with a metal casing, was anathema to hobby rocketry. Both sounded like the work of the basement bomber of the 1950s. The Code for Model Rocketry and nearly thirty years of hobby rocketry in America had settled the question of hobbyists assembling their own motors. It was not allowed by Tripoli or the NAR. And a hobby motor made primarily of metal was also forbidden. Yet Rosenfield believed he had an opportunity to create something that would not only satisfy the safety concerns of the hobby but also allow flyers the ability to move into areas so far unexplored in rocketry.

"I believed reloadable motors would be accepted for a number of reasons," Rosenfield said later, not the least of which was a massive reduction in the retail price of a rocket motor. "I thought we could bring them out at a lower price because the customer didn't have to throw away the casings. Also, if the labor costs of assembly [of single-use motors] were removed and if epoxy resin weren't used, the costs should be even lower, and we would be able to pass the savings on to the customer."

Once again and despite some risk, Rosenfield began work on a product that would revolutionize hobby rocketry.

The concept of a reusable rocket motor was not unprecedented. In fact, when he was younger, Rosenfield had occasion to use a similar product. It was called JETEX.

JETEX was the brand name of a miniature reloadable jet engine developed for model airplanes in the late 1940s. JETEX motors went on sale in 1948 and were marketed to people who wanted to give their flying aircraft "an effect of jet propulsion." A small JETEX motor affixed to a balsa airplane provided extra boost in flight. An early promotional brochure explained that JETEX engines were intended to produce jet-type thrust rather than rocket thrust. The starting thrust was low, building up slowly over time.

The JETEX motor had a metal case in which tiny fuel pellets were inserted for flight. The user filled the motor case with other components such as spacers, insulator disks, and a wick. During flight, the fuel pellets and small parts were used up. The metal case was completely reusable.

"They would produce 1 to 3 ounces of thrust for up to 15 seconds, and then you could reuse them," said Rosenfield. "The motors would get smoking hot. They were not like a modern reloadable. But it was the idea."

The JETEX line continued through the 1970s, but it was never marketed as a hobby rocketry product.

"The JETEX motors, even with their thrust augmenter tubes, could not even lift their own weight," observed one NAR member at the time. "They were better suited to being strapped under a hand-launched glider than for launching rockets."

Rosenfield's knowledge of the JETEX model airplane device, his experience with rocketry propellant, and his work with the ultralight parachute motors at ISP led him to believe he could develop a similar product that might work for hobby rocketry.

"I thought we could make a composite version of the JETEX, only it would be better suited to model rocketry. We could put a liner in the casing, and in addition to the fuel, we would supply a plastic nozzle that could be discarded after flight," he said.

Dan Meyer had computer design experience, and so he and Rosenfield worked on engineering the parts needed for a reloadable rocket motor. By 1988, they had a working prototype.

"It was a case of downloading civilian aerospace technology into the hobby market," said Meyer.

"We tested it as a glider motor first because that was already acceptable," said Rosenfield. "We started there because it was more politically correct. One day I was coming back from an NFPA meeting, and I mentioned to Bill Stine that we were thinking of making these reloadable motors for big gliders, and he was very supportive. It was just my first toe in the water, but I got a positive response."

Today, with a quarter century of hobby experience with reloadable motors, it is difficult to appreciate the problems associated with the creation of reusable technology from scratch. But Rosenfield faced several engineering hurdles before reloadables were ready for use by the public. One of the first issues was the composition of the reloadable case. What should it be made of? At the time, single-use motors were made of fiberglass, phenolic paper, or similar materials. However, for a reloadable motor, metal was the best choice; anything less durable would not last long enough to be reusable. So Rosenfield picked metal as the casing for his new rocket motor.

But what type of metal should be used? Steel was possible; so was iron. Both were durable and cheap. Both were also heavy and could fragment into shrapnel in the event of catastrophic motor failure. Overpressurized homemade motors made of steel pipe were the cause of serious injuries in hobby rocketry in the 1950s. Also, hardened steel's higher ballistic coefficient allows the metal to travel farther in the event of a case rupture. Rosenfield knew he had to find something safer. He chose aluminum, the same metal used by ISP for its ultralight products.

"When it fails, it does not create shrapnel like steel does. It just kind of opens up like a flower opens up," he observed. "Aluminum would just flare open, and the nozzle would come out."

Another hurdle in the design of a practical reloadable motor was the creation of a motor-based ejection charge. In the 1980s, motor-based ejection charges were the standard in both model and high power. Occasionally, a creative flyer would place a mercury switch or some other homemade device in a rocket to trigger an ejection charge in flight; others used remote-control devices borrowed from RC airplanes to fire ejection charges. But for the most part, flyers relied on the ejection charges built into their single-use motors to initiate deployment and recovery of their parachute gear.

In early 1989, Rosenfield began working with a lathe to modify a reloadable glider motor prototype into a configuration that would include a delay and ejection charge. The trick was keeping the burning propellant gases in the case from getting around a piece of pyrotechnic material that served as the delay grain. If the gases escaped around the delay grain and reached the black powder, the ejection charge would ignite too early. This would cause early deployment of recovery equipment while the rocket was under thrust; such an event usually led to destruction of the rocket. Rosenfield solved the problem of escaping gases by using O-rings that fit tightly in between the delay element in the motor and the case.

"The delay grain then butted up against the front of the motor case," he said.

This sealed the combustible ejection charge from the delay element and burning propellant below. Additional O-rings and a liner between the fuel grains and the case prevented the burning propellant from damaging or even melting the aluminum case. The first commercial reloadable motors built by Rosenfield were 29 mm in diameter and were static-tested—meaning they were ignited in the ground and not on board a flying rocket—near Las Vegas. The first flights of the new technology were in Southern California.

"This was all very secretive," said Rosenfield, who conducted his tests at Lucerne and a few other remote western launch sites. "I was at Lucerne in the spring [of 1990], and Steve Buck was there, and he was an AeroTech dealer, and I asked him, 'What would you think of a rocket motor with a metal casing that you could fire and then open up and put parts in and then reload the motor?'"

Buck's reaction and the reaction of others in whom Rosenfield confided his plans was enthusiastic.

Rosenfield brought several prototypes of his reloadable rocket motor to the historic Black Rock Two launch in the summer of 1990. He confirmed that flyers liked the motors, which were H-impulse–sized 29-180s and 29-240s. At Black Rock, he took his first commercial preorders for the new product, to be delivered at the upcoming LDRS-9 launch at Hartsel, Colorado.

"Dan Meyer and I literally sat for days and days working out the final designs for these motors," he said, and the pair arrived at LDRS-9 with the first commercial reloadable motors ever sold to the public. "I introduced it at a Tripoli launch because Tripoli had basically no rules about any of this stuff. I am not sure that they were even certifying motors yet. But the NFPA exempted Tripoli from everything, so we thought everything that we did through Tripoli was also exempted . . . It was like a show-and-tell. I had samples, and we did demos. At the [flyer's] meeting, I held up the hardware and told them how it came about. I showed them a reload kit—it was just a pack of parts—and I said, 'You load it, screw the ends on, fly it, and then take it apart and clean it up for use again.'"

Rosenfield also pitched the price difference between his reusable motors and traditional single-use rocket motors.

"A single-use J motor was about $120 at the time," he said. "And I told them that this would now cut that price in half. There was a lot of interest in that. Yes, the [reusable] hardware cost a lot. But after four or five flights, it paid for itself. Beyond the cost factor, I believed the high-power rocket people would enjoy assembling these motors. I felt the hobby could benefit from some increased participation by the rocket customer. Rather than just taking a premade motor and shoving it into a motor mount, now you would have more involvement in the usage."

Here, AeroTech's founder had tapped into a new source of personal challenge in hobby rocketry, one that had previously been forbidden by Tripoli and the NAR: people *assembling* their own rocket motors. This was an exciting concept to some flyers and, unbeknownst to Rosenfield at the time, would ultimately lead to flyers making the entire motor themselves. In many ways, the reloadable rocket motor developed by Rosenfield was a milestone marking the beginning of the experimental motor revolution in high-power rocketry, which was to follow only a few years later.

After LDRS-9, Rosenfield and Meyer took their new reloadable motor product to NARAM-32 in Dallas in 1990. NARAM was the biggest NAR event of the year. Not long afterward, NAR member John Pursley wrote an enthusiastic review on the new motor technology for *American Spacemodeling*. For Pursley, it was obvious the reloadable motor was the future of the hobby. Even though users handled the propellant themselves and the motor case was made of metal, "the prototypes were of a much higher precision and quality and offered greater safety benefits than almost any available nonreusable motor," wrote Pursley.

The NAR board of trustees, initially slow to come around to high power in the early 1980s, was also excited about ISP's reloadable motor.

"ISP's innovation may well be the mod roc shot heard 'round the world," wrote *American Spacemodeling* editor Larry Shenosky in late 1990. "The concept has heads turning and NAR officials scrambling to assess if and how the new technology might be incorporated into the hobby."

In October 1990, the NAR board created a Tiger Team headed by G. Harry Stine to evaluate safety codes and regulations for the new technology. Pres. Pat Miller advised NAR members that reloadable technology may be incorporated into NFPA 1122 as early as 1992. In other words, the NAR was not opposed to the new technology; they were embracing it. Others were taking notice too, wrote Shenosky, who claimed that at least one AeroTech rival was already telling flyers that it was going to market reload kits patterned after the technology in which ISP was now seeking patents.

"We're looking at it for high power first," said Rosenfield in a 1990 interview on his reloadable products. "We'll gauge reaction [and] then possibly work on regulations and specifications with reference to the smaller stuff." Rosenfield was soon working on reloadable *model* rocket motors too.

"ISP knows that no matter who eventually buys the product, making the technology successful means making it accessible," observed Shenosky. "That's why it's no mystery the company's [reloadable] propellant segments weigh 62.5 grams each. Under present rules, that means the reload kits can be shipped via UPS as a Class C propellant. The motors will come right to your front door."

While Gary Rosenfield was working out the details of ISP's reloadable rocket motor, another California high-power enthusiast was attempting to do the same thing. His name was Frank Kosdon. Franklin "Frank" D. Kosdon was born in Los Angeles on November 14, 1941. He lived there until he was six, and then his family moved to California's Central Valley, where he developed a love of science and the night sky.

"My dad and my uncle purchased some ranch property near Buttonwillow, California, which is west of Bakersfield," Kosdon told

Extreme Rocketry magazine in a 2000 interview. "I became interested in astronomy and constructed several telescopes. After looking at things through these telescopes, I decided I wanted to visit them. That is what first spurred my involvement in rocketry."

Kosdon was also deeply influenced by the Space Race between the United States and Russia and the launch of *Sputnik*. "I remember the newspaper headlines, which read, 'Russian Moon Circling Earth at 540 Miles Out,'" said Kosdon, who was already experimenting with his own rockets and making his own fuel when the satellite launched in 1957.

Kosdon was lucky he was never injured during his motor-making experiments. "The first rocket I designed contained a mixture of sodium chlorate, charcoal, and sulfur for propellant—an analog for black powder," recalled Kosdon. "I constructed a casing that had a washer brazed over the aft end for a nozzle. I laid a trail of powder coming from out of the nozzle washer and placed the rocket on a gallon [paint] can. Fortunately, it was at right angles from me. I was back about 5 or 10 feet when I lit it. When the burning powder trail entered the casing, a large amount of flame issued forth, which produced a whistling sound that became increasingly louder. Suddenly, the rocket just disappeared, and there was a big indentation [on] the top of the paint can. I never found the rocket, and fortunately, none of it found me."

After high school, Kosdon enrolled at the Massachusetts Institute of Technology (MIT), where he studied physics and aerospace engineering. At MIT, he joined the Rocket Research Society and conducted experiments with propellants such as potassium perchlorate and polyvinylchloride.

"The Rocket Society maintained a test cell at MIT, [and] during the course of my sophomore, junior, and senior years there, I conducted hundreds of firings of these propellants in that test cell. We developed test motors in which the propellant was castable into metal cylinders. These were perhaps the earliest reloadable motors," he said.

While at MIT, Kosdon and fellow student Ronald Winston were the recipients of a National Undergraduate Student Award presented by the American Rocket Society and the Chrysler Corporation. In addition to the prize money of $1,000, the pair was honored at a ceremony in New York City attended by then vice president Lyndon Johnson and astronaut John Glenn.

"I've only worn a tuxedo twice in my life so far," said Kosdon. "The first time was when I won this award."

Kosdon went on to Princeton University and earned his master's degree. His thesis was on the ignition temperature of various solid propellants. He then returned to California, where he earned his doctorate in fire research.

"When I got out of school, I took about a six-month trip around the world," he said. "When I came back, I was faced with unemployment. That was when the *Apollo* program and the Vietnam War were winding down. It was the first major crash of the aerospace industry. So I went back to what I had been doing during my summers when I was an undergraduate—selling *Collier's Encyclopedias* door to door. Actually, I found that a lot of fun."

In the mid-1980s, Kosdon discovered high-power rocketry.

"I wanted to get back into aerospace engineering, and I didn't want to work for somebody else," he explained. "So I decided I would try to be self-employed in a rocketry-related business. I didn't really know I was going to go on to make hobby rocket motors at that point. But at least that was the part of aerospace engineering that I found most exciting."

He found the Tripoli Rocketry Association in 1987. Soon thereafter, he began making single-use composite motors for himself and his rocketry friends.

"At first, I thought that these motors composed of phenolic or epoxy fiberglass tubing with the ends epoxied in were really neat," he said.

But over time, he also noticed that many of the parts in single-use motors were perfectly usable after flight.

"I would examine some spent motors, especially the ones with fiberglass tubing," said Kosdon. "I noted that if they had been constructed properly, just about all the parts in them were still good, sans the propellant."

So like Rosenfield, Kosdon began to consider the feasibility of a reusable rocket motor based on a metallic motor casing that could be flown, cleaned up, and reloaded again for additional flights. By virtue of his education and training, he already had extensive experience with metal-cased rocket motors. He also knew metal casings were taboo in commercial hobby rocketry.

"In the late 1950s, when I was first involved in the hobby, several people had been killed, which resulted in the prohibition of metal motor casings," remembered Kosdon.

However, since that time, he had learned that aluminum cases, unlike cases made of steel, were safe.

"With aluminum, you could see a tremendous amount of stretching, which resulted in ripping and tearing. But it all stayed in one piece. It was more ductile," he said.

Kosdon spent a couple of years perfecting his own version of the reloadable motor, and like Rosenfield, he had engineering and design hurdles to overcome on the way to a successful design that was both practical and safe.

"In 1989, I started making what I would later call a modern generation of reloadable motors," he said. "They consisted of aluminum motor casings with threads at both ends. Some of them were actually case-bonded motors."

In the spring of 1990, an acquaintance suggested to Kosdon that instead of threaded end caps to secure the contents of his motors, he should try using snap rings instead.

"I had a vague recollection of what snap rings were. I acquired some of them and designed a 1-inch [-diameter] motor and fired some of them right away," he said. "My first one failed because it contained bad propellant. The second one actually contained PBAN propellant. That worked very well."

Kosdon chose to go with snap rings over threaded closures. This was one of the primary distinctions between his motors and the reloadables being developed at AeroTech.

"Actually, I thought there might be a problem with [motors with] threads," he explained. "Bob [Baker] and I thought that the guys that bought the product could possibly cross-thread the motors, or they would get grit on them, which would result in damage to the threads."

However, this issue never materialized in practice, admitted Kosdon later. Threaded closures worked fine for AeroTech, but he decided to stay with cases that were grooved on either end for snap rings, which he found easier to use.

"The only drawback is that somebody needs a set of snap ring pliers," he said.

Another distinction between the reloadables being developed by Kosdon and those developed by Rosenfield was the material used for the rocket motor's exhaust nozzle. AeroTech/ISP made its nozzles from a phenolic-type material. Kosdon thought aluminum nozzles would be better for his motors. But the testing of his aluminum nozzles revealed new problems; he chose graphite instead.

"I had an idea that an aluminum nozzle might work because even though it had a low melting point, the conductivity would be real high, so I concluded that perhaps it would conduct the heat away rapidly enough so nothing bad would happen," explained Kosdon. "This worked well for about half of the burn, and then [the rocket] vectored off as the nozzle reached its melting point. After that, I used graphite, and it worked [really] well."

By the middle of 1990, Kosdon was testing his prototype motors at Lucerne. He made the motors, and friend Bob Baker built the rockets to fly them in. Kosdon knew Gary Rosenfield was also working on a reusable motor at the time. And they each used Lucerne to test their reloadable wares in secret at the Summerfest launch in 1990.

"I would load the motors up in Bob's trailer or motor home and then put them in a rocket, and no one else except for a couple of my buddies were allowed to see what was going on," said Kosdon. "At that launch, we took off to some mound on the other side of the lake bed, and Gary Rosenfield went off another way to test his motors."

It wasn't long before Kosdon had a viable reloadable motor ready for commercial use.

In the end, Rosenfield's AeroTech/ISP won the race to be the first commercial manufacturer to bring reloadable high-power rocket motors to the nationwide consumer market. But Frank Kosdon made it a close contest.

AeroTech/ISP published its first national advertisement for reloadable motors in the December 1990 issue of the *Tripolitan*. The professionally designed announcement introduced hobby rocketry to ISP's Reloadable Motor System, called "RMS" for short. The ad depicted a photograph of several shiny black motor cases with aft closures anodized in gold. There were promises of reloadable motors through M power, and there was a photograph of a typical reload kit. With the RMS system, said the ad, flyers could enjoy AeroTech's White Lightning, Blue Thunder, and Black Jack propellants at about half the price of current "throwaway motors." The advertisement took the back cover of the *Tripolitan* and was in color.

"We had a revolutionary product, and I wanted to put it in the best possible light," said Rosenfield. "We got hooked up with a graphic design firm, created a logo, and also a whole new catalog. It was all happening at the same time."

As a result of the advertising revenue generated from ISP's reloadable motors that month, the *Tripolitan*, the immediate forerunner of *High Power Rocketry*, had the first full color cover in its history. ISP's advertisements and color back covers continued over the next several issues.

Kosdon's reloadable motor advertisements were not far behind. His "Truly Recyclable Motors" appeared nationally several months later in a black-and-white full-page advertisement inside the April 1991 *Tripolitan*. All at once, high-power rocketry had two manufacturers from which to choose for reloadable motors.

The Tripoli board of directors addressed the new reloadable motor technology at LDRS-9 in Hartsel in August 1990. This was at the same event where Gary Rosenfield introduced his H- through K-impulse reloadable motors under the ISP label. The board had no opposition to reloadables per se. But they wanted safety rules for the new technology as soon as possible. The board decided that reloadable motors must be based on only ductile cases that would not fragment. In other words, no hardened aluminum, iron, or steel would be approved for motor cases as these materials could lead to shrapnel-type injuries in the event of catastrophic motor failures. Another issue was how Tripoli would incorporate reloadables into its almost-obsolete confirmation process. President Rogers proposed that only currently confirmed members be allowed to fly reloadables for the next six months, while the board performed a study to establish a more permanent policy. However, some board members thought Rogers's proposal would curtail the spread of the new technology, and so the board rejected it. Rogers then amended his motion, this time proposing that confirmed members not only be allowed to fly reloadables but also confirm other members for reloadable use. To confirm a new flyer for reloadables, the proposed rule would also require the signatures of two confirmed members. This allowed for some control over the new technology without limiting reloadables to only a small handful of flyers. That motion passed.

The board also agreed that for the next six months, reloadable motors would be studied on a probationary status by a committee composed of board members Rogers, Gary Fillible, Gary Price, Ken Vosecek, and Bill Wood. Finally, the board agreed it was time for Tripoli to restructure its confirmation process. They charged the new committee with the task of creating a multilevel confirmation procedure that would include not only traditional single-use motors but also reloadable motors. At the end of the probationary period, the Tripoli board agreed that reloadable motor technology was safe. The board decreed all commercial reloadable motors to be sold in high power be submitted to Tripoli's Motor-Testing Committee for testing and certification.

"This board has made decisions that will steer the course of advanced high-power rocketry for years to come," said board member Vosecek. "The decision to accept this new technology may have been the single most important move ever made in sport rocketry. This board is not timid!"

Like Tripoli, the NAR formally accepted reloadable motor technology almost immediately. Rosenfield was soon marketing reloadable model rocket motors, and full-page color ads for reloadable motors in the D- through G-impulse range began appearing regularly in the NAR's *American Spacemodeling* too.

The second revolution in hobby rocketry composite propellant had begun. Over the next ten years, single-use motors would virtually disappear from high-power rocketry, much like the compact disk replaced the cassette tape and LP record album. And once flyers were assembling their own motors, they began to think about manufacturing the rest of the internal motor components too.

No one knew it at the time, but high power had taken another big step toward the upper atmosphere. In a little more than a decade, flyers would be using reloadable motor technology to reach altitudes of 40,000 feet and more. The climb to 100,000 feet continued.

16

Showtime

Reloadable motors swept over high-power rocketry like an unstoppable wave in the early 1990s. And high-power consumers voted for the new technology with their wallets. The result was a steady decline in the use of single-use motors and a rapid rise in the number of people using reloadables. Prices drove the new technology. The cost difference between reusable and single-use motors was not imaginary. And it was most dramatic when one compared the upper ends of the impulse ladder.

In the 1980s, single-use J motors cost as much as $110 each. AeroTech and Vulcan single-use K motors cost at least $200 each, and a single-use L motor listed between $300 and $400. The cost of an M-class motor was in the stratosphere. Single-use M motors were between $795 and $1,000 each. A $1,000 M motor bought in 1990 is the equivalent of paying nearly $2,000 in today's dollars. And this was for a single flight. Very few hobbyists could afford the luxury of an L- or M-motor launch in the 1980s. And even the cost of J or K motors was prohibitively expensive for many flyers.

After the reloadable revolution, motor prices in every class declined, and more people than ever could afford higher-power motors. AeroTech/ISP led the way with other companies, such as Kosdon's Truly Recyclable Motors, in hot pursuit.

By the mid-1990s, a reusable J motor cost as little as $40 per flight. K motors were under $100, and L motors were less than $200 each. Even the cost of an M motor was within reach of the average flyer. By the end of the 1990s, AeroTech M motors would drop to as low as $250, a whopping 75 percent price reduction in the cost of the premier impulse motor class. As a result, the total number of M-motor flights in rocketry increased every year. The initial cost of reusable hardware was significant—up to $300 for a complete set of M motor hardware. But the value of this one-time investment was returned over the course of many subsequent flights. Some people even pooled their resources and purchased reloadable hardware to be used by several flyers. And by the early 2000s, the single-use high-power rocket motor was a curious oddity at the field, a reminder of the way things used to be.

Not everyone was excited by the prospect of the reloadable rocket motor. In April 1991, a representative of Estes Industries wrote NAR leader G. Harry Stine, urging him to refrain from revising the Code for Model Rocketry to include the new reloadable technology for model rockets. The representative raised several concerns in the letter, including the view (1) that there was a lack of any in-depth testing of reloadables, (2) that reloadables did not fit the definitions of the CPSC or the FAA of a model rocket motor, and (3) that reloadables involved metal casings—a longtime model rocketry taboo. The writer acknowledged the reloadables produced by AeroTech were "of good quality and probably are safe" but questioned the reliability of motors made by unnamed manufacturers with less stellar reputations.

The Estes writer did not call for a ban on reloadables. Instead, she urged Stine to use his influence to restrictre loadables to high power, at least for the time being. Allowing metal casings in high power would provide an excellent testing ground for the new technology. In the meantime, she said, reloadables should not be part of model rocketry. What's more, she added, the NFPA should specifically divide model and high power into two separate safety codes.

"It is our belief that this would eliminate confusion among legal regulatory officials that would result from the different requirements of the two activities," she wrote. "These differences include the ages of the consumers, the expertise required, the size of the model being flown, the power of the engines being used . . . and the heightened responsibilities of those involved in high-power rocketry."

Shortly thereafter, Gary Rosenfield wrote to the NFPA Committee on Pyrotechnics to propose a Tentative Interim Amendment to the Code for Model Rocketry, asking the NFPA to amend the code to specifically include reloadable motor technology.

Then on July 3, 1991, the Estes representative revised her position on reloadable technology and advised Stine that reloadables were appropriate for the Code for Model Rocketry. Part of the reason for this change may have been a decision by Estes to pursue their own reloadable model rocket motor technology. Another factor in this change in attitude may have been the testing of reloadables by the Aquarius Commission, a joint effort by Tripoli and the NAR to explore potential safety issues with the new reloadable motors. The commission was led by Harry Stine.

The Aquarius Commission conducted motor tests on June 6, 1991, on a dry lake bed near Las Vegas, Nevada. The purpose of these tests was to assess potential hazards, real or imagined, regarding reusable motors. The motors tested included AeroTech/ISP and Kosdon reloads. By their own admission, commission members intentionally misused reloadable motors all day long.

For example, several AeroTech/ISP reloads in the G and H range (both 29 mm and 38 mm) were assembled with either the wrong parts or missing some parts. The motors were mounted on a stand in the vertical position and were fired with the nozzle facing toward the sky. On several motors, the wrong nozzle was even installed. Predictably, some of the motors failed. The result of such failure was usually just a blown nozzle. There were no catastrophic failures.

On several other motors, the nozzle's forward O-ring was intentionally left out of the motor. Many of these motors still functioned normally. In another test, the commission mixed motor and casing parts from different manufacturers; the motor still functioned properly. In another example, Blue Thunder grains from an ISP reload were used in a LOC 38/240 casing. This motor sustained a CATO upon ignition. The metallic casing simply opened up like a flower. There was little fragmenting, and the pieces stayed close to the motor. There were other CATOs during the day. But they too resulted in no flying fragments, just case fissures.

The Aquarius Commission tests revealed there was nothing inherently dangerous about reloadable motors, even when they were incorrectly assembled. In a letter written by Stine to the Office of Compliance of the CPSC, dated November 8, 1991, Stine offered his support and that of the NAR to reloadable motor technology.

"[T]echnological progress has now made possible reloadable, reusable rocket motors with ductile metallic casings that are just as safe as expendable rocket motors that use paper and plastic casings," explained Stine. "The NAR has conducted tests of consumer abuse of this new technology and has more than a year's experience with consumer use of this new technology in the field . . . [M]etal-cased reloadable rocket motors currently meet all NFPA and NAR standards for performance reliability, shipping, and storage conditions."

With both Tripoli and the NAR supporting reusable motors, there seemed little doubt the NFPA would approve their use in all of hobby rocketry. Then the forward momentum of the new technology ran into a brick wall.

In early 1992, a videotape circulated in the rocketry community that was highly critical of reloadable rocket motors. The video was submitted to NAR officials, members of the NFPA Committee on Pyrotechnics, and the CPSC. It was developed at the testing grounds of the Colorado Springs Fire Department on December 17, 1991. The creator of the video was somewhat nebulous. But it had its supporters in the rocketry community. Foremost among them was motor maker Scott Dixon of Vulcan Systems, a longtime friend and supporter of high-power rocketry.

The video depicted a series of fire tests purportedly revealing certain inherent dangers associated with reloadable motors, especially in the context of commercial shipping. For example, one of the tests consisted of placing twenty-one reloadable kits into a cardboard shipping box, installing an igniter (presumably connected to one of the grains; this is not shown on the video), and then electrically lighting the igniter. The box went up in a brief and fiery mess that consumed the contents in seconds. Similar tests were depicted with larger boxes of propellant, with the same or even more spectacular results. These tests, said the on-film narrator of the video, demonstrated the danger of shipping reloadable packages containing fuel grains.

The film was more than just a critique of the means by which reloadable motors were shipped; it was an assault on the technology itself. The video was accompanied by a twenty-seven-page report entitled "Reloadable Model Rocket Motors: An Accident Waiting to Happen." The introduction to the written report further described the video's purpose as a demonstration of the dangers of reloadable high-power motors. Reloadables posed hazards to retailers, carriers, and the general public and was a "very real threat to the future of the hobby in these lawsuit-prone times," said the report.

The report recited some of the history of model rocketry and the development of the disposable hobby rocket motor in the 1950s. The manufacture and construction of the disposable rocket motor, noted the author, was an area best left to professionals. To allow consumers the opportunity to assemble motors on their own was to place hazardous materials in the hands of those without the pyrotechnic skills and training to handle them:

> Rather than progress, as it is often labeled, the author feels that reloadable motors are a retrograde development that takes rocket motor assembly and the handling of raw propellants out of the hands of the professionals, where they should be, and puts the onus for safe transport, storage, and assembly on the untrained but willing public. This is not right . . . Model rocketry is a safe pastime enjoyed by enthusiasts the world over. The reasons for its excellent safety record are clear and easily demonstrated. Should the endeavors of one or two manufacturers to sweep the market with a new and potentially dangerous product threaten the future of a well-established hobby through the legislation that could undoubtedly result from a single serious and probable accident? The answer is a resounding no!

The presentation was more than just a plea to find a safe and reliable way to ship reloadable motors, which was a legitimate concern. It was a call to action to eliminate reloadable technology altogether.

Perhaps in an effort to further legitimize its assertions, the report repeated a remark made by Stine years earlier regarding the unsurpassed safety of the factory-made model rocket motor. What the report did not say was that Stine's quotation was made at a time when there was no safe reloadable model rocket technology. The paper said nothing about Stine's then current and very public support of the new reloadable motor technology. The report also made no reference to the safety testing of reloadable motors by Stine with the NAR/TRA Aquarius Commission the previous summer.

Stine ultimately described the Colorado burn tests and the surprise video as "irrelevant." Reloadable motor technology was safe, he declared. And the NAR was likewise unequivocal in its support for the new technology. The NAR wanted the technology out in the open and regulated rather than seeing the new motors "being driven into an uncontrolled underground market where standards cannot be maintained." The NAR supported AeroTech's request that reloadables not be classified as a banned hazardous material. Unfortunately, not everyone shared this view—at least not yet anyway.

The NFPA met in Phoenix, Arizona, on April 6 to 8, 1992. One of the subjects for discussion by the Committee on Pyrotechnics was the new reloadable motor technology. The video and report were shared with committee members prior to their meeting on April 5 and 6. Scott Dixon was also on hand to answers questions regarding the video. Supporters of reloadable motors were also present at the meeting to share their views. The committee was shown videos of more recent burn tests conducted by the NAR that called into question the reliability of the Colorado video results and demonstrated the safety of reloadable motors in assembly and shipping.

The new NAR burn tests were conducted in late March in Phoenix and were again overseen by Stine. The purpose of the tests was to reconfirm the testing done for model rocketry motors that resulted in the DOT shipping classifications in 1978 and to also compare the behavior of model rocket motor reloading kits under the same conditions. Prior to testing, Stine wrote to commercial manufacturers such as Estes, Vulcan, and AeroTech, requesting products to be donated for testing. No manufacturers were to be directly involved in any of the testing, said Stine, although they were invited to observe.

Estes agreed to donate single-use motors for the cause. However, an Estes general manager who was reportedly opposed to reloadables warned Stine that he did not believe the tests were being validly designed. Among other things, the Estes representative said AeroTech was submitting motors that were atypical of shipped reloadables. He also suggested that the safety of disposable motors—or lack thereof—was of little doubt.

"Everyone knows the only way to ignite a disposable is via an open flame," he wrote, "while reloadable [sic] can ignite in many different ways, i.e., friction, impact, electrostatic discharge."

Vulcan Systems agreed to supply motors to Stine for his burn tests too. But Dixon warned the testing could be manipulated by motor manufacturers. He also echoed the concerns of the Estes general manager regarding the danger of open fuel grains being subject to, in exactly the same order that Estes had stated, "friction, impact, and electrostatic discharge."

Dixon also believed that AeroTech's reloadable motor design was nothing new at all. He said reloadable prototypes for hobby rocketry had been around since at least 1968 and that he had been making reloadables for the military for years. The reason this technology was not brought to the public before, said Dixon, was that it was simply too dangerous.

"[O]ne to 50 pounds of raw propellant stored under a kid's bed [or] in a closet, home workshop, or basement poses a significant hazard to the public," he wrote. "In short, reloadable motors are not new. All that's new (in our opinion) is an apparent willingness to 'gloss over' the issue of safety."

Stine rejected claims that reloadables were a danger to hobby rocketry. The NAR burn studies revealed "no increased fire hazard attributable to existing model rocket motor reloading kits in comparison to those of expendable model rocket motors." The NAR tests also confirmed that accidental ignition of a reloadable motor was not as easy as some people claimed.

"Follow-on accidental ignition tests resulted in failure of burning cigarettes to ignite the propellant grains of model rocket motor reloading kits," wrote Stine. "In aggregate, these tests showed no increased level of hazard in shipping, storage, or consumer use exists for current model rocket motor reloading kits in comparison to expendable model rocket motors that have been shipped, stored, and used for at least twenty-five years."

The NAR burn tests were provided to the NFPA at the April 1992 meeting, along with the video produced the previous December. For the Committee on Pyrotechnics as a whole, there was too much contradictory information to absorb at one time. According to one commentator, there was much confusion created by the conflicting videos. The non-rocketry members of the committee were simply not yet ready to vote for such a significant change to the Code for Model Rocketry. It would take more time.

The attempt to process Gary Rosenfield's Temporary Interim Amendment to NFPA 1122 to include reloadable motors in hobby rocketry was withdrawn. Things were moving too fast for the NFPA, and the proponents of reloadable motors would have to work harder to convince committee members that reloadable technology, which had been in use for nearly two years at model and high-power rocketry fields all over the country, was safe. The Committee on Pyrotechnics decided that it was more appropriate to include reloadable motors in discussions of the complete revision of the code in 1994. It was only a temporary setback. Reloadable motor technology would not be stopped.

The Colorado burn test video captured the attention of not only the CPSC but also other federal regulators who seemed to have a newfound interest in hobby rocketry. Both the CPSC and the Bureau of Alcohol, Tobacco, and Firearms (ATF) had nonvoting members on the Committee on Pyrotechnics for years. Not long after the reloadable video was released, one senior member of the NAR noted the ATF "has begun to exhibit a renewed interest in consumer rocketry regulations, especially in the field of high-power engines, propellant materials, and consumer storage of these items."

Both the CPSC and the ATF would soon take a more active role in the regulation of rocketry, both model rocketry and also high power. Eventually, the ATF's interest in the hobby would have long-term consequences for Tripoli, the NAR, and rocketry in America.

17

Black Rock, Argonia, and *Down Right Ignorant*

The success of the tenth anniversary LDRS in 1991 led to a return visit of the national launch to Northern Nevada the following summer. LDRS-11 was held at Black Rock on August 14 to 16, 1992. Nearly 150 flyers attended.

Unfortunately, range temperatures at LDRS exceeded 100 degrees every day. Some attendees claimed the heat rose above 110 degrees at times along the range. The midsummer launch introduced rocketry to another desert weather phenomenon too. A powerful sandstorm pummeled the rocketry encampment Friday evening. Gale winds lifted up particles of sand, dust, and debris from dozens of square miles of desert floor, transforming the air into a squall of tiny projectiles that permeated everything in their path. Flyer Alan Cooper witnessed the storm racing in from the west that afternoon.

"There it was, billowing over the mountains toward us," said Cooper. "Within 15 minutes, this enormous dust cloud had rolled onto the playa and proceeded to surround us. Then it hit. One minute, [it] was hot and calm. The next minute, it was hot and very windy and dusty. My tent bucked and swayed in the swirling dust but stayed up. After an hour of this wind, I realized sadly that it would be impossible for me to set up my Coleman camp stove and cook a hot dinner. I consoled myself to a dinner of salami, gorp, and cold tortillas."

Those who failed to stake down their tents, lean-tos, or makeshift shelters were left with crumpled poles, torn canopies, and missing gear; anything that was light enough to catch the wind and not tied down was taken away by the storm.

"Campers tell me they had their hands full just huddling in tents to keep things from being airborne," said one observer, adding, "Now there's an irony for you—rocketeers holding things down."

The weather Saturday was calmer. But it was still hot. Cluster-powered flights, camcorder payloads, and glider-carrying rockets were standard ticket items at LDRS in the early 1990s. One flyer obtained confirmation using a radio-controlled ejection charge that deployed a main parachute barely 100 feet from the ground. Flyers were not calling this dual deployment, at least not yet.

Utah flyer Randall Redd entertained everyone with scratch-built creations of all shapes and sizes. He reminded everyone that although LDRS had some serious rockets, the gathering was never intended to be only a serious event. LDRS was fun, and there was plenty to laugh at in Redd's rocketry world. Redd's collection included the *Dizzy Dog Rocket Ranch*, with *Penelope Pig* and *Pigs in Space!* He also launched *Old Fire Face*, a green dragon with an F14 motor hanging out of the mouth that gave the appearance of a fire-breathing dragon in flight.

Jim Schoenberg arrived from Hawaii with a half-scale *Patriot Missile* that flew well, and John Cato Jr. flew a one-third-scale *Nike Tomahawk* perched atop a *Nike* booster loaded with a K1100. Several flyers from Kansas brought a scratch-built full-scale *Patriot Missile* nearly 20 feet tall. The launch of the bigger-than-life *Patriot* was one of the most dramatic moments of LDRS-11. The rocket was powered by three AeroTech K1100s and three I284s.

"Everybody assisted the LCO in counting down to launch," said one flyer.

The rocket roared off the pad but did not go very high. There was a good ejection charge, and the rocket separated, but the booster broke free of the main parachute and tumbled back to the desert floor for a hard landing.

"As the rocket hit, the thud was felt in the stomachs of everybody," said another flyer.

There were several high-altitude flights on Sunday. Tom Binford returned with another version of his *Cloudbuster*, powered by a Vulcan O3000 motor. The rocket streaked to more than 24,000 feet, had deployment issues, and, like most of the ultrahigh flights of this era, was destroyed on impact on its chuteless return.

New England flyer Martha Sienkiewicz flew a *LaserLoc 3.1* rocket on a single-use long-burning K250 to more than 14,000 feet. Frank Kosdon and Bob Baker launched a rocket with a reloadable M3700 that tracked to 15,525 feet.

"The rocket accelerated so quickly that there was a large *crack!*" said one witness. "No one is really sure if that was a sonic boom or just the noise from the motor."

Deb Schultz of LOC launched a *LaserLoc* 2-inch-diameter rocket, also on an AeroTech K250, to a tracked altitude of 21,500 feet. That's more than 1,000 feet higher than the tallest mountain in North America.

On Sunday night, another storm blew through the playa. But in place of sand, this time, there was a fiery lightning show.

"We all stood out on the playa and watched the bolts touch down on the surrounding ranges," said Wayne Anthony of New York. "Mother Nature really put on a great show for us. It lasted for several hours before dissipating."

LDRS was still a three-day event in 1992, and the launch ended Sunday afternoon. There is no flight count that survives, but it is likely that three hundred rockets took to the air in three days.

On Monday, August 17, 1992, the second Fireballs launch was held. The launch was not yet a Tripoli-sanctioned event, and attendance was much lower than LDRS. It was sponsored by Bill Lewis and the crew from AERO-PAC and was intended for special projects or "experimental" flights that went beyond the limits of the Tripoli Safety Code. Among the highlights at Fireballs-2 was a drag race of seven LOC *Magnums*, each powered by K250 motors. All seven rockets flew well and deployed colorful parachutes for successfully recoveries, a rarity for the times.

Other flights at Fireballs included Chuck Mund's L1000-powered rocket, which was recovered 4 miles away in perfect condition, and a two-stage project from Johns Hopkins University that purportedly carried an M2000 to a K300.

Until 1992, the largest rocket ever flown in high-power rocketry's short history was Gary Fillable's *SK3J*, a scratch-built design by Fillable and named for the children of Vulcan Systems' Scott Dixon (Vulcan supplied the N3500 Smokey Sam motor for the flight). The rocket took more than three years to build and blasted off from a 900-acre Oregon farm in 1990. Fillable's rocket contained a number of unique features for its day, including a sounding rocket nose cone he molded out of fiberglass and explosive bolts that held the airframe sections together during flight. The vehicle was more than 22 feet long and 20 inches in diameter at the base. In addition to a 4-inch-diameter N3500 motor, the *SK3J* also carried three I283 motors.

To get the rocket into the air, Fillable manufactured a rotating launcher that he mounted in the bed of his pickup truck. On September 29, 1990, his 249-pound rocket thrilled a crowd of onlookers as it roared into the sky for a perfect flight to 1,800 feet, followed by a good recovery on the field. The flight made Fillable an instant star in the hobby and was probably the first mega-rocket flight in Tripoli history. The *SK3J* later appeared on the cover of *High Power Rocketry* magazine.

Now, barely two years later, Fillable's heavy-rocket record was about to be eclipsed by the star of Fireballs-2: a leviathan of a rocket built by Tripoli members Dennis Lamothe, Mike Ward, and Chuck Sackett. The trio called their record-breaking rocket *Down Right Ignorant*.

The reason for the name of the rocket was not hard to figure out, explained Lamothe's wife, Terry, another Tripoli member: "If you think about it, you've got three guys who are not in the aeronautics industry building a 35-foot-high rocket in one of their father's machine shops. It's kind of scary. I mean, what if they get pissed off?"

Down Right Ignorant weighed more than 800 pounds on the pad, which is just slightly less than the weight of a grand piano or an Arabian riding horse. The rocket was, appearance wise, a super-scale version of the popular LOC *Esoteric* rocket. It was 24 inches in diameter at its widest point and taller than a three-story building. The rocket was powered by a central O-motor, surrounded by thirteen more K and L high-power motors.

The project was built over the course of several hundred hours, most of the construction taking place in a machine shop owned by Chuck Sackett's family in Florida. The rocket's airframe was made of two 10-foot-long fiberglass tubes that were each 2 feet in diameter and served as the booster and payload/parachute sections of the vehicle. The upper end of the rocket, including the nose cone, would be deployed by the firing of an ejection charge in the form of a J460 motor installed upside down. The J motor would power a piston to separate the rocket at apogee and release the parachutes. Once the parachutes deployed, the main body would descend on three 24-foot-diameter chutes, with the nose cone and upper end descending on its own parachute.

"We originally attempted to set up Sunday night after LDRS," said Sackett, "but when an electrical storm rolled in, we decided it was best not to be playing with a 30-foot metal tower. We left the tower, trailer, rocket, and all out on the desert floor to make a fresh start the following morning."

On Monday, the trio was joined by several other Tripoli members to erect *Down Right Ignorant* on its launch pad. The crew also began assembly of the fourteen high-power motors that would propel the vehicle.

"I had loaded the O motor the night before, and it had to be fixed to the motor plate," said Sackett. "The five L motors were then epoxied into the thrust plate around the O motor. The original plan called for a reloadable O motor surrounded by four Energon L1100 motors. But when it was realized that the weight of the rocket was going to exceed 800 pounds, we decided that more thrust would be needed. One more L motor and eight ISP K1100s were added."

The *Down Right Ignorant* team planned for a noon launch on August 17. However, the flight was delayed by technical issues for several hours and was not ready for launch until just before 7:00 p.m. It was quiet, and the air was calm. The sun was low in the sky. The lone rocket standing in the midst of the Black Rock Desert could be seen for miles. On ignition, a huge fireball rushed out of the aft end of the rocket. Spectators thousands of feet away saw the bright yellow flame and watched the rocket moving up the launcher, but they heard nothing. They were too far away; the fury of the sound of all fourteen motors lighting at once had yet to reach their ears.

In a moment, they heard the roar of *Down Right Ignorant* as it climbed away from its pad straight and true, albeit slowly, trailing flame and smoke to 3,500 feet above the desert floor. The combined burn time of the motors was more than 10 seconds, and after burnout, the rocket coasted briefly. Then the bright flash of the J-motor ejection charge was seen, separating the rocket into two large pieces.

"We all jumped around, congratulating each other under the light of a crimson sunset," recalled one observer.

The main booster dropped quickly as one of its parachutes was damaged by the ejection charge. But the bulk of the rocket landed successfully under chute.

"Although it did not plummet out of the sky, it did land hard enough to crack the fiberglass joints where the [body] tubes were connected," said Sackett. "Meanwhile, the nose cone landed without a scratch."

Down Right Ignorant was a successful launch and the first mega-project in the history of the BALLS launch events. It was Tripoli's largest rocket ever flown—at least for the time being.

One of the issues discussed by the Tripoli board at LDRS-11 was where LDRS would be held in 1993. The launch had been in Nevada for two years in a row, and it had been several years since LDRS was east of the Rocky Mountains. Kansas Tripoli member John Baumfalk addressed the board on August 14 at the Gerlach Community Center, and he proposed LDRS-12 be held on August 13 to 15, 1993, at a little-known rocketry venue just south of Wichita, Kansas. No decision was made by the board that night.

The board took up the issue again the following evening. This time, Baumfalk was joined by Tripoli vice president Lamothe, who also lobbied for Kansas. The pair reported that the Kansas location offered an FAA waiver of 23,000 feet, with windows up to 50,000 feet. There were plenty of hotels available in Wichita, which was only an hour away from the launch field. The field itself was located just outside the town of Argonia, which, according to Lamothe and Baumfalk, would lay down the red carpet for the Tripoli Rocketry Association if it chose to hold its national launch in their community.

On a motion by President Rogers and seconded by board member Bruce Kelly, the Tripoli board of directors voted unanimously to take a chance on Kansas and an unknown rookie prefecture. They called themselves the Kloudbusters.

Today Argonia, Kansas, is synonymous with LDRS and high-power rocketry. By 2019, LDRS was held on this parcel of ground seven times, more than any other location in America. The Kloudbusters organization is recognized as one of the most experienced groups in high-power rocketry. They set the standard for a well-run national launch. But in the early 1990s, the Kloudbusters were unknown outside of Kansas.

The Kloudbusters were organized by John Baumfalk and Lyle Christ in the late 1980s. Both men wanted a regional club near their homes. To spread the word, they printed business cards for the Kloudbusters and distributed them at hobby shops in the Wichita area. Soon, they had plenty of paid members in their ranks. The Kloudbusters' first launch, put together for the benefit of a Boy Scout troop, was held on November 10, 1990, near Goessel, Kansas. The club became Tripoli's thirty-fourth prefecture the following spring, on May 5, 1991. They organized their first large-scale event, dubbed Kloudburst I, later that summer. The launch was a success, and plans were made to repeat the event in 1992. Kloudburst II was held at a regional airfield in Argonia, about 40 minutes southwest of Wichita. Local residents were welcoming, and they allowed flyers to house their rockets in airfield hangers the evening before the launch. The Argonia Historical Society stepped forward to supply food during the weekend.

Kloudburst II had the usual assortment of model and high-power rockets for its day, with one large exception: a scratch-built full-scale *Patriot Missile*. This was the same *Patriot* that would fly at Black Rock at LDRS-11. The *Patriot* was 16 inches in diameter and stood nearly 20 feet tall. On the pad, it weighed over 100 pounds, built primarily of unfiberglassed Sonotube purchased at a home improvement store.

"We chose the *Patriot Missile* design because this was about the time of the first Gulf War, Desert Storm, and the *Patriots* were in all the news," said Baumfalk, the group leader for the project.

The *Patriot* launch at Kloudburst II was scheduled for Saturday.

"Just as more help was called for to assist with the *Patriot*'s setup to the pad, a cold front rolled through, and the rain came," said flyer Allen Sawyze. "We hurriedly covered the bird with tarps and waited for Sunday. The remainder of Saturday night was spent telling old war stories of flights past and declaring war on the frigid weather."

Sunday's weather was not much better.

"Around noon, we got a break in the rain and cloud ceiling," said Baumfalk. "It took five of us to get the *Patriot* on the pad." The rocket was powered by a K1100 and several outboard I motors. "It went up about 50 feet on the outboard I motors, but the central K failed to ignite," he said. "The rocket went horizontal and immediately blew out the parachute."

Tripoli board members Lamothe and Ken Vosecek were on hand at Kloudburst II as observers. Both leaders were impressed by the organizational skills of the new prefecture and the support the club had from the community of Argonia. Lamothe called Argonia "rocketry heaven."

"There are friendly people, good country food, and there were few restrictions," observed Lamothe, who was very impressed with the high-altitude FAA waiver obtained by the Kloudbusters.

Not long after Kloudburst II, the club began to discuss the possibility of a mid-America LDRS.

"It was a dream of Lyle Christ and [me] to bring the national launch here after we started the Kloudbusters," recalled Baumfalk. "We started out our launches by using the actual airport at Argonia, which is about an hour south of Wichita. We used the airport through [Kloudburst] II, and then we moved to a field a few miles east of the airport. A local flyer of ours invited a landowner by the name of Rick Nafzinger to one of our launches, and he enjoyed the launch. Mr. Nafzinger later told us that he had a parcel of land in the area where we might fly, and he invited us out to take a look. One area of his land was pasture land surrounded by flat lands that were open. We held Kloudburst III there. And we picked that site for the location for LDRS-12."

"My neighbors were a little skeptical at first," Nafzinger said at the time. "They couldn't believe that people would come all the way from California just to shoot off rockets."

But the locals were won over when, by one account, they saw a 25-foot-tall rocket climb into the sky on a "giant plume of smoke."

"It was the most beautiful sight," said Nafzinger.

The proposed field for LDRS-12 was a huge pasture surrounded by miles and miles of well-maintained and equally flat farmland. The location of the Nafzinger farm offered flyers a standing waiver of 29,000 feet, with windows up to 50,000 feet, a generous waiver even by today's standards.

At the Tripoli board meeting at Black Rock at LDRS-11, the Kloudbusters made their bid for LDRS. Baumfalk, Christ, Tripoli member Art Markowitz, and board member Lamothe helped make the final push to secure the bid. The Kloudbusters promised emergency medical services for the event as well as water and electrical power on site. They even obtained the support of the local Civil Air Patrol, whose members would help flyers locate and retrieve rockets after flight.

"I just told them that I thought that Central Kansas was one of the best places there is, anywhere, to launch rockets," said Baumfalk. "I told them that we had good cooperation from the landowner and the surrounding landowners and we have good flat wheat fields for recovery."

So LDRS-12 went forward on a wheat field in Kansas on August 13 to 15, 1993. It would become the largest LDRS up to that time, with at least 260 registered flyers and hundreds of spectators. There would be more than five hundred flights in three days of flying, a far cry from the first LDRS in Medina.

Argonia in 1993 was much like it still is today: a small farming community of several hundred people in Sumner County, Kansas. Argonia purportedly elected the first female mayor in American history, Susanna M. Salter, in 1887. The community rests at an elevation of just over 1,200 feet. The town worked closely with the Kloudbusters for LDRS, and the locals were excited about the possibility of hosting the national event in their backyard.

Flyers arriving in Kansas were greeted by a public sign in downtown Argonia: "Welcome, National Rocket Shooters!" *Discovery Magazine* and the Associated Press were also purportedly on hand, and on the first day of LDRS-12, the local newspaper carried a bold front-page story with the following headline:

> Rocketeers Blast into Argonia!

The typical modern high-power rocket launch functions much like a rifle range. At the range head, a boundary line separates the launch field from the spectator area. Behind this line—which, at some events, is delineated by waist-high bright yellow contractor's tape stretched between metal posts—are the people: their cars, pop-up tents, tables, barbecues, and rocket preparation areas. Here also sit the spectators and the commercial vendors in their trailers and trucks hawking their wares.

Beyond the yellow tape and beginning 100 feet or so in the distance are rows of individual launch pads arranged in parallel lines. The bigger the motor, the farther out the rocket is placed on the pads. If a rocket holds an H-impulse motor, the row is 100 feet away. K-motor pads are 200 feet away. The M-powered pads are at least 300 feet away, just under the length of a football field. If the motors get bigger, the pads can be a quarter of a mile away or more. At most launch events, there are several additional model rocketry pads up close.

A high-power launch pad may take many physical forms. Some pads are tripod-like contraptions that secure a stainless-steel rod extending skyward from a circular blast shield; others take the form of a complex tower resembling a rectangular exoskeleton that surrounds the rocket and holds it vertical as it awaits launch. At some launches, a pad is nothing more than a metal pole stuck in the ground. Some clubs have mechanized launchers that sit on flatbed trailers that can handle rockets weighing many hundreds of pounds.

Whatever shape a launcher may take, the purpose of the pad is to control a rocket's flight long enough to allow the vehicle to become aerodynamically stable. A pad may be slightly angled but not more than 20 degrees from vertical. Every pad is assigned a number, and that number is posted on a piece of cardboard or metal near the launcher so it can be easily read from a distance.

The predominant sound at a rocket launch besides the roar of the motors is the voice of the LCO. The LCO sits at a table along the range head boundary line, usually in the middle of the line. He or she is like a ringleader in a rocketry circus. Through a PA system, he barks commands over the range, ordering people back and forth, ensuring no one ventures into any dangerous areas (i.e., the field when pads are active). Among other things, the LCO ensures that the master firing circuits for any pad are not armed until a rocket is upright on the pad and ready for launch. He also performs the final countdown and pushes the button to launch individual rockets. The LCO warns of undeployed incoming rockets or other dangers; he also declares when the range is open or closed. When the range is open or "safe," there is no launching of rockets. Flyers swarm the field, looking for their rockets or hauling new missiles to empty preassigned pads.

The pads are assigned by the RSO. The RSO decides whether a rocket will fly at any given field. He or she is usually an experienced flyer who sits at their own table and who gives a rocket a visual inspection to ensure it is safe for flight. The RSO may check the fins for proper alignment, turn the rocket over to inspect for proper motor retention, and generally evaluate the rocket for airworthiness. If a flyer brings a scratch-built rocket to the table, the RSO may ask for the position of the center of gravity and the center of pressure to ensure that the rocket is stable. When the RSO is satisfied that the rocket *looks* like it can fly safely—no one knows for sure until the launch button is pushed—he assigns a pad, and the flyer waits for the range to open.

At each pad, there is an electric cable with two alligator clips at the end. After the rocket is on the pad and vertical, the flyer runs one end of a small igniter wire into the aft end of the rocket motor through the nozzle and attaches the other ends of the igniter wire to the alligator clips. The cable runs back to the LCO's launch button.

After all flyers place their rocket on the pad and hook up their igniters, the LCO declares the range closed. The flying field is cleared of all people. It's launch time. The LCO arms the firing circuits and then proceeds down each row of pads launching every rocket, one at a time, after reading briefly from the owner's flight card. The cards are filled out with the name of the flyer and a brief description of the rocket, its motor size, height, weight, and whatever other detailed information may be required. The LCO then announces the countdown over a hushed crowd—usually a 5-second count or maybe a ten count for really big rockets—before pushing a button on a master controller in a box in front of him on a table. The electric signal from that controller sends a current from the box out over the ground in cables and into the rocket motor via an igniter at the pad.

If all goes as planned, the igniter creates a hot spark inside the hollow core of the APCP motor, and the solid rocket fuel in the case ignites. The combustion of the fuel melts the igniter wire, severing the connection to the main cable. The motor case is pressurized by the burning and expanding fuel, which is then expelled through the nozzle in a fury of fire and smoke and noise. The rocket lifts off.

Things do not always go as planned. Sometimes the igniter fails to light the motor. There is a vague puff of smoke out the nozzle, and nothing happens. The rocket sits there. It needs a new igniter. Other times, the igniter lights the motor just fine, but improper assembly of the motor leads to a fire at the pad, or the ejection charges fire early, and the rocket lifts off with the nose cone already uncorked, the entire assembly spinning around in the air like some airborne Fourth of July pinwheel.

This is part of the thrill of all high-power rocket launches. It is dangerous here, no matter how many rules are applied and even if all safety codes are followed and everyone does their job. A high-power launch is not a phony reality show or part of some scripted performance. There is no way to make this hobby completely safe. When rockets come apart at the launch pad or in the sky and large airframes and flaming fuel grains rain down from the heavens, these are real hazards. People need to pay attention; danger is literally in the air. That danger is part of the attraction of high-power rocketry. Flyers and spectators do not come to high-power events to be fooled into believing they are taking real risks. If they want fakery, they go to Disney World.

Of course, when it all works properly, a high-power rocket roars off the pad like a NASA spacecraft heading into space. And high-power rocketry spectators all over America (and flyers) love that action too.

In this sense, not much has changed in amateur rocketry since the 1930s. As writer David A. Clary wrote in his biography of Robert Goddard, spectators who witnessed Goddard's flights remembered them for the rest of their lives. In August 1938 in New Mexico, Goddard launched a liquid-fueled rocket to 6,565 feet. "Of all the observers [of that flight]," wrote Clarey, Marjorie Alder was the most descriptive:

> At about six thirty, Dr. Goddard threw the switch; there was a metallic click followed in a moment by an explosion and then a steady roar. The flame, first yellow, became white, and in about 3 seconds after the contact, Dr. Goddard released the anchoring weights, and the rocket began to rise. Its rise from the tower was slow, and after clearing the tower, it turned into the wind. Its great acceleration was easily apparent, and after a moment, its direction corrected to vertical, altered into the wind, and corrected at least once again and then remained almost vertical. The rocket continued to move with increasing speed and left a slight trail of bluish white smoke. Its size diminished so rapidly that by the time the fuel was exhausted, it was almost invisible. It continued to rise vertically [and] then turned horizontally, and the cap came off, releasing the parachute. The flight, with the parachute release, was the most thrilling sight I ever witnessed.

For LDRS-12, the Kloudbusters had an onsite electrical generator powering a public-address system heard up and down the flight line and over much of the range. There were three rows of color-coded pads, eight per row, for a total of twenty-four launchers. There were model and low-power pads, mid-power pads, and high-power pads. The plowed wheat field in Argonia was not as flat as the Northern Nevada desert, but it was easy to navigate and spacious enough for rockets of any size.

"For modelers used to the stark barrenness of the Black Rock Desert terrain of the previous two LDRSs, the flying field secured by the Kloudbusters prefecture was downright civilized," said rocketry videographer Earl Cagle Jr.

The Kloudbusters opened the first morning of LDRS with a parachute jumper leaping from a plane who landed right in the middle of the range. He carried a bright blue flag bearing the Kloudbusters logo. One of the larger projects at LDRS-12 was Jim Cornwell's upscale version of the Estes *Mosquito*. Cornwell's project was 12 inches in diameter and weighed more than 75 pounds. The yellow-and-black rocket stood 7 feet tall and roared off the pad with an AeroTech M2000. The *Mosquito* delighted the crowd with a spectacular flight and a nice recovery. But it was severely damaged after landing as it was dragged by gusty winds pushing the parachute through the field, turning the three-finned vehicle into a rocketry plow.

Another exciting flight was an American-flag–themed LOC *Top Gunn* rocket, built by Bruce Markielewski and powered by three Vulcan J250s and four H220 Rocketflite Silver Streak motors. Flyer John Cato's two-stage rocket carrying a K1100 in the booster and an I65 in the upper stage had a great flight, successfully lighting both stages and deploying parachutes high in the Kansas blue sky. There was also the *DC-3 from Hell*, a former radio-control airplane that had its wings removed and was converted into a rocket. The plane stood over 6 feet tall from nose to tail, and it ripped off the pad with an AeroTech K1100. The parachute deployed, yet it refused to unfurl, trailing behind and above the rocket as it plummeted to the ground.

"Open! Open! Open!" chanted the crowd as the rocket headed toward the field.

It looked as though the vehicle would be destroyed on impact. But at the last second, the tangled parachute snagged on a high wire line, allowing the recovery harness to stretch out just enough so the aft end of the *DC-3* gently touched the ground—before being yanked back into the air again, where it hung suspended from the line.

One of the biggest rockets of the weekend was John Baumfalk's full-scale *Patriot Missile*. This particular *Patriot* was a familiar sight in high-power circles, having flown at LDRS-11 at Black Rock and several local Kansas events. It had been damaged, repaired, damaged, and repaired again. At LDRS-12, the 16-inch-diameter rocket was loaded with multiple motors. Gary Rosenfield donated a new 98 mm AeroTech M1939 as the central motor for the flight. Rosenfield personally assembled the powerful reloadable motor, which was surrounded by three single-use J700s and four Silver Streak motors. The rocket did not fly until Sunday, but when it launched, the 200-pound missile had a spectacular liftoff. It rose slowly, trailing a bright yellow flame and lots of smoke. The airframe rotated as it rose, and then at a fairly low altitude, there was an unexpected deployment event. The rocket halves detached, and while the upper end was drifting back safely under chute, the heavy booster with its four big fins careened headfirst into the ground, where it was destroyed.

The largest rocket launched at Argonia in 1993 was Dennis Lamothe's red-and-white *Aerobee*, which stood 25 feet tall and weighed more than 400 pounds. Lamothe's rocket was 14 inches in diameter and carried five motors: a central O4659 made by Lamothe that was flanked by two K550s and two K1100s. The metal O-motor casing was the same one Lamothe and his fellow team members used in *Down Right Ignorant*. At the time, the only Tripoli rocket ever flown that was bigger than Lamothe's *Aerobee* was *Down Right Ignorant*. The *Aerobee* was thus easily the biggest rocket to ever fly at any LDRS thus far.

The *Aerobee* was prepared for flight on Saturday but was not ready until late Sunday afternoon. When it was finally finished, the rocket was carted to the away cell in a U-Haul trailer and assembled in sections on the launch pad. The final countdown came just minutes before the expiration of the FAA waiver for LDRS. As the countdown reached zero, the crowd held its collective breath. Nothing happened. There was no smoke, and there was no fire. The rocket just sat there. Members of the *Aerobee* team raced to the pad and corrected what turned out to be an electrical problem. The FAA waiver expired, and the launch was scrubbed.

Kloudbusters president John Baumfalk called the FAA in an attempt to extend the waiver but to no avail. There would be no extension. At the conclusion of the third full day of LDRS-12, the biggest rocket at LDRS sat forlornly at the pad.

On Monday, the Kloudbusters arranged for the FAA waiver to be back in place, just for the launch of the *Aerobee*. For the first time in its history, launching at LDRS officially went into a fourth day. This time, the countdown yielded immediate success. The four-finned *Aerobee* roared off the pad, kicking up big chunks of Kansas soil and leaving a huge trail of smoke and flame in its wake. The 400-pound rocket climbed high into the sky to the cheers of the crowd. Then it slowed, reached apogee, and gracefully turned over. There was no deployment event. The *Aerobee* had a sizeable ejection charge connected to a timer that was supposed to light the onboard black powder exactly 17 seconds into flight. Yet nothing happened. The big rocket plummeted to Earth, nose first, where it furiously impacted the ground, sending a mushroom-type cloud of dust, dirt, and rocketry debris high into the air.

"Unfortunately, it had a recovery problem," said Lamothe later. "But who hasn't had a problem with recovery?"

"Not very much of it is above ground," wrote one observer. "Everything went underground except for the fins, the bulkhead, and the tops of the motors."

The only item visible of the remains of the former two-story-tall rocket was a few inches of the aft end of the airframe and the same amount of the four fins. A broken motor case or two lay amidst the wreckage. The rest of the *Aerobee* was buried under the Nafzinger farm. With that flight, said videographer Earl Cagle, "Dennis became one of the biggest planters in Kansas."

Another exciting flight at LDRS-12 was the launch of Frank Kosdon's all-metal rocket, *Full Metal Jacket*. This 4-inch-diameter aluminum missile was little more than a rough airframe constructed around an O10,000 motor built by Kosdon. It was a crude-appearing rocket, with an ugly, blunt nose cone and three equally blunt fins, with no sweep to them at all. The airframe was designed to simply carry and test the motor; it was not engineered for maximum altitude. The fins were part of a welded fin can held in place by set screws on the motor case. The rocket was just over 10 feet in length and weighed 67 pounds on the pad. Most of that weight was the fuel in the 6-foot-long motor. The motor had a neck-snapping burn time of a little more than 2 seconds.

"It is the biggest reloadable motor I have made so far," explained Kosdon as he prepared the rocket for flight. Kosdon added a smoke charge he claimed would burn for 55 seconds into the flight, to aid in tracking the highflyer. "You'll have to be blind not to see it—at least eventually," he added with a smile.

"They made everyone get way back because it was going to be so loud, it would hurt our ears," recalled Tripoli member Bruce Lee. "It screamed off the pad, but it wasn't as loud as everyone expected. But you could hear it go through Mach, and then it disappeared, never to be seen again."

Full Metal Jacket quickly disappeared, although the blue smoke trail hung in the air a long time. Optical trackers at the field claimed the rocket reached a maximum altitude of 35,407 feet, which was declared a new Tripoli altitude record. (These were still the days when altitude records did not require *recovery* of the rocket.) However, there was no deployment, or as Kosdon described so many of his rocket flights, "there was a failure of the recovery system." It was no secret that the renegade California motor maker was primarily concerned with the performance of his motors; the rocket was an afterthought.

"I have to wonder if 'failure of recovery' is just a euphemism," said Kosdon friend Ray Dunakin years later. "When I first met Frank, he only cared about the 'up' part. He didn't bother with recovery."

Indeed, Kosdon's approach to rocketry was mirrored by other budding motor builders over the next decade. Their rockets or the rockets built by their friends were just a means to an end; the end being the successful performance of a homemade rocket motor.

"Success occurs at motor burnout," echoed another Kosdon friend, Mark Clark, who assisted Kosdon in the construction of several of his infamous O-impulse projects.

For such flyers, the question of what happened to a particular rocket after motor burnout was answered in two words: "Who cares?"

Full Metal Jacket was thought to have impacted a mile downrange. But it was never found, and there was no electronic record of the maximum altitude. (Over the next several years, Kosdon would rebuild two more versions of *Full Metal Jacket* for flights at Black Rock to almost 38,000 feet.) The rocket became part of the lore of Argonia, LDRS, and high-power rocketry. At the general assembly on Saturday night at a hotel in Wichita, everyone voted to return to Argonia in 1994 for LDRS-13.

The Kloudbusters hosted LDRS again in 1994, and in many ways, it was a repeat of the success they had enjoyed in Argonia the year before. The event was held on August 11 to 14, 1994. There were more than two hundred registered flyers from at least thirty states. This was a little less than the number who had attended in 1993, yet the total flight count increased to more than six hundred rockets launched. And the Kloudbusters continued with their efficient pad management, with twenty-four pads available, in addition to several away cells that were brought by flyers with very large projects.

"We're all children of the *Apollo* program," explained flyer and rocket builder Rick Wills at LDRS-13. Wills was a former cockpit designer for the air force. Now he was the owner of Midwest Rockets Inc., a commercial rocketry retailer in Dayton, Ohio. "Shooting rockets is egotistical. Only in rocketry can you fill every sense you have completely. It deafens you. [And] then there's the smell. You can taste the stuff in the air. You can feel the vibration. After experiencing that, it's hard not to do. Rocketry doesn't draw the meek."

"Back when I was a kid in the early 1950s, when the Space Race seemed to be the thing of the future, I began making my own motors from black powder," said Robert Robinson of Indiana, the owner of Robby's Rockets, selling all things related to ejection charges in high power. "I was lucky to get them up 50 feet. Mostly, it would just burn up my rockets. I would have loved to have been an astronaut, but as time went on, it became too late."

For LDRS-13, Robinson brought a half-scale model of the United States Army's *Honest John* missile. He had worked around the real thing when he was in the army and stationed in Germany.

"The *Honest John* has always been my first love," he explained. "The roar, the excitement—it's hard to describe the ground shaking [and] massive roar of this thing lifting off the launch rail."

Robinson's scale version of the *Honest John* was 18 inches in diameter and nearly 20 feet tall. Unfortunately, high winds at Argonia knocked the rocket off its launch pad. It would fly one day but not at LDRS-13.

The biggest rocket star of LDRS-13 was again built by Dennis Lamothe of Florida, who returned with a new version of the *Aerobee* that was destroyed in 1993. This time, the big rocket carried three N motors. Lamothe built his huge rockets at home in a garage-turned-rocketry-shop.

"I'd kind of like to stay away from talking about that though," he explained. "I live in a residential area, and I don't want to alarm anyone. How many people in the world can go into their garages, build a 25-foot-tall rocket, take it out somewhere, and fly it legally? Only in the United States of America."

Lamothe worked on the *Aerobee* all day Friday and Saturday. This was the biggest rocket to fly at any LDRS through 1994, and at 6:00 p.m. on Saturday, the rocket was finally raised into vertical position. The lifting of the rocket on its 30-foot-tall pad took almost twenty men. At 7:20 p.m., the rocket was still sitting at the pad, waiting for something to happen.

"I was supposed to be at a wedding at seven o'clock," said one farmer who helped assemble the rocket on the field. "But I'm not about to leave now."

No one was leaving the field. This was the biggest rocket at the biggest launch in America. Finally, it was launch time, and under clear skies, the end of the countdown yielded a spectacular takeoff as all three N motors in the *Aerobee* came up to pressure. But as the rocket rose, flames erupted out of one side of the lower airframe, just forward of the fins. One of the N motors had come apart inside the rocket. The burning fuel from the doomed motor was spilling out inside the rocket, melting everything it touched. What it did not melt burst into flames.

Within seconds, fire engulfed the entire aft end of the rocket as forward momentum continued to push the vehicle higher into the sky. When the *Aerobee*'s forward momentum ended, the rocket rolled over gently and was momentarily parallel to the ground. At this brief apogee, which might have been 500 feet, the parachute deployed, pulling the nose cone and some of the rocket's electronics out of harm's way. The rest of the vehicle returned to Earth in a flat, fiery spin. Flaming chunks of propellant dropped separately from the open airframe in free fall, starting a small brush fire. As a local reporter summed up the spectacle:

> Once the rocket struck the ground, what remained of the Aerobee's airframe burned up and was destroyed. The heat was so intense the metal motor casings were reduced to what one observer described as "pools of molten aluminum." The cause of the N motor CATO was unknown, at least initially.

Up at the spectator area, there is complete silence. After a long minute, a chubby teenager gets up the nerve to approach Dennis Lamothe.

In what appears to be heading toward a movie scene out of a bad baseball movie, he plaintively asks, "What happened, Dennis?"

Lamothe though will have none of it. "Well, how the f—— should I know?" he replies.

Later, video taken of Lamothe's assembly of the three N motors exposed the problem.

"The film revealed that while assembling the N2700 Dragon Breath motor, I failed to insert the final spacer ring prior to placing the O-ring in and sealing the motor with the end closure," said Lamothe.

As a result, the burning propellant inside the motor case escaped by a route other than the nozzle, leading to the disaster. For Lamothe, it was another hard lesson learned—the second year in a row—and afterward, he wrote an article for *High Power Rocketry* reminding flyers to pay attention to what they were doing on the field.

"After countless hours of labor in Florida and thousands of dollars spent, I now fully understand and demand for myself the need for having an accurate checklist," wrote Lamothe. "All complex procedures require this effort, as I will on future projects. I will also require that when doing extra-critical assembly procedures, I will work in a secluded, roped-off area to allow me the time and lack of distractions to complete these tasks."

18

The Visionary

The burden of creating a national publication dedicated to high-power rocketry has traditionally been a thankless one. Just ask Tom Blazanin.

When Blazanin took over the *Tripolitan* as volunteer editor in 1985, Tripoli's newsletter was a black-and-white newssheet with an audience of a few dozen people. He quickly transformed that newsletter into a high-power journal that was almost but not quite a magazine. And he did so using primitive layout and formatting tools, at least by today's standards.

"Creating text on a Commodore VIC 20 to cut and paste for copying at Kwik Kopy was not the way," recalled Blazanin, who soon switched to the then new IBM 286 computer, which gave the *Tripolitan* a more professional look. "The first issues I had done were typical low-tech products. A maximum number of one hundred copies per issue were printed and distributed. As the Tripoli membership grew, the number of issues published grew. The first issue with a heavier cover appeared in October 1985, and about five hundred copies were printed."

Blazanin published twenty-five issues of the *Tripolitan* between 1985 and 1989, and by the end of the decade, he was printing nearly a thousand copies per issue. He personally wrote many of the stories, solicited commercial advertisements, hounded writers for articles, and begged everyone for adherence to deadlines, which were rarely met. The journal was mailed to every Tripoli member and was also handed out free at regional high-power events around the country. In the days before the rise of the Internet, the *Tripolitan* was a lifeline connecting far-flung high-power users to Tripoli, to commercial rocketry vendors, and to one another.

"I tried to maintain a bimonthly publication schedule for the *Tripolitan*," recalled Blazanin. "However, being president of the national organization and family activities left the *Tripolitan* on the old 'sometimes' schedule. Every attempt was made to keep a regular schedule, but it seemed I always had the odds against me. I wanted so much to deliver an on-time magazine of value to the members."

Blazanin stepped down as president of Tripoli in late 1987. One of the reasons for his departure was his plan to devote more time to the chores required to publish the *Tripolitan*. He warned Tripoli of the burden of having one individual serve not only as Tripoli president but also as publisher of their journal. It was too much work for one person, he said. After stepping down as president, he continued to publish the *Tripolitan* on a haphazard basis. He also increased its length, exceeding ninety pages for some issues.

"I think the *Tripolitan* was very important because people were joining Tripoli to buy Class B motors, but they needed another benefit to look forward to," explained Blazanin. "They needed to be updated with what was going on. I also realized it was important for people to see their pictures in a magazine. It made the members feel important, and it gave them notoriety, and with that, they would renew their membership."

For most high-power businesses, the *Tripolitan* was the only way to reach a national rocketry audience in the mid-to late 1980s. Yet for his efforts in creating, editing, and publishing the *Tripolitan*, Blazanin was paid nothing. The membership and apparently Tripoli's leadership expected him to do the job well and do it for free. Praise for Blazanin's efforts on the *Tripolitan* every year was limited. But if there was a mistake or clerical error in any issue or an unpopular opinion was expressed, there were plenty of critics. In rocketry, no good deed goes unpunished.

After Blazanin resigned as president in 1987, he remained on the board of directors. He was also one of the Tripoli representatives with the NFPA's Committee on Pyrotechnics. Eventually, someone else would have to take over the *Tripolitan*. But there were no takers.

"I apologize for the inconvenience you have had in not having all the news that stimulates and inspires one to tread on in rocketry," Blazanin told readers with a late issue of the *Tripolitan* in October 1988. "[B]ut as you have been told, it's hard for just one person to put this thing together."

The next issue was also delayed by several months. Blazanin advised members he was doing all the work on the *Tripolitan* himself, and now he was really finished. He penned the first in a series of "farewell editorials" that would appear in print over the next two years:

> I am finished as editor of the *Tripolitan*. I have done twenty issues since I joined Tripoli right as it swung into a national organization... This last year has been miserable for me because I have no time to relax and do the things I want to do.
>
> As everyone knows, I started my own little rocket business. It turns over dollars, not the greatest, but my time is productive in a monetary way. After years of doing things for nothing, dollars begin to count. Many of you have called to tell me what a great job I was doing. I really appreciated it... But words alone are not enough. Help was more important. To those who helped, I cannot thank you enough. The sad thing is there were many more "helpers" who dropped the ball. This is worse than anything when someone is counting on things.

The *Tripolitan* was supposed to be published bimonthly. It did not come out every other month in 1988 or in 1989 either. Things were so bad that only two issues were created with a 1989 date. In late 1989, Pres. Chuck Rogers lent a hand to the process. But Tripoli really needed a new editor to take over the journal. And they needed that editor immediately. In the February 1990 issue of the *Tripolitan* and then again in the August issue that year, Blazanin wrote two more farewell editorials.

"I can't do this anymore," he warned the membership.

Still, no new editor stepped forward. To many observers, the *Tripolitan* was about to fold.

In the summer of 1990, the Tripoli board authorized the creation of a newsletter called the *Tripoli Report*. The primary reason for the new publication was the *Tripolitan* was falling behind schedule, and the board needed an alternate method to keep in touch with its members.

The first issue of the *Tripoli Report* was dated June 1990. It was printed on a single sheet of 11-by-17-inch paper, front and back, which was folded in half to create four pages of type. It was mailed to all Tripoli members. There were no ads, no pictures, no launch stories. Instead, the *Tripoli Report* was limited to internal Tripoli business and direct communications between Tripoli and the membership.

"With this publication, we will better be able to direct serious [members'] concerns to everyone, and no one will be out in the dark, and a better overall picture of our organization will be presented to *all* members," said the board.

Other reasons for creating the *Tripoli Report* included the perception that the *Tripolitan* had become too much of a political publication and/or that the new publication was created because the board did not want the general public privy to Tripoli's financial statements. This last reason was curious since the financial statements were only printed in the *Tripolitan* and not that often and only Tripoli members received the *Tripolitan*. Its circulation outside the membership was virtually zero except for copies given away at launches.

The creation of the *Tripoli Report* did not end the board's search for a new editor of the *Tripolitan*.

"It's farewell time again," wrote Blazanin in the August 1990 issue of the *Tripolitan*. "This can't go on forever, and it won't. There'll be no more *Tripolitans* from me anymore. I know I've said it before, but goodbye for really, really, really, 'why doesn't anybody ever take me serious' real. Farewell, editorial."

"Yes, it's official," echoed President Rogers. "Tom Blazanin has fired himself again as editor of the *Tripolitan*."

Despite the advent of the *Tripoli Report*, the *Tripolitan* remained Tripoli's most important connection to its members, to commercial vendors, and to the world at large. Yet the board made no effort to offer a salary or compensation or even a token stipend for the job of running the publication. And the board was apparently willing to let the journal fail unless a suitable volunteer was found. For nearly three years, no one among more than one thousand members besides Tom Blazanin was ready, willing and able to rescue the club's magazine. And if the *Tripolitan* failed, the young high-power rocketry organization could fail too.

Then someone stepped up to the plate. That someone was a relative newcomer to high power who had discovered the hobby at a Southern California rocket launch in 1987. His name was Bruce Kelly, and he was about to change the image of high-power rocketry forever.

Like Tom Blazanin before him, Bruce Kelly had no publishing experience when he took over the *Tripolitan*. However, what he lacked in experience, Kelly made up for in personal energy and, even more importantly, a vision for the future of high-power rocketry. That vision included the expansion of the hobby across America.

Kelly had been a Tripoli member for barely three years when he volunteered to run the *Tripolitan*. In that short time, he had already made a name for himself in the hobby. He attended launches from Southern California to Ohio, and he often took his young family with him.

Kelly was an avid builder who enjoyed clustered rocket projects. He was also interested in the politics and inner workings of the hobby. He traveled to NFPA meetings at his own expense so he could observe firsthand the procedures that governed hobby rocketry. In 1990, he obtained a seat on Tripoli's board of directors. Kelly knew the *Tripolitan* was in trouble for some time, and he volunteered to be editor long before he was finally tapped for the job. But he was quickly rejected because he had no experience. Fortunately for high power, he was persistent.

Kelly was given his chance to try and run the *Tripolitan* in the fall of 1990. Under his direction, the *Tripolitan* would be completely redesigned and would soon be running full-color covers. Kelly would also rename the *Tripolitan* so it could be marketed to retail outlets nationwide.

In a few years the *Tripolitan*, renamed by Kelly to *High Power Rocketry*, would be printed in production runs exceeding twenty-thousand copies per issue. By the mid-1990s, *High Power Rocketry* could be purchased at almost every hobby store in America. The magazine would not only become the undisputed national voice for high power but also bring thousands of new members to Tripoli's ranks over the next ten years.

In a few years and second only to perhaps AeroTech's Gary Rosenfield, Kelly would be one of the most well-known figures in all of high-power rocketry.

Bruce Kelly was born in 1953 in Memphis, Tennessee. His parents, John and Geraldine, met in the late 1940s. Geraldine was living in Memphis and traveled to California for vacation. John Kelly was in the United States Marine Corps stationed at Camp Pendleton, just north of San Diego. He met Geraldine at the beach near Camp Pendleton. They married in 1951.

John attended barber college after he left the service, and the couple settled in Memphis, where John remained a barber for many years while Geraldine worked for Western Union. They had three children. Bruce lived in Memphis until he was nineteen. He went to high school in nearby Whitehaven. He had few hobbies as a child. Then in his teens, he discovered fast cars and drag racing.

"We had a world-famous racing service in Memphis—Race Head Service," recalled Kelly. "They would build and manufacture racing heads and engines for famous racers. It was only a mile from my house."

One of Kelly's first cars was a 1964 Chevrolet Impala. He bought a motor from the racing company.

"It was a special 327 ordered by someone else who never picked it up," he recalled. He took the block and the heads and assembled the motor on his own. He was sixteen.

"Ever since I got my driver's license, I had a heavy foot," said Kelly. "The times I raced, I was never beaten in my class. Before I was called to serve a mission for my church, I was building a Chevy II, a lighter car that would have taken me up several rungs in the classes. It was an early Nova before they called it a Nova. It was going to have more of a pro-stock engine in it, a very ambitious project for a kid my age."

"I didn't have any experience with model rocketry as a kid," Kelly said in an interview in 2002. "However, like most people, I was fascinated with the space program. I wasn't one of the fortunate ones to see *Sputnik* when it passed overhead. But I did watch just about every rocket launch from then on."

Kelly recalled America's rocket disasters and failures in the early days of the Space Race.

"They didn't really understand stability yet," he said, "and a lot of people didn't see much success coming out of the program. Then suddenly, Russia had something up there. Here was science fiction being realized. But from the American point of view, it was strange watching some other country doing it, and it was a country that we considered as our enemy."

After high school, Kelly pursued religious studies. "In the Mormon Church, when you turned nineteen, you get called to serve and do your mission if you are willing to accept the call," he explained. "I served in Oregon, Idaho, Northern California, and Southern Washington from 1973 to 1975."

Kelly's service included speaking to thousands of strangers about Mormon teachings. Public speaking and knocking on doors forced the shy teenager to learn how to walk right up to and talk to complete strangers.

"I think that without that experience, I wouldn't be nearly what I am now as far as being outgoing and willing to step in and help out," said Kelly. "I probably never would have done that with Tripoli. It was a real turning point in my life. It changed my personality."

Once his mission was completed, Kelly went back to school.

"My dad always wanted me to get an education, and nobody in my family ever graduated from college," he said.

He enrolled at Brigham Young University in Utah, where he met his wife, Phyllis.

"We met at a church activity and married three months later," he said.

The ceremony was in November 1975 at the Salt Lake City Temple. Bruce and Phyllis would have five children. Kelly left college early to support his family. He went to work for a company called Service Master. Among other things, Service Master did carpet cleaning and fire restoration work. At twenty-three, he decided to start a business of his own.

"It evolved into eventually just doing carpet work, and I was pretty good at it, and somebody in the church found out about it, and by the late 1970s, I was doing a lot of commercial work for the church. This work allowed Phyllis to stay at home, so we had a very traditional family."

Kelly's commercial work with the Mormon Church started in Provo, Utah. Over the next ten years, he worked all over the United States and beyond, installing carpet in temples and chapels from Utah to Hawaii, California to Tahiti. He was on the road constantly.

"I could knock out a job in a couple of weeks," he said, "but I stayed very busy, working on the road and raising my children when I was back at home. There was no time for hobbies of any kind."

Bruce Kelly discovered high-power rocketry while he was working on a carpeting job in Central California.

"I had a friend and neighbor by the name of Gary Price, and from time to time, he would talk to me about his rocketry activities," said Kelly. He and Price worked together installing carpets for the church. "Gary came back from LDRS-5 with copies of the *Tripolitan*. It looked interesting, but I still could not really fathom it. I knew that there was a difference between little rockets and large rockets. But I just couldn't grasp it until I saw it in person."

In early 1987, the pair traveled to California to a job near the state college in Fresno. Prior to leaving Utah, Price told Kelly about a Southern California high-power launch they might be able to attend while working in Fresno.

"We talked about it before we left, and we also talked about the possibility of getting confirmed. Gary wanted to get confirmed for high-power motors. So he built a rocket in Orem for a Vulcan H130 motor, and we brought the rocket with us when we went to California for the job."

Over a long weekend while in Fresno, they drove just south of Barstow to Lucerne, California, about 5 hours away, where Kelly witnessed his first high-power rocket launch.

"This was an exciting time for me as I watched for the first time Class B–propelled rockets hurl skyward," he said. "It's like when you see it, you know it, and I *knew* it. I said to myself, 'This looks interesting!'" Kelly saw several large rockets fly that weekend, including a LOC *Esoteric* owned by Steve Buck.

"It flew on a K1500 motor airstarting two I140s. The rocket also carried a camera payload. I was impressed with this rocket, its size, and all the nuances around it. It brought back all kinds of memories from when I was a kid, watching the big rockets on TV. When I saw Steve Buck launch his rocket, I was hooked."

Kelly and Price joined Tripoli that weekend. The two of them got confirmed using the same rocket Price had built. After the rocket was launched and recovered, both men obtained confirmation cards.

"We gave our membership money to Deb Schultz, and I then bought two rocket kits right on the spot," said Kelly, who soon discovered that he loved minimum-diameter rockets and clusters.

One of the first kits he had built was the LOC *EZI-65*, designed by Ron Schultz to fly on an AeroTech I-impulse motor.

"Most people who would fly those on J motors would find they would collapse around the time they hit Max Q," said Kelly, referring to the point in a rocket's flight path where the airframe undergoes maximum mechanical stress. "I figured that I could slide bulkheads down the tubes to strengthen the airframe and the shoulder of the payload section."

Kelly's success with modifying the *EZI-65* led to a Tripoli altitude record of nearly 9,000 feet on a long-burning J125 motor. He eventually captured other Tripoli altitude records too. He and his wife and five children attended Summerfest at Lucerne that year; every member of the family joined Tripoli.

Over the next few years, they traveled together to rocketry events all over America, turning launches into family vacations, staying at places like the Green Tree Motel near Lucerne, a local rocketry hot spot where the children would spend their nights cooling off in the swimming pool after flying rockets all day in the hot desert sun. The Kelly children were not just bystanders either. They built and launched their own rockets. At LDRS-9 in Colorado in 1990, Lisa Kelly, Tripoli member 1130, became the first person to obtain Tripoli confirmation with one of Gary Rosenfield's brand new reloadable motors.

From the very beginning of his involvement in high power, Kelly's enthusiasm for rocketry went far beyond that of the average Tripoli member. He enjoyed the camaraderie among flyers almost as much as the actual launching of rockets.

"The highlight of these launches became the evening social gathering at the motel," Kelly would later write. "All of us would migrate from room to room and look at the craftsmanship of the rockets. It was interesting to sit and chat with those prepping for the following day. It was also a more relaxed atmosphere to visit vendors out of the sun and away from the business of the launch site. There was always someone with video playing scenes of the days' flights or of previous launches held months before. And some of the manufacturers and dealers even served refreshments in their rooms. I realize now that it was the social times that endeared me to not only the hobby but the people as well."

Kelly was curious as to how Tripoli functioned as an organization. He met Blazanin in 1988, and he learned how Tripoli worked by watching Blazanin in action.

"I met Bruce through Tripoli," recalled Blazanin. "It might have been at Hartsel, or it was at Lucerne. But I had known him over the phone because he was always calling me and asking me rocket questions!"

"At my first LDRS, I attended all the board meetings," said Kelly. "At that time, I found out about the NFPA and about that self-regulating body and its influence on the hobby."

Kelly then traveled to several NFPA meetings just as an observer. Through such personal efforts, he met not only pyrotechnic experts from around the country but also the movers and shakers of amateur and model rocketry, including Harry Stine, Vern Estes, and then NAR president J. Pat Miller.

"Most of the people who had a part in pioneering the hobby, I met because I had an interest in attending the NFPA meetings," said Kelly.

Kelly knew the *Tripolitan* was in trouble in 1990. At one point, a Tiger Team of six Tripoli members, including Blazanin, was created to put out the magazine as a joint effort. Each member of the team would serve as the editor of at least one issue. Four of the Tiger Team editors produced nothing. One editor produced two issues that were almost finished but were pulled at the last minute.

"When asked to produce what he had completed to date, it was found that he had turned Tripoli's magazine into a publication featuring himself on both covers and his products throughout both issues," recalled Kelly.

The Tiger Team was another abysmal failure at keeping the *Tripolitan* alive.

Kelly was elected to the Tripoli board in the summer of 1990. Among other things, he pledged to improve the *Tripolitan*'s sporadic publication schedule. At Black Rock that summer, he and Blazanin approached Pres. Chuck Rogers regarding the possibility of Kelly becoming new editor of the *Tripolitan*. Kelly had already been turned down for the position once before. Yet by now, Rogers was truly desperate—Blazanin had resigned again—and as Kelly later recalled, Rogers was ready to take a chance on just about anyone.

"In total frustration because of magazine woes, no doubt the leader's heaviest burden, Chuck threw up his hands," recalled Kelly.

"All right, go ahead!" exclaimed Rogers. "You're the new editor. No one else is going to do it, so you might as well give it a try. Good luck!"

Kelly assisted Blazanin with the very next issue of the *Tripolitan*. The collaboration was really a test to see if Kelly could actually run the magazine.

"I had purchased my first computer and got it set up," said Kelly. "I put some of the publishing programs on there. Tom actually did his last issue [August 1990] at my house on my computer. I think he did one of the articles, and he sort of looked over my shoulder while I did all the rest of it. This was actually my first experience laying it up."

Kelly did a fine job on this first issue. He was named editor and chair of Tripoli's Publications Committee in the fall of 1990. He introduced himself to the entire Tripoli membership in the October 1990 issue of the *Tripoli Report*.

"As the new editor, I hope you will be patient with my shortcomings," said Kelly. "We have had problems getting the *Tripolitan* out on time, and doing this issue of the *Tripoli Report*, I can understand a lot better the scope and magnitude involved in doing this stuff. I am, as Tom was and still is, a very busy person. My work takes me all over the country, and my time to devote to this is limited, but I know it needs to be done, and I am trying. I cannot guarantee how many issues I will do, but I promise you I will do everything in my power to maintain a schedule as close as possible to what is required by the organization—four *Tripolitans* and eight *Tripoli Reports* per year."

Kelly took sole control of the *Tripolitan* with the very next issue, December 1990.

"My first goal was to improve the magazine a little at a time," he said.

He quickly got the chance to do so. AeroTech motor maker Gary Rosenfield wanted to debut his new reloadable motors in an advertising campaign, and he had hired an advertising agency to create something entirely new for ISP. He was desperate to reach a national rocketry audience.

"This was a professionally made ad, and I was very proud of it," said Rosenfield. "We had such a revolutionary product, and I wanted to put it in the best possible light. We got hooked up with a graphic design firm, created a logo and a whole new catalog. It was all happening at the same time."

Rosenfield approached Kelly regarding a full-color back-page ad in the *Tripolitan* to debut ISP's new line of reloadable motors. Kelly jumped at Rosenfield's offer. He knew with this new stream of advertising revenue from AeroTech and ISP, the front and back cover of the magazine, which were printed on the same sheet of stock paper, could both be in full color. Thanks to Rosenfield, Kelly offered Tripoli members their first ever full-color cover for the *Tripolitan*. That cover appeared in the December 1990 issue of the *Tripolitan*, and it featured a group photo of the participants of the breakthrough Black Rock Two event organized by Steve Buck the previous summer.

ISP and AeroTech continued to buy full-color back-page ads in the magazine for the rest of the 1990s, interrupted only a few times by other vendors eager to do the same thing. From this issue forward, Kelly's rocketry publications would always have a color cover. But Kelly wanted more than just the insertion of a color cover to the publication.

"I thought that we could start making the magazine look more professional," he said. "If you look at my first dozen issues, you can identify from issue to issue the small changes. You can see something different in every issue."

Kelly found new design ideas by perusing other magazines at bookstores, where he spent hours studying different layouts and editing techniques.

"When I first started the magazine, I also came across a publication about publishing that had an interesting article by someone who did the layout for *Time Magazine*. He talked about the importance of word spacing, justification of paragraphs, gutter spacing, and cover layouts."

Kelly also learned what drew people to certain publications and the importance of continuity between issues of a magazine.

"One of the things that I settled on after a while was the logo for the magazine and the color of the logo," said Kelly. "I didn't want to change the logo much. I made style changes on the inside, but I tried to keep the outside the same. *Time Magazine* had a red border around the cover on every issue, and *National Geographic* has the standard yellow cover on every issue. These magazines are easy to recognize, even from a distance. I wanted to have that with a rocketry magazine too."

Within seven months of taking over as editor, Kelly published four issues of the *Tripolitan*, bringing it current for the first time in years. In addition to color covers, Kelly switched to a glossier, more attractive paper for the inside pages of the magazine. He chose a type style that was easier to read, and he began selling full-color ads in the interior of the magazine too. Two more advertisers joined with AeroTech in June 1991 to purchase color ads inside the magazine. Later that summer, he began running color photos in feature articles and launch stories too. Soon, the magazine bore little resemblance to any of its predecessors. Still, Kelly had bigger things in mind for the *Tripolitan*.

"I wanted to get word out not only that we had this hobby but that there was an organization that supported the hobby," said Kelly.

His plan was to put the magazine on store shelves across America, from supermarkets to hobby shops to bookstores. He wanted something that no high-power publication had ever achieved: national retail distribution. He wanted to place the hobby of high-power rocketry directly into the American stream of commerce. Yet how would he do it?

"I have been communicating with national distributors who are guiding this effort," he reported to the membership in the spring of 1991. "Soon, you will see a bar code on the cover. This is a federal requirement as well as distributor required for the retailer's benefit. The *Tripolitan* will need to become as fit as any other magazine on the newsstand. It will have to be printed on a laser printer [and] have better graphics, interesting articles, and so forth."

Despite his progress, Kelly soon discovered a roadblock to widespread distribution of the magazine. The problem was the *Tripolitan* name itself.

"I sent copies of the *Tripolitan* to every major distributor I could find," he said, but all he got back in return were polite rejections. "The most promising contact asked me a question. 'We know that *Car and Driver* is about cars and their drivers. We know that *Sports Illustrated* is about sports. But what is a *Tripolitan*?'" *Good question*, thought Kelly.

The *Tripolitan* was a fine name for a local club newsletter in Pittsburgh. It also worked as the title of a semiprofessional journal for a small high-power organization. But for nationwide distribution of a high-power magazine, the name had to go. It was too esoteric and parochial to grab the attention of anyone looking over a newsstand. It communicated nothing about the contents of the magazine. Kelly suddenly understood the importance of having a title people could understand, a title that would capture their attention in seconds.

Kelly could not change the name of the *Tripolitan* at will. It was not his magazine—not yet, anyway. He was just another editor, a link in a chain that reached back more than twenty years to Francis Graham and others in the Tripoli Federation of the late 1960s. So he approached the Tripoli board of directors and later the members and explained the problem he was facing.

"The word *Tripolitan* means nothing to the casual observer scanning the shelves for magazines. As you know, magazines that are stacked on racks are obscured by the magazines in front of each other. The feature photos that 'sell' the magazine are not seen until one picks a copy off the shelf. Distributors and consultants asked us to make a complete title change to better market our product."

Kelly asked the board to grant him permission to rename the *Tripolitan*. He then suggested a new name: *High Power Rocketry*

The Tripoli board of directors agreed. However, they were yet not ready to abandon the *Tripolitan* logo altogether. Kelly therefore created a subheading for the new title in smaller print, combining both titles in a single heading:

> The Tripolitan ... America's *High Power Rocketry* Magazine

This first issue of *High Power Rocketry* was dated August/September 1991, less than a year after Kelly had taken over as editor. The *Tripolitan* subtitle remained on the cover for several more issues until November/December 1992, when the magazine dropped all subheadings and became simply *High Power Rocketry*. Prior to these name changes, the *Tripolitan* had reportedly been carried by only twelve hobby stores nationwide. With the new format and easy-to-understand title, Kelly found a distributor to take the magazine nationwide into more than 4,500 retail outlets.

When Bruce Kelly took over the *Tripolitan* in late 1990, the print rate for the journal was approximately one thousand copies per issue, distributed primarily to Tripoli members who received the journal as part of their paid membership. By mid-1992, the circulation level of the magazine had more than *tripled*, reaching 3,400 copies per issue. For many, the improvements brought by Kelly to the association's magazine played a central role in a 33 percent increase in Tripoli's membership in just a year and a half. The Tripoli board was so impressed with Kelly's performance that on August 16, 1992, at LDRS-11, they voted to turn the entire magazine over to Kelly as the owner/publisher.

Yet this was only the beginning. By mid-1994, *High Power Rocketry* reached a production rate of twenty-three thousand copies per issue. It was in hobby shops, bookstores, and grocery markets. You could find the magazine at Barnes &Noble, HobbyTown USA, and everywhere in between. People around the country were suddenly discovering *High Power Rocketry*. For many flyers in the 1990s, the *High Power Rockery* magazine was their gateway to the hobby. Their anecdotal stories are legion.

Child psychologist Richard King took his nephew to a hobby store in Fresno, California, where a copy of *High Power Rocketry* caught his eye. As he turned the pages, King was amazed at what he saw. As a result, he joined Tripoli. In a few years, he would set the Tripoli N-altitude record at more than 32,000 feet.

David Leninger of Minnesota was at his favorite brew stop in Minneapolis in 1996 when he wandered into the hobby shop next door.

"To my surprise, there were some of the biggest rocket motors I had ever seen," recalled Leninger, who was looking at a few E, F, and G motors. "And wouldn't you know it? On the shelf and right next to them was a stack of *High Power Rocketry* magazines." Leninger bought a 24 mm motor and a copy of the magazine. "After reading that magazine cover to cover, I quickly discovered that there are even bigger motors yet, and I had to get my hands on them."

Randall Ejma of Virginia bought an electric remote-control car at a local hobby store and went back the following week for some parts.

"While the guy was looking for whatever it was that I was asking for, I was perusing the magazine rack," said Ejma. "I picked up a copy of *HPR*, and it instantly blew my mind . . . I was amazed that this underground hobby even existed."

"I knew of E motors from an Estes catalog from my childhood," recalled Tripoli member Matt Jones of Nebraska. Then Jones walked into HobbyTown USA in La Vista, where he found a *High Power Rocketry* magazine. "I had no idea that you could fly anything larger than 5 pounds!"

Tripoli high-altitude record-setter Bill Inman had a similar experience. He hadn't discovered high-power rocketry until he came across a copy of Kelly's magazine at a retail outlet in 1995. "I was amazed by the monsters in there. I had no idea that motors this size existed, so I started buying the magazine regularly." Eventually, Inman would hold every Tripoli altitude record in the F- through J-power impulse range.

Flyer Tom Swenson discovered high power at the Hobby Horse in Madison, Wisconsin.

"There was a copy of *HPR* with *Down Right Ignorant* on the cover," said Swenson. "I've never managed to replicate the feeling I got when I opened those pages and discovered high-power rocketry for myself."

In Appleton, Wisconsin, there was a hobby store called Galaxy Hobby. Longtime Tripoli member David J. Miller used to browse the aisles there, looking at plastic models or trains.

"I was in giving the magazine rack a spin when I found a copy of *High Power Rocketry*," said Miller, who was surprised to learn that Tripoli even existed. "Now I have never been an altitude junkie, but the realization that I could lob larger rockets with the power of something more than an F and that they could weigh more than a pound got my creative juices flowing right away. I do believe that I wrote for a catalog from every vendor in that magazine!"

Curtis Scholl found Tripoli in 1994 after visiting a hobby shop in San Antonio, Texas, looking for model rocket motors. And Tim Eiszner, who would go on to become the launch director of the first East Coast LDRS held in Orangeburg in 1996, found the hobby through a copy of *High Power Rocketry* he had purchased at a Barnes &Noble in Atlanta. Tripoli board member Bruce Lee found his first copy of *High Power Rocketry* in a hobby store in Nebraska. It was the 1991 issue, with Gary Fillable's *SKJ3* rocket launched out of the back of a pickup truck.

"I bought it, brought it home, and joined Tripoli the next day," said Lee, one of the longest-serving board members in Tripoli history.

These stories and a thousand more like them were the direct result of Bruce Kelly's determination to bring high-power rocketry into the mainstream.

By 1994, Bruce Kelly's success with *High Power Rocketry* made him one of the most well-known figures in all of hobby rocketry. His hard work and personal vision for the magazine—that it would serve as a gateway for people everywhere to discover the hobby—served as a catalyst for rapid growth in the hobby. And as the subscriptions and purchases of the magazine grew under Kelly, so too did the ranks of the Tripoli. With his accomplishments as the creator of rocketry's premiere publication and his leadership role on the Tripoli board, Kelly was on his way to another milestone.

Soon, he would be elected president of the Tripoli Rocketry Association.

19

A Code for High Power Rocketry

In his 1960 work *Rocket Manual for Amateurs*, United States Army captain Betrand R. Brinley wrote that in the field of amateur rocketry, safety is the leader behind which even science and the advancement of knowledge must follow.

"Rocket experimentation can be a frightfully dangerous business, and there is no sense in risking your future happiness and well-being in an ill-conceived and badly executed attempt to show your friends and neighbors how smart you are," warned Brinley.

In the mid-1980s, Tripoli told potential members that one of the reasons they should join Tripoli was "to gain access to safe, preloaded rocket motors, technical information . . . and the advantages of a large responsible voice in matters dealing with all the various agencies involved in regulating rocketry." Tripoli's Articles of Incorporation in 1986 affirmed this claim. The stated purpose of the corporation included "[establishing] regulations and procedures for testing, training, and authorizing members of the corporation to acquire and use commercially available Class 'B' solid propellant rocket motors in a safe and legal manner."

Tripoli leaders and their counterparts at the NAR could use their influence to affect safety code legislation on a state-by-state basis. In the alternative, they might create a code on a national scale. They chose the latter method, and the document that was ultimately created to effectuate a national safety standard would be NFPA 1127, also known as the Code for High Power Rocketry.

Tripoli published its first safety code in February 1987. It was created by a three-person committee: Tom Blazanin, Edward Tindell, and Gary Fillible. It was called the Advanced Rocketry Safety Code. In 1988, this code was renamed the Tripoli Safety Code.

The safety code governed motors, rocket construction, launch and ignition systems, and field conditions. The code sounded a lot like the NAR's Code for Model Rocketry, which is not surprising since its creators were former NAR members.

The NAR's safety code and Tripoli's early codes prohibited metal in rockets and required users launch only certified commercial motors. Both codes prohibited alteration of commercial motors. And both codes shared a similar language for launch field safety. The NAR prohibited flights in winds that reached 20 miles per hour. And so did Tripoli. Both organizations said no rocket could be launched at an angle of 30 degrees from vertical. The NAR published a launch site dimensions table for its rockets based on engine size up through G power. In 1987, Tripoli was not ready for such tables—they would come later—yet the Advanced Rocketry Safety Code set forth safe distances from the launch area to the pad. The minimum distance for a high-power motor was 150 feet. For rockets that exceeded a J-class impulse level, that distance was increased to 200 feet.

The most significant difference between the Code for Model Rocketry and Tripoli's 1987 code was enforcement. The Tripoli code was not mandatory. It was offered as a guideline so as not to "stifle creativity" on the field—whatever that meant. In practice, organizers at some Tripoli launches required the Advanced Rocketry Safety Code be observed, even if Tripoli as an association chose not to.

One of the first moves toward an enforceable safety code came in late 1987 when G. Harry Stine called Tom Blazanin and invited him to observe an upcoming meeting of the NFPA's Committee on Pyrotechnics in Missouri. Stine was chair of the NFPA's Committee on Pyrotechnics, a position he held for many years. The committee made recommendations to the NFPA related to hobby rocketry (and fireworks) in the United States. In 1987, this meant model rocketry. There were no rules on high power yet.

Stine likely realized that sooner or later, the steady expansion of commercial high-power rocket motors would be addressed by the NFPA. In his position at the NFPA, he was in a good place to influence the new hobby. If he chose to, Stine could have been a formidable roadblock to high-power rocketry. Instead, he offered Tripoli's president the chance to participate with the Committee on Pyrotechnics. Tom Blazanin's first NFPA meeting was in St. Louis in 1987. The experience was an eye-opener for him.

"One was surprised to see that at no time were the powers of anyone overshadowed by another," wrote Blazanin. "Also surprising were the open attitudes of those representing the government agencies. At the time, those agencies included the ATF as well as the CPSC. Manufacturers, consumers, enforcement agencies, and experts are brought together, and each input their specialty to create a balanced system to decisions that are acceptable to those that the final result will affect . . . When we look at the NFPA codes, we automatically see rules that 'control' that which we want to do. [Yet] these controls were set up by persons knowledgeable and intelligent to the needs of the end users."

Shortly after the meeting in St. Louis, Tripoli was extended a formal invitation for a seat on the Committee on Pyrotechnics. Blazanin was Tripoli's first NFPA representative, with Tripoli vice president Ed Tindell as his alternate. Still, Blazanin was not the first high-power enthusiast on the committee. That distinction, like so many other firsts in the history of high power, belonged to Gary Rosenfield, who already had a seat as a result of his model rocketry and motor-manufacturing expertise.

"While we have only one seat, we join with Gary Rosenfield, who holds the seat representing manufacturers," said Blazanin. "Our interests are common with regards to high power, and Gary has acted in the past with the interests of both manufacturers and users in mind."

By the late 1980s, it was known that NFPA 1122, the Code for Model Rocketry, was soon going to be revised by the NFPA. Many hobbyists assumed this revision would include incorporation of high power into Section 1122. This seemed an easy path to follow and was certainly simpler than creating an entirely new NFPA code for high power alone.

"A complete rewriting of NFPA 1122 will take place in 1989, and it is imperative that Tripoli have a voice in this procedure," reported Blazanin to the board.

Blazanin was among those who thought that NFPA 1122 would be amended to include high power. Blazanin, Tindell, and Rosenfield—along with Bruce Kelly, who was often on hand to observe—attended NFPA meetings throughout the late 1980s. They got along with other members of the committee, and Blazanin, in particular, attributed some of their success to Harry Stine as well as Mary Roberts of model rocketry maker Estes Industries.

"It seems that working together pays off," wrote Blazanin in 1989. "In another year, NFPA 1122 goes up for a rewrite. Part of that rewrite will cover advanced rocketry. It is hoped that this united concern by all right now will continue and grow and be an asset where we, as advanced rocketry consumers, are concerned."

At the start of the 1990s, Chuck Rogers, Tripoli's president and a working professional rocket scientist, brought additional credibility to high power's position on the committee. At their joint meeting in Dallas that year, both Tripoli and the NAR agreed to participate as partners in the rewrite of the NFPA codes.

"I've actually had discussions with NAR board members asking if Tripoli would support raising the limit of high-power rocketry from our proposed full 'O' power limit to the limit of the U.S. Space Act, which is equivalent to 1,000 pounds of propellant," said President Rogers that spring. "They've also asked for Tripoli's position on allowing metal airframes. No kidding. I'm serious . . . At the Dallas meeting, an old NAR member who shall remain nameless—let's just refer to him as the original cowboy rocketeer—said that as far as high-power rocketry was concerned, and I quote, 'anything, anywhere, anytime' was fine with him as long as it was addressed in the NFPA code. I guess from now on, we will have to call him 'Anything, Anywhere, Anytime Stine.'"

Rogers believed Tripoli and the NAR working together would result in positive changes in the hobby.

"There are no 'winners' or 'losers' here, except the big winner was *you*, the sport rocketry consumer. What you will get out of all this is unified positions on the NFPA pyrotechnics committee that will get the kind of responsible, progressive codes we can all live with."

For Rogers, this meant more opportunities to launch high-power rockets from more fields and better services from each organization because they were now in competition with each other. The importance of a good working relationship between Tripoli and the NAR and of presenting a unified front to the NFPA was recognized by leaders in both clubs.

"Many of the public safety officials on the NFPA pyrotechnics committee are going to be surprised and shocked when they find out that a thousand people are out there flying high-power rockets of considerable size," explained Rogers. "They are going to require some convincing to write a code that will allow the activity to grow to its full potential."

According to Rogers, Tripoli and the NAR needed an effective alliance to bring a safety code to reality.

"As I and others on the NFPA pyrotechnics committee can attest, if the rocket members of the committee can't come out of the upcoming NFPA pyrotechnics committee rocket caucus with a unified position on high-power rocketry, then frankly, we don't stand a chance of getting the kind of code we want passed by the full committee. The NAR also realizes this."

Despite some differences, Rogers believed the NAR and Tripoli were, in reality, quite close on how they viewed the rewrite of the safety code. Together, said Rogers, the two organizations should work out any differences in their proposals for high-power rules before the presentation of those rules to the full meeting of the committee. This way, the two clubs would have the greatest chance for success in getting high-power rocketry formally approved by the NFPA.

In 1991, Tripoli and the NAR published drafts of a revised NFPA 1122 for their respective members. These drafts incorporated high power into the model rocketry code. The proposals disparaged the private manufacture of rocket motors and set an upper limit on high-power motors at the O level of impulse or 40,960 Newton seconds. No weight limit was set for a high-power rocket. The proposal also included "suggested field and range site dimensions for conducting high-power activities."

During this time, Tripoli and the NAR also revised their own respective safety codes to incorporate much of the anticipated NFPA 1122 revisions. Then in late 1991, a decision was made to split up model and high power within the NFPA codes. Section 1122 would be revised and would continue to be known as the Code for Model Rocketry. But an entirely new section, designated NFPA 1127, would be drafted. This new section would be called The Code for High Power Rocketry.

The precise reason why model and high-power rocketry were split into two separate codes is nebulous. But it was later stated that the committee determined "a separate NFPA code should be adopted for high-power rocketry because of significant differences in operations and to prevent confusion of both model rocketry and high-power rocketry in the minds of public safety officials."

In other words, high power should be segregated from model rocketry to ensure that any potential problems with the former would not lead to local safety officials imposing restrictions on the latter, a long time fear of the model rocket community and the NAR. If there was an accident or a tragedy involving high power, the consequences could be confined to high power. Model rocketry would share none of the blame. This was probably a good idea from model rocketry's point of view. And from the view of high-power enthusiasts, it was what they wanted too: national recognition for high power and a separate, standalone safety code of their own.

The first draft of NFPA 1127 was reviewed by the Tripoli board of directors in the fall of 1991. On October 26, 1991, at a board meeting at the Danville III launch in Illinois, the board adopted the draft version of Section 1127 as Tripoli's new safety code. The author of the draft, which contained the essential elements of the code that would be adopted, was likely Harry Stine.

In January 1992, the NAR published an interim "NAR High Power Rocketry Safety Code." The NAR's interim code was less than two pages long, set forth basic launch safety, flying conditions, and recovery rules, and recognized high-power motors up through 40,960 Newton seconds of impulse, or an O motor. The code stated that the NAR was in the midst of developing a motor-testing program for high-power motors. In the meantime, the NAR said it would accept motor-testing certification conducted by Tripoli. The NAR interim code also contained a safe distance table. Although the interim code recognized the existence of high-power motors through O power, the NAR's newly implemented user certification process allowed for certification through J power only.

In February 1992, Tripoli published the draft of the proposed Code for High Power Rocketry adopted by the board the previous October—NFPA 1127. A final draft would be ready for adoption by the NFPA in two years, in 1994. The Tripoli board asked its members for comments regarding the proposed new safety code, but these comments were for Tripoli eyes only. The NFPA "official public comment period" was still in the future. Tripoli board members and NFPA representatives wanted to obtain the pulse of the membership in advance of any public comment period, probably to head off any strong opposition that might be transmitted directly from high-power users to the NFPA. In soliciting member comments, the board asked people to be patient and realistic.

"Please be aware that high power is still a touchy subject in many states. The NFPA pyrotechnics committee is composed of people from the rocket industry, the fireworks industry, various fire marshals, the Bureau of Alcohol, Tobacco, and Firearms, and other experts concerned with public safety. Please consider the areas you may wish to disagree with and see changed. Ask yourself if you can live with it the way it is or are . . . bothered enough by a certain regulation that you must comment. Do not nitpick these codes! The Tripoli representatives on the committee would have a difficult time getting 1127 through the process if we all wanted to launch our *EZI-65*s in school yards (this example is extreme, but you get the point). The majority of the people on the committee are not rocket people. They have their ideas as well. Much time has been spent by all concerned to get rocketry this far. Having a high-power code at all is quite an accomplishment."

In 1992, there were twenty-four voting members on the NFPA Committee on Pyrotechnics. Seven of them were directly involved in rocketry: Chuck Rogers (Tripoli), J. Pat Miller (NAR), Gary Rosenfield, Michael Platt of the High-Power Rocket Manufacturers and Dealers Association, Vern Estes, Mary Estes, and Dane Boles of Quest. Other voting members on the committee included state and national fire marshals and pyrotechnics and fireworks experts from around the nation.

There were also a number of alternates to the committee, including rocketry notables such as Bruce Kelly, Mark Bundick, and Trip Barber. Nonvoting members included a representative of the U.S. CPSC, the U.S. Occupational Safety and Health Administration (OSHA), and Harry Stine. Stine was semiretired and was designated member emeritus of the committee.

The draft of NFPA 1127 shepherded through the committee stood alone, with no dependence on the Code for Model Rocketry. Its stated purpose was to "assure the availability of high-power rocket motors" to certified users and to "establish guidelines for [the] reasonably safe operation of high-power rockets to protect the user and the public." The foreword to the draft, which was left intact in the final version of 1127, contained a brief recital of the history of high-power rocketry. That recital began with the launching of higher-powered motors in the late 1970s and the observation that there had now been more than a decade of operational experience with commercially made solid-fuel motors and rockets with larger and heavier airframes.

"Indeed, the experiences gained thus far [exceed those] amassed by model rocketry when the first NFPA Code for Model Rocketry [NFPA 1122] was adopted," observed the NFPA.

Both the draft and the final version of NFPA 1127 were broken down into seven chapters covering most aspects of high-power activities. This included safety issues, requirements for commercial high-power rocket motors, safe distance tables, and prohibited activities. The draft required high-power users be a minimum of eighteen years old.

This 1992 draft elicited many comments from high-power flyers. For example, Tripoli board member Bill Wood disagreed with the draft's call for a 10-foot-wide clear area around high-power launch pads to prevent fires. This was inadequate, said Wood, who argued that 50 feet was more reasonable. Wood also took issue with the safe distance Table. The minimum safe distances were too short, he argued. He believed any high-power safety code should have safe distances more in line with those established for explosives by the Department of Defense. Wood even suggested that Tripoli require hard hats at launches to protect flyers from fragments, not necessarily from motor cases but from airframes that held motors subject to catastrophic failures.

Despite some negative comments, a majority of the Tripoli board agreed with the draft, and a motion to accept it passed easily. In the majority's view, although the proposed code could be improved, it was better than Tripoli's then current safety code. They adopted the draft and resolved to make it better. At the same time, the board decided that Wood should become Tripoli's alternate member to the NFPA's Committee on Pyrotechnics.

As for the membership comments, most were concerned with proposed launch site dimensions. Some flyers, especially those who launched rockets on the East Coast, believed the site dimensions were too conservative; the code called for too much overall space for a high-power launch, they said. As reported in the *Tripoli Report* in early 1993, the proposed site requirements might "shut down half the launch sites in the eastern part of the country." Indeed, one board member said that the upcoming launch at Danville, Ohio, one of the most well-known events in high power at the time, would have to be shut down if the Section 1127 site dimensions were to be followed.

The launch site dimensions in the draft called for safe distances that ranged from half a mile of space for an I-impulse motor to 3 miles for an M-class motor, all the way up to 5 miles for an O motor. One board member asked that Tripoli vote to rescind their support for Section 1127 in its entirety based on the launch site dimensions table. Fortunately, cooler heads prevailed. Tripoli continued its support for NFPA 1127, with the exception of the launch site dimensions table.

"The board decision was that this was the best compromise to continue adoption of the draft NFPA 1127 code without adversely affecting East Coast launches, which would have been shut down with little or no warning," reported the *Tripoli Report*. It was expected that with some additional negotiation, the launch dimensions would be reduced prior to the final version of Section 1127.

On January 21, 1994, at a meeting in New Orleans, the Tripoli board of directors adopted the latest draft of NFPA 1127 as Tripoli's safety code. At this same meeting, the board also created Tripoli's Future Directions Committee, charged with studying motors that were specifically prohibited by Section 1127, namely, homemade, experimental motors. The following day, the board reaffirmed its position that unapproved and experimental high-power motors shall *not* be flown at Tripoli-sponsored launches. The final published version of the first edition of the Code for High Power Rocketry was released by the NFPA in 1995. It went into effect on August 11, 1995.

NFPA 1127 applies to the design and construction of high-power motors, vehicles, and launch operations. Its stated purpose is to provide for safe and reliable motors, establish flight operation guidelines, and prevent injuries. Among other things, the code sets forth surface wind-launching rules, certified user guidelines, spectator safe distance tables, launch angle requirements, and related safety regulations. Some of the rules reflect simple common sense. The consumption of alcohol when preparing a rocket for flight is prohibited. There were warnings to never insert an igniter into a motor before the rocket is vertical on the pad. A flyer must be eighteen before he or she can be certified, or the sale of uncertified motors is prohibited.

"It has been occasionally argued that Tripoli should have never participated in the NFPA process," observed Bruce Kelly years after the code was adopted. "However, if we had not become part of the process, we would have allowed others to write codes that would have had an impact on our activity without our input. This, of course, would have been foolish on our part—to be the largest high-power rocketry organization and yet to have had no say whatsoever in how we are regulated by others. We entered the code writing arena with this philosophy as our guideline—'It is better to act than to be acted upon.'"

In the spring of 1991, the NAR released its first official high-power safety code. This was the first set of written rules promulgated by the NAR allowing its members to fly high-power rockets, at least through the J-impulse level. The code was soon expanded to allow NAR members to fly rockets K and above. In the near future, said the trustees, this interim code would be replaced by a new safety code that was being created by the NFPA.

Along with their interim safety code, the NAR announced a new system by which members would be allowed to fly higher-impulse motors. This system would not be as easy as Tripoli's confirmation program. Under the NAR program, flyers did not progress from an H motor to an N motor based only on a single flight. They had to wait. NAR flyers would have to successfully launch and recover a rocket at multiple intervening impulse levels as they worked their way up a motor ladder. NAR members were required to successfully fly and recover using an H motor before they could use an I-impulse motor, they had to successfully launch an I-impulse motor before they could then purchase a J, and so on. The NAR called their new stepped program "certification."

Meanwhile, Tripoli leaders were preparing to scrap their motor eligibility system too. The board long realized change was needed in the Tripoli confirmation system. Many Tripoli members agreed the current system was inadequate.

"Too damn easy!"

"Our current confirmation program is obsolete!"

"Flying a rocket on a single H proves only one thing—that you can fly an H."

"Too many people are being confirmed who should *not* be confirmed."

In early 1991, the board proposed replacing its confirmation program with a stepped certification system. This draft proposal was more complicated than anything the NAR was using or proposed to use at the time. To reach Level 1, which would allow a flyer to launch H-powered rockets, the user was required to fly and recover any rocket successfully. There was no motor requirement at all. But he or she would also have to pass a written test regarding the high-power rocketry safety code.

To reach Level 2, a flyer had to not only successfully launch a rocket with an H motor but also demonstrate proficiency—the proposal did not specify how—in Newton's fundamentals of the relationship between the center of pressure (CP) and the center of gravity (CG). There is some suggestion that this meant proficiency in Newton's laws demonstrated with knowledge of his specific equations and how they were used to determine the CP and the CG. (Such a requirement in the absence of computer programs on the subject probably would have made Level 2 unattainable for the majority of high-power users.) A Level 2 flyer could then use an I- or J-impulse motor.

Both Level 3 and Level 4 of the proposed system required increasing knowledge of Newton's laws and the ability to formulate additional equations related to the aerodynamic drag and motion of rockets. At Level 3, a flyer could launch up to a K motor. To reach Level 4, the user would launch a K or L motor at supersonic speeds, and then they could buy M, N, and O motors. Level 5, which would mean you were an "acknowledged expert in high-power rocketry," would be granted to those who launched and successfully recovered an M-, N-, or O-powered rocket. There was even an "R" certification level for flyer proficiency in reloadable motors.

This early board proposal was never implemented. But it was an early step, albeit a little behind the NAR, on the road to an effective program. Although this 1991 draft was not adopted, the idea of a stepped certification program slowly percolated through the Tripoli ranks. At LDRS-11 at Black Rock in 1992, the Tripoli board formally replaced the term "confirmation" with "certification" in all Tripoli materials. The next step was to find someone to take the laboring oar to create a high-power certification program from scratch. Tripoli found their man in board member Scott Bartel of California.

"Rocketry had been interesting to me ever since I watched the *Wonderful World of Disney* when they had the Werner von Braun *Man in Space* series," Scott Bartel said in an interview in 2001.

The *Man in Space* shows were first broadcast in 1955. With documentary footage and special effects animation, the history of rockets and the anticipated effects of space travel were taught to an eager audience on the brink of space exploration. Pres. Dwight D. Eisenhower was so impressed with the show, he requested a print of the film to be screened at the Pentagon. A scale model of the space station built for the series is housed at the Smithsonian in Washington.

"I was completely overwhelmed by the idea of men in space and space flight in general," recalled Bartel. "When the first satellite, *Sputnik*, flew overhead, I was watching it."

Bartel bought his first rocket kit in 1962 after seeing an Estes ad in *Boy's Life Magazine*. The rocket was the Estes *Astron Scout*.

"I straightened out a coat hanger for a launch rod and took the rocket out in the back yard," he said. "I found the largest motor in the box and put it in the rocket. I flew it and never saw it again! However, it was all history after that. I started flying everything I could."

Bartel graduated high school and enlisted in the United States Air Force in 1969. He was trained as a radar technician and worked on some of the navigational systems that were the precursors to the Global Positioning System. He also spent a year in Iceland, where his duties included loading rockets into the munitions bays of Convair *F-102 Delta Dart*s.

In 1976, Bartel went to work for defense contractor Martin Marietta in their quality assurance department. Some of this work involved early computer systems for NASA's up-and-coming space shuttle.

"I inspected all the print wiring boards and assemblies, and I assisted with the functional checkouts of these really old computers," said Bartel. "These computers were big, really slow, and very touchy. You couldn't run them for very long without them failing."

After Martin Marietta, Bartel held similar jobs with the United States Air Force, this time as a civilian, and other aerospace companies, such as Hughes Aircraft. His experience included working on satellite programs and quality assurance on the *Hubble* spacecraft.

Bartel discovered high-power rocketry in the early 1990s while attending a meeting at the Johnson Space Center in Texas. While there, he wandered into a local hobby shop.

"There on the shelf were black cylinders, composite motors. Of course, I had heard about Enerjet composite motors as a kid but hadn't seen or heard of them since. Here were several AeroTech motors in the E, F, and G size. Well, it was all over. I saw them and said, 'Oh my gosh. I've got to do this again!'"

His first high-power rocket was a LOC *Graduator*. He took it out near his mother's home in Victorville, California, and flew it on an F motor.

"This was back when you had to use Thermalite and wrap wires for igniters," he said. "I built a rocket launch controller and used that. About that time, I found out about Tripoli and joined."

Bartel became Tripoli member 975. He became active with his local Tripoli prefecture, and he also wrote stories for the *Tripolitan* and later *High Power Rocketry*. He took a leading role in learning the ropes of legally launching high-power rockets in his home state, California, which had not adopted any NFPA codes related to rocketry. He was instrumental in publishing explanations of California's byzantine pyrotechnic rules and regulations. He joined AeroTech's Gary Rosenfield in presenting the California fire marshal with a detailed written proposal to incorporate high-power motors into the state's pyrotechnic regulations. Bartel was selected to the Tripoli board of directors in 1993.

Bartel volunteered to create a multistep certification program for Tripoli in 1993. He presented the board with his first draft of such a program in October that year. As much as Tripoli should strive to be egalitarian and trust the technical ability of all Tripoli members, the board must be realistic, he said.

"It is my observation that any Tripoli member is fully capable of successfully constructing a 'conventional' (Kraft, phenolic, plywood, or fiberglass) rocket and flying it with H-, I-, J-, [and] K-rated motors. However, not all Tripoli members are capable of building an L-, M-, N-, [or] O-class or complex rocket without advice or guidance.

"In the professional world, I have the dubious distinction of being a systems engineering and technical assessment (SETA) contractor. What I do is provide advice to Defense Department program offices on whether their other contractors are doing the right job. I deal primarily with complex space shuttle experiments and hardware. My tasks consist of assisting with requirements development, systems engineering studies, engineering reviews, safety evaluations, and 'nosing about.' What I have found is that even the most skilled contractor (Hughes, TRW, Rockwell, Lockheed, etc.) sometimes misses, overlooks, or simply isn't aware of an operational, safety, or reliability requirement. I believe that this happens with our [rocketry] activities as well."

Bartel presented the Tripoli board with four alternatives for a new safety program. The first alternative was really just rhetorical in that he suggested Tripoli could simply ban motors larger than a K. This would eliminate the increased risks inherent in L through O motors, he said. No one was in favor of this option, including Bartel. The second alternative was the creation of a new set of regulations covering design rules, redundancy requirements, and operations safety for L, M, N, and O motors.

"Tripoli members would have to certify their rocket meets requirements by completing a detailed checklist and providing design details to support their contentions," explained Bartel. "This is sort of like getting a building permit for a house," he added. "You don't build if you don't design to code."

Bartel's third option was a three-step certification system that he called the "Advanced Tripoli Consumer Program." Under this scenario, there would be a "general" certification level for H to J motors and a "high impulse" rating for L to O motors. There would be a final rating called "complex." This would apply to high total-impulse clustered rockets and some multistage rockets. Bartel liked the idea of a three-part system as it reminded him of the ratings private pilots had to obtain as they went from a simple pilot's license through an instrument rating and then ratings for multiengine aircraft.

The final option did not involve certification at all, at least not for motors in every impulse class. Instead, Bartel suggested the creation of an advisory committee to support flyers of L- through O-powered rockets. Under this plan, several highly skilled Tripoli members would be appointed to what Bartel called the Tripoli Advisory Committee, and they would serve as inspectors of high-impulse rockets prior to flight. Bartel concluded his initial offering to the board with the suggestion that perhaps some combination of the second, third, and fourth alternatives would work best.

The Tripoli board made no final decision on a certification system prior to the end of 1993. Instead—and with Bartel at the helm—the certification discussion moved slowly forward over the next eighteen months. However, the board did take early action regarding Bartel's fourth option. In November 1993, the Tripoli Advisory Panel, called TAP for short, was born. On motion by Bartel and fellow board member Dick Embry, the board unanimously approved the creation of the TAP to oversee L through O flights in the future.

The rules governing the first TAP committee were somewhat different from those that guide that same panel today. For example, membership on the panel was originally small and very exclusive. TAP members were appointed directly by the board. They were to be persons with an engineering background or having a proven track record of competency in the design, construction, and recovery of high-impulse rockets. The submission of materials to the TAP was also only voluntary. That is, members who were planning on launching a rocket with an L through O motor were under no obligation to send anything to the TAP. Instead, flyers were encouraged to do so. (Over time, the TAP would expand in size and would be given teeth; the submission of preflight materials for Level 3 certification flights would eventually be mandatory.)

In 1994, there was little published discussion by the board regarding the rest of Bartel's proposed certification rules. In the meantime, Bartel, who was appointed to the board to fill a vacancy in 1993, successfully ran for a board seat and was elected in his own right in 1994. During the election campaign, he made his proposed certification system part of his platform. Bartel redrafted his certification proposal and the TAP rules in early 1995. The new certification system was unanimously approved at the board's meeting in Las Vegas, Nevada, on February 3, 1995. It would remain in place, subject to periodic modifications and amendments, up through the present time.

The program created by Bartel separated certification into three distinct classes: primary, advanced, and unlimited. Today, these classes are recognized as Levels 1, 2, and 3.

Level 1 or "primary" certification required a flyer to successfully launch and recover a rocket powered by an H motor. A successful Level 1 launch allowed the flyer to then purchase and fly motors up through the I-impulse class. The prior confirmation system, which allowed flyers to confirm with an H motor and then fly an O motor, was eliminated. From this point forward, high-power users would earn their way to the top.

Level 2 or "advanced" certification required a six-month waiting period after a flyer's successful Level 1 flight. The rationale for the wait was to require a flyer to gain experience at Level 1 before he or she moved to the next level. Level 2 also required the flyer to pass a written test regarding rocketry fundamentals and safety issues prior to the certification flight. This was a bold stroke by Tripoli. It helped ensure that members had some minimal level of rocketry knowledge as they moved up the impulse ladder. The test was not a difficult one. It was twenty-five questions out of a pool of fifty; questions and answers could be studied by flyers prior to examination. But it was the first time that any test was required for hobby rocketry, and it would become a cornerstone of Tripoli's certification program and similar programs offered by the NAR.

The questions developed for the Level 2 test ranged from Newton's laws to rocket stability (the center of pressure and the center of gravity), the proper use of launch equipment, and motor dynamics. Although the test questions have been revised and added to over the years, many of the questions initially developed by Bartel are still used on the Level 2 certification test today. Level 2 certification also required the successful launch and recovery of a rocket with a J, K, or L motor. Once certified to Level 2, flyers could launch rockets up through L power.

The final step was Level 3 or the "unlimited" certification class. Level 3 required a six-month wait after Level 2. It added the requirement that the flyer have a "detailed inspection of their rocket and discuss its construction with two TAP members prior to flight." This requirement made TAP members indispensable to the certification process. Among other things, a Level 3 certification rocket was also required to carry onboard electronics for deployment and recovery. If a flyer met these criteria and successfully launched an M-, N-, or O-powered rocket, then they were considered for Level 3. Final approval had to come directly from the Tripoli board of directors. The original rules stated board approval had to come before the Level 3 flight. This requirement was dropped when the ranks of Level 3 flyers swelled in the late 1990s.

At the 1995 Las Vegas meeting, the board revisited the new TAP system too. It was agreed that Level 3 flyers would automatically be placed on the TAP list. Eventually, when Level 3 became more common, membership in the TAP was modified to a system of nomination and appointment. In addition to being a requirement for a Level 3 certification, the TAP was also offered to Tripoli members as the "go-to" team for any flyers planning a large or high-impulse project. There was no requirement that such projects be reviewed by TAP members, but the system was in place for help as needed. At the board meeting held in Gerlach, Nevada, on August 9, 1995, Pius Morozumi was named as the first TAP chair.

"I think the TAP program is one of the big success stories [of] Tripoli," said former Tripoli President Chuck Rogers years later. "You get a TAP adviser identified for you. You are building your Level 3 rocket. You don't just show up with a rocket. You are sending plans and drawings. Your progress is being reviewed. A Level 3 rocket is a pretty serious rocket, and we expect people to have a certain level of expertise. In the old days, we didn't have that."

Tripoli published its first listing of certified members in 1996. The largest class was in the Level 1 category. In the first year of practice, there were nearly 1,300 Level 1 flyers. There were 130 Level 2 flyers. There was only one Level 3 flyer in 1996: Karl Baumann of California. Baumann was the first Tripoli member to receive Level 3 certification under Tripoli's modern system. Two years later, the ranks of the certified increased dramatically, especially in the Level 2 and 3 categories.

By mid-1998, the number of Level 2 flyers had nearly tripled, to more than 360. And there were now sixty-four Level 3 flyers. These people would become the new leaders in the field. They would know how to load and reload a motor, how to quickly assess the center of pressure and the center of gravity, how and when to arm electronics and insert igniters, how to secure a high-power rocket motor safely in a rocket, how to build an ejection charge, and how to dodge a falling rocket. They would teach new flyers how to strip igniter wire, secure a canopy to the desert floor, or search for a rocket in a cornfield. By example, they would show everyone how to walk away with dignity from the embarrassing mishaps that sooner or later happen to all experienced flyers.

The certification systems used by Tripoli and the NAR to qualify flyers for high-power motors and the NFPA's Code for High Power Rocketry are the twin pillars upon which the safety foundation of high-power rocketry now rests.

"NFPA codes for consumer rocketry protect all of us, to one degree or another," wrote former Tripoli president Bruce Kelly in 2004. "Our insurance coverage is based on the NFPA codes. We feel safe at our rocket launches because NFPA 1127 is also our Tripoli Safety Code. There is a greater degree of responsible control at our launches, and we feel comfortable about inviting guests, allowing walk-on spectators, and having our children around. The NFPA rocketry codes [also] provide safety officials a greater degree of confidence in our activity so they regulate us less on local levels."

20

Theodolite's Demise

> Now, friends, what is a solution to the altitude determination problem? First, it must be defined. The problem is altitude determination, not tracking or optics or baselines or calibrations. The solution is to come up with a system that will determine the true altitude achieved by a model rocket above its launching site within an acceptable set of limits . . . said system also being capable of operating and doing its job on about 95 percent of the models launched regardless of altitude achieved or sky conditions or previous conditions of servitude. There is a solution. I do not know what it is right now. If I did know, I would build it and tell you about it. Model rocketry is a very Edisonian hobby. By this, I refer to the famous saying attributed to Thomas Alva Edison—"There must be a better way to do this job. Find it." Let's start being Edisonian about the *altitude determination system*. (Emphasis added.)
>
> —Message to the NAR membership, G. Harry Stine, October 1969

Harry Stine's challenge to the NAR to find an "altitude determination system" for hobby rocketry would take almost another quarter century to fulfill. The device he was looking for in 1969 was the modern rocketry altimeter.

In the mid-1980s, a few hobbyists with electronics expertise built homemade devices that could purportedly track a rocket's peak altitude up to a mile or so. By 1988, there were at least two companies marketing commercial altimeters to the high-power community. One of the first was Transolve.

Transolve was run by John Fleischer, an electronics whiz from Ohio who was marketing his "A-1 altimeter," an altitude data logger that could record a rocket's peak altitude up to several thousand feet. The A-1 was powered by a 9-volt battery. After flight it was plugged into a separate unit that displayed a digital image of the rocket's altitude.

The A-1 altimeter was unique, yet it was also large and bulky and required some electronics know-how. It was almost a foot long and had to be built by the flyer. Fleischer wrote a how-to article on the A1 that appeared as the cover story for the October 1990 issue of *Radio Electronics*, a national technology trade magazine. According to Fleischer, the A-1 could be built for about $135. At or about this time, another version of Fleischer's altimeter was being sold as a complete unit by North Coast Rocketry for $225.

Another early entry into the commercial altimeter market was Flight Control Systems, based in Camp Hill, Pennsylvania. Flight Control began selling its "FP1 Model Rocket Flight Data Logger" no later than 1990. The Data Logger was an early version of an on board computer. The device required an understanding of computer command functions and required some software skills. It came with a technical manual eighty pages long. It could purportedly record altitudes to 12,000 feet, and it had the ability to fire a stage separation charge. Like the Transolve A1, the Data Logger was designed to be plugged into something else for data retrieval after flight, that something else being an IBM-style personal computer. Also, like the A-1, the Data Logger was bulky and expensive. It was almost two inches wide and nearly a foot long. The altimeter version was nearly $300. Additional software, cables, and lithium battery packs pushed the total cost to more than $400. (This would be like paying $750 for an altimeter in 2018.)

Both Transolve and Flight Control Systems were pioneers on the road to an easy-to-use, affordable, and reliable altimeter for high-power and model rocketry. However, these and other early manufacturers would not be the first to market a modern, self-contained, inexpensive altimeter for the hobby. That device would be built by Tommy Billings. The company he created was called Adept Rocketry.

Thomas "Tommy" L. Billings was born in Charleston, South Carolina, in 1945. His father was from Kentucky, his mother from Louisiana. His parents married in the early 1940s and had four children. His father served with the Seabees during World War II and then moved to South Carolina when the war ended. When Tommy was born, his father was a welder at the United States Navy shipyard in Charleston. Mrs. Billings was a homemaker.

"My plan was to be a medical doctor. A heart surgeon is what I was hoping to be," recalls Billings, who graduated from Hanahan High School in 1963. "I was really into biology and frogs and things like that."

He also enjoyed electronics. "My elder brother got interested in electronics, and then he got a HAM radio license, and because of that, I ended up getting a HAM radio license when I was in high school, and then it occurred to me that I also liked electronics," he said. "So for a number of reasons, I decided maybe I don't want to be a medical doctor."

Billings started college after high school. But he was not quite ready for academic life. He took a job at the naval shipyard in Charleston, where he learned how to repair sonar and radar on United States Navy vessels. His stint as an apprentice electrician was cut short by the Vietnam War. In 1965, he joined the army. He was twenty years old.

"I joined the army thinking that I would have better opportunities," he said. "I went to Fort Belvoir, Virginia, where they invented the starlight scope and the M16. They put me through an electronics school, where I worked on mine detectors and searchlights and a number of different things that the army used."

Billings eventually became an electronics instructor for the military. He left the army in the late 1960s and returned to the shipyard at Charleston to finish his electronics apprenticeship. Then he enrolled at Clemson University and earned his bachelor of science degree in electronic engineering in 1972.

"My first job was at Westinghouse in Baltimore, Maryland," he said. "I was a digital designer, and I designed a couple of components for the AWACS radar. I was trying to get back to the South, and I then took a job with IBM in Lexington, Kentucky. The integrated circuit was really getting going back then. The microcontroller was just coming out, and I just stayed with it."

Billings accepted a position with IBM, and he moved to Colorado in 1974. He remained with the computer giant for six years. Then he left IBM in 1981 to start his own company as a digital designer.

"I invented a computer using one of the early microcontrollers, and I joined a computer club at the School of Mines in Golden, Colorado," recalled Billings. "About thirty of those computers got built up. Then several of us decided that we wanted to spin off and do our own thing. I formed a corporation called TLV Controls Corporation and and several of us got together and formed a company making a low-cost logic analyzer because there simply wasn't anything on the market that was low cost."

Billings and his colleagues succeeded in making the analyzer. In 1986, a larger company bought them out. Billings made plenty of money in the deal, and he could have stopped working right then and simply retired. But he was only forty-one.

"I piddled around with another software company in Colorado," he said. "And I put together another corporation and dabbled with developing software for income tax but didn't really go anywhere."

Meanwhile, he was also doing some work for medical companies. As part of this work, he handled devices such as CPAP pressure monitors and other medical devices. Through this experience, he became an expert on pressure sensors and pressure monitors. He didn't know it yet, but his work on pressure sensors in the medical field in the mid-1980s would one day be to the benefit of high-power rocketry.

"My elder brother Jimmy was a rocket nut," recalls Billings of his childhood. "Jimmy and a friend used to make rockets attached to long sticks, like skyrockets or great big bottle rockets. Eventually, we were all making rockets. We made rockets out of metal or rolled-up cardboard tubes. Or the whole rocket was a motor with fins attached. We would use a lathe to turn a nose cone, and we would cut fins out of tin and then attach them with screws."

In the late 1950s, science fairs were common, and rockets were all over the place, recalls Billings. Homemade rockets in those days were constructed of wood, aluminum, or even cast iron. And like other teens of their day, Tommy and his friends made their own rocket fuel out of potassium nitrate, sulfur, and charcoal.

"The biggest problem people had with their own motors was making a working nozzle," said Billings. "There was one story where a guy filled a pipe with some type of homemade black powder, and he was hammering the end of the pipe to form a nozzle, and it blew him to smithereens. This happened to a number of people. They would go off into their house or the basement to build these things, and a few of them got killed. That is where the term 'basement bomber' came from."

Billings remembered the local government putting a stop to fuel making; he and his friends moved away from rocketry. When model rocket motors came out, he tried them but was disappointed.

"They didn't even have D motors yet," he said. "Our own rockets used to weigh 5 to 10 pounds, and it was all fuel."

Billings discovered high-power rocketry in the late 1980s. He launched a few rockets while living in Bloomfield, Colorado, and he joined a local club. With his electronics background, he soon found himself making rudimentary timers for the staging and deployment of his friends' rockets.

"I already had all the equipment, and I was an expert in soldering, so I built this stuff in one basement room of my house," he recalls. "I would then test it at the local launches in Colorado."

Billings eventually began selling his electronics to local rocketry flyers. He decided to form another business devoted to rocketry-related work. He called his company Adept Rocketry.

One of Adept's early projects was a timer that Billings constructed for David Gianokis, a NAR member in Colorado. Gianokis scratch-built scale models of America's most famous rockets. Some of his work would eventually be displayed in aerospace museums across America, including the Smithsonian.

In 1990, Gianokis was completing work on a one-twelfth-scale model of the *Mercury Redstone* rocket, the vehicle that propelled Alan B. Shepard into space in May 1961. The *Redstone* would fly on a Vulcan I283 motor. To make the flight more realistic, Gianokis wanted the capsule escape tower to be fired off in flight with an onboard Estes C6-5 motor. He asked Billings to develop the electronics to ignite the escape tower motor. Billings had already developed an acceleration switch for his rocketry timers that activated upon takeoff. It might work for the *Redstone*, he thought.

"It was a double fulcrum switch made with a tiny moving brass bar and a micro switch with a roller on it," he explained. "This switch could detect acceleration and liftoff."

The timer worked perfectly, and the escape tower was propelled off the vehicle in flight and without a hitch.

LARGE AND DANGEROUS ROCKET SHIPS

The following year, Gianokis built an even bigger rocket: a one thirty-third replica of the *Saturn IB* that carried a central Vulcan I motor surrounded by four G motors, all to be ignited at liftoff. The rocket would stage in flight, and the upper stage would be propelled by another H motor. Once again, Gianokis turned to Billings for help with electronics for the project. Billings designed four plastic fins for the sustainer of the *Saturn IB* that would be controlled by springs. The fins would pop out upon stage separation to help maintain stable flight for the upper stage of the rocket. He also installed an onboard sonic beacon locating device, tracking transmitters, and a flight computer. The computer controlled the ignition of the second stage H motor and also served as an altimeter for the flight. Once the computer was retrieved, the information it held, such as peak altitude and speed, could be downloaded to a personal computer.

The flight of the Gianokis *Saturn IB* was on December 28, 1991. The stages of the rocket separated perfectly during flight, and the sustainer reached nearly 2,500 feet in altitude. The electronics recorded the rocket's top speed at 365 miles per hour. After the flight, the rocket was shipped for display to an aerospace museum in Southern California. The *Saturn IB* and Billing's role in designing the electronics for the rocket was the cover story for the March/April 1992 issue of *High Power Rocketry*.

Billings soon began marketing his Adept timers, stagers, and sonic locating beacons to high-power flyers all over America. Along the way, he made some mistakes as he figured out the safest way to design his products.

"Pius Morozumi had one of my early timers in a two-stage rocket at Black Rock," recalls Billings. "He had already turned it on as he fiddled with it, and the timer was 'live' as it sat on his lap. He was trying to put the wires into a terminal block on the unit. Well, Pius accidentally hit a wire underneath the terminal board. The rocket suddenly ignited and took off, flying horizontally from his lap. It raced across the ground and literally pierced a tire on someone's car! That tire is now the famous tire on display in front of a house near Bruno's Country Club in Gerlach. And I thought, 'Holy crap, he was just trying to hook up the unit after he powered it up.' I was glad no one got hurt. But I decided right then to get rid of terminal blocks."

Billings also began developing an inexpensive self-contained altimeter to detect maximum altitude for a rocket's flight. The altimeter would have the ability to fire onboard ejection charges in flight too. It would revolutionize high-power rocketry.

French physicist Paul Cailletet developed the first altimeter in the 1880s. Cailletet's pressure altimeter was a simple device used to measure surface variations in land or the altitude of an object, such as a building, in relation to the land. Paul Kollsman, a German inventor, is credited for creating the first reliable barometric altimeter for aircraft in the 1920s.

In the early 1980s, there were no dual-deployment–type altimeters in high-power rocketry. Occasionally, inventive flyers such as Korey Kline used model airplane RC equipment to fire multiple onboard ejection charges to deploy one parachute at apogee and then another larger chute as the rocket came closer to the ground. Yet this was rare; if they were so inclined and proficient in electronics, people also built homemade timers or stagers that relied on mercury switches to fire ejection charges at apogee to deploy their parachutes.

A mercury switch is a type of switch that opens or closes an electrical circuit via a small amount of liquid mercury. The mercury is typically encased in a tiny glass vial through which electric contacts are placed. The small bead of mercury is pulled by gravity to the lowest point in the glass. However, in flight, that bead can jump up between the electrical contacts at the wrong time, completing a switch that actives a circuit, such as a stager or a timer.

"In the old days, people said, 'Use a mercury switch to activate your parachute,'" recalls Billings. "But I think that was wrong. The mercury in the switch is going to move immediately upon motor burnout. It becomes a free-floating object in the glass vial. If this occurs, you would immediately activate your onboard charges. If it were a deployment charge for your parachute, you could be deploying at 500 miles per hour because the rocket is still moving very fast at burnout!"

Billings learned about pressure monitors when he worked on medical devices in the 1980s. Based on that experience, he experimented with pressure sensors in altimeters specifically for high-power rockets.

"A pressure monitor is a piezoelectric device that is a little container that has a diaphragm in it, and if you bend that diaphragm, you change the resistance across it," explains Billings. "And you can inject a voltage into that little disk and obtain readings of how voltage changes. For an altimeter device, one side of the disk has a perfect vacuum on it, and the other side is open to ambient air, so now when you are at sea level, you are reading maximum pressure because you have a vacuum on one side and sea level air on the other. And as you go higher in altitude, the pressure goes down, changing the output voltage, so you get an indication of altitude."

Billings tested a prototype barometric pressure altimeter near Boulder, Colorado, in 1990 using a digital circuit board. He plugged the board into the altimeter, and he drove up into the mountains.

"I would simply watch the numbers climb a foot at a time as I gained altitude," he said.

Billings ultimately abandoned the digital readout because at the time, they were too bulky and expensive for rocketry. Instead, he settled on an innovative method to determine altitude: a series of high-pitched electronic beeps.

"I was already familiar with beepers through my medical device work, so I used a series of beeps to announce the altitude of the rocket after flight," he said.

Beeping tones emanating from the altimeter in a recording loop after flight would allow the flyer to obtain his rocket's altitude *immediately* upon recovery. As a flyer walked up to his or her rocket, they paused and listened for the sound. A string of beeps would end with a pause; the number of beeps represented a numerical value. So for example, the altitude of 2,312 feet would sound like this: *beep, beep,* pause, *beep, beep, beep,* pause, *beep,* pause, *beep, beep.* There was nothing to download and no need for a computer. The result was an altimeter that was less expensive and was small enough to place in almost any high-power rocket.

Adept's first catalog was released in 1992, and ads quickly appeared in rocketry magazines around the country. Adept altimeters were not kits; they were complete, ready-to-install units. There was nothing to assemble, solder, or build. Although the early units were large (approximately 5 inches long and less than 2 inches wide), they only needed a single 9-or 12-volt battery to operate. The battery plugged directly into the unit. There was no need for any computer knowledge or electronics expertise. And Billings did not offer just one altimeter; he had five different altimeters for sale. The prices were affordable. Adept's altimeters ranged from $59.95 to $129.95.

Adept altimeters were a wake-up call for a new generation of high-power users and manufacturers alike. The ALTS1A altimeter, for example, provided altitudes up to 59,000 feet in 1-foot increments and had the ability to fire two separate ejection charges during the flight of the rocket. The practical aspects of this unique function were real.

Typically, a drogue chute is deployed first soon after the rocket reaches maximum altitude and starts descending. Then at a predetermined altitude before the rocket reaches the ground, the main chute is deployed for a soft landing and minimum wind drift. The term "dual deployment" was still several years away. Yet here it was, in concept and design, in the advertisement for Adept's new line of altimeters in 1992.

Adept was not the first commercial company to espouse the advantages of a multiparachute deployment system. In the late 1980s, North Coast Rocketry was selling a timer built by Transolve called the Veet-1 that, in conjunction with a motor-based ejection charge, would deploy a small parachute first (by motor ejection), followed by a larger parachute deployment charge later that was controlled by the Veet-1. The Veet-1 was a unique product and was one of the forerunners of modern deployment systems. A similar timer was offered in kit form by another flyer in 1988.

Billings gave this technology a big push forward with his superior, compact, multideployment, altimeter-based ejection system.

"Well, you watched someone pop a parachute at apogee, and the wind could carry it for miles," he said years later. "And it just came up one day—I don't recall how—that someone said something to the effect that if you waited until the rocket came closer to the ground before you opened the parachute, it would be better."

It was more than just better. By the end of the 1990s, the dual deployment altimeter would be a standard ticket item for every sophisticated high-power rocket in America. It would be incorporated into the certification programs of the NAR and Tripoli, and the dual deployment altimeter, a term Billings would coin in the mid-1990s, would become the primary reason for the existence of not only Adept but also every other yet-to-be-formed altimeter manufacturer in the hobby.

As time passed, several other manufacturers would enter the altimeter market, including Missile Works, blacksky, Perfectflite, G-Wiz, and others. In time, mass-produced commercial altimeters would become more complex, they would handle more functions, and they would become smaller, much smaller, and less expensive. Flyers could choose between altimeters based on air pressure (barometric) or acceleration (accelerometer). Eventually, GPS was added to the mix too.

For hobbyists hemmed in on small fields, the dual deployment altimeter was a godsend. The altimeter ensured that rockets fell quickly from higher altitudes with little drift until the main parachute deployed only a few hundred feet above ground. For western flyers who had the luxury of wide-open spaces, the modern altimeter also allowed the opportunity to more accurately gauge flights to fantastic altitudes—20,000, 30,000, 40,000 feet and higher—but still have the main parachute deployed so low, the rocket would be returned within a short distance of the pad.

The mass-produced altimeter also changed the construction techniques of hobby rocketry. Altimeters required special housings in the rocket's airframe called altimeter bays that had to be vented to outside air as well as sealed within the airframe of the rocket to prevent ejection charge incursions into the bay. Ejection charge gases could damage an altimeter. Manufacturers began marketing commercial rocket kits complete with prebuilt altimeter bays; others specialized in the construction of the bays themselves, or flyers simply built then from scratch.

Drogue parachutes became common, and manufacturers would develop altimeters that allowed flyers the ability to program their main parachute deployment at almost any altitude they wished. On a windy day, a flyer might want to deploy the main at 400 feet, on a calm day perhaps 1,500 feet, or any combination thereof. Flyers learned the importance of arranging the drogue and main parachutes in separate sections of their rocket to avoid tangling. The term "shock cord" would be replaced by "recovery harness," and flyers would build all these things around the functions of their ever more sophisticated altimeters.

The introduction of the modern altimeter affected the evolution of the high-power rocket motor too. Until the early 1990s, the primary method to deploy recovery gear was by ignition of the ejection charge at the forward end of a single-use rocket motor. This was no longer necessary. Altimeters could fire ejection charges that had nothing to do with the motor. Black-powder charges could be fired by an altimeter at or near apogee, when the rocket was moving slowly, reducing the chance for airframe damage caused by high-speed early deployments. Flyers would no longer have to make educated guesses about the length of motor-based ejection charge delays. As the reloadable motor revolution swept through the hobby in the 1990s, many motor makers abandoned motor ejection charges altogether for their L, M, and N reloadable motors. Such charges were no longer necessary. Altimeters could handle the deployment duties for the rocket.

In 1992, when the first Adept catalog was released, flyers of model and high-power rockets were still using a theodolite to gauge a rocket's altitude. A theodolite is a precision instrument, typically a telescopic device mounted on a tripod, for measuring horizontal and vertical planes. It was invented in the sixteenth century and was used in the twentieth century by people such as Goddard and Von Braun to optically calculate the altitude of a rocket in flight.

If you were even a casual flyer in high-power rocketry in the early 1990s, you probably knew what a theodolite was. You also knew what "tracking" meant. Flyers spent hours at a time sharing the duties of optical tracking, usually sitting on a stool hundreds of yards from the flight line, off the edge of the field, peering through the telescope-like device, and recording numbers and angles to be used in calculating a rocket's altitude. Observers strained to keep their eyes on a rocket as it raced into the sky. Optical tracking was the only way to go in those days. It was how flyers determined the winner of altitude contests or the setting of altitude records. And even with the advent of the early altimeters in the late 1980s, optical tracking remained indispensable. However, the modern dual deployment altimeter caused the theodolite to disappear from high-power rocketry virtually overnight.

"What killed optical tracking was it was a lot of work," recalls high-power pioneer Chuck Rogers, who was an expert at optical tracking at Lucerne and Black Rock in the 1980s. Rogers and his fellow flyers once tracked a Frank Kosdon rocket to 39,000 feet, he said, using a 3.3-mile baseline and multiple people at the range taking measurements. "As soon as these altimeters started coming out, optical tracking just died. In hindsight, we should have had an overlap period where we flew rockets with altimeters and we optically tracked them to see how they compared."

By the late 1990s, optical tracking was nowhere to be found on the typical high-power range. The theodolite disappeared from the launch field and from the collective consciousness of high-power rocketry. Today the average high-power flyer has likely never even heard of a theodolite, let alone seen one on a rocketry range.

LARGE AND DANGEROUS ROCKET SHIPS

With the advent of the commercial altimeter, high-power rockets now had a brain, and every year that brain became more sophisticated, performed more functions, and reported more information than ever before. One day even more sophisticated altimeters would help track high-power rockets to the edges of space. People like Tommy Billings made it that way.

21

"My Own Private Hindenburg"

After two years in Kansas, LDRS returned to Nevada's high desert in 1995. AERO-PAC hosted the four-day launch at Black Rock on August 10 to 13, followed by BALLS on Monday, August 14.

There were more than 350 registered flyers from at least 26 states and several countries, including Japan, England, Canada, and Sweden. For the first time at any LDRS, more than a thousand flights were logged, more than twice the flights at LDRS-12 two years earlier. In terms of the number of flights and registered flyers, LDRS-14 was the first LDRS mega-launch. There was nothing else like it in all of hobby rocketry.

Launching at Black Rock started on Thursday, August 10. High winds that morning kept most people in their vehicles and trailers. By midafternoon, a full-scale dust storm halted all operations. That evening, after the wind died down, a few impatient flyers hooked up someone's car battery to a lone launch pad and fired several rockets into the night sky. The next day brought better weather. There were nearly two hundred flights, and at one point, the spectator line stretched out more than 700 feet along the flight line, several rows deep.

These were the days when an M-powered flight still brought everything on the range to a halt. Many participants had yet to see an M-powered rocket fly at their local launch. They saw plenty of M-powered rockets at LDRS. David Cotriss thrilled everyone with his *Dauntless*, a 48-pound 9-foot-tall rocket that streaked into the air on an AeroTech M1939, trailing fire and smoke for several thousand feet. Cotriss had remote-control deployment switches on board, which were becoming a rarity in 1995. The switches failed to fully deploy the vehicle's recovery gear. The *Dauntless* plummeted to Earth and was destroyed.

Other flights on Friday included single and multistage rockets. There were L750 Silver Streaks, AeroTech K550s, and Vulcan K2000 motors. There was also a night launch under a full moon. One observer described "a kind of hovering mist on the desert floor over near the mountains. It looked like an artist's conception of some remote moon on an alien planet."

The most spectacular flight that evening belonged to Jason Blatzheim, who launched a J200 Firestarter motor in a North Coast Rocketry *Tomahawk* rocket. The vehicle had stability issues and went awry immediately, "pin-wheeling and cartwheeling across the launch area, showering sparks in all directions."

The number of flights increased on Saturday with scores of launches on all motor sizes, A through M power. The star of the day was a two-stage scratch-built *WAC Corporal* created by brothers Bill and Damian Davidson of New Jersey. The *WAC* was more than 11 inches in diameter and stood almost 25 feet tall. It was powered by three M1939 motors in the booster and carried an additional M1939 in the upper stage. The rocket was equipped with six altimeters and stagers, all manufactured by Adept Rocketry. The black-and-white United States Army rocket weighed more than 230 pounds and was one of the most powerful rockets to launch at any LDRS so far, eclipsed perhaps only by Dennis Lamothe's multiple-N-powered *Aerobee* at LDRS in Kansas the year before.

The Davidson family, also known as "Team WAC," spent Thursday and Friday assembling the *WAC Corporal* and hauling it out to their own pad more than two miles downrange. The project was ready Friday, but winds picked up, so the team decided to wait until Saturday morning. The rocket and pad were covered up Friday night, and the coverings were lashed to the ground with straps and steel spikes, as if pinning down some great beast. Flyer Neil Young camped near the pad all night to ensure nothing was disturbed.

By midmorning Saturday, everything was ready. Electronics were armed, and three huge igniters—each made out of a 1-inch slug of AeroTech White Lightning propellant, a loop of Thermalite, and also

Magnalite—were inserted into the booster's multiple 98 mm motors. At 11:00 a.m., the LCO pushed the launch button, the igniters lit, and the *WAC Corporal* thundered off the away cell into the bright blue desert sky.

All three M1939s fired as planned, and the booster and sustainer separated cleanly in flight. The last M lit right on cue, carrying the upper stage of the *Corporal* to more than 13,000 feet. The booster then returned safely via two large parachutes, and the sustainer was lowered under a 26-foot-diameter canopy that allowed the rocket to hang in the air for nearly half an hour. In an era where most large rocket projects failed at some point along their flight path, this flight was a complete success. The rocket graced the cover of *High Power Rocketry* the following year.

On Sunday, more than four hundred flights took to the air, including thirteen Ls, ten Ms, and one N-powered rocket. Everything from simple H-powered rockets used to certify Level 1 all the way to a Frank Kosdon/Mark Clark minimum-diameter M3200 flight to an estimated 32,000 feet (which was never seen again) was flown. It was the kind of day that confirmed LDRS was the rocket launch of the year.

One of the most memorable launches Sunday was the two-stage *Spirit of America* by Ron Early and Patrick McConaughy. Their 7.5-inch-diameter rocket was 17 feet tall, and the booster held nine motors. The sustainer carried an additional L750. The rocket had a great flight and staged perfectly.

Among the many flyers on the desert playa this weekend were several new faces who, in the coming years, would leave their mark on the hobby, people like Paul Robinson, who launched an M-powered rocket to 26,148 feet—Robinson would one day run one of high power's largest commercial motor-making companies—or Alex McLaughlin and Mike Hobbs, who each obtained their Level 1 certification on Sunday. Both would become dominant figures in the experimental rocketry scene, and within a few years, they would each be launching O, P, and Q motors at Black Rock. A young flyer by the name of Ken Allen launched on Sunday too. In the 2000s, Allen would become the largest dealer of high-power rocketry parts on the Eastern Seaboard.

The fifth BALLS event followed the day after LDRS-14, beginning on Monday, August 14, 1995. The BALLS launch was growing in stature every year. By now, it earned the reputation as one of the most exciting launches in amateur rocketry, where rockets of incredible power and size were flown every year.

Black Rock veteran Tom Binford of Georgia launched another one of his *Cloudbuster*-designed rockets at BALLS-5. The rocket weighed 190 pounds on the pad, of which 118 pounds was propellant. The motor was a 6-inch-diameter 7-foot-long P motor made up of nine BATES-type propellant grains. Binford estimated the motor would generate 2,000 pounds of thrust for more than 8 seconds—in other words, enough thrust to pick up a Volkswagen and throw it across town.

The Georgia flyer was looking for an altitude of 60,000 feet. To track the missile at high altitude, he poured a gallon of bright red chalk dust into the payload bay. Unfortunately, while preparing the rocket at the launch pad, one of the ejection charges accidentally triggered. There was a loud bang, and Binford found himself covered from head to toe in powdery red chalk dust. He cleaned up quickly, and the rocket was made ready for flight again. As it turned out, the high-altitude tracking dust was unnecessary. On ignition, the P8000 motor suffered a mighty CATO, destroying his *Cloudbuster* right there on the pad.

"Unfortunately, the forward closure let go inches into the flight," explained Binford afterward, who said he lost all his onboard electronics when the forward bulkhead was shot forward like a projectile by the CATO into the unprotected electronics bay. "I believe all the 6-inch motors at this BALLS launch sustained a CATO."

It was true. There were several other disasters on Monday. Jim Hart's aqua-green *Progressive Propulsion* was topped off by a titanium nose cone machined from a 155 mm howitzer shell. Hart's rocket carried a Ron Urinsco O6700 motor. The rocket roared off the pad and looked great for a few moments. Then it sustained a CATO in flight, completely destroying the vehicle.

Scott Ghiz and John Heinze had a great flight of a two-stage M-to-L rocket on Sunday at LDRS. On Monday, they returned to test their skills with the launch of a Dynacom *Tarantula*, a very advanced commercial kit for its day, on an M1939. The *Tarantula* disassembled under thrust and was also destroyed in flight.

Another ambitious yet ill-fated flight was the *Thunderbolt 2B* by Mark Clark, Robin Meredith, and motor maker Frank Kosdon, who supplied the O10,000 motor for the all-metal minimum-diameter high-altitude attempt. The two-stage rocket was aiming for an altitude of 85,000 feet and carried timers by blacksky and tracking smoke to help follow the rocket to altitude.

"The boost was perfect, fast and straight," said Clark later. "The sustainer lit, and it was moving so fast that it left no smoke trial."

At 30 seconds into flight, tracking smoke from the upper stage appeared, and the sustainer was followed high into the sky. The booster, however, was nowhere to be seen. A short time later, said Clark, "there was a tremendous roar for several seconds, and the ground shook as the booster impacted the ground 150 feet on the other side of the launch tower."

"The booster payload was too deep to recover." said Clark, "So the cause of the recovery failure is not known."

And the sustainer for the rocket? It could not be tracked all the way to apogee, even with a discernable smoke trial. It took another full day of searching before all the sustainer's parts were found. The onboard deployment systems had failed.

Yet there were some successes at BALLS-5 too. Karl Baumann launched a half-scale replica of a *Phoenix* missile on an AeroTech M2400 Blue Thunder motor for a perfect flight and a successful recovery. It was a historic flight of sorts as Baumann became the first official Level 3 flyer in Tripoli history. And Walter Blanca set a new Tripoli altitude record for an N-impulse–powered rocket with his two-stage vehicle powered by dual Energon L1100 motors. The rocket reached 30,942 feet. The altitude was gauged by an onboard altimeter, and both stages and the altimeter were recovered intact. It was the first altitude attempt verified by Tripoli's new Contests and Records Committee. Blanca's N-altitude Tripoli record would stand for four years. High-power rocketry had now launched and recovered a rocket from an altitude taller than the peak of Mount Everest.

The premiere attraction at BALLS-5 was also one of the largest high-power rockets ever flown: Chuck Sackett's *Project 463*.

Until 1995, *Down Right Ignorant* was the biggest rocket ever flown in hobby rocketry. That rocket—a joint project by Sackett, Dennis Lamothe, and Mike Ward—was flown at Fireballs in 1992. *Down Right Ignorant* weighed more than 800 pounds and was successful in both launch and recovery.

LARGE AND DANGEROUS ROCKET SHIPS

Project 463 was bigger than *Down Right Ignorant*. Much bigger. Sackett's new giant was the size of a small building and was built around a base of 36-inch-diameter Sonotube, a product commonly found at construction and home improvement stores. Sonotube consists of a preformed casing made of laminated wax paper that is almost cardboard like in appearance and texture. It is used by building contractors to make concrete columns by pouring concrete into the tube, waiting for the mix to harden, and then removing the tube after the form is set. By the 1990s, Sonotube was used frequently in high power as airframe material, especially in large projects.

The design of *Project 463* was a combination of several rockets, explained Sackett, including a two-stage pseudo ICBM with some influences from the *Saturn IB*, the *Mercury Redstone*, and the Russian *Vostok*. Sackett also built a one-twelfth-scale version of the vehicle to test his design. From its 3-foot-wide base of Sonotube, *Project 463* transitioned to a 24-inch-diameter airframe in the sustainer and a custom-made nose cone on top. The rocket contained fiberglassed sections of not only Sonotube but also commercial airframe sections by PML and LOC. There was aluminum tubing in the upper stage, and the entire vehicle had an internal skeleton of tubular steel. It took Sackett eighteen months to build the rocket in his Orlando machine shop in Florida, he said.

Sackett designed the rocket to carry a P motor and two N motors in the booster and a third N motor in the sustainer. It also carried multiple parachutes, including a set of chutes in the aft end of the rocket that would carry the tail section back to the ground separately after flight. When completed, *Project 463* stood 43 feet tall and weighed 1,200 pounds. It was by far the largest high-power rocket ever built up to that time.

The rocket was transported from Florida to Black Rock alongside another high-profile rocket: Mike Ward's *Stratospheric Dreams*. Final assembly of the individual components was completed on the desert floor during the four days of LDRS-14. Sackett and his volunteers—including LOC founder Ron Schultz, John Cato, John Sicker, and others—worked on *Project 463* all weekend. On Monday morning at BALLS, the rocket was raised to skyward on its pad. Sackett designed a massive launch tower that allowed several people to raise the rocket and tower into vertical position with a winch. Even from a mile away at the rocketry range head, the rocket looked gigantic.

Project 463 was ready by noon on Monday. Then the four motors were installed: a P9300 and two N2750s in the booster and another N2750 in the sustainer. Mike Ward inserted igniters in the booster motors, and the range was cleared of all persons. The countdown commenced. Tripoli member Tim Eiszner, LCO for the launch, warned everyone to stay back more than a mile from the pad just in case. Sackett described what happened next:

> It was time. The flight line gave us the go-ahead for Mike [Ward] to start his countdown. When he reached zero, I pushed the button and held on for dear life! The motors lit, and *Project 463* leapt skyward. It was hard to believe after all this time that this beast is in the air! Then things went awry. The second stage failed to ignite, and as the rocket came over the top of its arc, I realized the upper stage was doomed. A few seconds later, when the ejection charges failed, it was obvious the whole rocket was doomed. At about 150 feet off the ground, the two stages separated because of wind resistance [and] then impacted less than 30 feet from each other. And that was it. I had to sit down.

The impact of the heavy rocket threw up an expanding column of dirt, dust, and rocketry debris. The cloud slowly rose above the playa in all directions. The rocket was entirely destroyed. Even at the launch pad, there was a crater. For several yards in every direction, there were boulder-sized chunks of playa littering the ground. They had been blasted out of the desert floor by the simultaneous ignition of the P and N motors on takeoff. Sackett and his crew searched for clues as to the cause of the mishap. But they never pinpointed the precise mechanism for the crash. They knew onboard electronics failed to separate the stages in flight, and there was no activation of the multiple parachutes carried in the rocket. Why the electronics failed was anyone's guess, said Sackett.

Project 463 failed to duplicate the success of *Down Right Ignorant*. Yet as of 2018, it remains the largest rocket to ever fly at any BALLS event and one of the largest rockets in high-power history.

"I did not plan to have the biggest plow-in in Tripoli history," said Sackett afterward. "Nor did I expect to blow a hole in the desert from the rocket motors, big enough to put a Volkswagen in. It is part of the hobby. You win some, and you lose some. *Project 463* will be my own private *Hindenburg*."

Even with the destruction of *Project 463*, Sackett and his launching partner Mike Ward were not finished flying at BALLS-5. The pair had transported another rocket from Florida to Black Rock, and this rocket was as ambitious as the first. Sackett's *Project 463* was designed to put a lot of weight into the air, low and slow. Their second vehicle, *Stratospheric Dreams*, was the exact opposite: a minimum-diameter vehicle aimed at the earth's upper atmosphere.

"I wanted to put up something that went so high, it would take a long time for anybody to beat it," said Ward, whose rocket was the most ambitious high-altitude attempt by any Tripoli member to date.

It was all built around a motor that also served as the airframe and bottom end of the rocket: an S24,000 that computer simulations estimated could propel the project to more than 100,000 feet. Sackett and Ward had worked on the rocket together. Ward designed the propellant, and Sackett performed the machine work on the motor case and the rocket's metal air frame. The case for the S motor was made of carbon steel. Empty, the case weighed more than 200 pounds.

"I wanted it to be safe," he said. "The PSI burst pressure of the steel case would be at least 4,500 pounds. The computer simulations predicted a PSI in the S motor that would not exceed 900 pounds. We would therefore be nowhere near the burst pressure of the steel case."

The motor case accepted three metal fins that were based on a United States Air Force *Nike Missile* fin. The propellant was composed of four Bates-type grains cast inside PML 11.41-inch-diameter coupler tubes. The grains slid into standard PML airframe tubes that served as the motor liner, separating the burning propellant from the metal case. Each fuel grain weighed 80 pounds for a total of 320 pounds of high-power propellant.

Sackett machined the nozzle for the motor out of steel. It was much cheaper than graphite, he said, "and if you were to drop it, it would not break." The nozzle weighed 85 pounds. The weight of the fuel, motor case, nozzle, and metal airframe pushed the liftoff weight of the rocket to 750 pounds.

Stratospheric Dreams was unpacked on Tuesday, August 15, the day after BALLS officially ended. The plan was to have everything ready for a flight early Wednesday morning. By now, there were few flyers left from LDRS-14 or BALLS. Yet several people remained to lend a hand for the launch, including Tripoli president Kelly and high-power motor guru Frank Kosdon. If this rocket performed as planned, it would be a truly historic flight: a high-altitude shot many years ahead of its time.

On Wednesday morning, August 16, 1995, the big rocket and its launch tower were raised together. The assembly was 17 feet tall. The rocket was topped off by a custom-made 85-pound nose cone made of pressure treated, 6-by-6-foot wooden timbers, and an all-steel tip to withstand the heat generated by the rocket's estimated top speed of Mach 2. The total impulse was 340,500 Newton seconds. At its peak, the S motor was expected to generate more than 5,000 pounds of thrust.

To alert everyone within earshot of the impending countdown, Kelly activated his truck's burglar alarm for 60 seconds before takeoff. After a ten count, Ward pushed the button to light one of the biggest motors in high-power rocketry history. At the aft end of the rocket, a bright yellow flicker of flame shot out of the throat of the nozzle and barely teased the ground below; the mighty S motor was alive and coming up to pressure.

Then the entire thing just blew up.

With a thundering roar, the S motor disintegrated as internal pressures from the burning gases pushed outward in every direction, exceeding the 4,500 PSI burst strength of the all-steel motor case. The instantaneous CATO not only destroyed the motor and lower end of the rocket but also obliterated the entire vehicle.

There would be no new high-altitude record today. There was nothing left of *Stratospheric Dreams*.

22

1996: Orangeburg and the OuR Project

Tripoli's national launch in 1996 was the first LDRS east of the Mississippi River in ten years. There were only forty-nine people at LDRS-1 in Medina in 1982; at LDRS-15, there were sixty launch pads and nearly a thousand flights. More than 1,100 rocket motors were burned. The setting for LDRS-15 was unlike any LDRS held before. There was no furrowed farmland, no hard-baked desert soil, and no Midwest winds to contend with; there were no Rocky Mountains to climb either. In fact, there was hardly any dirt on the ground. The launch was held on dozens of acres of luxurious green grass. Flyers could kick back on the ground in a T-shirt and shorts with no shoes on and watch their rockets return to Earth. This was Orangeburg, South Carolina.

Orangeburg was named for William IV, prince of Orange, who was the son-in-law of King George II of England. It was settled by Europeans in the early 1700s when an Indian trader opened up a trading post there. The town was the site of large cotton plantations in the nineteenth century, and eventually, Orangeburg was home to a large African American community. During the civil rights movement of the 1960s, the city was rocked by demonstrations. In 1968, several students at the local college were killed during a confrontation between police and demonstrators.

Today Orangeburg is the county seat of Orangeburg County. Geographically, the city is located off Interstate 26, about 40 minutes southeast of the state capital at Columbia and along the road to Charleston, one of the busiest seaports in North America. The population of the metropolitan area is sixty-seven thousand. It is the largest city to ever host an LDRS.

Orangeburg is also known as the "Garden City." The area is dotted with lush green landscapes, and it is home of the Edisto Memorial Gardens, a collection of all-American roses planted across 150 acres of meticulously cared-for soil.

Beginning in the mid-1990s and continuing through 2015, Orangeburg was one of the premier rocketry venues in America. It was the site of the annual Labor Day weekend Freedom Launch as well as numerous other rocketry events, including the NAR's National Sport Launch in 2008 and 2015. Orangeburg was also the home of both a Tripoli prefecture and a NAR chapter.

The seeds of the first East Coast LDRS were planted in Georgia. In the early 1990s, members of Tripoli Atlanta sought out a company called Super Sod to request permission to launch high-power rockets at one of their farms in South Georgia. Super Sod is one of the largest producers of grass in the South. Turf at Super Sod is grown on huge farms scattered throughout the Southeastern United States, and the sod is harvested as lawns for homes and businesses in Georgia, the Carolinas, and Florida. Super Sod farms are wide-open expanses of manicured short-cut grass. They are typically devoid of trees, plants, or other manmade obstacles. They are perfect locations for high-power rocketry.

Thanks to the efforts of Tripoli Atlanta, Super Sod granted high-power flyers the use of one of its farms in Douglas, Georgia, in the early 1990s. This flying area was later expanded to include another Super Sod farm near Byron, Georgia. Soon, there were multiple Tripoli prefectures in Georgia. From these oversized lawns and with FAA waivers ranging from 10,000 to 25,000 feet, high power experienced rapid growth in the Southern United States throughout the 1990s. Super Sod charged flyers nothing for the privilege of using their land.

Tripoli member Tim Eiszner was prefect of Tripoli Atlanta. As the club grew in the early 1990s to more than sixty people, Eiszner met Larry Smith, who, in turn, met a flyer by the name of Jim Conn, who was from South Carolina. The trio launched rockets at the Georgia sod farms every month and became friends. There were several other influential high-power flyers at these farms too, including Tom Binford, Steve Roberson, Rick Boyette, and John Cato. Cato was likely the original Tripoli contact with Super Sod, and he held the FAA waiver at the Douglas field for several years.

One day Jim Conn decided to do some research to see if Super Sod had any farms a little closer to his home in South Carolina. He learned the company had a sod farm at Orangeburg. He and Larry Smith drove to Orangeburg seeking permission to fly high-power rockets there. When they arrived, they met the vice president of Super Sod, who happened to be living on the property at the time. Fortunately for high-power rocketry, it was a productive meeting. Another new Tripoli prefecture was born, and within a short time, rockets filled the air above the Orangeburg Super Sod site. It was 1993.

By 1995, LDRS was in place for fourteen years. Yet there had been no LDRS east of Kansas since the last Medina event in 1986. Every year a handful of East Coast flyers made the trek to LDRS at Hartsel, Argonia, or even to Black Rock. But for the average flyer in the east, the trip was too much of a hardship.

"The problem was the distance was so great, it was difficult for the average East Coast flyer to get to even Kansas, let alone the Black Rock Desert," recalls Eiszner, who lived in Georgia. Eiszner and a few of his friends had made the trip to some of the western LDRS events. "But there were few others who would ever be able to experience an LDRS if something wasn't done to bring one east."

From time to time, there were searches for an East Coast LDRS location, but they always came up short. Then in 1995, Eiszner, Conn, and Smith decided to do something about it. The trio contacted Tripoli President Bruce Kelly, inviting him to Orangeburg in the summer of 1995 to survey the field and surrounding area during a monthly Tripoli-ICBM launch. Kelly flew to South Carolina, surveyed the field, and liked what he saw. At the same time, Conn and Eiszner approached Super Sod to ask if high power's national launch could be held on their sod farm in Orangeburg. Meanwhile, Smith went to the FAA to try and secure a special waiver for a possible four-day-long launch event.

During their LDRS discussions with Super Sod, Eiszner and Conn learned the sod company had a tradition of granting their Orangeburg employees a long weekend whenever the Fourth of July holiday fell on either a Tuesday or a Thursday. They noted that during the following summer in 1996, the Fourth of July was on a Thursday. Super Sod told Tripoli that they were welcome to use the sod farm for four days, July 4 to 7, 1996, for their national launch. LDRS was about to move east again.

The first Orangeburg LDRS began with setup of the 60 launch pads, 4 miles of caution ribbon, and the 4,000-square-foot tent that would serve as the meeting and eating area for the membership, reported Conn, who personally built forty of the launch pads needed for the big event.

Volunteers from several eastern prefectures stepped up to help staff the launch. Meanwhile, Eiszner, who would serve as launch director, set up a 1,000-watt public address system with sixteen speakers that could be heard all over the range. Flying began on Thursday, the Fourth of July. Overhead, Space Shuttle *Columbia* was two weeks into one of the longest shuttle missions ever. This was *Columbia*'s twentieth flight, having been launched on June 20. It would spend more than sixteen days in orbit.

"After having been approached with the notion of hosting the country's premiere rocket event at my local field in Orangeburg, South Carolina, I started having my worries," said Conn later. "Who would waste their time and money coming to a relatively small field with a mere 10,000-foot waiver?"

Conn received his answer immediately. On Day One, usually the slowest day at any LDRS, at least three hundred flights took to the sky as flyers from all over the East Coast converged with every type of model and high-power rocket available, A through M power. It was likely the biggest opening day in LDRS history.

The weather was perfect, with "hardly a breath of wind," said one flyer. Sometimes a light breeze carried rockets off the field and into a nearby pine forest that was dark, dank, and almost primeval in nature. Recovery in the dense cobweb-filled trees was tricky. This remained a problem throughout the event. But with minimal skill and a little bit of luck, it was an obstacle most flyers avoided.

"Lots of people were using Thursday to get their Level 2 certification flights launched and recovered," noted another flyer. "Most [people] were flying AeroTech RMS Js. Some [flights] were on Vulcan J250s and hybrids."

The biggest rocket on Thursday was a hugely upscaled LOC *I-ROC*. The red-and-white rocket was meticulously constructed and beautifully painted. It carried thirteen high-power motors: a central AeroTech M1939 surrounded by four K550s and eight I211s. The size of the rocket and its ambitious power plant grabbed everyone's attention as it was prepped and carried out to an away cell for launch. At ignition, the motors kicked out a long yellow flame and endless amounts of dense white smoke. The vehicle rose well initially, but not all the motors lit on cue. The rocket became unbalanced, and it crashed to the ground. The owner of the rocket was a flyer by the name of Steve Eves.

"My hat is still off to Steve, who I know will be back with projects that will be bigger and better," said one writer after the launch.

After this flight, Eves left high-power rocketry for many years. When he returned a decade later, he would launch the biggest high-power rocket in Tripoli history.

Prior to LDRS-15, local flyers did research to check on the historical weather patterns in the area. They learned it had not rained at Orangeburg over a Fourth of July weekend in 132 years. Until 1996, that is. Friday's weather was wet, very wet, with rain falling off and on for much of the day. Flyers were confined to their tents, cars, and nearby hotel rooms. The number of launches plummeted.

"Connecticut Tripoli was running the show on Friday, and [they] did everything they could to keep the flights going as long as the weather conditions permitted," recalled Conn. "[So] there were still ninety-one flights, eleven of them certification attempts."

On Saturday, the weather was still wet and gray—but not for long. The skies cleared late in the morning, and rockets once again took to the air in great numbers. More than 360 flights were recorded in the heaviest day of flying at LDRS-15. There were almost fifty Level 1 and Level 2 certification attempts on Saturday as all sixty launch pads on the field stayed busy most of the day.

"At one time, I was out in the field only to glance up and see four rockets floating down under canopy while the countdown was on for a fifth!" said one excited flyer.

Saturday was also a big day for M-powered flights. There were at least nine M flights, six of which were Level 3 certification attempts, the first Level 3 attempts at any LDRS. At the time, there were only a handful of Level 3 flyers in the country. The launching of an M-powered rocket at LDRS was still a showstopper.

One of the M flyers on Saturday was, fittingly enough, Tom Binford. Since 1989, the Georgia resident made several treks to Black Rock for LDRS and BALLS. He launched the first O motor at Black Rock and was a constant contender for Tripoli's high-altitude records. Now he was launching in his own backyard. Binford's *Mega-Motor-Eater* had a great ride up on an AeroTech M1939.

In addition to Binford's flight Saturday, Jim Cornwell sent up a black-and-yellow ultra-scale *Mosquito* on an M1939, and flyer Damian Russo launched a multistage rocket that held not only an M1939 but also fourteen additional motors (two Ks and twelve Hs). The exhaust from Russo's rocket left a large crater under the pad. Unfortunately, the failure of one of the K motors was fatal to his Level 3 success. Bill Davidson also flew a rocket on an M1939 with four K550s for a good flight and successful certification.

On Saturday night, the annual LDRS members' banquet was held outdoors on the grass field under a huge tent where Southern-style barbecue was served. It was a great evening, followed by another full day of flying.

There were several M launches on Sunday, including the *Flying Pyramid of Death* by Brent Wynn. This odd-roc design was a crowd favorite. Wynn's pyramid-shaped rocket was 42 inches wide at the base and almost 4 feet tall. It weighed 50 pounds with an AeroTech M1419. The pyramid had a spectacular takeoff and then arced over and veered, pointy end first, into the ground a couple hundred feet away. Flyers would eventually perfect the M-powered odd-roc–type flight—but not at LDRS-15.

More than 260 rockets were flown on Sunday, making the official LDRS-15 total 995 flights, although the actual number of flights may have been even higher.

"LDRS-15 was a smashing success and broke records for attendance and flights," declared Tripoli board member Bruce Lee after the launch.

The sixth annual BALLS launch was held a few weeks after LDRS-15, on August 16 to 18, 1996, at Black Rock. At the time, the Tripoli board of directors and the Future Directions Committee were trying to figure out whether Tripoli should authorize homemade motors outside of Black Rock. As the board pondered the future of research motors, several Tripoli members decided to take another big leap with experimental motors. They were going to launch a rocket carrying an R-impulse motor to the edge of space. They called their creation the *OuR Project*.

The *OuR Project* was conceived by Frank Kosdon, Paul Robinson, Jim Rosson, John Dunbar, and Phil Prior. Their goal was to launch a single-stage rocket to an altitude of at least 100,000 feet. The team spent two years creating their rocket, investing thousands of man hours and nearly $20,000 in personal funds into the project. Their final creation was an all-metal missile containing 300 pounds of APCP. The weight of the rocket on the pad was approximately 700 pounds.

The *OuR Project* was initially scheduled for launch at BALLS-5 in 1995. However, construction delays slowed the project, which, in some ways, may have been a blessing, said team member Phil Prior. The crew attended BALLS-5, and they observed that the majority of large-motor rockets were destroyed either on the pad or in flight. And the culprit was usually a catastrophic motor failure.

"It became apparent that simply upscaling the smaller [motor] designs wouldn't work," said Prior. "After BALLS-5, we had an entire year to work out a solution. We consulted every professional source we could find. Consensus among the propulsion engineers that would work with us was that the lower grain was likely collapsing into the nozzle throat under thrust and choking or overpressurizing the motor."

With an extra year to prepare, the *OuR Project* team changed their motor design, especially with regard to the lower grains and the nozzle configuration. They were now ready in 1996. The final motor characteristics were more than just impressive. The rocket was powered by an R17,542 with propellant by Kosdon. The aluminum motor case held five Bates grains stacked one on top of the other. Each grain was nearly 10 inches in diameter. The fuel would burn for 14 seconds, generating more than 4,000 pounds of thrust at its peak. It was the most ambitious high-altitude attempt since *Stratospheric Dreams*. The *OuR Project* captured the imagination of everyone in high-power rocketry, and it also attracted media attention on launch day on August 16, 1996, at Black Rock. There was a film crew from the BBC as well as writers and photographers from *Popular Mechanics* magazine.

On the pad, the rocket stood 21 feet tall. Its nose was carefully angled a few degrees away from the town of Gerlach and the spectator area just in case. At ignition, the rocket came to life. There would be no CATO of this experimental motor. Phil Prior described what happened next.

> There is no lag, just what appears to be almost an explosion under the rocket. The fireball grows rapidly, and there is a loud boom as the rocket shoots off the pad. This is the first launch I have seen with a blast deflector. Instead of mixing with the dirt, the initial motor blast is deflected out and up around the rocket. I am convinced for a split second that the motor has blown the nozzle. These fears are short-lived as the rocket comes out of the flash and dust cloud trailing the anticipated 30-foot flame. Jim [Rosson] and Paul [Robinson] watch silently, but I cannot contain my glee as the motor burns and burns on a perfectly straight trajectory.

"The rocket flew perfectly straight, and the motor functioned flawlessly, with a total [burn] time [of] around 12 to 14 seconds," said Robinson later. "The rocket coasted for about 80 seconds, slightly more than calculated, into the ozone layer."

Onboard video from the *OuR Project* relayed live images to a television screen on the ground.

"What was seen was a clear outline of the entire Black Rock Desert and the surrounding terrain, the curvature of the earth, and the great blackness of space," reported Robinson.

Computer simulations performed afterward suggested the rocket reached an altitude of more than 93,000 feet. This was a magnificent victory for high power: an R-powered rocket that left the pad and reached a height of more than 18 miles without a mishap, no CATO, no shred during ascent, no in-flight disaster, nothing that had not been planned.

It was recovery time. Like any high-power rocket, the *OuR Project* needed to fire its ejection charges, separate into two sections, and unfurl its multiple parachutes for recovery. Instead, the rocket turned over at apogee and plummeted straight back to the ground, intact and undeployed. A sonic boom heralded the missile's return to Earth. The rocket's impact site was almost 5 miles from the pad. There was little trace of it on the surface.

"Fortunately, someone had seen a dust cloud and located the impact site," said Prior. "There was no discrete hole, but [there was] an area 20 feet across [that] appeared to have been fragmented and lifted by underground shockwaves."

"The lake bed literally opened up and swallowed the rocket whole," explained Bruce Kelly, who was on hand to witness the important launch. "The earth was disturbed, but no discernible entry hole could be found, no 'fin slices' to indicate [the] angle of penetration. For a radius of approximately 25 feet, the ground moved like liquid for nearly 2 hours. It reminded me of an earthquake and the 'liquefaction' that sometimes occurs in the immediate area. After a couple of hours, the ground returned to the hard clay silt we are all used to."

The *OuR Project* was buried so deeply, ordinary shovels could not reach any portion of it.

"The kind folks attending BALLS wanted the rocket recovered and collectively donated over $400 to rent a backhoe for excavation," said Prior. "Paul [Robinson] rented a backhoe and dug down approximately 10 feet. He was still unable to touch the wreckage with [an] 8-foot-long probe."

In the end, nothing was recovered of the rocket save a few small pieces of twisted metal and some burned-up fragments of Nomex cloth. Today the rocket rests deep under the Black Rock Desert. Although onboard electronics in the *OuR Project* were never recovered and a definitive altitude cannot be ascertained, subsequent calculations by Tripoli board member Chuck Rogers revealed the rocket likely reached an altitude of 94,000 feet. This would never be an official record, however, since the vehicle was destroyed. But the flight excited everyone in high-power rocketry.

It would be many years before any flyer would reach an altitude anywhere near what was accomplished with this rocket. Yet the *OuR Project* demonstrated such heights were possible.

23

The Rise of the Basement Bomber

Part I

In the 1980s, the manufacture of an AP composite motor was a dark art involving closely held formulas, exotic chemicals, and other secrets unknowable to the ordinary hobbyist. For the average flyer, a rocket motor may as well have been a witch's brew of potions, toxins, and spells that, when mixed together just right, yielded a magic stream of fire shooting out of the nozzle. The conventional wisdom was that the creation of such devices was best left to the expertise of commercial manufacturers.

The advent of the Tripoli Rocketry Association in 1985 did little to alter the idea of who should and who should not make high-power motors. Tripoli, like the NAR, had no interest in homemade rocket motors. Such motors were banned at Tripoli events in the 1980s. The early years of high-power rocketry were about the spread of *commercial* motors, not private ones.

Don Carter and Mark Weber, among the cofounders of the modern Tripoli Rocketry Association, toyed briefly with the idea of circulating APCP motor formulas to the public in 1986. In reality, however, Carter believed the typical enthusiast would find it difficult to make his or her own motors from scratch.

"First, you have to find out who sells what and what the minimum order is," explained Carter at the time. "Often companies won't sell to you unless you are a business yourself. If they do sell to you, you often have to pay a minimum of $100 or more to get whatever you want. Now let's say you're able to get all the chemicals you need and they arrive at your doorstep, all packaged very neatly. What do you do with them? You know that the composite fuel is hard like rubber. You need to know how to get it like that. And just what percentage of what do you put in? Starting to sound a little harder now, huh? So let's suppose you do get a formula and you are able to get the fuel to harden. Now what? How much fuel goes into that tiny vessel? And how do you get it to stay put without falling out? And are you going to have a core burn motor, end burner, or some other exotic grain configuration? And how are you going to do all that? So you think to yourself, 'Maybe I can call someone in the motor business to help me.' Well, think again! There are very few people who will give you the time of day, let alone any information about rocket motor design."

Carter and Weber considered writing a book on homemade motors.

"But of course, there is the chance that some fool will blow themselves up with the information," said Carter.

So they never wrote their book. And neither did anyone else—at least for a while.

Then the reloadable motor revolution arrived in 1990. Suddenly, Gary Rosenfield and Frank Kosdon blurred the distinction between the basement bomber and the average hobbyist. Flyers were now assembling model and high-power rocket motors right there on the edge of the field. And neither Tripoli nor the NAR voiced any objections. Indeed, both organizations embraced the change. People could now assemble commercial high-power motors almost from scratch in a reusable metal case from a package of premanufactured chemicals, pieces, and parts.

"What's in the bag?"

"Everything you need to build your own motor."

"Is it hard to do?"

"Nope. Takes less than 5 minutes."

"What are those things?"

"A few O-rings, a nozzle, a liner, and the fuel grains."

"That's all there is to building a high-power rocket motor?"

Any mystery regarding the internal components of the high-power motor was now over. And it was only a matter of time before someone else would ask, "Why can't I make this whole motor myself?"

As it turns out, curiosity may be the real fuel of high-power rocketry.

In the late 1960s, with projects such as the *Gloria Mundi* rocket, early Tripoli Federation members experimented with the manufacture of their own fuel.

"We just weren't sufficiently skilled in those days, and no one was trying to make APCP, at least not yet," recalled future Tripoli leader Ken Good.

The idea that average hobbyists were capable of making their own APCP motors from scratch steadily gained support in the early 1990s. During the Tripoli elections of 1991, Tripoli board candidate Ken Vosecek of Illinois was among the first to call for broad-based support for homemade motors.

"The members that 'push the envelope' with these activities represent the true frontiersmen spirit that Tripoli was founded on," said Vosecek. "They must be allowed to continue that spirit of tradition."

Bruce Kelly's *High Power Rocketry* began expanding coverage of amateur rocketry—that is, rocketry where people made their own fuel—no later than 1993, when he added an "Amateur" section to the magazine. He published several articles on motor making over the next year, ranging from zinc-powered motors to experimental hybrids used by the Pacific Rocketry Society. He also printed stories on the formulation of AP for fuel. One of these articles, written by Randall R. Sobczak, outlined nozzle theory and design, propellant characteristics, grain geometry, binders, and particle sizes. It was less than a blueprint for the creation of homemade APCP, but it was a step in that direction.

That summer, Kelly renamed the Amateur section of the magazine. The use of the word "amateur" was connotative of the basement bomber days of model rocketry, he thought. He changed the title: amateur rocketry would now be called "experimental" rocketry.

Although a few Tripoli members were already making their own fuel and there was open dialogue in the hobby press on the subject, Tripoli as an organization was reluctant to commit to the new technology. In October 1993, the board reaffirmed its prohibition on the use of homemade motors at any Tripoli-sanctioned event. The board believed Tripoli's insurance policy and safety code mandated only commercial motors at such launches.

Not surprisingly, the board's decision was viewed by some members as a barrier to progress. The subject would become a hot topic in Tripoli elections over the next few years. In early 1994 and in response to calls for change, the board created the Future Directions Committee, charged with, among other things, investigating methods for Tripoli members to legally and safely manufacture their own motors. At the same time, the board reaffirmed its position that members were prohibited from using their own motors at any Tripoli-insured events. Later that year and then again in 1995, additional board candidates expressed support for members who wanted to make their own fuel.

"I consider myself prochoice," declared one candidate. "I use disposable, reloadable, and my own homemade rocket motors."

Another candidate demanded the taboo of the homemade motor be removed by Tripoli immediately and the board's recent decisions on the subject of experimental motors be reversed. "This activity is currently quite widespread and growing," he said. "Do we continue to turn our backs on it?"

President Kelly, by now one of the most influential voices in hobby rocketry, also weighed in on the issue of experimental motors. Kelly had originally joined with other board members who voted to prohibit homemade motors in late 1993. But times were changing.

"As a board member, it has been one of the most difficult decisions that I had to support," he said later. "Some have argued that the number of members interested in this activity is small in comparison to the whole organization. True as this is, the number is growing, and this issue must be addressed."

The Future Directions Committee was charged with not only looking into the feasibility of solid-fuel homemade motors but also studying another new source of commercial high-power rocketry propulsion: the hybrid motor.

A hybrid is a motor that uses propellants in two different states of matter, one solid and the other either gas or liquid. A hybrid motor consists of a pressure tank, usually containing the liquid or gas to act as an oxidizer, and a combustion chamber that contains solid propellant. A valve separates the tank and the combustion chamber. When it is time to light the motor, the valve is opened, and the gas (or liquid) in the tank flows into the combustion chamber. An ignition source is introduced into the chamber, and the fuel ignites, providing the thrust needed for rocket flight.

Amateur rocketry enthusiasts had been experimenting with hybrid motors for some time and the essential technology had been around for many years. Yet in the early 1990s, there was little, if any, mention of the subject in the high-power community. At the time, the reloadable motor revolution was still fresh in everyone's mind.

Then in 1994, a Florida-based company asked the Tripoli board of directors for permission to demonstrate hybrid rocket motors at Tripoli launches around the country. The company was called Hypertek, and one of its cofounders was a longtime high-power rocketry enthusiast by the name of Korey Kline.

Korey Kline was among the pioneers of high power in the early 1980s. He built his own composite motors, manufactured a popular line of rocketry kits, and participated in several of the early LDRS events at Medina. By 1985, Kline was a household name in the high-power community. Yet as the decade wore on, Kline adopted a lower public profile, and in some ways, he dropped out of sight. But he never stopped thinking about new ways to improve rocketry.

Kline was born in 1958 in Encino, a small suburb north of Los Angeles in the San Fernando Valley. Kline's mother was an artist; his father was an engineer.

"That's where my curiosity and dream of space flight originated," Kline said in an interview in 2003. "When I was eleven years old, my father would turn on the TV, and we would watch liftoffs and landings of the rockets and spacecraft. My father always encouraged me to participate in hobbies and model airplanes and things like that. His generation was more into planes. But I always felt the new and upcoming hobby was rocketry."

Kline discovered model rocketry when he was ten. His first rocket was an Estes *Streak*, which he launched on a B motor.

"As soon as I pushed the button, the *Streak* disappeared," he said. "I was so amazed that the rocket took off that fast and disappeared that I was hooked at that moment."

Kline built model rocket kits made by Estes, Centuri, FSI, and Enerjet.

"I started doing my own scratch-built rockets when I was twelve or thirteen," he said. "When you look at the costs of buying a kit and what you could get by buying the components and designing it yourself, it only made sense . . . I liked the idea of emulating the NASA space program, and going to the moon or Mars was probably the deepest motivation at the time. I had hoped to become an astronaut. That was my dream."

After high school, Kline sought to pursue his dream of a career in the space program. He applied for admission to the United States Air Force Academy, one of the best pathways into a space-related career.

"I wanted to be a test pilot and then later go to astronaut school," he said. "The way you get into the Air Force Academy is to be nominated by a senator or a congressman. Each state only gets a handful of candidates. Because California is a very large state, it had hundreds of candidates. I tried for two years in a row to get in, but it didn't work out. Since the Air Force Academy wouldn't have me, I decided to start my own space program. At that point, I decided to get serious about learning about rocket motors and propellants."

He enrolled at California State University, Northridge, and studied mechanical engineering. Through a mutual model rocketry acquaintance, he was introduced to another young man in Southern California who was also interested in rocketry. His name was Gary Rosenfield. The two became friends, and Rosenfield helped Kline land a job at Bermite, where Rosenfield was already employed as an engineer working with solid-fuel propellant.

Kline and Rosenfield worked together at Bermite in Santa Clarita for about a year and a half. Among other things, they worked on reduced smoke propellant for the military's *Sidewinder*, a state-of-the-art air-to-air missile carried by military jets. Kline eventually moved on to other defense contractors in Southern California, including Space Ordinance Systems and Special Devices Incorporated. In the late 1970s, he and Rosenfield were among a handful of model rocket flyers launching bigger and bigger rockets out in a lonely desert spot near the Lucerne Valley. At Lucerne, Kline was introduced to several other model rocketry enthusiasts who were also pushing the limits of the hobby, including Chuck Rogers, Jerry Irvine, Bill Wood, and Roger Johnson.

"We were developing what was called model rocket technology, which was the precursor to high-power rocketry," said Kline, who was a NAR member at the time. "We believed in all the NAR safety rules, except for the maximum propellant and takeoff weights."

When Rosenfield and John Davis teamed up to form Composite Dynamics in 1978, Kline was among the first to buy one of their higher-powered motors.

"They were such a deal back then—only $13!" he said. "If you bought ten of them, it was $100! I remember thinking at the time, 'A hundred dollars for rocket motors! What is this hobby going to come to if you have to spend a hundred dollars to fly a rocket?'"

In 1979, Kline started his own company called Ace Rockets. He marketed his products locally and through advertisements in Jerry Irvine's *California Rocketry*. And in 1983, Kline attended LDRS-2 at Medina.

"We had been networking with Chris Pearson for a couple of years," said Kline. "A lot of it had to do with the NAR cracking down on the high-power guys. California was pretty much developing the technology, and the East Coast was using the technology, but they were very much in touch with the politics of the NAR."

Kline recalled the first few LDRS events as small and informal, where clusters of model rocket motors and the occasional H or I motor would steal the show.

"I do remember that if you flew a three-motor G cluster, everyone would stop and watch," he said. "A four-G-motor cluster would shut down the launch, and everyone would stare in awe and amazement."

Kline built his own rocket motors in the early 1980s, which he distributed under the name Visijet. The Visijet was among the first, if not the first, high-power special effects motors. It produced a purplish flame, sparks, and lots of smoke. He introduced his Visijet motor at LDRS-3.

Kline was just the ninth member to join Tripoli when it became a national organization in 1985. At the time, his rocket kits and motors were already a staple at launches around the country. But he eventually grew weary of making kits, so he entered into a licensing agreement with Chris Pearson and Matt Steele, where they took over the Ace line as part of Pearson's North Coast Rocketry.

"They carried the kits for a couple of years [and] then dropped the line after developing their own designs," said Kline. "By then, I had moved onto other things, and it all just sort of fell apart."

Kline was also one of the early pioneers of large composite motors. This included the N1940, which Kline said was the result of a joint effort between him and Bruce Kelly and Tom Blazanin. In the mid-to late eighties, this N-impulse motor was among the largest commercial high-power rocketry motors available; for years, it was one of the only commercial N-class motors certified by Tripoli.

"The motor was 4 inches in diameter and 4 feet long," said Kline. "We built ten of these motors, which, in theory, could obtain an altitude of 45,000 feet. Several people attempted to fly the motor, but it was too powerful, and we shredded a lot of rockets."

Korey Kline's interest in hybrid motors began in the 1980s.

"Back then, I was mixing propellant by hand, and I was concerned about the legal ramifications and other issues," he said. "I felt I had pushed the limit with the N motor and couldn't see building larger motors than that, so I lost interest in that and moved to pure research into hybrid rockets. At that point, I was completely alone on developing these motors. This was back in 1981, when I was working on gaseous oxygen [GOX] motors. It was a brand-new technology, and nobody was doing it."

Kline conducted secret tests developing hybrids at Lucerne over the next several years. "The first five or six years were spent designing grain geometry, injector design, flame-holding capabilities, and the nozzle end of the technology."

His early work centered on using gaseous oxygen as the oxidizer for the system. Then at the suggestion of mentors John Krell and Bill Wood, Kline switched to a higher-density oxidizer: nitrous oxide, which would become the key to the development of his most successful design. In 1989, he successfully fired his first nitrous hybrid motor. Over the next four years, he continued improving the design and performance of his hybrid until he was ready to go public with the technology in 1993. Kline's first public demonstration of his hybrid was for the Reaction Research Society.

"By 1994, I think my record was about 500 pounds of thrust, and my maximum burn time was about 50 seconds," he said. "By that time, the motors started getting so large that I stopped thinking in terms of letters of the alphabet as size designations. I began thinking in *pound seconds* total impulse. The letters really became irrelevant to me."

In early 1994, Kline was contacted by two Florida men, Kevin Smith and Tom Bales, who recently sold a biomedical company and were interested in pursuing other endeavors. Smith and Bales had come across a copy of *High Power Rocketry* and were fascinated by the size of hobby rockets and an article written by a member of the RRS concerning hybrids. Smith learned that Kline was the developer of the hybrid technology. He called Kline to find out more.

"At the time, I used to get lots of calls from people interested in hybrids," recalled Kline during an interview in the early 2000s. "I used to give them a list of things to do and to call me back with the results. If they called me back and the results were correct, I could tell they were serious, they knew how to build things, they did the test, and they were worth giving more information to. A lot of people would call back a week later and literally lie to me about the results. I was astonished by that. Kevin Smith was one of the first guys to call me back and give me the right answer."

Over the next few months, Kline sent Smith several experiments. He received good results in return. He decided it was time to meet Smith in person, so he flew to Miami. The pair hit it off, and shortly thereafter, Kline took a big chance. He quit his job in California and moved to Florida. Kline, Smith, and Bales then formed the Environmental Aeroscience Corporation. The trio were going to build hybrid motors. Their first commercial product would be the Hypertek system created for high-power rocketry.

The Tripoli board of directors approved the first demonstration Hypertek motor at a Tripoli launch in the summer of 1994. The demonstration was conducted by Kevin Smith under the auspices of Florida's newest prefecture, Tripoli Tampa. The board required Hypertek to post a $5,000 performance bond prior to the flight to cover Tripoli's insurance deductible in case anything went awry. Both the static test and rocket launch were successfully conducted on September 4, 1994. This was the first time a hybrid was officially launched at any Tripoli launch.

Two weeks later, Smith and Kline took Hypertek's system on the road, this time for a demonstration at BALLS-4 at Black Rock. The first flight of a hybrid at BALLS was on Saturday, September 17, 1994. The demonstration motor contained 2 pounds of nitrous oxide for the oxidizer along with approximately a half pound of polymer solid-fuel grain in the combustion chamber.

"It was interesting to hear the unique sound of the throttled hybrid motor," said one observer, who thought that the motor's sound was reminiscent of the pulsejet motor used by the Germans for their *V-1 Buzz Bomb* in World War II.

The hybrid motor worked perfectly. At ignition, the 7-foot-long rocket left the pad cleanly and climbed to more than 6,800 feet. Unfortunately, the rocket's parachute and recovery system failed. A recovery harness made of bungee cord, typical for high-power rockets at that time, tore apart during deployment, and the rocket plummeted in from altitude sans the parachute, taking a core sample about a foot deep from the desert soil. The next day, the Hypertek crew tried again. This time, they sent up a larger motor in a 9-foot-tall rocket that arced over at 10,000 feet, followed by a good recovery.

The following month, in the October 1994 issue of *High Power Rocketry*, a cryptic two-page-wide advertisement appeared in the centerfold. Both pages of the ad had an entirely black background. In the center of the pages, running from one side of the magazine to the other in 1-inch-tall uppercased green letters, read the following:

THE FUTURE OF HIGH POWER IS NOT SOLID

There was nothing else on these two pages except for a tiny illustration of a small green rocket in the upper right-hand corner of the second page. The advertiser was not identified. No product was listed. There was no one to contact about the ad—there were no addresses or telephone numbers—nor was there any other printing anywhere on the two pages. It was a mystery designed to capture the attention of the high-power rocketry community, and it did just that.

The advertisement in *High Power Rocketry* preceded any coverage of the demonstration high-power hybrid launches. Indeed, coverage of the demonstration launches in September would not appear for a few more months. Yet word of the Hypertek high-power motor was already getting out.

In December 1994, the same two-page ad was repeated again in *High Power Rocketry*. There was still no mention of any product; nor was the advertiser identified in the advertisement. Bruce Kelly revealed in his editorial that the name of the advertiser could be found elsewhere in the magazine. The advertiser, said Kelly, was a new propellant company by the name of Hypertek, and with its "The Future of High Power Is Not

Solid" slogan, Hypertek embarked on a marketing campaign to bring the hybrid into the mainstream of rocketry.

Gary Rosenfield and AeroTech were also pursuing hybrid motor technology in 1994. For AeroTech, it was a game of catch-up. Still, by the end of 1994, Rosenfield had a working prototype. He gave flyers a preview of his new motor system at the Danville Eleven launch in Illinois in November. He explained the theory of the hybrid, brought samples for people to see, and produced photographs and preliminary thrust curves of test firings AeroTech had recently completed. The following month, AeroTech followed Hypertek's lead in aggressive advertising for its new product. Rosenfield took out a full-page back-cover advertisement for his hybrid, the "RMS/Hybrid," in *High Power Rocketry*.

Although the Hypertek and AeroTech hybrid motors were based on the same underlying technology, the two systems were put together differently. The AeroTech hybrid had three primary components: the solid-fuel case, a tank for the oxidizer, and a release valve that connected the case and oxidizer tank together. The solid-fuel case was like any other AeroTech reloadable 54 mm motor casing. Once the tank, already filled with nitrous oxide, was attached to the forward end of the motor case via the release valve, the entire assembly was inserted into the aft end of the rocket, much like any other motor. The fuel grain for the AeroTech system was based on a cellulose (i.e., paper) derivative.

The Hypertek system, on the other hand, used a nitrous tank and a separate disposable fuel grain that were joined together by an injector bell that contained a valve that allowed the nitrous oxide to combine with a single piece fuel grain at the time of ignition. The Hypertek system operated with additional parts at the pad, including a remote tank that allowed the user to fill the motor with nitrous oxide at the time of launch. In other words, the tank is initially empty when the rocket is placed on the pad. The Hypertek solid fuel was thermoplastic.

On February 4 and 5, 1995, the Tripoli board convened a special meeting at the El Dorado Lake Bed near Las Vegas, Nevada, to observe flight test demonstrations of the hybrid systems developed by both Hypertek and AeroTech. The tests were a success, and it was noted that neither the AeroTech nor the Hypertek systems appeared to require any special permits for flight or shipping. All the components were essentially inert until the moment of ignition. There was no shipping issue, and there were no potential explosives issues.

Hypertek was granted permission to begin immediate beta testing of its motor system after the February demonstration. This meant that Hypertek could recruit prefectures around the United States to test its product. AeroTech, on the other hand, was unable to give the board full disclosure regarding its product, apparently because of patent issues. AeroTech would have to wait for permission to begin beta-testing its hybrid until later in the year. By the end of 1995, both companies were testing hybrid motors around the country. Hypertek would ultimately get their hybrid to consumers first. Yet by 1997, both Hypertek and AeroTech were selling hybrid motors to flyers all over America.

Korey Kline's development of the Hypertek line of rocket motors heralded his return to a leadership position in hobby rocketry. In the spring of 1995, he was elected to the Tripoli board, a position he would hold for the next three years. On the board, Kline joined with other board members who would push for a broadening of Tripoli's rocketry purpose, not only with hybrids but also with other rocket motor technology, including experimental solid-fuel motors. In Kline's opinion and in the view of a growing minority of Tripoli members, homemade propellant was the future of the hobby.

"Tripoli should be an all-encompassing rocketry organization," he urged, "from large model rockets to an amateur space shot. We are in the Wright Brothers days of rocketry, where people can build spaceships in their garage. Welcome to the future."

24

The Rise of the Basement Bomber

Part II

In the fall of 1994, Tripoli's Future Directions Committee mailed every Tripoli member a questionnaire to gauge interest in homemade rocket motors and the emerging commercial hybrid market. The results of the survey, compiled by Art Markowitz, revealed widespread interest among Tripoli members in hybrid technology. The committee recommended hybrids be included in Tripoli's official motor certification listings, which was approved by the board.

On the issue of homemade solid-fuel motors, a different consensus was reached. Although the survey revealed members were interested in experimental motors, the committee recommended against the approval of such motors, and a majority of the Tripoli board of directors appeared to agree. Over the next few months, there were various reasons expressed by either committee or board members for their stance.

"Tripoli should not become an amateur rocket organization."

"Many members probably do not realize the potential complexity of law applicable to the making of solid-fuel rocket motors."

"Those seriously interested in the manufacture of their own motor(s) should have to jump through the same hoops as the commercial manufacturers."

"What is the liability if a member burns his house down and then the rest of the neighborhood? Is Tripoli responsible because we said it was okay?"

In February 1995 and by a vote of seven to zero, the Tripoli board voted against member manufacture of their own propellant for use at Tripoli events. The board also voted members be prohibited from using homemade propellant at Tripoli-sanctioned events unless they had previously certified said motor(s) with Tripoli's Motor-Testing Committee. In other words, they would have to comply with the same requirements as a commercial manufacturer. This was an onerous task for individuals as it would require, among other things, the submission of detailed paperwork and multiple sample motors to the certification committee for testing. The practical effect of this stance by the board was a ban on experimental motors at Tripoli launches, at least for the time being.

LDRS-14 was held at Black Rock that summer. The subject of experimental motors came up at both the annual board meeting and at the members' meeting. Board members Korey Kline, Sonny Thompson, and Scott Bartel were tasked with creating experimental motor guidelines for the board's future consideration. The board also authorized another study of experimental motors under the auspices of the Future Directions Committee. As the NAR has done with high-power motors in the mid-1980s, Tripoli was proposing more study before experimental motors could be used at their launches.

Meanwhile, a growing number of Tripoli members continued to make motors for their own use. Like the reloadable revolution a few years earlier, this was another technological tidal wave that could not be stopped.

On January 20, 1996, and by a vote of seven to zero, the board revoked their 1995 position on experimental motors and announced Tripoli now "explicitly recognizes the need to support solid and hybrid motor research." The board also directed Bartel, Kline, and Thompson to draft protocols for homemade motors ready for discussion at LDRS-15 in Orangeburg that coming July.

The next day, on January 21, 1996, the board took an even bigger step and, for all intents and purposes, lifted the ban on homemade motors at high-power rocketry events. The board granted five different prefectures permission to hold experimental launches. In return, the board required each of these prefectures to come up with its own set of safety rules and required organizers of each launch to collect data on all experimental flights and to forward that data to the Future Directions Committee for analysis.

One of the first prefectures to submit experimental guidelines was Tripoli Middle Tennessee at their Barrens Test Range site near Manchester, Tennessee. Among the new rules adopted at Manchester was that the experimental portion of any launch be separated from regular Tripoli launches by at least 24 hours. Any rocket carrying homemade propellant was to be flown on a designated experimental or "EX day." The rule was reminiscent of the NAR's 3/48 Rule of the late 1980s, wherein the NAR allowed high-power flights as long they were separated in time and distance from model rocketry launches at the same location.

The separation rule used in Manchester would later be adopted for all Tripoli EX events. It would be followed, with some local modifications, at almost every other Tripoli location for the next ten years. The Tripoli board of directors published the Tennessee "EX Rules" in March 1996 as an example of what the board was looking for in terms of procedures for experimental launches in the future, although the board had not yet formally adopted any distinct set of rules anywhere.

At LDRS-15 at Orangeburg in July, Kline and Thompson presented the board with an outline for the use of experimental motors, now being called "Tripoli Research." It was a lengthy proposal, suggesting the creation of multiple certification levels (novice, intermediate, and advanced), the creation of a permanent research committee in Tripoli, and detailed instructions as to how flyers should present requests to the new committee for using experimental motors at any Tripoli EX launch. Among other things, the outline required any Tripoli member who wanted to launch his own motors sign a "hold harmless agreement," exonerating Tripoli from liability should anything go wrong.

The complicated outline proposed by Thompson and Kline was not adopted by the board. Over the next several months, the board struggled with what to do formally about experimental motors at Tripoli launches. At the same time, Tripoli was also working with the NFPA's Committee on Pyrotechnics on finalizing the Code for High Power Rocketry. At one point, Tripoli provided NFPA committee members with a first draft of a proposed safety code for homemade motors. Not long afterward, it was suggested that Tripoli abandon its homemade motor proposal as it might endanger Tripoli's relationship with the NFPA. Tripoli President Kelly rejected such a suggestion.

"Contrary to what you may have heard from 'Internet sources,' the NFPA committee was not opposed to the December presentation of our first draft outline [for research motors]," Kelly told members. "This does not mean that there will be no opposition in the future. Anything you heard about the committee being opposed to the concept at this time is not true."

Kelly went on to state that Tripoli's exploration of homemade motors, while new and perhaps revolutionary, would not signal the beginning of the end for high power. "There are some 'Chicken Littles' who believe experimental activity will endanger the 'legitimate' side of the hobby. This is no [truer] than when [high power] was born or the development of reloadables or the introduction of hybrids. This is the typical attitude when anything new comes along. The bottom line is that anything can be unsafe if practiced in an unsafe manner."

In June 1997 and after reviewing data from several experimental launches around the country, the board presented Tripoli members with its first enforceable experimental rules. The rules, dubbed the "Tripoli Research Interim Launch Rules," were similar to the first set of research rules suggested by Tripoli Middle Tennessee a year earlier.

The Tripoli Research rules authorized homemade motors if they were hybrids or composites only; neither black-powder nor liquid motors were included. They prohibited steel motor cases and set new safety distance limits for homemade motors. The rules allowed field officers the right to refuse to launch any rocket with experimental motors that he or she deemed unsafe, and they required a special application be made to Tripoli by prefectures who wanted to hold a research event. The interim code limited the upper end of research motors to a P motor or the equivalent of a Q in rockets launched with clusters. All rockets powered by a research motor were required to have electronic-based deployment systems as the primary means of parachute deployment.

A significant aspect of Tripoli's new rules was what they did not require. There would be no requirement for flyers to submit homemade motors to Tripoli Motor Testing. Flyers were free to make their own motors for personal use at Tripoli Research events without having to participate in the formal Tripoli motor certification process.

Finally, the new rules established a time boundary between experimental and normal launches. There could be no experimental motors launched at any ordinary commercial Tripoli launch and no commercial motors at any research event. If the same location was used for both a regular launch and an experimental launch, there must be a one-day separation between the two events. There would be no commingling of commercial and homemade motors at the same launch on the same day. Some members objected to this restriction, but on this subject, there was no room for interpretation.

President Kelly reported that the NFPA's Committee on Pyrotechnics wanted strict separation between research launches and other high-power events. The reason was simple. The committee did not want local authorities confused as to what was an accepted codified activity by the NFPA (i.e., model or high-power rocketry pursuant to 1122 and 1127) and what was not (i.e., research/experimental motors). Kelly made it clear he expected this rule to change in the future once research motors gained wider use. NFPA 1122 and NFPA 1127 are still the generally accepted safety codes, noted Kelly.

"Tripoli Research [rules] do not enjoy this status and most likely won't for a long time. However, acceptance usually begins with a level-headed, well-thought-out approach to the activity. Acceptance will also only come with documented proof that the activity is safe and is conducted legally and by the guidelines. This is our first step toward that direction . . . The major complaint that many of you will have is about this separation issue. Don't waste your time complaining. It is not likely to go away. If we get rid of that rule, you can eventually expect the whole activity to go away from Tripoli. We are not in a position to jeopardize high power, the focal point of our charter, for the sake of a few who insist on experimental activity."

The Tripoli Research rules continued to evolve over the next few years, and by the end of 1997, homemade motor technology would be fully incorporated into Tripoli's launch structure.

The increased use of homemade motors was accompanied by a new vendor in high-power rocketry: the fuel designer.

As soon as experimental motors received coverage in high-power media, advertisements began to appear promising to teach flyers how to safely mix their own fuel. So even before Tripoli approved research motors, the decades-old stigma of the basement bomber began to disappear from hobby rocketry. One of the first modern ads for homemade rocket fuel appeared as a half-page ad in the March 1995 issue of *High Power Rocketry*:

Take the Next Step in High-Power Rocketry Make Your Own Motors!

The ad was created by John Wickman of Wyoming. Wickman's CP Technologies offered videos and booklets on the manufacture of ammonium nitrate motors in the G- through O-impulse range. The ad was repeated in the June issue of *High Power Rocketry* (and every month for years thereafter). Soon, Gary Rosenfield at AeroTech was also advertising consulting services on all kinds of motor issues, including technical services on solid motor propellant design and rocket motor component sourcing for businesses or individuals. It was the beginning of a marketing gold rush to bring motor-making technology to the average flyer.

In August 1995, a company from Oklahoma called Propulsion Systems began selling booklets describing the formulation of APCP motors. For less than $25, flyers could purchase design information for at least six pretested formulas for APCP motors. In October 1995, *High Power Rocketry* ran its first ads for liquid propellant—in this particular case, gasoline—from a manufacturer in Canada. The ad promised blueprints instructing flyers how to create motors powered by regular pump gasoline with more than 3,000 Newton seconds of total impulse—in other words, an L motor.

Through the end of 1995 and continuing throughout 1996 and while the Tripoli board busied itself with debating whether to allow research motors at Tripoli events, advertisements for homemade propellant in the rocketry press increased. If there was an internal debate on whether Tripoli should allow experimental motors at its launches, one wouldn't know it from the solicitations in publications such as *High Power Rocketry*:

"If You're Using Reloads, You're Already Assembling [Your] Own Rocket Motors—Take the Next Step."

"Do It Yourself—and Save!"

"Our Books Show You How to Do It the Right Way!"

"Uses Regular Gasoline as Fuel!"

"Watch Your Rocket Lift Off on a Motor You Made!"

These ads carried no disclaimers that experimental motors were prohibited at Tripoli-sanctioned events. It was taken as a given perhaps that eventually, they would be allowed.

The formulation of homemade propellant was the cover story for the January 1997 issue of *High Power Rocketry*. The article was written by David Crisalli, and it followed several flyers as they made their way through a propellant-making course offered by the RRS in the Mojave Desert. The story contained photographs outlining nearly every step in making composite fuel, from mixing to casting to igniting the motor on a test stand at the end of the class. The article was followed by the society's full-page advertisement offering motor-making courses for APCP motors. For $475, flyers would spend three days in class, and they would build and test-fire their own L motors. They would get to keep the motor hardware for their own personal use in the future.

When Tripoli finally approved research motors in 1997, the number of books, videos, and instructional courses on homemade motors increased. Perhaps the most influential of these courses was Thunderflame. An entire generation of Tripoli flyers anxious to make their own propellant would learn how to do so safely through Thunderflame. And it all started in Tennessee with a man by the name of James Mitchell.

James "Jim" Mitchell was born in February 1960 in Hayti, Missouri, a small town in Pemiscot County along the southeastern corner of the state. His father, James, was skilled in everything having to do with his hands, especially plumbing, electrical work, and welding. In the 1960s, the elder Mitchell worked at shipyards building gigantic barges that would be floated down the Mississippi and Missouri Rivers. Later, he specialized in the repair of diesel engine heads and cast-iron welding. Mitchell's mother, Carolyn, was a hairstylist from Tennessee. Both sides of the family lived in Missouri or Tennessee going back several generations. Mitchell was the eldest of five children. He discovered rockets when he was twelve.

"It was a small town and among the best times in my life. I'd mow lawns for a couple of bucks and send the money in to buy rocket motors. Back then, we didn't have all this electronic stuff to entertain us, and there were only three television stations," said Mitchell. "I read a lot, including comic books, where I discovered an ad for Centuri Rockets. I asked around at school if anyone knew anything about this stuff, and one guy did, and he showed me some model rockets. We launched one in his backyard, and I really liked it."

Mitchell certainly enjoyed the thrill and excitement of launching rockets. But he liked the idea of making rocket fuel even more. At the age of thirteen, he came across a book in a local library about some kids who were making their own motors out of potassium nitrate and sugar. Mitchell was able to get potassium nitrate for a dollar a pound from the corner drugstore. He dove into motor making.

"I got the chemicals and mixed them and started cooking them down on the stove using an empty tuna fish can to hold everything," recalled Mitchell in 2014. "When it got to the point of caramelization, it flashed off, meaning it caught fire. It was like lava. It burned the top of the stove and burned holes in my shirt. It also filled the whole house with smoke. There was a 3-foot-high cloud of smoke all the way through the house. My mother said that my daddy was going to kill me when he got home!"

Afterward, the young chemist moved his rocketry experiments outside the house. It took him almost a year, but he finally found the right formula.

"It was just random luck," he said later.

Mitchell was almost fifteen. Soon thereafter, he discovered motorcycles and cars, and he left hobby rocketry.

The Mitchell family moved to Tennessee in the 1970s, where Jim finished high school. He took factory jobs in textiles and furniture making and eventually went to work in his father's business performing diesel engine repair and welding. Mitchell remained completely unaware of any new developments in rocketry, including composite motors, high power, or LDRS.

Then in 1991, he came across a copy of *High Power Rocketry*. Turning its pages, he saw photographs of rockets launched with K motors and advertisements for high-power rocketry supplies. He contacted one of the magazine's advertisers, High Sierra Rocketry, and soon he was launching rockets with 29 mm reloadable motors on a friend's farm in Tennessee.

"It was awesome!" he recalled. "I liked these motors much better than the old black-powder motors, and they were louder too."

Mitchell's interest in the chemical side of hobby rocketry was immediately rekindled. Working on his own and not yet a member of Tripoli, he began taking apart single-use and reloadable motors to see if he could figure out how they worked.

"There were a few people who were making their own motors back then," he said. "But for the most part, the formulas were kept secret unless you were part of the 'in' crowd."

One day Mitchell came across an advertisement for a company marketing its own motor formulas and selling bulk AP. He purchased the formulas and supplies but did not have much luck with the motors. However, by reverse-engineering motors, he eventually created several of his own high-power formulas.

To determine the thrust characteristics of his motors, Mitchell built a MacGyver-like thrust stand in his backyard using only a barbecue grill motor, a popcorn container, some springs, a fish scale, and a ballpoint pen.

"I needed something simple at the time, not a load cell with all sort of electronics," he said. "I just made it with stuff lying around the house. It looked kind of like an EKG or an earthquake monitor."

Mitchell joined Tripoli in 1994. He attended his first high-power launch near Manchester, Tennessee.

"Bruce Kelly was actually at that launch, and he even signed off on my certification," he said.

While there, Mitchell met Terry McCreary, another new high-power flyer, and the two discovered their shared interest in motor making. Over the next two years, Mitchell increased his knowledge of motors through trial and error and long discussions with McCreary, who was a chemistry professor in Kentucky.

Like any good scientist, Mitchell was a serial note taker. He compiled copious pages of his motor-testing results, which he kept in a well-worn spiral notebook. He became an expert at curatives, chamber pressures, nozzle performance, and grain geometry. He built a sophisticated thrust stand on his property and a 24-by-40-foot shop where he built his motors and rockets. Initially, Mitchell was not allowed to launch his larger K- and L-impulse homemade motors at Tripoli events. They were still prohibited by the Tripoli board. His big break came not at a Tripoli launch but at NARAM-38 in Evansville, Indiana, in 1996. At NARAM, he was allowed to launch a *Phoenix* rocket carrying a homemade L motor.

"Everybody went and hid behind their cars because it was a homemade motor," said Mitchell. "They also demanded that I be the person to push the launch button."

The flight was a success, and a picture of the big rocket even made its way into *Sport Rocketry*—sans any mention of the origin of the experimental L motor.

Mitchell's increasing proficiency at motor making came at a time when Tripoli's leadership was struggling with how to deal with an increasing number of flyers interested in making their own fuel. Several board members were vocal in their support for experimental motors, including Sonny Thompson and Bill Davidson. Thompson reached out to Mitchell, buying several motors and flying them in his own rockets. Thompson liked the performance of the motors, so he approached Mitchell with a proposition.

"Sonny wanted to put together a project where we would go around the country teaching people about experimental motors," recalled Mitchell. "He said it was mostly for safety reasons. We could teach people the right procedures for putting together all the chemical components for a high-power rocket motor."

Mitchell agreed, and the flyers combined their efforts into a joint venture, and Thunderflame was born. Thunderflame courses were initially limited to Tripoli Level 2 flyers and higher. Thompson would usually do the lecture part of the two-day course, while Mitchell would lead the laboratory study, where students were shown how to mix the chemicals.

"I put together the kit, which was all the metals and liquids and oxidizer, everything that was needed for a motor," said Mitchell. "We wanted people to do everything right, so we made the propellants very user-friendly in a complete kit with less work, making it as simple as possible."

In one of their first full-page ads that appeared in the spring of 1998, Mitchell and Thompson offered hobbyists their vision of the future of high-power rocketry:

> Wish you could make your own propellant? *You can!* Our 1.5-day *basic* propellant-making course at $235 is reasonably priced, and . . . it works! We invite you to ask any of the people who have already taken our course. Our course was designed by flyers for flyers. We teach a propellant system that is as safe as it gets, and [it's] fun. The course is designed

for people with little or no propellant-making experience. Course includes lecture, propellant formulas (2), mixing, casting, grain processing, and . . . reusable hardware for a 6-grain 38 mm J750 motor, which you can keep. We will also show you how to make five other types of propellant, including a more powerful red that makes the others look pink . . . Can you fly these motors at experimental launches? *Yes*, and the number of Tripoli Research experimental launches is growing each year. Ten (or more) Tripoli Research launches have already been approved for 1998. And in 1999? By then, there will be well over 100 Tripoli members who know how to make their own propellant. You can be one of them.

Thunderflame was offered at LDRS-17 in 1998 at the Bonneville Salt Flats. This was the first time flyers had the chance to learn how to make their own motors at an LDRS event. The course was limited to twenty-six students and promised each flyer the opportunity to not only build their own motor during the lab part of the course but also launch it in a rocket at LDRS on Sunday.

"This is an elementary course," Thompson explained to his students. "[It] is intended to teach the basics of composite propellant manufacture."

Thompson believed there were many Tripoli flyers anxious to learn how to make their own motors. "The problem was that you could buy and read every book you could find on propellant making and still not know how to do it. Words are good, and pictures help, but it is just not the same as watching someone do it."

Thunderflame courses focused on teaching the fundamentals of propellant manufacture, to teach students how to "make and cast propellant the same way, every time," said Mitchell. The classes covered mathematical calculations, particle size, burn rates, and the dynamics of heat and pressure. Students learned about the relationship between thrust produced by burning a given amount of propellant over time, also called specific impulse. They learned that some types of propellant had a higher specific impulse than others. The basic mechanics and safety issues related to mixing, pouring, and casting fuel were discussed, and students were provided formulas for high-power fuel with names like "Diamond Back" and "Stinger." Basic courses took students through

J-motor formulation. Advanced courses were created later, providing instruction for K, L, and even M motors.

The Bonneville Thunderflame class in 1998 was well attended. An orientation at the Stateline Hotel in Wendover, Nevada, was held on Thursday, followed by motor-making sessions at the range on Friday and Saturday. Tripoli flyers from all over the country signed up for Thunderflame at LDRS-17, people like Ken Shock of Washington, Lannie Cross of Iowa, Charles Webb of Texas, and Gary Logan of New York. Tripoli member Rolf Oren, who came from Sweden to obtain his Level 3 certification at LDRS, also signed up for Thunderflame. In early registration, more than thirty-five Tripoli members registered to launch their own experimental motors on Sunday. Most of them were disciples of the Thunderflame course, taken either at Bonneville or elsewhere.

Over the next few years, Thunderflame courses, fueled by Mitchell's expanding inventory of homemade propellant formulas, became the most well-known high-power motor class in America. The list of people who attended these courses included many of the future leaders of high power, including Derek Deville, Neil McGilvray, Robert Utley, and even Tom Binford, who was already making his own motors in Georgia. Using the skills that they obtained from Thunderflame, these and other flyers would build their own N, O, P, and even larger motors that would be launched at high-power events all over America.

"The experimental bug hit hard and dug in," said Deville after he attended a Thunderflame course in 1999.

The stigma of the basement bomber disappeared from hobby rocketry. Homemade fuel was now on the cutting edge of the hobby. And one day Deville and others like him would be mixing their own fuel for high-power flights to the edges of outer space.

25

Return to Colorado

LDRS-16 was held on August 7 to 10, 1997, at Hartsel, Colorado, the same site where the event was held from 1987 and 1990. More than 275 registered flyers launched 700 rockets into the air.

"Thunder in the Rockies" got off to a good start on Thursday with fine weather at the field, perched 8,800 feet above sea level in the Rocky Mountains. The waiver was 23,000 feet, with windows to 28,000. There was a herd of grazing bison adjacent to the rocketry field all weekend, about a mile from the range head.

"The rocket flights over the next four days didn't seem to bother them at all," noted one flyer.

The flying began at 6:30 a.m. on Thursday, and several flyers took advantage of near-windless conditions during the early-morning hours. Newcomer Jeff Taylor, who would one day create his own successful high-power motor manufacturing company, achieved Level 2 certification on an AeroTech J350. Ray Bryant flew a 35-pound *X-Wing Fighter* on a J800 "but had an overly energetic ejection charge, which disassembled the rocket into many pieces." And Scott Meinhardt, who handled Tripoli's liability insurance policies for years, set a new Tripoli altitude record for J-class motors with his *Long Gone II* on a J700 AeroTech single-use motor. Single-use motors were a rarity in 1997 as they were eclipsed by reloadables. Meinhardt's single-use J powered his rocket to 10,584 feet, as determined by his onboard altimeter.

On Friday, Jim Rosson launched his *Big Screaming Banana* on a Kosdon East 76 mm N3200 "Green Gorilla" demonstrator motor. The 75-pound rocket was 11 feet tall and painted bright yellow. It lifted cleanly off the pad, trailing a bright green exhaust flame out of the nozzle. Phil Prior launched a Kosdon East M2800 in his rocket *Purple Haze*. It shredded in flight. Later that weekend, James Denney launched a rocket with a Kosdon East M4000 Skidmark motor that "put out a shower of sparks as it burned."

Rocket activity on Saturday was cut short by a powerful thunderstorm that rolled in that afternoon. Yet there was still plenty of action and lots of large flights, including several M-powered launches and hybrid-powered rockets.

One of the biggest rockets at LDRS-16 was a group project called *Nebraska Heat*. The rocket had an airframe made from 11.4-inch-diameter Public Missiles Systems phenolic tubing and was built by several Tripoli members in Nebraska. It stood 14 feet tall and held an AeroTech M1419 surrounded by eight I and J motors. The plan was to lift the 104-pound rocket on the four J motors initially and then light the central M and remaining I motors in midair. It can be dangerous having smaller motors lift a big rocket off the pad. If the smaller motors are not up to the task or are too slow to light, a big rocket can turn into a cruise missile in a hurry. But *Nebraska Heat* worked fine. The rocket lifted easily off its launcher, trailing distinct flames from the four J motors. The M and remaining I motors fired on cue in midair. *Nebraska Heat* had a great ride up and was successfully recovered on multiple parachutes.

Another big flight Saturday was Greg Daley's Level 3 certification in his *Return to Sender*. The red-and-black vehicle was a four-finned design with a 9-inch-diameter airframe that was 16 feet long. Daley used an M1939, which lifted the 110-pound rocket high into the air before returning for a successful certification via a 24-foot-diameter military parachute.

There were a number of flying contests haled at LDRS-16. The majority were geared toward altitude events. LOC/Precision sponsored a 5,000-foot challenge. This contest tested a flyer's ability to get closest to 5,000 feet without going over the mark. The winners of the contest were Scott Carter and Vern Knowles, who teamed up to launch a PML *Cirrus* rocket to 4,955 feet on an H123. Public Missiles Systems sponsored a 10,000-foot challenge. First place went to flyer Ken Goldstein, who put his *Oh K* to 9,702 feet on a long-burning AeroTech K185.

The AHPRA also sponsored a highest altitude contest, but it was reserved for black-powder motors only. The trophy went to a PML *Cirrus* rocket that achieved 4,389 feet on an H330 Silver Streak motor. Rocketman Enterprises sponsored a "closest to the pad" contest. The winner hoisted his rocket to 8,700 feet on an H motor (in another *Cirrus* rocket). When the rocket returned, it landed barely 70 feet from the pad.

The largest rocket of LDRS-16 was *Total Obsession*, launched by Damian Russo, Larry Zupnyk, and Bill Davidson. This was the third year in a row that Team WAC was a star at LDRS. *Total Obsession* was 18 feet tall and was powered by the relatively new AeroTech N2000 and four APS L1100 Redeye motors. It weighed 165 pounds on the pad. The big rocket was hauled out to a special pad on Sunday morning while a crowd gathered in anticipation. At ignition, the AeroTech motor failed to light, but the four L motors lifted the heavy rocket to a couple thousand feet. Despite the failure of one of its motors, it was a majestic flight and a crowd favorite at LDRS-16.

The weather on Sunday threatened rain much of the day; many flyers packed up for home by midday. As a result, the total number of flights this year was lower than expected. Still, the numbers at Colorado's fifth LDRS were impressive, with 730 launches in four days. There were at least twenty-three M-powered flights at LDRS-16, the most Ms yet at any LDRS.

Flyer Tony Cochrane, who authored an article on LDRS-16 for *High Power Rocketry*, even wondered if the days of M-motor flights stopping all action on the field were drawing to a close. Cochrane asked readers if one day the mighty M flight might even become blasé.

"Speaking of M motors, remember back, only a couple of years ago, when a rocket flying an M motor was rare?" asked Cochrane. "The whole launch used to come to a standstill whenever someone would fly one of these. This year saw twenty-three M-motor flights, nearly doubling the number flown just a year ago in South Carolina at LDRS XV. It was almost becoming routine. 'Oh, just another plain-Jane M flight.' However, to the person building and prepping the M-powered rocket, these flights are anything but routine. Most of the flyers are a bundle of nerves. The costs of large motors are high enough that they are still once a year . . . for the majority of flyers. The future will be quite interesting, seeing if the power cross section of motor sizes flown will keep increasing or if a plateau of sorts has been reached."

It was an interesting question. One day M motors at LDRS would be common. But for now, the M-powered rocket was still king at most fields around the nation, even at LDRS. And the national launch was still the place to see big projects powered by big motors.

Model rocketry legend and NAR cofounder G. Harry Stine passed away at his home in Phoenix, Arizona, on November 2, 1997.

Stine grew up in Colorado and embarked on a career in rocketry as soon as he graduated college as a civilian scientist at White Sands Proving Ground in New Mexico. At White Sands, he was involved in testing both solid- and liquid-propellant rocket motors for the United States Army.

"In 1955, I went to work for the U.S. Naval Ordnance Missile Test Facility at White Sands as head of the range operations division," he said later. "I became familiar with flight operations and flight safety [with] rockets such as *Viking* and *Aerobee*."

Stine wrote science fiction in his spare time, and in 1955, he branched out to write the first in a series of nonfiction articles on space flight for a magazine called *Mechanix Illustrated*. He also agreed to reply to letters written to the scientists at White Sands on the subject of rocketry.

"I would gladly answer letters from young people who wanted instructions on how to make and fly their own rockets. I also found myself talking to youth groups and amateur rocket clubs visiting the proving grounds. As a result, I became acutely aware of the 'youth rocketry problem,' as it was known in those days. One of my responses to this problem was to devote one of my *Mechanix Illustrated* articles to the subject of safety rules practiced by professionals and a list of safety rules for amateur rocketeers that I derived from this. This list of safety rules was the ancestor of today's NAR Safety Code. The article was entitled, 'The World's Safest Business,' and appeared in the February 1957 issue of *Mechanix Illustrated*."

Years later, Stine said that one of the purposes of his *Mechanix Illustrated* story on safety was to "help stem the growing number of accidents by basement bombers."

On January 27, 1957, Stine received a letter from a Nebraska shoemaker by the name of Orville Carlisle. Carlisle had seen Stine's *Mechanix Illustrated* story. He told Stine that he and his brother had invented a model rocket with a replaceable solid-fuel engine. The rocket was reusable and was recovered by a small parachute for multiple launches, said Carlisle, adding that his model rockets might be the solution to the safety problems Stine had discussed in his recent safety article. Carlisle asked Stine if he was interested in seeing such a rocket.

"I accepted his offer with some trepidation because I had become very skeptical of nonprofessional rocket experiments and devices," said Stine.

Still, he told Carlisle to go ahead and send some samples:

> A great big box with fireworks labels all over it showed up at my home in Las Cruces, New Mexico. And here were these lovely little model rockets . . . He sent me a launch pad too. So I went out in the cotton field one cold February morning, and I lit the fuse. I turned around and looked at this thing going up. *Whoosh.* You've seen it. You know what they do. Out comes a parachute. Hey, that's neat! Let's see if it will do it again. I go back home, get another motor, pack the parachute, and go back to the cotton field. *Whoosh.* I did it again and again. What fun! This is neat. This is wonderful. So I called up my White Sands buddies, particularly the guys who worked in the range safety office. We had a ball with these things. We couldn't figure out how this shoe store owner from the middle of Nebraska had achieved this fantastic reliability on these things when it was far better than what we could do.

Stine launched his very first model rocket on February 5, 1957. His enthusiasm for Carlisle's product led to a full-length magazine article entitled, "Shoot Your Own Rockets," which appeared as the cover story for *Mechanix Illustrated* that October.

"I got three young teenagers to build some of the *Mark II* models from kits sent to me by Carlisle," said Stine, who took the flyers out to the desert east of Las Cruces to test their rockets on March 24, 1957. "We launched about fifty models that afternoon, got some great photos, and confirmed my belief that this was indeed the answer to the youth rocketry problem."

The White Sands engineer approached several professional colleagues that spring with a proposal to create a model rocketry company. Stine wanted to call the company "Capital Rockets." However, his coworkers were not interested.

"They all declined to become involved because they felt the risk was far too great," said Stine. "The major high-risk factor was the fear that model rockets could be classified as 'fireworks' by all but six Southern states. The dictionary defines a firework as a 'device which produces a spectacular display of color, sound, light, or a combination thereof.' So did most state fireworks laws, and those laws were, at the time, not at all precise in their terminology or definitions."

Stine left White Sands that summer and went to work for the Martin Company in Colorado, working on Martin's *Titan* missile. He continued his correspondence with Carlisle, and the two finally met in person at Denver's Union Station on July 8, 1957. That weekend, they forged a business agreement and came up with a name for their soon-to-be rocketry endeavor: Model Missiles Inc. or MMI.

"We decided to call the firm 'Model Missiles' because we felt the word 'rocket' was too closely allied with fireworks," said Stine. "The word 'missile' was an 'in' word at the time."

A few months later and on October 4, 1957, the Soviet Union launched *Sputnik 1*. This event "fired the imaginations of young people all over America," said Stine. "The time was ripe for the model rocket."

Stine had previously written a book about the Space Race with the Soviets. In that book, published in early 1957 and entitled *Earth Satellites and the Race for Space Superiority*, Stine accurately predicted the coming of the Space Age.

"[W]hether we like it or not, we're going into space to explore the last frontier, the vertical frontier to which there is no ending," said Stine. "We will do it for many reasons. Because we are curious and want to find out the answers to the questions we have about the universe we live in, because we want to build better lives for ourselves and our children... The impact of the launching of manmade Earth satellites is already being felt in our civilization. In fifty years, it will alter our way of life beyond all recognition."

In his book, Stine also opined that if the Russians were to launch a satellite first, America could suffer a loss of international prestige. On the night of the *Sputnik* launch later that same year, Stine was called by a local reporter for his thoughts on the momentous event. Stine quoted right from his book, and when the story appeared in the newspaper the following day, he was told by his boss to clean out his desk. He had been fired for talking to a reporter without getting clearance from Martin's public relations department. The dismissal had little effect on Stine's views. Later that month, he provided *Life Magazine* with some colorful quotes regarding his personal opinions of America's now second-string role in the Space Race.

"We lost five years because no one would heed rocket men," he told *Life*. "We're a smug, arrogant people who just sat dumb, fat, and happy, underestimating Russia."

For Stine, the closing of the door at Martin also opened up another door that would have a long-lasting impact on the young scientist's life.

"When Martin Company fired me on October 5, 1957, there was only one thing left that I really had to do. I had to go out and get this model rocketry company started and get these kits on the market because the newspapers at the time were absolutely full of stories of kids packing matches into steel pipes and putting odd chemicals and fertilizers and stuff into pipes and blowing their hands off and killing themselves."

Beginning that month and continuing over the next two years, Stine and his colleagues tested model rocket products on a 560-acre plot of ground west of Denver on the slopes of Green Mountain. During this hiatus, he decided to create an organization to promote model rocketry nationwide. On December 7, 1957, Stine's Model Missile Association was granted corporate status in Colorado. In 1959, that name was changed to the National Association of Rocketry.

"We ran this thing out of our basement, literally," said Stine. "I found an old mimeograph machine, and I started turning out a monthly newsletter called the *Model Rocketeer*."

Stine was also writing regularly on the subject of model rocket safety for *American Modeler*, often outlining what he saw as the need for education and safety in the new hobby.

"There is plenty of rocketry literature around [but] it so highly technical that it is difficult, if not impossible, for the amateur to decipher into useful information," wrote Stine. "As a result of being ignored, using trial and error, and having no good information . . . an amateur rocketeer isn't getting the kind of results he wants. He works with the wrong tools, the wrong methods, the wrong techniques, the wrong materials . . . and just seems to do everything the hard way. You can't blame him either because rocketry is a complex subject."

Stine's career took many twists and turns in the years after the NAR was formed. He ultimately went back to work in the aerospace industry. For a time, he was involved in the testing of the escape pods for the Convair *B-58 Hustler* strategic bomber.

Yet he never left hobby rocketry. In the 1960s, Stine led the charge to legitimize model rocketry nationwide. With his guidance, the NAR established range education and safety standards that were later adopted by the NFPA and incorporated into legal codes all over the country. The NAR worked with federal agencies such as the FAA and the CPSC to bring model rockets into the mainstream of America. In 1965, Stine published the first edition of *The Handbook of Model Rocketry*. Now in its seventh edition, the handbook is often referred to as the "bible of model rocketry."

Stine was not an early advocate—and many flyers felt he was a fierce opponent—of high-power rocketry when the technology first appeared in the late 1970s. His reluctance to venture into high power is perhaps best understood in light of the government resistance he encountered during the creation of model rocketry. In the early days of high power, there were few, if any, safety rules in place. Many high-power enthusiasts preferred it that way. For Stine, this lack of safety guidelines threatened the entire hobby rocketry community, including his beloved model rocketry. However, as the years passed and high power established its own safety record, Stine became a high-power advocate within the NAR.

Stine was the chairman of the NFPA's Committee on Pyrotechnics in the 1980s, and using his considerable influence, he could have blocked the fledgling high-power community from membership in the NFPA. Instead—and at a time when Tripoli's leadership was not even aware of the NFPA's importance—Stine lent his hand. It was Stine, with the help of J. Pat Miller, who extended to the young Tripoli Rocketry Association the invitation to become a member of the NFPA's Committee on Pyrotechnics in 1987. This was the early foundation of the Code for High Power Rocketry.

"As Tripoli grew, he recognized the potential of high power and helped with the first draft of NFPA 1127," wrote Tripoli president Kelly shortly after Stine's death. "He, in fact, gave Tripoli credit for the growth and acceptance of high power as a hobby in the code's introduction."

Stine also took action that protected both model and high power in the future. Along with representatives from the Reaction Research Institute (RRI), Stine helped free hobby rocketry from the restrictions of the Space Launch Act of 1984. That act required, among other things, payload and launch permits for all rockets launched by U.S. citizens. Together with George James of RRI, Stine secured an exemption to the act's requirements for all of hobby rocketry, including high power.

"High-power manufacturers and consumers must understand how close they came, along with model rocketry and amateur rocketry, to being effectively shut down," wrote *Tripolitan* editor Tom Blazanin in praise of Stine's work in 1986.

Stine was a leading force behind the NAR's entry into high power in the late 1980s. And he was counted among those leaders in the hobby who stood up to government regulators and private manufacturers opposed to the introduction of reloadable motors in the 1990s. He did not always see eye to eye with those running high-power rocketry, and Stine was no shrinking violet. He had strong opinions, and he was always a forceful advocate for his positions. Yet without his considerable efforts over the course of several decades, it is possible that neither model nor high-power rocketry would exist in any form today.

"Think what you will about G. Harry Stine's role in high power," said President Kelly, "but always remember his influence that laid its foundation. As for me, I considered him a friend, and I respect the role he played in the history of our hobby."

Stine authored more than fifty fiction and nonfiction titles on subjects that ranged from science to history to the future of man in space. His pen name was Lee Corey.

Harry Stine was sixty-nine.

———◆———

Not much has changed at the Bonneville Salt Flats since the Donner Party passed near here on the way to their Sierra Nevada destiny more than 160 years ago. The flats are as close to an alien landscape as anywhere else on Earth: a densely packed salt pan of more than 44,000 acres in Northwestern Utah, just over the Nevada border, named for United States Army captain Benjamin Bonneville, who explored this desolate part of the American West in the nineteenth century.

Prehistorically, the salt flats were a small sliver of what is now called Lake Bonneville. Geologists believe this once-massive body of water covered nearly all of Western Utah, at least until the end of the last Ice Age, more than 15,000 years ago. Like the Black Rock Desert a few hundred miles to the west, the salt flats are a living testament to the fact that climate change is nothing new; it is an immutable feature of the earth's dynamic environment.

Speed demons have set records on Bonneville's hard salty surfaces since the 1930s, when some nut from England streaked across the desert in a primitive automobile at 300 miles per hour. A world land speed record of 622 miles per hour was set here in 1970 by a rocket-propelled vehicle known as *The Blue Flame*. That record stood for a decade. Subsequent world speed records moved on to Black Rock in Nevada and then to North Africa, where they are set today. But Bonneville remains the ancestral home of speed in America. Enthusiasts from around the world still gather here every August for several weeks of racing contests involving every type of vehicle, from cars, trucks, and motorcycles to golf carts, minibikes, and even riding lawn mowers.

The flats are free of any obstacles—no trees, no rocks, nothing. In August 1998, this location was chosen for LDRS-17, cohosted by Tripoli Utah and the Utah Rocketry Club (UROC). Both clubs had been launching rockets at Bonneville for several years, and the annual Hellfire event, one of the oldest in high power, was hosted here by UROC since 1995. For LDRS-17, both clubs combined efforts to produce an outstanding and well-organized launch, attended by at least three hundred registered flyers who launched more than nine hundred rockets (and burned 1,100 motors) in four days, August 6 to 9.

The landscape at the salt flats is bright white, blindingly so, and visitors are encouraged to apply sunscreen on all exposed skin, even under the bridge of their nose, to prevent sunburns that occur from reflected light bouncing off the surface of the desert floor. The salty ground seems a bit squishy beneath one's shoes, like walking on a fine layer of crushed ice. But beyond a depth of a few inches and depending on the year, the soil of the ancient lake bed is like concrete to incoming rockets with deployment issues. Some years, there is no penetration of the surface by diving rockets from any altitude. Unlike Black Rock or Lucerne, undeployed rockets at Bonneville do not get swallowed up by the earth. Instead, they crush themselves against an imaginary brick wall, forming unrecognizable, accordion-like shapes of cardboard, fiberglass, Kevlar, or even carbon fiber. Some rockets shatter on impact, leaving barely a mark in the ground.

The majority of flyers at LDRS-17 were relatively new to high power. Most had Tripoli numbers between 4,000 and 6,000. If you had a number between 1,000 and 2,000—like Mark Clark, Robin Meredith, Jim Rosson, or Neal Stephenson—you were a high-power old-timer. Less than a dozen flyers had membership numbers under 500. Only a few flyers from the 1980s were on hand for the seventeenth anniversary of LDRS. One of them was Gary Rosenfield, who had been present at every national launch since 1983.

The Stateline/Silversmith Hotel & Casino, located on Interstate 80 only 15 minutes from the range head, served as the LDRS headquarters. More than two hundred rooms were reserved by flyers. The setup of fifty launch pads was completed on Wednesday, August 5, and the pads were laid out in several rows with two separate banks on either side of the range, by now an LDRS custom. This allowed rocket launching from one bank of pads while flyers on the other side of the range were simultaneously loading rockets. There was an elevated LCO tower constructed to allow club officials a view of the entire rocketry range, which rested just over 4,200 feet above sea level. Several vendors sponsored contests again this year, including closest to the pad (Rocketman) and the 5,000-foot (LOC) and 10,000-foot (Public Missiles) high-altitude challenges.

For the first (and perhaps only) time in LDRS history, there was no FAA waiver granted for Tripoli's national launch. The FAA officials in Salt Lake City told launch organizers not to worry about it as the salt flats were within the jurisdiction of the nearby Hill Air Force Base.

"Because we [are] in restricted air space, the normal waivers aren't required here," explained launch organizer Jay Reitz to all flyers on the first day of the launch.

Reitz and other local club members learned of the FAA's disinterest in their launch location some months earlier. They then contacted air traffic controllers at the air force base to see if Tripoli could launch rockets at Bonneville. The air force said sure; flights up to 20,000 feet required no notice to anyone, they said. Over that altitude, at least from Thursday through Saturday, the controllers asked that they be given a heads-up if a particular rocket was expected to clear 20,000 feet.

However, on Sunday, when there would be no air force activity in the restricted area, the maximum altitude was as high as anyone wanted to go. This made for the highest waiver in LDRS history.

Day One of LDRS-17 brought a beautiful blue desert sky, free of any humidity. But it was terribly hot. The peak temperature at the range reportedly reached 115 degrees. Yet there was still some flying, especially with H-, I-, and J-powered rockets. There was also the usual assortment of skywriters, lake stakes, and CATOs.

Charles Simpson certified Level 3 on a Kosdon M2240. Chris Cummings had a good flight on his Public Enemy *Performer* loaded with a 75 mm AeroTech L1120 motor. Another impressive vehicle was a *Saturn V* owned by Brian Weese of California. Weese's scratch-built creation was a near-perfect one-thirty-second-scale replica of the biggest rocket ever built in America. It was 12 inches in diameter and stood almost 12 feet tall. Weese spent 240 hours building his *Saturn V*, and he traveled to the salt flats to obtain his Level 3 certification. The rocket was displayed on its side all day Thursday, resting on a blue plastic tarp stretched out against the snow-white ground. Next to the rocket was Weese's successful Level 2 rocket: a one-fiftieth-scale *Saturn V* that looked just as nice as the big one, only a lot smaller. Weese's rocket did not fly Thursday, but he would be ready soon.

Friday brought overcast skies and cooler temperatures. By midmorning, there was a barrage of rockets taking to the air. As soon as one rocket deployed its parachute, another was ignited on the pad. LOC founder Ron Schultz was attending his fourteenth LDRS in a row. This year, he introduced his latest creation: aone-twelfth-scale replica of the infamous German *V-2* rocket. The demonstration flight on an AeroTech I motor was a success and was the beginning of hundreds of similar flights of LOC *V-2s* that would continue long into the future.

Another *V-2* flyer on Friday was Brad Overmoe of Utah. Overmoe's rocket was scratch-built. The 9.25-inch-diameter rocket brought the entire crowd to its feet as it blasted off on an M1419. At ignition, the rocket was completely engulfed by flame and smoke, momentarily disappearing from view. It then shot out of the billowing smoke and climbed to 7,000 feet for a great flight and good recovery.

New Mexico flyer Bill Cordova was next up at the away cells with a Level 3 certification on an AeroTech M845 hybrid. Not long afterward, there was a series of demonstrator flights of a new motor Gary Rosenfield had dubbed the "Warp 9." This high-power motor was all thrust in a big hurry. The L3790 had an advertised burn time of barely a second. As the crowd held its breath in anticipation of this first L flight, the "go" button was pushed. The result was a massive CATO that destroyed the rocket. In the meantime, another one of AeroTech's demo motors, this time a 54 mm K620, blasted off in a rocket built by Wayne Anthony of Massachusetts for a great ride and recovery.

Just after 1:00 p.m., the LCO announced that lightning storms were headed toward the flats.

"You could see the dark clouds coming from the north," reported one observer. "About an hour later, they announced once again. There was talk of electricity in the air, and we had to leave."

But not before one last large rocket was sent skyward: Dave Wysack's upscale version of an Estes *Alpha* model rocket. Wysack's *Alpha* was 10 inches in diameter and 10 feet tall. At ignition, an M1419 easily lifted the heavy rocket skyward, trailing columns of smoke as it slowly rotated on its climb. The *Alpha* turned over at apogee, but there was no deployment of the recovery gear. The big rocket plummeted back to the salt flats.

"Stay alert, everyone!" shouted the LCO over the public-address system. "We've got no chute! We've got no chute!"

With this last exclamation, the deployment system miraculously activated a few hundred feet from the ground. The rocket floated gently to the ground for a perfect landing. This was the last flight of the afternoon. As the thunderstorm approached, people gathered their gear and retreated to their cars, tents, trailers, or hotel rooms a few miles away in West Wendover. The rain and winds came and went quickly. The weather cleared a few hours later, and there was a brief attempt at a night launch. Then the storms returned. A hoped-for evening launch was scrubbed after only a couple of flights.

Later that night, a much more powerful storm marched through the rocketry site.

"When we arrived at the site on Saturday morning, it looked as if a massive tornado had blown through the parking area," said photographer Nadine Kinney, attending her first LDRS with her husband, Neil, a Tripoli member. "Almost every tent and awning [was] crumpled in some way. Most suffered minor damage, and a few were totally destroyed. There were bent legs [and] torn covers, and some of them were just blown clear away from where they stood the day before."

There was a silver lining to the carnage. The bad weather was gone for good. Saturday was the biggest launch day of the event.

The big rockets started early that morning with a successful Level 3 certification flight by Lyle Vaughn on an AeroTech M1419. Juerg Thuering from Sweden claimed his Level 3 certification with another M1419 in a half-scale *Patriot Missile*. Thuering was among the first European flyers to achieve Level 3. And Tom Gosner of Seattle fired off his *Pinky and the Brain*, a two-stage rocket powered by AeroTech and

Kosdon M motors. Gosner was hoping for 13,000 feet and his Level 3 certification on this flight. The 80-pound rocket had a great boost and a clean separation at first stage burnout, but the sustainer ripped apart high in the air at Max-Q. The booster dropped gently under chute and "stuck the landing" in the ground. In other words, it landed standing up.

The biggest star of LDRS-17 was, for the fourth year in a row at LDRS, built by Team WAC. Team WAC was Damian Russo, Larry Zupnyk, and Bill Davidson. In the mid-1990s, they were as close as one could get to achieving celebrity status at LDRS. Every year they arrived with large—and oftentimes the largest—rockets of the event. For LDRS-17, they created an upscaled version of LOC's three-finned *Magnum*. With the help of anyone who could lend a hand, their 150-pound *Mad Max* was placed into a special launch tower the team had constructed for this flight.

"It took six guys more than two hours just to put the tower together," said one team member. On Saturday, *Mad Max* ripped off the pad on a trio of AeroTech M1939 motors. The flames from the three motors beat the ground furiously as the rocket roared off its away tower. The rocket had a picture-perfect ride to an estimated 11,000 feet in altitude. *Mad Max* was a dual-deployment rocket, and the plan was to deploy the main parachute at approximately 1,200 feet. The ejection charges fired on time, and the parachute and recovery harness spilled out of the rocket for a seemingly ideal recovery. Yet the parachute failed to fully unfurl; for some reason, it could not catch any air. The parachute simply trailed the big rocket like a long piece of windblown ribbon on a car antenna. The hard impact against the salt flats reduced the 12-foot-long airframe into crumpled-up Sonotube. Inspection of the crash site, noted one observer, confirmed the salty surface at Bonneville "is only slightly softer than asphalt."

The festivities at LDRS-17 moved from the salt flats to West Wendover for the annual banquet and Tripoli members' meeting on Saturday night. By vote of the Tripoli board, it was decided that Sunday, August 9, the last day of LDRS, would be designated both an experimental and a commercial launch day. For the first time at LDRS, homemade motors would share the same field at the same time as traditional commercial motors. The board also waived the requirement that flyers could not launch rockets with experimental motors above their certification level.

There were at least three dozen rockets launched on homemade fuel on Sunday. Most of the fuel was from the Thunderflame classes held at LDRS that very weekend. There were several H and I experimental motors as well as more than a dozen Js and several Ls. There was an M1200 and even an O-impulse motor, the latter being flown by Robert Justus. Some of the experimental motors performed poorly—there was obviously a steep learning curve in making your own propellant—yet many of the research motors seemed as reliable and exciting as any commercial motor in its class.

Future Tripoli president Stu Barrett of Washington launched his former Level 3 rocket on a new AeroTech L1900 Blue Thunder for a good flight, and Mike Vaughn of California followed suit with his *Long Burn Express* with a ten-year-old AeroTech K125 single-use motor he had been storing for a special occasion. Vaughn's rocket streaked into the sky on its 20-second burn time to an estimated 13,000 feet. Jeff Taylor launched an L5000; new Tripoli member Ken Biba achieved his Level 3 on M power.

The last day at LDRS brought some heartbreaking flights too. Brian Weese's *Saturn V* was slated for flight on Saturday and was set up on the pad and ready to go. But some of his ejection charges were radio-controlled, as was common in the mid-1990s, and an errant radio signal fired those charges as the rocket was prepared for launch. The vehicle came apart at the pad, and the launch was scrubbed. On Saturday night, Weese repaired the damage, and on Sunday, he and a friend carried the 56-pound rocket back to the away cell.

"This is thought to be the largest-scale model flight of a *Saturn V* ever attempted," said one observer just before launch.

At ignition, the M1419 in Weese's rocket came to life, and the *Apollo*-era vehicle had a perfect liftoff. Moments into the ride, however, the 11.5-foot tall rocket veered sharply off course and did a huge loop in the sky, followed by an equally spectacular cartwheel. The rocket remained unstable and smashed directly into the ground. Although some electronics and a few hard parts were salvaged, the airframe was destroyed. It was later determined that a clear plastic fin had broken off in flight, leading to the rocket's instability and crash.

Bill Davidson's Thunderflame-powered M1700 rocket with four J motors sustained a mighty CATO in midair, raining down motor parts, fins, nuts, bolts, and all kinds of tiny airframe parts and pieces.

The most spectacular mishap of the day and maybe of the entire event was reserved for Scott Raumberger. His full-scale *SAAB Missile* replica was powered by five motors: a central experimental M1400 and four Sparky I-impulse motors. The plan was to light the M motor on the pad and airstart the outboard I motors. However, at ignition, the M failed to light, and the I-class motors fired early. The rocket rose lazily, almost in slow motion on I power; then it became unstable, flipped over, and faced nose first to the ground. It hung there in the blue sky for a moment, pointed straight down.

Suddenly, the M1400 came to life. The rocket screamed nose first into the ground, striking the flats hard, bouncing high into the air, and then striking the ground again and again, careening out of control and still under power in the midst of the away cells. The air-to-ground dance continued for several seconds as the M motor finally consumed all its fuel. The rocket finally came to rest, a smoking hulk on the bright white desert floor.

26

The Showman

In high-power rocketry, he was a force of nature. "If you can dream it, you can achieve it. That is my philosophy of life," declared Ky Michaelson in 1997.

One of Michaelson's dreams was to launch an amateur rocket into outer space:

> When I was eight years old, my father showed me *Collier's Wonder Book*, which was published in 1920. It showed a man sitting on top of a rocket with a leather helmet on his head. In another picture, it showed him lying on the ground, smoldering. The caption read, "And he lived to tell about it." Those two pictures played a big part in my life. They gave me the spirit of adventure and the dream of actually launching a rocket into space. Fifty years later, I remember that dream . . . a dream that I will do everything possible to make into reality.

Michaelson was born in Redwing, Minnesota, on the eve of the Second World War. When he was twelve, his parents gave him a chemistry set.

"It had a lot of experiments in there, and as a result, I became interested in black powder. Later, I got a bigger set and started doing things with black powder, like putting it in containers to make a rocket motor. I was building rocket motors before I was even building a rocket."

As a teenager, Michaelson learned to make his own form of rocketry fuel. "I would cut off match heads and use them as propellant for rocket motors. We tried putting the match heads in paper tubes, but that didn't work too well. So we tried metal pipes, but we didn't make the nozzle big enough. We also made a rocket car that ran on a string that was powered by match heads. After that, we started experimenting with zinc and sulfur, and I had my first successful rocket flight."

Michaelson also loved fast-moving vehicles. "Since my very early childhood, I've always felt a great need for speed, which is why I believe the good Lord blessed me with my life in this era versus, say, the horse-and-buggy days. Although I'm half certain I could have concocted some way to make those old buggies go just a bit faster."

Michaelson joined the National Hot Rod Association in the 1950s; he raced motorcycles and dragsters through the early part of the 1970s. He also built rocket-powered bicycles, motorcycles, snowmobiles, go-carts, and dragsters. He had even built a rocket-powered backpack.

One day he took a rocket-powered motorcycle to a local racetrack, and the announcer roared, "Here comes the Rocketman!"

The nickname stuck, and Michelson would one day use the Rocketman label as the brand for his own rocketry-related business. His lust for speed eventually led him to Southern California and a career as adviser to stuntmen in Hollywood, where he designed rocketry-related stunts and equipment for movies and television shows into the 1990s.

Ky Michaelson was not a model rocketry enthusiast. Nor was he involved in the high-power rocketry movement of the 1980s. Then one day in 1994, a friend dropped by to show him a copy of a magazine he had picked up at a local hobby store in Minnesota. It was Bruce Kelly's *High Power Rocketry*.

"Man, look at the size of these rockets!" exclaimed Michaelson.

The particular issue in his hands that day had a Frank Kosdon advertisement for reloadable motors. Michaelson read the caption: "A new altitude record of 37,793 feet." The idea of setting an altitude record suddenly stirred something deep in his soul.

"After that, I went to my very first rocket club meeting and took my rocket backpack," recalled Michaelson, who became Tripoli member number 3567.

Soon, he was flying high-power rockets. Then he started a small business selling parachutes for high-power rockets. Michaelson's first advertisements appeared in *High Power Rocketry* in 1995 with his product line Rocketman Recovery Systems. He marketed parachutes of all sizes featuring parabolic cups and four shroud lines. His ads were glossy full-page billboards impossible to overlook. It was clear that Michaelson had good instincts for marketing products in a manner rarely seen yet in high-power rocketry.

"We'll Pay You to Set Records Using Our Chutes!" proclaimed Michaelson in a Rocketman ad in June 1995.

He promised cash rewards to flyers using his parachutes on record-setting flights: $100 to the first person to set any Tripoli altitude record in the F- through O-motor impulse range; $500 for the first high-power rocket recovered from 40,000 feet using a Rocketman parachute; and $1,000 for the first flyer to achieve 75,000 feet with one of his parachutes. Soon, Rocketman parachutes were showing up at launch events all over America. Michaelson was quickly making a name for himself in high power.

Not long after releasing his parachute products, Michaelson announced the creation of a rocketry video business. He traveled to launches all over the country, such as Danville in Illinois, Summerfest in Nevada, Culpepper in Virginia, and LDRS wherever it was held. At each launch, he would interview flyers on the field, shoot copious amounts of film using a camcorder, and then edit VHS tapes into one-hour programs for every event. He sold his videos for $20 each under the banner of *High Thrust Video*.

In late 1995, Michaelson began marketing a line of high-power rocketry kits with splashy names like *Firefly*, *Hybrid Mama*, and his largest rocket: the *Big Kahuna*, which was 11 inches in diameter and nearly 20 feet tall. The *Big Kahuna* retailed for $850 and at the time was the largest commercial high-power rocket kit on the market.

Michaelson did more than just market rocket products to consumers. He also launched rockets—lots of them. In 1995, he and his wife, Jodi, made their first trek to the Black Rock Desert to attend LDRS-14 and BALLS-5.

"We had to pack up over 1,000 pounds of parachutes, video equipment, rockets, and support equipment and have it flown to Reno from Minneapolis," said Jodi later. "Ky and I flew into Reno and rented a U-Haul truck. Then we went back to the airport and picked up our support equipment. Next, we were off to Black Rock. What an awesome place to launch rockets! For the next four days, we sold parachutes and T-shirts and shot video for our *High Thrust Video* series."

Michaelson was one of the rocket stars at BALLS launch that year. Combining his machining and rocketry-related skills with a deep wallet, he built an all-aluminum minimum-diameter missile packing a Frank Kosdon P25,000 motor. The rocket was 6 inches in diameter and more than 18 feet tall. Michaelson declared he was seeking an altitude of 100,000 feet, which would set new Tripoli record. He named the vehicle after a friend and former Hollywood stuntman, the late Dar Robinson. He called it *DR Hero*. At ignition, the big aluminum rocket left the launch tower in a bright flash of fury and flame. Barely a second into its flight, the rocket disintegrated, scattering huge chunks of burning propellant and debris all over the desert floor.

Jodi brought her own rocket to Black Rock, and it too was impressive: an all-aluminum 3-inch-diameter missile housing a Ron Urinsco N4300 motor with a 3-second burn time. On ignition, Jodi's rocket raced into the air and was optically tracked to more than 31,000 feet. Unfortunately, there was no deployment of the recovery gear. The rocket plummeted back to the ground and was lost deep in the lake bed or, as Jodi put it, "somewhere between Black Rock and China."

These early setbacks for Michaelson and his wife at their first Black Rock event were temporary, and they seemed to take them in stride. He jokingly referred to his BALLS-5 flight as "one of the world's biggest amateur CATOs."

"Was I disappointed?" asked Michaelson. "Yes. Did I learn from it? The answer is still yes."

Michaelson's lesson was a simple one: "Never launch a rocket unless you have first hydrostatically tested the motor casing and then test-fired the motor."

Despite his setbacks in 1995—or perhaps because of them—Michaelson was becoming somewhat of a celebrity in high power. In the spring of 1996, he static-tested a P motor at a rocketry range near Delamar, Nevada. The P motor worked fine, and while they were there, he and his wife launched another all-aluminum rocket, this time with an O2800 motor that propelled the rocket to more than 50,000 feet. Safe recovery eluded them one again. The nose cone drifted off and was never seen again. The rest of the airframe came in ballistic.

"I learned two things at Delamar," said Michaelson. "One, we could build a P motor that would not blow up. Two, never try to recover a rocket and the nose cone with the same parachute, especially if you have a large ejection charge in it. The aluminum nose cone snapped a 2,000-pound tensile-strength stainless-steel cable. We had to recover the rocket with a shovel. I think the nose cone landed someplace over in Area 51."

Over the next few months, Michaelson attended several regional high-power events. He and Jodi flew rockets at LDRS-15 in Orangeburg, South Carolina, where he launched one of his *Big Kahuna* rockets, likely the largest rocket at LDRS that year. The rocket sustained a CATO on the pad and was destroyed. It was another big disappointment.

"I thought to myself, 'This is the last *Big Kahuna* that I am ever going to build,'" he said.

Then later that summer, Michaelson received a telephone call that reenergized his enthusiasm for rocketry. The call came from Nicholas Graham, the CEO of the Joe Boxer Corporation, one of the largest men's underwear makers in America. Graham told Michaelson he was coming out with some new rocket underwear and that he wanted to get some national exposure.

"I told him about the altitude project we were working on," said Michaelson. "He said that he was very interested in that, but he wanted me to build a large rocket so everyone could see the Joe Boxer logo, and he wanted to put a pair of Joe Boxer underwear and a pair of Russian underwear in it. I thought to myself, 'Oh man, now I have to build another *Big Kahuna*.'"

With the Black Rock launch only weeks away, Michaelson told Graham he could build a Joe Boxer rocket, and he got right to work. Over the next month, he put the finishing touches on a P-powered altitude rocket while simultaneously constructing another *Big Kahuna* for Graham's underwear company. This time, the *Big Kahuna* would be adorned in a splashy black-and-yellow paint scheme and $4,000 worth of high impact graphics. The Joe Boxer logo would run up and down both sides of the huge rocket.

Michaelson had his hands full at the BALLS-6 event with not only his two large rocket projects but also plenty of media attention, some of the first media attention ever garnered by high-power rocketry at the Black Rock Desert.

"Before I left, I made a number of phone calls to friends of mine in the media. I didn't know that they were all going to show up for the record attempt. I was trying to prep the Joe Boxer *Big Kahuna* and the Joe

Boxer altitude rocket at the same time, trying to give interviews to the press—the Learning Channel and the Sci-Fi Channel, Discovery, NBC, CNN, and the BBC—plus do the commercial for Joe Boxer."

Fortunately, said Michaelson, many of his fellow Tripoli flyers came to his rescue and helped him prepare the rockets and also set up the launch tower. The first Joe Boxer launch at Black Rock was of the *Big Kahuna*. Prior to flight and as part of the television commercial, a dozen flyers clad in sheer white body suits, each suit bearing the *Joe Boxer* label, hoisted the 20-foot-long rocket on their shoulders and carried it across the range.

"We marched around the desert in those hot suits for about 15 minutes until the producer said it was good enough," said Michaelson. The image was captured on film and it remains one of the most iconic high-power photographs ever taken, a surrealistic vision of the hobby that later appeared in a photo spread in *High Power Rocketry*.

The massive rocket was loaded on the pad with an M1939 and fired off into the air. The rocket had a good liftoff but deployed its recovery gear too early. The landing was very hard.

"Let's put it this way—if I ever want to fly the *Big Kahuna* again, I'm going to have to build another one," said Michaelson afterward.

Next up was his high-altitude P-powered project, and this time, Michaelson's hard work and the lessons he had learned from several prior disasters paid off. The missile roared off the pad for a flight to an estimated altitude of nearly 10 miles. Recovery was near perfect, with the rocket landing under parachute not far from the launch pad.

"Finally!" exclaimed the LCO over the PA system. "A high-power rocket recovered from over 50,000 feet!"

It was among the most impressive high-power flights of the year. Michaelson was now among the first high-power flyers in history to launch and safely recover a rocket from 50,000 feet.

Michaelson's high-profile rocketry presence continued to grow over the next few years. In 1998, he was contacted by Rick Perry, a Hollywood special effects coordinator who had stumbled across a Rocketman catalog on the Internet. Perry told Michaelson he was looking for someone to build rockets for a movie based on the Homer Hickman Jr. book called *The Rocket Boys*. The story was set in rural West Virginia in the 1950s, explained Perry, and focused on several high school students who built their own rockets. One of the students, who seemed destined to follow in his father's footsteps and work in the coal mines, instead went on to work as an engineer for NASA. Michaelson was sent a copy of the script. He quickly read it, and then he agreed to manufacture thirteen rockets for use in the film. The movie was called *October Sky*.

In the late 1990s, Michaelson ran for a seat on Tripoli's board of directors.

"I think it is important to promote the sport at a national level so the public and government officials understand that high-power rocketry is a family sport and we are not a bunch of terrorists," said Michaelson, who promised to support the expansion of experimental rocketry.

He called for additional money to challenge laws he believed were hindering the entire hobby. His election platform also included a call for better insurance coverage for rocketry, increased production of *High Power Rocketry* to twelve issues per year, and more media involvement in the hobby.

Michaelson was unsuccessful in his election bid to the Tripoli board. But as the decade came to a close, he found himself pursuing another dream: assembling a team of flyers to launch a rocket into space.

LDRS returned to Argonia for five days of flying from July 29 through August 2, 1999. There were nearly 375 registered flyers and more than 1,100 flights at LDRS-18, making this the biggest LDRS to date. There was also an increase in the number of big projects.

At least forty-seven M-powered rockets were launched at LDRS-18; there were twenty M-powered flights on Saturday alone. For the first time, the Discovery Channel was on hand, looking to perhaps one day create a show dedicated to high-power rocketry. This was also the first LDRS to have its own dedicated web page on the Internet.

Unfortunately, LDRS-18 also began in the midst of a terrible heat wave. Field temperatures Thursday were well over 100 degrees, and that was in the shade. Still, despite excessive heat on Thursday and windy conditions on Friday afternoon, more than four hundred rockets took to the air in the first two days of LDRS-18.

Among the rockets launched was Doug Gerrard's *My Mind's Eye*, a scratch-built vehicle launched with eleven onboard cameras. Gerrard entered high power in 1988 to fly camera rockets. By 1999, he was one of the leaders in the field. His LDRS rocket held eight Olympus and three Minolta cameras. In the era before digital photography, that is a lot of film. Gerrard engineered intricate relay and timing circuits to fire all his onboard cameras during flight. This was also Gerrard's Level 3 certification flight. His rocket was more than 17 feet tall and weighed 110 pounds on the pad with an AeroTech M1939. It carried altimeters and timers by blacksky and Adept. This was truly a state-of-the-art high-power rocket in 1999. *My Mind's Eye* roared off the pad, sailing high into the sky on the big M motor. A good deployment brought it all back in one piece for a successful certification flight.

The beginnings of the upscale *Fat Boy* rocket craze might be traced to LDRS-18 too as numerous scale versions of the Estes kit started appearing in all sizes this year. On Friday, two M-powered *Fat Boy*s were hauled out to the M pads. One was built by Darren Owens of Kansas. It was 16 inches in diameter and was powered by an Ellis Mountain M1000. The rocket had a good flight and recovery. Later that morning, Andy Schecter of New York launched a 14-inch-diameter *Fat Boy* on an M1939.

Gary Rosenfield introduced a new M-impulse motor at LDRS-18: the 75 mm M1315. In a demonstration flight on Friday, the new M motor carried a rocket high into the air. It was the first flight of an M1315 at any LDRS. In a few years, this motor would become one of the most common and affordable M motors on the market.

Nearly 450 rockets would leave the pads on Saturday in one of the biggest single-day launches in LDRS history. This worked out to nearly one flight per minute. Gene Nowaczyk of Kansas City launched his version of the *Starchaser*, a rocket made famous by Stephen Bennett of England. This was Nowaczyk's Level 3 certification flight. His 18-foot-long rocket held an M1419 and altimeters by Adept and Olsen. The flight was near perfect.

Another crowd favorite was a flying tetrahedron built by John Ritz of Nebraska. The big odd roc roared into the air on an M1419 for a flight to about 1,000 feet. Derek Deville returned for his second LDRS to launch a minimum-diameter missile on an L3000 motor. Duane Wilkey launched a beautiful *Nike Smoke* on an M1419, and John Bolene of Oklahoma let loose with a Level 3 attempt on an Ellis Mountain M1000.

On Saturday night, Tripoli members made the trek back to Wichita, about an hour from the field, for the annual banquet and members meeting. More than three hundred dinners were served. The guest speaker was Quentin Wilson, one of the true-to-life characters portrayed in the now released hit movie *October Sky*. Wilson had recently joined a Tripoli prefecture in Texas, and he obtained his Level 1 certification that weekend in Argonia. He entertained Tripoli members and their families with stories of his rocketry adventures with his West Virginia classmates in the 1950s.

Sunday morning brought cloudy and breezy conditions to the range. Although this was a welcome respite from the heat, the wind cut the flying day short. Yet it was still a good day for flying, and upward of two hundred rockets left the pads, including two of the largest rockets of LDRS-18.

Ray Halm Jr. saw a photo one day of the upper stage of a *Minuteman II* ballistic missile. He liked that picture, and the New York resident thought a scale model of the missile would make for a great high-power rocket. Halm began his project, which he dubbed *Aries*, after he found an airframe tube that was 24 inches in diameter. This would make for a perfect half-scale version of the rocket, he thought; he was off and running.

When Halm arrived at LDRS-18, the *Aries* was almost done—but not quite. He and his team spent the first few days in Kansas completing the paint job and otherwise preparing the rocket for flight. When it was ready, it was a sight to behold. The *Aries* was 2 feet in diameter and 13 feet tall. There were four massive fins that were removable via slots in the airframe and a floating aft centering ring that supported a custom aluminum thrust plate. The thrust plate would be needed for this rocket. The power plant for the vehicle was four AeroTech M1939s. The motors weighed 80 pounds.

"This is the first AeroTech four-M-motor cluster ever flown," said AeroTech employee Ed LaCroix, who was on hand to help Halm and his team with last-minute motor preparations before flight.

The total weight of the *Aries* was 350 pounds, ready to go at the pad.

As the *Aries* was being carried to the away cells on Sunday, another huge project was being prepared at the other end of the flight line. This was Tripoli Oklahoma's *Shamu the Killer V-2*, a 14-foot-long black-and-white *V-2* replica powered by a central N2000 and four K motors. The 22-inch-diameter *V-2* garnered special attention from the Great Britain–based Discovery Channel film crew. During the Second World War, hundreds of *V-2* rockets rained down on England during the last year of the conflict. The film crew pored over every inch of the replica at LDRS, aiming their camera this way and that during final preparation for flight. They also installed an expensive camera in the side of the rocket's airframe to capture video from the air.

When it was ready, the *V-2* was lifted by several men into a pickup truck for a ride out to the away pads. Along one side of the rocket was the word "Magnum," designating one of the rocket's primary sponsors. In the 1990s, Magnum was one the largest high-power vendors in America. Meanwhile, back at the *Aries* pad, everything was near ready. Final assembly of Halm's rocket was performed at the edge of a waist-high-tall cornfield. The launcher and rocket were raised in unison, and igniters were inserted in the aft end of each of the four 98 mm motors to ensure that ignition of the M1939s was simultaneous. Quentin Wilson was asked to push the launch button for the *Aries* when the countdown hit zero. When he did, the rocket lit up like a Roman candle.

"In my opinion, there was nothing better," wrote flyer Gene Nowaczyk after the launch. "Just try to imagine four M1939s lighting at the same time. The smoke and fire that poured out from the business end of the *Aries* was amazing. I have never seen anything like it." (One day, in the distant future, Nowaczyk would set a new standard for high-altitude flights in high-power rocketry.)

The heavy rocket rose easily from its Unistrut launch pad, turning slowly as it increased in speed and gently nosed into the wind. It climbed a few thousand feet before reaching apogee, and then it deployed its multiple parachutes as planned. The *Aries* descended in two sections. One of the three chutes attached to the main airframe did not quite unfurl, but it all landed gently on the ground. The crowd roared with cheers and applause.

It was time for the *V-2*. The 165-pound rocket was quickly raised up and brought to vertical on its pad. The bottom edge of each of the four fins rested on the ground. The wind was picking up slightly, but it was within safe parameters for flight. The Discovery Channel crew finished filming at the pad. Then everyone retreated for cover. At ignition, the World War II replica lifted perfectly, trailing a gigantic column of smoke and fire. The *V-2* climbed high above the field, and like the *Aries* just moments before, it deployed its parachutes on cue and began a long and slow descent back to Earth for a great recovery. Under a 28-foot-diameter parachute, the rocket drifted off, landing gently two miles away.

Despite the wind, it was a great day of flying. That night, the Tripoli board met at a hotel back in Wichita and decided that as a result of the weather delays on Sunday, commercial flying would be allowed to continue to Monday, originally designated as an LDRS experimental day.

Every now and then, there comes a special project that captures the attention of the entire rocketry community. It is something that everyone wants to see, and they will come a long way to see it. In 1999, such a rocket made its debut in Argonia on August 2 at LDRS-18. It was a one-third-scale replica of the famous *Mercury Redstone* rocket.

On May 5, 1961, the *Mercury Redstone* lifted astronaut Alan B. Shepard into space. Shepard was the first American to achieve that feat, and 90 days later, another *Redstone* propelled Gus Grissom into space. It was the beginning of America's manned space flights, culminating in the moon landing in 1969.

"Sitting around with friends at LDRS last year at the [Bonneville] Salt Flats, I was talking about wanting to build a 30-foot-tall model," said Tripoli Nebraska member Bruce Lee. "A tribute to something that NASA did would be cool. 'How big was the *Mercury Redstone* rocket?' I asked. 'Around 100 feet tall,' someone replied. And I thought, 'Let's see . . . 100 feet . . . hmmm . . . One-third scale would put it around 30-plus feet . . . That's about right. A scale model of Shepard's flight would be awesome!'"

Lee and twenty-seven friends from five different states began work on a project that would ultimately tip the scales at more than 600 pounds, loaded and ready to go.

"We broke up the project so people in different geographic areas could work on different subassemblies in parallel with the main construction being handled in Omaha," explained Lee at the time. "The *Mercury Redstone* would not see a complete assembly and integration of all parts until it was on the pad at LDRS. This was an extra complication and meant everything had to be designed and built exactly as required— the first time."

Most of the rocket's airframe construction was built by a team of flyers in Omaha, Nebraska. This included cutting internal components such as centering rings, motor mounts, and bulkhead attachments. The airframe alone was more than 20 feet long. The application of fiberglass to the 24-inch-diameter airframe was completed in Lincoln. After the fiberglass was applied and cured, the airframe sections were driven to Omaha for subassemblies, detailing, and finishing.

Construction of the detailed *Mercury* space capsule and its escape tower was undertaken by Ky Michaelson, working on his own in his shop in Minnesota. A sample piece of the 24-inch Sonotube airframe was shipped to Michaelson so he could use it as a fitting sample to ensure the capsule assembly would mate perfectly with the main body of the rocket. The capsule was created from scratch by Michaelson. He made the escape tower by welding steel tubes together with plates on the top and bottom of the tower. The tower would house three I-impulse motors. Michaelson also supplied the parachutes for the entire rocket. Electronics for the *Redstone* project came from a package put together by another team, this one led by Missile Works owner Jim Amos of Colorado. A relative newcomer to high power, Missile Works was gaining a foothold in the altimeter and timer market. Amos supplied three of his MRC2 remote-control systems as well as an electronic timer for backup for the rocket's ejection charges. An Adept electronic timer was also installed to fire the ejection charges and activate the motors in the capsule's escape tower. These motors would pull the capsule assembly off the main airframe for a recovery under chute. All told, the *Redstone* would carry more than fifty separate black-powder ejection charges. The *Redstone* was finished in authentic colors, complete with NASA insignias.

"It had taken eight people working eight hours to get the decals on," said team member Arley Davis. "But they were perfect and to scale."

Meanwhile, television learned of the rocket, and according to Lee, the medium wanted to be there.

"The Discovery Channel found out about our project, and the *Extreme Machines* show contacted us about filming. They flew in from England and arrived at LDRS on Friday. They wanted to know if we could fly a couple of their onboard video cameras, and the answer, of course, was yes. At the field, we cut a hole in the payload section and fiberglassed in one of their cameras. The second camera was attached to the escape tower motor casing. Both had dramatic downward-looking views."

The *Redstone* was displayed horizontal and near the flight line from Thursday through Sunday. Everyone walked over to see the mighty replica as it was being prepared for its flight on Monday.

"We spent over 1,500 man hours in planning and construction in Omaha and probably another 500 hours of work outside of Omaha on this project," said Lee, who estimated that the total cost of the rocket was $10,000.

The launch tower was assembled on Friday and Saturday. It was positioned three-quarters of a mile east of the flight line. Final setup for the flight began at 6:00 a.m. on Monday. Discovery Channel cameras were installed in the airframe, the upper and lower airframe sections were fitted together while the rocket was vertical on the pad, and the capsule was set in place at the top of the rocket, 30 feet above the ground. It took all day, but the *Redstone's* army of volunteers made ready the 600-pound rocket for flight. For power, a 6-inch-diameter P13,000 motor was installed in the aft end. The motor was more than 6 feet long. It was engineered for a 6-second burn time, which would propel the big rocket several thousand feet into the air.

Final setup was completed just after three o'clock on Monday afternoon. Two minutes before ignition, a fire truck from the Argonia Fire Department turned on its sirens to alert everyone on the range of the impending launch. In attendance was Laura Shepard-Churchley of Kansas, daughter of astronaut Alan B. Shepard. This would be the biggest rocket ever launched at LDRS up to this time, and after a ten count, the daughter of America's first man in space pushed the go button for launch. What happened next was, as one commentator noted, "a heartbreaking disaster."

The igniter fired, and the fuel grains inside the P13,000 came to life. The motor pressurized, and momentarily, a "white-purple-ish glow" emerged from the aft end of the *Redstone*. The rocket began to climb up the tower like a slow-moving freight train chugging up a steep hill. But unlike the Shepard launch in 1961, this *Redstone* halted suddenly midway up the rail and burst into flames, suffering a motor CATO before the rocket cleared the launch pad. It happened so quickly that the aft end of the rocket seemed to literally explode, the motor ripping open inside the airframe and propelling fiery chunks of propellant directly out of the lower left side of the vehicle. Simultaneously, there was a scattering of rocket parts, big and small, in every direction, extending outward from the pad in a circle of calamity. Oh, the humanity!

As the lower half of the *Redstone* burned furiously on the pad, the upper half of the rocket separated and was carried skyward by the initial momentum of the P motor. The upper airframe continued on a weak trajectory straight up into the air for perhaps 50 feet or so. Then the airframe, complete with its escape capsule, fell forlornly straight back down again, impaling itself on top of the launch tower. As the upper section of the *Redstone* dangled from the peak of the 30-foot-tall tower, two of its recovery harness lines fell into the fiery mess burning directly below like the dangling tentacles of some wounded beast. The booster was totally consumed by fire.

It was one of the most spectacular failures in the history of LDRS, and for several minutes, as the fire consumed the rocket, the *Redstone* Team and a thousand spectators could do little more than look on in stunned silence while the Discovery Channel cameras kept rolling. This is the hardest part of the high-power rocketry learning curve: watching a carefully built and well-thought-out scientific creation as it is utterly destroyed. There are no participation trophies in rocketry, and no matter how hard one prepares for a rocket's flight, there is never any guarantee of success.

Thankfully, no one was hurt in the disaster, and after the fires calmed down, the crew was picking up the pieces, trying to figure out what had gone wrong. Postflight inspection of the wreckage revealed the aluminum motor case likely had a metallurgic defect. When the P13,000 reached full pressure on the pad, the defect gave way, the case split wide open, and the superheated flaming contents inside the motor spilled out into the airframe, destroying the rocket.

"The next morning, our *Mercury Redstone* project was the lead article on the cover of the *Wichita Eagle* newspaper," reported team member Richard Burney.

"An incredible shot of it blowing up had been taken by one of the newspaper's photographers."

27

The Biggest LDRS Ever

The national launch returned to the sod farm in South Carolina over the Fourth of July weekend in 2000. Four hundred registered flyers launched nearly 1,400 rockets in four days. In terms of the logged flight count, this was the biggest LDRS in Tripoli history.

"I need therapy," wrote Tripoli board member Bob Schoner after the launch. "I have been thoroughly desensitized to the thrill of an M motor . . . [D]o you remember not so long ago when phone lines and email were abuzz with word that an M motor was going up? The words were usually announced in whispered tones of half awe and half fear. Oh, how things have changed."

Schoner was right; things had changed. The sixty-five pads at LDRS-19 launched at least sixty motors in the L- and M-impulse range; forty were M motors. This was the Super Bowl of high-power rocketry.

One of the early M flights at LDRS-19 was a 12.5-foot-tall *Iris* replica built by brothers Paul and Andy Bonham. The 57-pound rocket blasted off Saturday morning, July 1, with an M1419 to more than 6,000 feet, followed by a perfect recovery on the Super Sod farm. Moments later, Al Stone of Illinois stepped up to the pad with his *Black Brant II*. This rocket tipped the scales at 120 pounds. An AeroTech M1939 carried it more than 4,000 feet high. It landed 100 yards from the pad. Up next was Maryland flyer Kevin Mitchell, who launched another *Iris*-based scale rocket, this one with a Kosdon M1130 motor. The flight was near perfect; Mitchell walked away with his Level 3 certification.

It was like this all day on Saturday. There were AeroTech White Lightning Ms and Kosdon East Green Gorilla Ms, M motors that emitted bright red flames, and Skidmark Ms that showered the launch area with colorful titanium sparks. The spark-producing motors sometimes started small fires in the dry grass. Fire crews rushed out with ATVs or on foot with water and other fire-suppression devices. The crowd loved all of it.

Ray Halm Jr. of New York launched the multiple-M-powered *Arias* rocket at LDRS-18 in Argonia in 1999. On Saturday at Orangeburg, he sent up another big rocket to 8,000 feet on a Kosdon M2240 motor. And Tom Binford launched his 100-pound *Mega Motor Eater* on four K550s and a central M2400 for another exciting flight.

A fast-moving thunderstorm arrived late in the afternoon, cutting short Saturday's range activities. On Sunday, flyers quickly picked up where they had left off. Bruce Lee and Lyle Woodrum of Tripoli Nebraska launched their *Bomb Pop*, a 120-pound rocket version of the popular frozen ice cream. The rocket carried two 75 mm M1315 motors.

"Knowing that a two-motor cluster is the most difficult to fly, I decided to make a more reliable motor igniter," said Lee. "Taking a page from the NASA shuttle booster ignition system, I came up with a similar 'motor within a motor' igniter. An Estes A10 motor was installed with an electric match at the top of the core of each M1315. At ignition, the A10 generates a column of flame that reliably starts the whole core."

Lee's system worked perfectly, and *Bomb Pop* roared into the blue skies for a good flight and recovery. Mark Roszell had a successful Level 3 certification with his *Ready to Rumble* on a Kosdon M2240, and Andy Schechter had a fun flight with his *Pinhead*, a rocket that was basically just a nose cone with four fins attached. It was propelled by a fast-burning M2400. At Max Q, the fins tore off the airframe, and *Pinhead* tumbled from the sky to a hard landing.

Prior to LDRS-19, Frank Kosdon delegated some of his motor manufacturing to New England Tripoli member Paul Robinson, who was making motors under the name Kosdon East. At LDRS-19, there were plenty of Kosdon East flights, including a new Skidmark motor flown by flyer Jim Scarpine of North Carolina. Scarpine's 60-pound rocket served as a demonstration vehicle for the new motor. The rocket blasted off amidst a shower of sparks trailing a dense, oily black plume of smoke high into the air. The Skidmark would become one of the most popular motors on the Eastern Seaboard. Robinson also launched a rocket of his own on an M4150 for a nice flight on Sunday; flyer John

Taylor sent up a 40-pound rocket on a Kosdon M1130 for another good ride.

One of the more memorable flights of LDRS-19 was Florida flyer Rick Boyette's upscale *Maxi-Alpha*. Boyette's giant version of the tiny Estes model rocket was powered by an M1939, which lifted the 80-pound *Maxi-Alpha* high into the sky. It was a great flight up, but deployment issues led to a less-than-optimal recovery.

There was at least one hybrid N motor on Sunday: Bruce Kilby's *Obsession*, a 100-pound Hypertek-driven vehicle launched from a custom-made wooden tower. The big rocket was 12 inches in diameter and stood 15 feet tall. It carried a central N motor and multiple K and I motors. All were lit on the pad. The N1310 hybrid emitted a bright flame as it rose several thousand feet into the air. Recovery was near perfect.

The annual Tripoli banquet and members' meeting was held at the sod farm on Sunday evening. Organizers set up a large white party tent under which hundreds of Tripoli members and their families enjoyed Southern-style barbecue. The weather was cool and pleasant, and the tent was ringed with lights. The speeches, awards, and alcohol-fueled rocketry revelries went on late into the evening.

On Monday, there was more terrific weather, along with the biggest rocket launch of the holiday weekend: Derrick Deville's red-white-and-blue *Freedom Phiter*. Deville's rocket was built by the future Tripoli board member at his home in Miami. It weighed 165 pounds on the pad with three Kosdon M3200 motors. The rocket's airframe was 12 inches in diameter and stood 15 feet tall. It had flown once before, said Deville, at a local launch on an M and a couple of Ks. The three Kosdon M motors for LDRS-19 put the combined impulse range of the motors at about an O9600, he said.

"It was a poster child for raw power," said videographer Earl Cagle Jr. of *Point 39 Productions*.

This flight brought to a halt all other activity on the range. At ignition, all three of the 75 mm Kosdon motors fired immediately. The rocket left the pad so quickly, it caught some people by surprise. As it climbed up and away, three long flames filled with Mach diamonds trailed in its wake. The rocket thundered straight up, said Schoner.

"Actually, 'thundered' isn't the right word to describe 9,600 Newton seconds of Kosdon power," he added. "Every one of your senses gets whacked. Your ears and eyes are assaulted from something like that. My eyes hurt. The image is still burned into my retina. My ears hurt. The roar still rings in my head days later. My hair hurts, but I don't really know why."

Freedom Phiter almost disappeared at apogee, but it could be seen turning over and splitting in half just shy of 9,000 feet. The two airframe halves, now connected by a long tether, dropped quickly, followed by a successful main parachute deployment directly above the rocketry range. When the rocket set down gently on the grass under chute, Deville was treated to a loud round of applause for his perfect ground-pounding launch and recovery. *Freedom Phiter* graced the cover of *High Power Rocketry*'s LDRS-19 issue in 2001. The rocket would also appear on the cover of a new high-power rocketry magazine that was only on its third issue. That magazine was called *Extreme Rocketry*.

Another impressive flight on Monday was Mark Oullette's *Draco*, named for the constellation bearing the same name. The Montreal resident had spent two years building the 17-foot-tall vehicle. It looked like a museum piece. Oullette received the Best Technical Award of LDRS-19, a contest sponsored by LOC. Although a beautiful rocket, on the pad, *Draco* was all business. It weighed 125 pounds with an AeroTech M2500. It took several men to hoist the rocket into vertical position on its launcher; when it was ready to go, the entire range came to a halt again.

Draco climbed effortlessly ahead of the big AeroTech M. The rocket turned over a little under a mile in the sky and then deployed multiple parachutes, bringing down the nose cone and the rest of the rocket separately. The parachute for the nose cone was tangled a bit, resulting in some bruises on landing. The rest of the rocket landed nicely on the green grass.

LDRS-19 continued through Tuesday, the Fourth of July. By the time it ended, there had been 1,379 logged flights—the most recorded flights at any LDRS ever. The Orangeburg launch in 2000 remains the largest LDRS ever held, a remarkable feat considering the event lasted only four days.

"All in all, it was a fun four days," said flyer Jerry O'Sullivan of Maryland. "I've never experienced such a concentration of cool flights and great disasters!"

MARK CANEPA

For the first time in LDRS history, a three-day Tripoli experimental launch followed LDRS at a separate location with a day off (July 5) between events. The launch was called "Small Balls," and it was held at Tripoli North Carolina's field at Whitakers, a half-day's drive from Orangeburg into the Tar Heel State. The flying range was an 800-acre cow pasture surrounded by 1,000 acres of lush farmland with twice the waiver altitude of Orangeburg. It was a fine spot to launch rockets powered by motors similar to but not quite as large as those featured at the annual BALLS event at the Black Rock Desert in Nevada.

Many of the homemade motors at Whitakers were based on designs by Thunderflame owners Jim Mitchell and Sonny Thompson. Mitchell was on hand teaching research motor-building skills. Also present was Kentucky chemistry professor Dr. Terry McCreary, who had just released his first book on the manufacture of composite propellant for high-power rocket motors.

Neil McGilvray was part of a contingent of Maryland flyers who had traveled to North Carolina to take advantage of the research event, launching at least two rockets, including *Rancor*, carrying an N2890 motor designed by Jim Mitchell. This was the second N motor built by McGilvray; his first research motor sustained a CATO only 400 feet into the air in Maryland. This time, McGilvray's homebuilt motor worked perfectly, carrying his 95-pound rocket 5,300 feet above the pasture for a nice flight and recovery.

"It worked!" exclaimed McGilvray as he headed out to collect his rocket. "Big improvement over my last N attempt. Thank you, Jim!"

Maryland flyer Kevin Mitchell also had success with his N700-powered research rocket called *N-Sane*. At 200 pounds, this was the biggest rocket of the event. It had a great flight on a fuel called Super Blue Propellant.

Flyer Fred Wallace entertained the crowd with the flight of his *Big Wahoo* rocket, although the vehicle did not survive the show. This was the ninth M-powered flight of the rocket, which Wallace loaded with a research M150. Takeoff was great, but 1,500 feet into the air, there was a *boom* as the motor sustained an in-flight CATO. The rocket descended in several pieces, some parts of the airframe under parachute as the still-burning propellant provided for a bright red flame during descent.

McGilvray described what had happened next to his friend's rocket: "Then to add insult to injury, every bit of the rocket, with the exception of the burning section, floated into the shallow pond that the entire herd of cows had used earlier in the day to cool off and God knows what else!"

"You can imagine the stench he had to deal with," added photographer Nadine Kinney. "So he spent $10 in quarters and hours at the local car wash cleaning up his rocket."

Flying at Small Balls went through Saturday, July 8. On Friday, there were numerous launches, including the early evening flight of a rocket by Dave Bullis called *Redemptive Power*, holding a Thunderflame N4070 Super Blue motor. Other flyers included Kathy Gilliand of Maryland, who had launched several rockets, Mark Lloyd, who had flown a 102-pound missile on an M2200, and Kelly Mercer, who had launched a full-scale *ASP* on an M3000. There were no statistics retained after the event, but there were at least forty-five flyers, and they sent up close to a hundred rockets, with many of them in the L-, M-, and N-impulse range. Experimental motors were becoming more popular every year, and the Whitakers event made it clear that eastern flyers were just as anxious to formulate their own high-power fuel as any of their brethren out west.

"Kudos to the North Carolina boys for running a first-class event," wrote McGilvray after the launch ended. "I, for one, will be looking forward to the next one as the next launch can never come around too soon."

28

Return to Lucerne

At their annual meeting in Orangeburg in 2000, the Tripoli board took up the issue of who would host the twentieth anniversary of America's biggest rocketry event the following summer. The primary contenders were the Black Rock Desert in Nevada and the Lucerne Valley in Southern California. Member support for a return to Black Rock was strong. A Black Rock LDRS would be hosted by AERO-PAC, one of the most experienced clubs in the hobby. Black Rock also had the highest waiver in high-power rocketry at 100,000 feet. In contrast, the flying range at Lucerne was smaller, and the waiver paltry in comparison—only 5,300 feet above ground level, with windows to 10,000 feet.

A Black Rock LDRS also had the advantage in motor size selection. There were no restrictions in the Silver State on impulse size for high-power rocket motors. By 2001, flyers at Black Rock were launching N, O, and P motors with some regularity. However, in California, restrictive state regulations prohibited motors larger than an M, with a slight exception for clusters, which could equal the total impulse of a single N motor. Any LDRS at Lucerne would also be governed by the Golden State's pyrotechnics regulations developed for sparklers, pinwheels, and skyrockets. The freedom to fly would certainly be curtailed. Not surprisingly, at the board meeting in Orangeburg, there was discussion in favor of dumping any proposal that LDRS-20 be held in California.

"We cannot fly a motor larger than an M at Lucerne. That won't do."
"The waiver is way too low for high power!"
"Let's avoid California. There are too many legalities and regulations."

Lucerne seemed outclassed by Black Rock at every turn. Yet Lucerne had one factor in its favor: sentimentality. Few people doubted that

Lucerne was the ancestral home of high-power rocketry. In the 1970s, composite rocket motors were being flown at Lucerne, likely before any other location in America. During the 1980s, Lucerne was also the site of several high-profile annual launches, including Summerfest and Octoberfest. The first reloadable and hybrid motors were tested at Lucerne, and many high-power pioneers had launched their first big rockets here too; people such as Gary Rosenfield, Chuck Rogers, Jerry Irvine, Chuck Mund, Korey Kline, and Steve Buck, to name but a few.

Lucerne had also seen a fair amount of controversy. In the twenty years prior to 2001, the range was shut down from time to time by state authorities opposed to high-power rocketry. Tripoli and NAR members were sometimes disciplined and even expelled from their respective organizations for violations of safety rules at Lucerne. Still, despite periodic problems, Lucerne was a legendary rocketry range. But it had never hosted LDRS.

After careful consideration and debate and with more than just a bit of history in mind, board member Chuck Rogers made a historic motion. It was Rogers who had led the charge to hold the tenth anniversary LDRS at Black Rock in 1991, a time when the remote Nevada location was relatively unknown to most flyers. Now more than ten years later, Rogers moved the board to vote that the Lucerne Dry Lake Bed be chosen for LDRS-20 in 2001. Black Rock could be the alternate location if any problems arose in California, said Rogers. The Tripoli board unanimously agreed; the motion was passed. LDRS prepared to make its first trip to California in 2001.

"When I started this club several years ago with the other guys, one of my dreams was to become a Tripoli prefecture and host an LDRS," said LDRS-20 organizer Rick O'Neil as he looked out over the well-planned range on Day One of the national launch. "This is a dream come true."

The twentieth anniversary of LDRS went forward on July 19 to 22, 2001, at Lucerne. Pundits predicted it would be the largest LDRS ever, with four hundred flyers expected to launch more than one thousand rockets. The hosting prefectures, Tripoli Anaheim and Tripoli Los Angeles, did a good job of creating what was one of the largest high-power ranges ever seen. There were at least eighty pads spread over three adjacent but separate sections of the dry lake bed. As in Argonia, Kansas, this arrangement allowed for the nonstop launching and recovery of rockets throughout the day. The local clubs also had an outstanding launch control system capable of handling almost any combination of simultaneous flights.

David Reese of California launched one of the first rockets of LDRS-20. Reese's bird was a LOC *Vulcanite* kit that screamed off the pad on an AeroTech I211 for a good flight and recovery. Flyers Steve Whitney from Bangor, Maine, and Andy Woerner from San Diego each launched M-power rockets on Thursday as vendors worked nearby to set up their temporary shops behind the flight lines, waiting for the weekend crowds to arrive.

The first day of LDRS had fewer flights than the rest of the weekend. But there was plenty of socializing as flyers set up camp and got reacquainted with friends from all over the country. People arriving from the East Coast either drove out in their own vehicles or flew into nearby cities such as Los Angeles or Las Vegas and rented cars for their trip to the range.

"Easterners like me are frequently amazed at the landscape that reveals itself on the drive from Las Vegas to Lucerne," wrote Tripoli board member Ken Good of Pittsburgh. "I've seen it before, but the 'otherworldliness' of this environment, nearly Martian by nature, never fails to impress me. It's desolate, with flat or rolling high desert, dotted by dry scrub and punctuated by barren mountains and outcropping of rugged brown rock. You don't play games in this country. It doesn't tolerate fools."

The first day was a good opportunity for people to stroll the flight line, parking lot, and camping areas searching for the extra special projects that would fly at Lucerne. There were several rockets worthy of their attention. One was a scale version of the *N1*, a Soviet-era missile conceived as the Communist answer to America's *Saturn V*. This was the rocket that was going to bring the Russians to the moon. The scale rocket was a nearly perfect one-sixteenth-version of the original and was built by several Tripoli AERO-PAC members, led by John Coker of California. The rocket stood 22 feet tall and was nearly 4 feet in diameter at its base. It was painted in Soviet-style colors of dark green and white. The rocket was powered by forty-three motors and was admired by onlookers over the next two days as it was assembled for its anticipated launch Saturday morning.

LARGE AND DANGEROUS ROCKET SHIPS

Not far from the *N1* was an equally impressive rocket transported to California all the way from Europe by a team of Tripoli flyers from Switzerland: the ARGOS group. The rocket was a museum-quality scale version of the *Arianne 4*, one of the more successful space vehicles produced by the European Space Agency. The *Arianne 4* was first flown in 1988 and would ultimately conduct more than a hundred missions to space, primarily for satellite insertion into Earth orbit. Like the *N1* group, the Swiss team had not just constructed a high-power rocket but also built a work of art, prompting the same wide-eyed stares and comments by passersby, over and over, all weekend long.

"That rocket is too nice to fly."

"The *Arianne* belongs in a museum, not out here in the desert."

"That's the finest high-power rocket I have ever seen."

Yet the Swiss team clearly built their museum piece to fly, and led by Juerg Thuering, they were determined to launch their rocket at LDRS-20. The final product was a one-thirteenth-scale version 15 feet tall and 10 inches wide at its base. For power, the rocket would use five motors: a central M1939 with two K550s and two J350s in the outboard boosters.

There were several other scale projects also waiting in the wings, including a large number of *V-2* replicas of almost every size all the way up to a 285-pound behemoth of a rocket scheduled to fly Saturday. And then there were the rockets belonging to a pair of high-power newcomers, soon to be known simply as the Gates Brothers.

Dirk and Eric Gates had been in high power less than two years, during which time they both achieved Level 3 certifications. Yet beyond their local launches in Lucerne, they were all but unknown in 2001. LDRS-20 was their coming-out party. In just four days, "Gates Brothers Rocketry" would become a national high-power phenomenon, one that would last for nearly a decade. The Gates Brothers would conduct a nearly flawless high-power seminar, launching and recovering at least three mega-projects that burned nothing but K, L, and M motors. Their construction techniques were perfect, their motor selection was perfect, and their flights were perfect. And it all began midday on Thursday, July 1, 2001, with the flight of *Aramis*.

Aramis was a two-stage rocket that connected the lower half of Dirk's Level 3 rocket with a sustainer from Eric's Level 2 rocket. The combined creation was 6 inches in diameter and 17 feet tall. The rocket held an M1315 in the booster and a long-burning K250 in the sustainer. Dirk and Eric rose at 6:00 a.m. on Thursday to ensure they got an early start for their first big show.

"This was our first two-stage rocket," said Eric, "so prepping her went slow. We had two cameras to load and many electronics items to get into place, armed and ready to go. We wanted it to work."

At 11:00 a.m., *Aramis* roared off the pad. The M1315 left a giant trail of white smoke as *Aramis* climbed high into the air. The sustainer lit seconds into the flight, and the K250 took over, lifting the upper stage to more than 7,000 feet before returning under parachute for an excellent recovery.

"They made that look pretty easy," said one observer.

For the Gates brothers, this was only the beginning.

Unfortunately, the range closed early on Thursday afternoon as strong winds picked up and blew through the valley, carrying lots of dirt and sand and everything else that was not tied down. By midafternoon, most of the flying was over, an inconvenience that would repeat itself every day, significantly reducing the overall flight numbers for LDRS-20.

"The early mornings were just beautiful and also were the best time to launch," said NAR member Chuck McConaghy. "The first rockets were launched every day around 6:00 a.m., which was good because the high afternoon winds and blowing sand arrived like clockwork and shut things down by about 2:00 p.m. every day."

There were many more flights on Friday, July 2, culminating with the flight of another Gates Brothers mega-rocket, *Porthos*. The *Porthos* rocket was physically the largest vehicle flown by Dirk and Eric at LDRS. It was 12 inches in diameter and 16 feet tall; the rocket weighed 160 pounds without any fuel. For propellant, *Porthos* carried a central L1500 surrounded by six K1050s. The K motors were all single-use motors, a real rarity for high-power 2001.

"We were thankful for single-use motors," said Dirk. "We only had one L1500 to pack!"

Porthos weighed more than 200 pounds fully loaded. The big rocket lifted off Friday morning for another perfect flight and recovery to 6,700 feet. The rocket carried multiple video cameras that returned some great flight film of the entire ride.

There were several other memorable flights on Friday, including a Kurt Gugisberg nine-motor rocket called *Flying Space Cat of Death*, a successful 7.5-inch *V-2* flight by Ron Phillips, and an M-powered rocket launched by Kimberly Harms of Washington. Local flyer Wedge Oldham obtained his Level 3 certification with his *Necessary Evil* on an M1315.

Saturday was the biggest day of LDRS-20, and the big projects took center stage immediately. The Soviet *N1* team worked on their rocket throughout the night on Friday. They were ready to go Saturday morning. Launch conditions were perfect for the finless space vehicle. There was no wind and nothing but blue skies above. Just prior to countdown, the PA system sounded a rendition of the Russian national anthem, which was followed by a slow countdown in Russian by the LCO.

At ignition, the 230-pound 21-foot-tall *N1* blasted off, with virtually every one of its booster motors, all thirty of them, successfully lighting. The rocket left the pad cleanly amidst a huge cloud of smoke and in rapid succession went through its two additional stages in the air. The rocket was not going high (it did not reach 2,000 feet), but it looked like the real thing every step of the way. Recovery of all three stages was complicated by some parachute troubles, yet most of the rocket survived intact. The big crowd at Lucerne cheered the *N1* team for their efforts.

For the third day in a row, the Gates Brothers were at it again. This time, they prepared a full-scale replica of a *Jayhawk* rocket, a United States Navy target drone. Along with the Swiss *Arianne 4* and the Russian *N1*, the *Jayhawk* was among the most admired and photographed rockets of the entire weekend. It was finished in a bright red paint scheme complete with United States Navy insignias. The scale rocket stood more than 14 feet tall, had a wing with a 40-inch wingspan, and weighed 220 pounds dry.

"The *Jayhawk* was our LDRS-20 special project," said Eric. "When we were young, our favorite Estes kit was the one-fifth-scale *Jayhawk*, so we decided to build a full-scale *Jayhawk* for LDRS-20. This was our *Jayhawk* on steroids."

To power the rocket, the Gates Brothers turned again to AeroTech, picking two 98 mm M2500 motors for power. With both motors installed, the total weight of the rocket was nearly 250 pounds.

As the Gates brothers were getting their rocket ready, at another pad a short distance away, Andy Woerner and his crew from San Diego were preparing a mega *V-2* rocket, a one-third-scale replica of the German missile that had become the first rocket to reach outer space. Woerner's *V-2* was painted yellow and black and stood more than 15 feet tall, with a diameter of nearly 22 inches. His team wrestled the 285-pound vehicle onto the launch pad, and then everyone backed away to admire the rocket as it sat ready to go. The excitement in the crowd, which had grown to nearly three thousand people, was building. As the *V-2* and *Jayhawk* would be launched almost back-to-back, most people left their camps, vehicles, tents, and trailers for a better look at the two big rockets.

The Gates Brothers were ready first. The *Jayhawk* was electronically raised on a mechanized launch pad. With the rocket vertical, the LCO performed a ten count. The firing button was pushed. The smoke from the two M2500s kicked up clouds of dirt and debris that momentarily obscured both the launcher and the rocket. The bright red winged craft then appeared, moving quickly up its rail launcher, roaring out of the dust and smoke to more than 3,000 feet. At apogee, the *Jayhawk* turned over and descended in two pieces, with the main body dropping under two 26-foot-diameter parachutes. It all landed perfectly on the range.

The crowd's attention then turned to the Woerner *V-2* missile. A split second after ignition, the 285-pound rocket chugged up the 1-inch-diameter launch rod somewhat lowly, propelled by a central M1939 and four K560 motors. Briefly, everything looked fine. But as the rocket cleared the pad, it seemed to be struggling to achieve altitude, like a great beast running low on air. The rocket lost its vertical ascent and suddenly began to arc over at only a few hundred feet, and that arc quickly grew more pronounced. Disaster was in the air. After an apogee of less than 750 feet, the rocket turned completely over and was inbound again, heading for the desert parking lot, where hundreds of cars were located. There was no deployment at all. Trailing thick smoke, the heavy *V-2* picked up speed as it passed directly over the crowd, continuing its path for the parking area, illustrating well the rocketry truism: "Crashing is a lot like landing—only faster."

At the last moment possible before impact, the rocket suddenly deployed some of its recovery gear. The nose cone slipped away from the airframe and dropped under its own chute. But it was too late for anything else. The rest of the rocket made a sickening thud on impact. The entire range came to a halt. Had anything been hit in the parking lot?

"Then hundreds of people ran to the remains of the crashed *V-2*," recalled one flyer. "It was like a scene out of a movie where the alien spacecraft crashes. Civil Air Patrol cadets, who were to keep people at a safe distance, surrounded the crash site, which revealed a large crater with debris scattered everywhere. We watched the owner salvage the main chute, still in its deployment bag, and the fin section, which was sort of intact."

Fortunately, no one was harmed by the crash, and no vehicles were struck by the incoming missile. The rocket impacted in a small open space surrounded by cars and trucks. The *V-2* was utterly destroyed. Postflight inspection led to several theories for the crash, including the possibility that the heavy rocket was underpowered and/or that one of the K motors did not light on time. For some reason, the rocket had also carried its steel launch rod high into the air.

Once the range activity resumed, another beautiful *V-2* met its maker on the playa. This rocket was owned by the Arizona High Power Rocketry Association. It was a one-quarter-scale black-and-white replica that looked terrific on the pad. The rocket was 16 inches in diameter and almost 12 feet tall. It held an AeroTech M2000 Redline demonstrator motor. This *V-2* had no trouble clearing its pad. It roared away on a textbook-like trajectory, trailing a bright red flame high into the sky. It was a picture-perfect ride up. It turned over at apogee and began its descent. Yet there was no deployment at all. Instead—and just like the real thing—the *V-2* screamed back to Earth for quite the authentic landing. The rocket hit the dirt nose first and disintegrated, reducing the vehicle to rubble not far from its pad.

This second big crash interrupted launching for a bit of time. But soon, range activities resumed, and for the first time at LDRS-20, lines for launch pad assignments began to form as hundreds of people brought forward rockets of all shapes and sizes. Even with eighty pads scattered over three adjacent ranges, it was difficult for the RSOs and LCOs to keep up with the traffic. The wait in line grew to more than 45 minutes, and once rockets were at the pad, they too might sit for half an hour waiting for the go button to be pushed. All at once, it seemed everyone at Lucerne wanted their turn to fly.

By midday, the sky was raining rockets. Four, five, six, and even seven parachutes at a time could be seen in the sky at the same time at different altitudes, lowering their cargo. As is absolutely critical at any large launch, as soon as one parachute opened, another rocket took to the air. New smoke trails zipped among descending vehicles. There were scores of good flights and a few bad ones too. One 10-foot-tall rocket carrying seven K motors, dubbed *The Seventh Element*, shot into the air on three K550s, and then as the vehicle coasted for a moment prior to airstarting two more, it separated into two pieces connected by a long recovery harness. Two more K550s lit anyway. The rocket careened all over the sky like a crazed skywriter before settling down, still connected together, to land under chute.

By early afternoon, dozens of downed rockets were scattered all over the playa. In every direction, rockets were lying on the ground, their owners making the trek to retrieve them. Still more rockets sat on their pads, electronics armed and beeping, waiting for their turn. One hundred, two hundred, and then three hundred rockets took to the air while the crowd continued to grow in size. Vendors were selling every conceivable rocketry-related item, from motors and airframes to lapel pins, bumper stickers, and hats. There were raffles and announcements for lost children. It was like some giant carnival, or perhaps a three-ring circus with the LCOs as ringleaders; the entire scene was the antithesis of the first LDRS held twenty years earlier. Who would have thought a tiny launch on a farm outside of Cleveland, Ohio, would end up in the Southern California desert with all this?

Almost four hundred rockets had taken to the air, and it was barely two o'clock in the afternoon. Perhaps this would be a record-setting day after all. The crowd was surging forward, and more and more rockets were being set up on pads.

Then, right on time, the desert wind came up suddenly and with great force. Flyers saw it moving toward the range from the western edge of the playa. A long cloud of dust, low to the ground at first, rose higher and higher as it moved across the desert toward the rocketry range like a dirty tsunami. The cloud reached the launch area, and now people were scrambling as untethered tents were blown away, pop-up canopies collapsed, and anything not tied down and light enough to fly was carried into the air. In only a few minutes, it was all over.

Saturday's flying at LDRS-20 came to a sudden halt. Rockets were left standing on their pads; people retreated to their cars or campsites or back to their motels in Victorville. Or they just went home. This may have been the busiest Saturday morning in LDRS history. But there would be no flight count records set at LDRS-20. The winds had seen to that.

LARGE AND DANGEROUS ROCKET SHIPS

That night, the annual banquet was held at the Ramada Inn in Victorville. At the members' meeting, awards and honors were bestowed. Gary Rosenfield and Korey Kline received Tripoli lifetime memberships. It was announced that the state of Texas would see its first ever LDRS the following summer in 2002, near Amarillo.

There were far fewer flights on Sunday. Carl Delzell launched a 165-pound *Army Hawk* Missile on an M2000 Redline, and there were several rockets that launched with the new Hypertek M1000 Armageddon motor. Wedge Oldham impressed the crowd with the flight of his scratch-built one-third-scale *Nike Hercules*. The *Nike* was launched from a mechanized pad that had one of the longest launch rails ever seen at a high-power event. The two-stage rocket, which weighed 200 pounds, roared up the rail on four K700s and staged to an M1939 in the sustainer. It was a fantastic flight up as the sustainer reached more than a mile in the air. Unfortunately, the booster failed to deploy its recovery gear properly, and the lower half of the rocket impaled the hard desert floor. But the sustainer had a fine dual deployment and touched down gently in the distance.

29

"The Fuse Is Lit!"

The ATF had a seat on the NFPA's Committee of Pyrotechnics for years. Yet for all intents and purposes, the agency took little notice of model or high-power rocketry. Then in 1994, the agency suddenly expressed interest in exercising power over the hobby. At the heart of the ATF's assertion of authority was the bureau's claim that APCP was an explosive, thereby subjecting high-power rocket motors to federal explosives regulations.

When this issue first arose, rocketry leaders tried to work with ATF bureaucrats to reconcile the freedom to fly high-power rockets with federal regulations. This effort was initially expressed through attempts by rocketry to carve out exceptions for rocketry motors to federal explosives rules. The Tripoli board of directors and the NAR were confident that a reasonable outcome could be achieved by working together with the ATF. All that was needed to reach a compromise with the government was time and patience, they thought.

Unfortunately, they were wrong.

———◆———

In its online history, the ATF makes the absurd claim that it has been around in some fashion since the late 1790s. In reality, the ATF was established in 1972 as a bureau within the U.S. Department of Treasury. After the World Trade Center attack in 2001 and the creation of the Department of Homeland Security, the ATF was renamed the Bureau of Alcohol, Tobacco, Firearms, and Explosives and was put under the auspices of the Department of Justice. By 2012, the ATF had become yet another federal bureaucratic leviathan, with 5,000 employees and an annual budget exceeding $1,000,000,000.

Although the ATF had been a member of the NFPA for many years, there is little, if any, record of significant ATF involvement in amateur, model, or high-power rocketry through the 1980s. It has been suggested by the ATF that one of the reasons for this disinterest, at least as far as high power is concerned, was that until the early 1990s, high power was too small to notice. This seems like a possibility. Another reason for the ATF's disinterest in high power may have been a lack of opportunity. Although the ATF was involved with the NFPA prior to the 1990s, published revisions to NFPA safety codes prior to 1993 do not show any ATF member officially on the Committee on Pyrotechnics until the mid-1990s. However, by 1995, the ATF had at least two members on the committee: Larry J. McCune as primary voting member and David S. Schatzer as an alternate member. Both were part of the committee's work on the Code for High Power Rocketry (NFPA 1127). Presumably, they were among the first ATF employees to get a clear look at high power in its modern form.

In 1993 and during the course of NFPA committee meetings on the proposed Code for High Power Rocketry, the ATF suggested, seemingly for the first time ever, that federal explosives rules must be applied to high-power rocket motors. This was startling news to the rocketry community. At the specific suggestion of ATF officials, requests for clarification on the applicability of explosives laws to high-power rocketry were submitted to the agency from rocketry representatives. These requests included a letter to the ATF from AeroTech's Gary Rosenfield, dated October 17, 1993.

In his letter, Rosenfield asked the ATF whether it believed rocketry was regulated by federal explosives laws. At the time, Rosenfield and other hobby leaders were confident such laws did not apply to high power or, if they did apply, hobby rocketry was exempt. It was reported that a former ATF representative who had been a member of the pyrotechnics committee for many years had previously stated that the agency was not interested in high power. Rosenfield sought to affirm this belief through his letter to the ATF.

Several months passed, with no reply by the ATF to Rosenfield's inquiry. Then on April 20, 1994, the ATF provided a response that set the high-power community on its heels. The bureau announced that in its view, high-power rocket motors were *explosives*, and as explosives, they were subject to federal regulation. In its reply to Rosenfield's inquiry, the ATF claimed its authority over rocketry derived from the federal government's power to create an explosives list, created by the Organized Crime Control Act of 1970. That act defined an explosive as follows:

> Any chemical compound, mixture, or device, the primary or common purpose of which is to function by explosion. The term includes but is not limited to dynamite and other high explosives, black powder, pellet powder, initiating explosives, detonators, safety fuses, squibs, detonating cord, igniter cord, and igniters.

Indeed, in 1971, APCP had been placed on the federal explosives list. But since that time, the ATF had expressed no interest whatsoever in hobby rocketry as an explosive. The ATF was not involved in the regulation of model rockets, which they believed posed no safety hazards to the public.

"This exemption for model rocket motors, common fireworks, and propellant-actuated industrial tools was intended to cover explosive items that, because of the small quantities involved, would not likely be a source of explosives for a bomb or be a hazard during storage situations," said the ATF.

What about high-power rocket motors? In the spring of 1994, the ATF announced there would be no similar exemption. From this point forward, high power would be governed by explosives laws. When asked whether motors might be excluded under exemptions that applied to "propellant-actuated devices," the bureau initially suggested the exemption might apply to a fully assembled motor. Yet it would not apply to the possession of propellant prior to assembly. Furthermore, said the agency, an exemption was never meant to apply to large high-power rockets. ATF representatives also told high-power leaders that the agency's newfound interest in rocketry was not new at all. It was merely a "clarification of existing law" that would soon be written into an upcoming revision of ATF guidelines and would be presented to the NFPA as part of the revisions to proposed NFPA Section 1127.

Tripoli President Chuck Rogers and Michael Platt, president of the High-Power Rocket Manufacturers and Dealers Association, advised hobbyists of the ATF's new interpretation of federal explosives laws. In an open letter addressed to the rocketry community on April 25, 1994, Rogers and Platt suggested a pragmatic approach to the ATF's new pronouncement. First, they assured users that they did not expect any federal enforcement actions would be undertaken any time soon. Next, Rogers and Platt also told members that plans were already underway for rocketry leaders to work with government regulators to minimize the impact federal explosives rules might have on the hobby:

> The consumer groups and the trade association have already initiated the process to exempt high-power rocket motors, reloads, and related items from the federal explosives laws. It is also our intention to investigate the possibility of obtaining an amendment to federal explosives law. Unfortunately, both these processes will take a considerable amount of time, and the high-power rocket community has to bring itself into full compliance with the law while these other avenues are pursued.

For rocketry leaders, this seemed like a reasonable approach: work within the law while taking reasonable steps to have explosives regulations revised so as to minimize their impact on high power. In the interim, what did compliance with federal explosives laws actually mean for the average high-power flyer?

According to Platt and Rogers, flyers who purchased and then stored a high-power rocket motor must (1) follow applicable federal guidelines related to storage containers for explosives and (2) obtain a federal low explosives user permit, also called an LEUP. To obtain an LEUP, applicants were required to pay a federal license fee and submit fingerprints. They were also required to fill out an application that contained, among other things, a waiver of any licensee's rights regarding search and seizure. High-power flyers would have to grant consent, ahead of time, to the federal government to enter their homes at any time without a warrant to conduct a search of the premises. In other words, possess a high-power rocket motor, and you waive your Fourth Amendment rights against unreasonable search and seizure guaranteed by the U.S. Constitution.

The reaction of high-power flyers to the news of the sudden need for an explosives license varied. Many flyers complied with the new rules. They picked up the federal forms, filled out the LEUP application, turned in their fingerprints, and got their license for $20. They built or bought special containers that complied with federal storage rules.

Other flyers refused to comply. Some people decided it was better to forego an LEUP and stay off the ATF's radar, especially since neither Tripoli nor the NAR were planning on enforcing federal explosives laws on America's rocketry ranges. Still other flyers called the ATF directly to ask questions about the new rules or to complain about the fact they were suddenly being applied to hobby rocketry. As reported in the *Tripoli Report* in mid-1994, the board discouraged calls directly to the ATF's office in Washington, D.C.

"One and a half to two hours per day have been consumed answering the same questions every day since the original posting," the board reported. "They give the same answers to everyone, and it is a waste of their productive time that could be spent working on the solutions that we all want and need."

Board member Kelly told flyers that the ATF acknowledged that federal explosives laws were not written with high-power rocketry in mind and that agency representatives had personally assured high power they were going to be as flexible as they can with the hobby.

"There is room for some exemptions, [but] in what form they will come, we (and they) do not know," reported Kelly. "There is a possibility that they may set their policy according to 1127, which they are considering, which has storage allowances built in . . . If we back off for the time being, the ATF will assemble the data they have and provide high power with a ruling that will allow us to continue what we are doing . . . They have no plans for immediate enforcement. Warren Parker from enforcement said we could quote him on this."

Tripoli Motor Testing chair John Cato also urged a flexible approach to the ATF's newfound interest in high power.

"It would be the height of stupidity to give these guys the impression that we do not want to work with them as they really do want to work with us," said Cato. "It is further true that we, as an organization, can seek exemptions in the future. But to seek exemptions, we will first have to be in compliance . . . We are talking about devices that are classified as Class B explosive, and everyone must be realistic and acknowledge that fact."

At the Tripoli board meeting at LDRS-13 in 1994 at Argonia, Mike Platt discussed recent talks he had had with the ATF, and he took questions from members regarding the ATF's involvement with rocketry. There was not much new to report, said Platt, but meetings between rocketry representatives and the ATF were being scheduled.

By late 1994, federal explosives laws were common topics of discussion at Tripoli board meetings and among members around the country. At the leadership level, Tripoli board members made little progress in convincing the ATF to grant rocketry any exemptions. In fact, rocketry leaders were having trouble getting any response from the ATF at all.

In 1995, Kelly and Gary Rosenfield spoke with an attorney experienced in working on ATF-related matters. The purpose of this contact was to obtain preliminary advice on rocketry's options in the dispute and to discuss a possible course of legal action if necessary. Kelly suggested to the board that they at least consider legal action against the ATF if no progress was made by March 31, 1995. The board unanimously approved Kelly's suggestion. However, that decision was later rescinded when the board learned ATF's hands were allegedly tied because the agency did not have the legal ability to grant an exemption to rocketry. This claim came at an NFPA meeting in Huntsville, Alabama, in April 1995. During that meeting, Kelly, Platt, and NAR president J. Pat Miller met with several ATF representatives, including agents from enforcement, compliance, and legal counsel. Platt reported that the ATF even apologized for taking so long to respond to questions from the rocketry industry. He also said ATF representatives claimed they were in favor of making changes to explosives regulations to accommodate rocketry, but they were legally unable to do so.

"It was concluded by [the] ATF's counsel [that] the agency does not have the authority to do so while remaining within the scope of the law," reported Platt.

This remark was later set forth in the *Tripoli Report*, which reported further that the ATF could do nothing more for high-power rocketry as it "is following a congressional mandate and cannot effect changes requested by Tripoli, the NAR, or [manufacturers] without regulatory or statutory changes approved by Congress."

The ATF was willing to help rocketry in the drafting of proposed legislation to address the issue, added Platt. To help accomplish this, the NAR and Tripoli were preparing a "wish list" of proposed changes to the current law. In the meantime, he added, a Tentative Interim Amendment to the draft of NFPA 1127 was proposed by the bureau. The amendment incorporated federal explosives laws and storage requirements into the new Code for High Power Rocketry.

In another open letter to Tripoli members, President Kelly later explained that during the ATF meetings at Huntsville, there was some clarification of federal storage rules. He also reiterated his belief that flyers who purchased and immediately used high-power motors at local events would not need an LEUP. If there was no storage of fuel involved, there was no requirement for a federal license, said Kelly.

Kelly also used his letter as an opportunity to chastise a commercial motor vendor who had refused to sell motors to flyers who did not have an LEUP. Tripoli had no requirement that its members have LEUPs, declared Kelly, and flyers who had bought and flown motors locally did not need a license in the first place. Kelly made it clear that although Tripoli members were expected to comply with federal laws, Tripoli was not in the business of enforcing such laws. If flyers chose to ignore the law, said Kelly, they were on their own.

In the summer of 1995, the Code for High Power Rocketry was approved. In August, the Tentative Interim Amendment incorporating federal low explosive rules was made part of the new code. High-power rocketry finally had a safety code approved by the NFPA. Yet for some high-power users, the federal licensing and storage rules transformed this new code into a federal ball and chain, and they were not going to stand for it.

In October 1996, Tripoli published LEUP filing instructions in the *Tripoli Report*. A template prepared by Platt took flyers through every question on the federal application form and provided advice as to how to complete the form as it applied to rocketry. By now, many Tripoli members were obtaining an LEUP; a few local Tripoli clubs were even creating community storage lockers for all members of a prefecture to use.

Then in January 1997, President Kelly reported the ATF was proposing additional rule changes that would affect high-power users. These changes were published in the Notice of Proposed Rulemaking (NPRM) posted by the ATF in the *Federal Register*. The first change was a proposal requiring persons who stored explosives to notify all local law enforcement and fire officials where such items were being stored. The second change was a proposed increase in the initial LEUP fee from $20 to $100.

Kelly and other Tripoli board members viewed the requirement that some local officials be made aware of explosives storage as reasonable, especially for the protection of firefighters and other first responders. However, the proposal was vague and ambiguous as worded, Kelly told members. As written, the new rules could require rocketry users to needlessly notify all kinds of local officials, such as building inspectors or municipal executives. Kelly and others recommended to the ATF that the proposed rules be narrowed to emergency responders only.

Kelly viewed the fee increase as simply unreasonable; he encouraged Tripoli members to oppose it. He asked members to write the ATF and express their opinions on both the proposed new regulations. To help keep members focused, Kelly printed a sample letter to the ATF in the *Tripoli Report* that contained essential objections he believed should be made as well as the appropriate address where the objections should be mailed. If members chose to send their own objections to the ATF, Kelly asked that they stay on subject, or their arguments might be tossed out.

"Remember," cautioned Kelly, "mixing items not relevant to the NPRM or making comments about the ATF being a bunch of 'jackbooted so-'n'-sos' will not help our cause either . . . Therefore, it is important that our public comments be levelheaded, measured, and without questioning the ATF's parentage or political leanings. Having said that, I trust all of you to prove you are the responsible and trustworthy adults that I have been representing to the ATF you are."

ATF received more than four hundred written objections to the proposed rule changes. These objections resulted in a narrowing down of the local officials who needed to be notified of a member's motor storage. However, there was no stopping the fee increase. The LEUP rose from $20 to $100. And there was still nothing new with regard to the grant of any exemption for high power to federal explosives laws.

For the time being and after three years of dealing informally with the ATF, the APCP remained on the federal explosives list, and high-power rocketry continued to be regulated by the ATF.

It had been ten years since Frank Kosdon had discovered high-power rocketry. During that time, he had become one of the most respected albeit sometimes controversial figures in the hobby. The MIT graduate had played a significant role in the reloadable motor revolution, and he had created a high-power motor company that developed into two distinct branches: Truly Recyclable Motors out of his home in Southern California and Kosdon East, run by Paul Robinson and Jim Rosson in the Eastern United States.

Kosdon was also a positive force in the research and experimental rocketry movement. In the 1990s, Kosdon motors propelled several rockets to altitude records, including the *OuR Project*, which reached an estimated altitude just shy of 100,000 feet in 1997. Attired in his beach-like short pants and whatever tie-dye shirt was available and looking like he was a beach bum instead of a rocket scientist, Kosdon had become a fixture at western desert launches. He was enthusiastic about hobby rocketry and was especially responsive to new flyers with questions about rocketry—or anything else, for that matter. He loved science; he loved to talk to anyone about rocketry. But Frank Kosdon hated conformity. In fact, he despised regulations of all kinds, even rocketry rules imposed by Tripoli, and he had nothing but disdain for regulations imposed by the government.

"I have a bad history of not liking licenses and regulations going back to my days selling *Collier's Encyclopedias* door-to-door in 1960," he said during an interview in 2000. "Collier's used to take $1 out of each commission to go into the 'fine fund.' If you ran into [government] trouble for not having a solicitor's license, your fine was paid through this fund."

"Don't let the bastards grind you down" was one of Kosdon's favorite descriptions of government authority. It was a phrase favored by many other high-power flyers too. Kosdon believed the ATF's exercise of authority over high power was without foundation and was premised on the government's desire to control the hobby rather than any hard science that rocket motors were dangerous, let alone "explosive."

"There has never been a good explanation of why they came up with the 62.5-gram stipulation," he said in reference to the ATF's exemption for model rocket motors. "It's already obvious that the bigger propellant grains are much safer than the equivalent weight of smaller grains. The DOT and the ATF will say whatever they want to be able to maintain their control."

By 1997, Tripoli and the NAR had spent four years trying to figure out a way to challenge the ATF's assertion that high power was covered by the federal explosives laws. Early on, Gary Rosenfield had suggested one means of exemption. Rocket motors were exempt from explosives laws because they were "propellant-actuated devices," also known as PADs. ATF explosives chief James Brown seemingly agreed in a 1994 letter that a fully assembled motor would be exempt as a PAD. However, he followed up this admission by asserting that prior to the assembly of the motor, propellant would not be exempt. Using this Catch-22–like argument, high-power motors were therefore not exempt, he said. The ATF was holding firm. They were in rocketry to stay; a PAD exemption would not be recognized.

Then in the summer of 1997, someone suggested an alternate approach to the federal government's intrusion into hobby rocketry. At the Tripoli general assembly meeting in Hartsel, Colorado, on August 9 at LDRS-16, the Tripoli board of directors opened up the floor for member comments on the then-current state of high-power rocketry. Frank Kosdon put up his hand, and he asked to be recognized. He then stood up to address the board and gathered membership. He said he wanted to talk about the ATF and its regulation of the hobby. The room went silent. What he said next was a revelation. Kosdon did not attack the ATF as a meddling governmental agency trying to control the hobby, he did not discuss the agency's flip-flop on PADs, and he did not argue that an exemption to federal explosives laws be created for high-power motors. For Kosdon, getting rid of governmental interference was simpler. An exemption wasn't necessary, declared the MIT-trained rocket scientist, because the government's classification of APCP as an explosive *is wrong*.

The foundation of the ATF's authority to regulate high power stemmed from the inclusion of APCP on the explosives list created in 1971, long before high power had even existed. So far, high-power leaders challenged the ATF's regulatory actions on two fronts. First, they argued that the ATF waived its right to insist high-power motors were regulated because they had taken no action on the hobby between 1971 and 1994. Second, they argued that high-power motors qualified for the exemptions that applied to PADs. Both arguments were entirely valid. The ATF rejected them nonetheless.

Kosdon and a few others like him were suggesting an entirely new course. High power should attack the foundation of the ATF's source of power: the inclusion of APCP on the explosives list itself. If composite propellant was not an explosive, Kosdon said, it could not be on the federal explosives list. No exemptions were needed. It was that simple.

———⋆•⬥•⋆———

On February 4, 1999, representatives from the NAR and Tripoli met with the ATF in Washington, D.C., to discuss the ongoing regulation of high power by the agency. Present at the meeting were NAR President Mark Bundick, Bruce Kelly, Michael Platt, and former NAR President Miller, now chairman of the NFPA's Committee on Pyrotechnics. At least two subjects were covered during the meeting. First, rocketry leaders continued to ask for clarification of the basis and extent of the agency's regulation of the hobby. Second, the ATF was told APCP was *not* an explosive, and on that basis, it should be removed from the federal explosives list. Among other things, the rocketry representatives presented the ATF with engineering and technical information supporting the fact that APCP "does not propagate an adequate detonation wave, a key characteristic of a good high explosive." The 3.5-hour meeting was described by rocketry leaders as "cordial, productive, and open." Yet ATF representatives were unmoved.

"They pointed out to us that many of the items on the annual list are lousy explosives," wrote Bundick after the meeting. "But they have no room to change the list without legislative relief from Congress." In other words, if rocketry wanted things to change, they needed to call their congressional representatives and write new law.

A few weeks later, Tripoli and the NAR made a historic decision. They jointly retained legal counsel to assist hobby rocketry in its struggle with the ATF. The Washington, D.C., law firm of Egan & Associates, led by attorney Joseph Egan, was hired to help obtain relief from the ATF. Egan was a NAR member. He assisted the hobby on regulatory matters in the past. To pay for its lawyers and on behalf of Tripoli, President Kelly signed an association check for $30,000 to Egan's firm on March 29, 1999. The NAR wrote a similar check, and a request was made for additional funding from Platt on behalf of the dealer and manufacturer's association.

"In my opinion, there is nothing more important to the hobby right now than an acceptable resolution to our regulatory situation," said Kelly that summer. "A few weeks ago, I received a phone call from an upset

Tripoli member. He made a couple of minor mistakes, and he was afraid he had hurt himself and the hobby. A visiting ATF agent went out of his way to intimidate him. Just last week, another member was being challenged by a field agent in another part of the country. In both cases, the agents contradicted themselves. It is as if the ATF is making up rules as they go along. Phone calls to Washington resolved these issues, but we should not have to call Washington every time a field agent gets it wrong. It's time for the federal government to act responsibly and train their agents to follow the same script."

In August, Kelly, Bundick, and attorney Egan met with the ATF in Washington. Egan presented a summary of the legal reasons why high-power flyers should not be subject to ATF regulation. He asked the ATF for clarification of its positions on motors and propellant. The ATF did not offer any direct input on legal issues at the meeting. Yet agency representatives agreed to review and then respond to a legal memorandum to be submitted in writing by Egan.

The memorandum the agency had requested arrived shortly thereafter in the form of a letter from Tripoli's legal counsel to the ATF, dated September 7, 1999. In that letter, Egan demanded that the ATF reverse its position on the regulation of high-power rocket motors, first announced by the agency in 1994 and reaffirmed by the ATF in 1997. He asserted that APCP did not function by explosion or explode when ignited. Therefore, rocketry propellant could not be on the federal explosives list.

Egan also argued, in the alternative, that high-power motors were "propellant-actuated devices" and were therefore exempt from regulation anyway, even if APCP was included on the explosives list. Finally, Egan argued that the inclusion of APCP on the ATF's most recent explosives list was not properly noticed and had not been left open for public comment. Therefore, the list was both procedurally and substantively defective, he said. To rectify these problems, concluded Egan in his memorandum, the ATF needed to remove APCP from the federal explosives list immediately.

ATF representatives rejected Egan's written claims the following month, in October 1999. Then on December 22, 1999, the bureau put its final position in writing. There would be no change of the ATF's regulation of high-power rocketry, said the agency. The ATF reiterated its view that APCP was an explosive, it would remain on the explosives list, it was not necessary for the agency to go through any additional rule making, and high-power motors did not qualify for the PAD exemption. Rocketry users must comply. Nothing had changed.

It had been almost six years since the ATF announced its regulation of high-power rocketry. Since that first letter to AeroTech in 1994, both Tripoli and the NAR had met with, reasoned with, and negotiated with government agents on numerous occasions in an attempt to ease federal regulation of the hobby, all to no avail. Flyers who purchased and stored high-power motors needed an LEUP. They needed to be fingerprinted, they needed to agree in advance to warrantless searches of their homes, and they were required to pay a fee to the government for these privileges. And without an LEUP, high-power users could become *criminals*. The ATF was firmly planted in the hobby, and there was no way—short of suing them—to get them out. For many flyers, it was all too much.

"Since 1994, every time we have met with regulators, we have come away with more regulation," said President Kelly afterward. "We are being treated like criminals, and we haven't done anything wrong."

On February 11, 2000, the Tripoli Rocketry Association and the NAR filed a suit against the ATF in the United States District Court for the District of Columbia in Washington, D.C. Among other things, the complaint asked a federal court judge to declare that APCP was not an explosive, that it should not be on the federal explosives list, and that rocket motors were propellant-actuated devices within the meaning of such laws.

"The fuse has been lit!" declared Tripoli member Ky Michaelson at a rocketry event held on the eve of the suit being filed. "I have taken action on big guns before. This will be a long drawn-out battle. It won't be quick or cheap. The ATF has lawyers on their payroll and don't care if they are sued . . . We must be in this for the long haul!"

Indeed, the fuse had been lit. High-power rocketry was challenging a powerful bureaucracy within the U.S. government. And unbeknownst to anyone at that time, the ensuing legal battle would burn for almost another decade.

16

Bruce Kelly of Utah heads to the pad at LDRS-7 in Colorado with his clustered LOC *Top Gunn* rocket. Kelly joined Tripoli in 1987 and brought to the hobby his personal vision of high-power with the release of his iconic publication, *High Power Rocketry*, in 1991. The magazine was available at almost every hobby store in the United States, attracting thousands of new members to Tripoli. Kelly was Tripoli's fourth and the longest-serving president, overseeing the organization from 1994-2002. During that time, Tripoli's membership numbers and annual budget increased dramatically. During Kelly' tenure, a national certification program was adopted; LDRS grew into the largest amatuer rocketry event in the world; and high-power flyers made their first real attempts to reach for the edges of outer space. *High Power Rocketry* was published from 1991-2005.

Ky Michaelson was both showman and high-power rocketry enthusiast. "I think it is important to promote the sport at a national level so the public and government officials understand that high-power rocketry is a family sport and we are not a bunch of terrorists," he said. The former stuntman created the rocketry company, Rocketman, and was later tasked with supplying rockets to the movie *October Sky*. In 2004, and with private and corporate sponsorship, Michaelson would lead a group of flyers in an attempt to reach space with an S-powered amatuer rocket at the Black Rock Desert.

19

20

James "Jim" Mitchell of Tennessee was an early proponent of the homemade experimental motor movement of 1990s. Together with Sonny Thompson, Mitchell created Thunder Flame, a hands-on course where students learned the basics of APCP motor-making. By the early 2000s, flyers using skills they obtained at Thunder Flame classes were constructing and launching their own high-power research motors, H through Q power.

LDRS XIII

ARGONIA, KANSAS
AUGUST 11, 12, 13 & 14, 1994

25,000 FOOT Ceiling

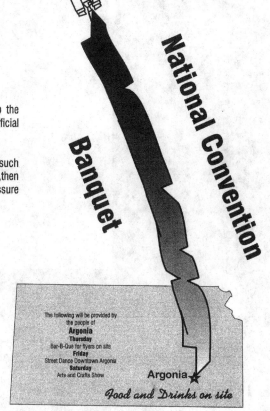

National Convention

Banquet

Official **LDRS** Hotel
Ramada Inn Airport
in Wichita, KS
(316) 942-7911
1 800-835-2913

There are other motels closer to the launchsite, however, this is the official motel for all Tripoli Activities

Manufacturers please identify as such when you make your reservations, then confirm with Scott LaForge to assure placement in manufactures row.

Sponsored by:
TRIPOLI ROCKETRY ASSOCIATION
AND
KLOUDBUSTERS
ADVANCED ROCKETEERING

FOR MORE INFORMATION CONTACT:

SCOTT LAFORGE
10809 HASKELL CIRCLE
WICHITA, KS 67209
316-729-8330

The following will be provided by the people of
Argonia
Thursday
Bar-B-Que for flyers on site
Friday
Street Dance Downtown Argonia
Saturday
Arts and Crafts Show

Argonia

Food and Drinks on site

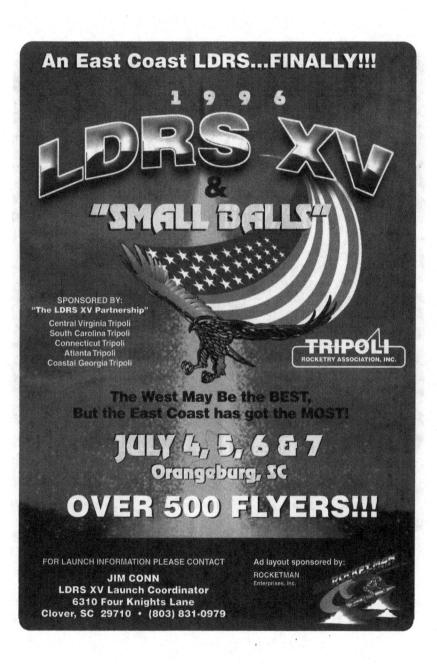

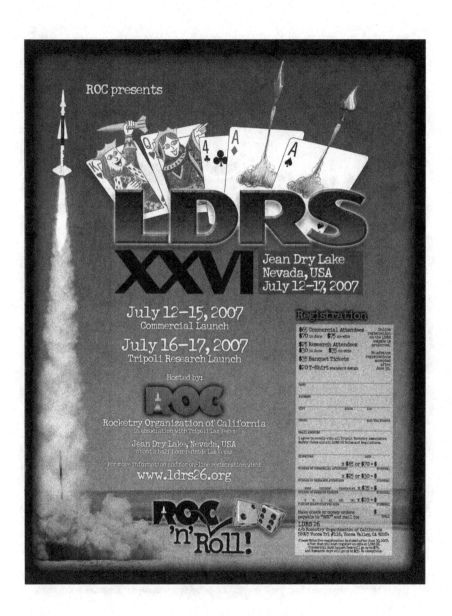

BOOK III
Here Is Your Temple
(2001–2018)

30

Texas

The first Texas LDRS began on a soybean farm in Kansas in 1999, when a cowboy by the name of Patrick Gordzelik helped run LDRS-18 in Argonia. Gordzelik had volunteered to help the Kloudbusters run Tripoli's national launch that summer, and the experience left a big impression on the Lone State native.

"They let me be launch control officer," said Gordzelik, who was still a relative newcomer to high-power rocketry. "I thought this would be killer if we could host an LDRS in Texas. I mean, why wouldn't anybody want to be surrounded by a bunch of people that love something as much as you do? I love to entertain people, and I thought what better way to pay back to these guys I love and this hobby I love than to get them all to come to Texas and throw a party for them?"

Three years later, Gordzelik would get his wish. In 2002, Texas was given the opportunity to host LDRS-22. The launch became one of the biggest LDRS events ever held, and it propelled Gordzelik into a position of leadership in high power that he would retain for years to come.

Patrick "Pat" Gordzelik was born in May 1955 in Amarillo, Texas. He likes to say his elder twin brother, Michael, beat him into the world by only 8 minutes.

The Gordzeliks were originally from Poland. The family fled religious persecution in Europe in the nineteenth century—a czar was out to kill them, said Gordzelik—and they eventually settled more than 5,000 miles away in Texas. Gordzelik's father, Edward Joseph, was born in White Deer, a small town in the Panhandle. Edward was also a twin, one of seven straight generations of twins in the Gordzelik family.

Pat Gordzelik's parents met in college. His mother, Alene Stewart, was of Irish descent and was from Clayton, New Mexico. Her family was in the water well drilling business, and Stewart was enrolled at Draughons's Business College in Amarillo.

"She was poor and didn't have any money for the university, so she went to this business school instead—like a tech school today," said Gordzelik. "Dad went there when he got out of the army. They met in class—they were sitting right next to each other—and it was love at first sight." Edward and Alene were married and had their twin boys and also a daughter, Myra.

Edward Gordzelik earned his living as a postman. In his spare time, he was fascinated by falconry and astronomy. He even built homemade telescopes to peer into the Texas night sky. Edward died at only thirty-one, when Pat was seven years old. The family left Texas and moved to Alene's family home in rural New Mexico, where they lived with Pat's grandmother, Mary Elizabeth Stewart.

"My grandmother was a postmistress, and she had a little store with a gas station in Steed, New Mexico," recalled Gordzelik. "She had 160 acres and a house with no electricity and no running water and an outhouse and a windmill that pumped water into a barrel."

There wasn't much work in Steed. Gordzelik's mother, Alene, moved back to Amarillo to attend nursing school, while her children remained behind in New Mexico. Over the next several years, Gordzelik and his siblings were raised primarily by their grandmother, who taught her grandchildren the values of discipline, self-reliance, and hard work. Gordzelik recalled, "If you get served lemons, you make lemonade, she used to say. She tried to give us a good background as to what we could do, and she always said that we would do more than she ever dreamed. I killed my first antelope when I was seven years old, and she taught me how to gut it. We put up barbed wire fences for ranchers, we had horses, we had rifles, we fished, and we learned to live off the land."

Mary Stewart recognized her grandson's interest in science too. "We didn't have a television," said Gordzelik. "But my aunt Alice married a wealthy rancher—he seemed wealthy to us because he had electricity—and we'd go to Sunday dinner there, and my aunt would ask after dinner if we wanted to watch cartoons, and I said no. I only wanted to watch the Walter Cronkite show about space travel."

Gordzelik's grandmother also encouraged her grandson to read, and he did so intensely, thanks, in large part, to a wandering bookmobile that traversed the back roads of rural New Mexico in the early 1960s. "It would come into town and allow you to check out seven books every two weeks," he said. "I was about nine years old, and I would check out the maximum, and I would have them all read by the time it came back around, and then I would check out seven more. I did this for three years."

During this time, Gordzelik built his first model rocket. It was an Estes *Scout* he bought after seeing an advertisement in *Boys' Life* magazine. He also joined the NAR. Later, he discovered he liked making rocket fuel even more than building rockets. "I got into motor-making when my grandma bought me a chemistry set, and I made up a batch of black powder-like rocket fuel," recalled Gordzelik during an interview in 2014. "I can remember going to school, and they let me give a demonstration of the fuel in class. I had a little pile of it on some aluminum foil. After it burned up, the teacher said that 'it makes a pretty flame.' I was insulted. 'No, it doesn't make a pretty flame,' I said. 'It's rocket propellant!'"

Soon, Gordzelik was conducting chemical experiments in his grandmother's tiny bathroom. "I had a half a gallon of this chemical set up, and I was testing it. I had a little spoon out, and I lit the chemical on the spoon and saw this tiny spark, and it then went to the container with all the chemical, and it ignited, and the whole house filled up with smoke. The nearby glass shattered, a wood table was set on fire, and my grandmother took away my chemistry set for a while, but then she allowed me to move my experiments out to the barn. I was about nine."

In 1969, and after his mother completed nursing school, Gordzelik moved back to Amarillo at the age of fourteen. "Amarillo was huge to us, and I did not want to go," he recalled. "I was afraid that if I went to Amarillo, I would become a 'cat daddy.' That's a hippie who wore beads and ran around smoking pot. My brother and I were ranchers. We wore cowboy hats and boots, and we rode horses in New Mexico. We vowed to all of our friends that we would not go to Amarillo and become a catdaddy."

Amarillo turned out to be a fine place for the Gordzelik brothers. "I found people there who liked rockets, and within a year, I was flying with two guys I met at Bonham Junior High School." Gordzelik continued launching rockets in high school, where he met more like-minded students interested in space and rocketry.

On afternoon in the early 1970s, Gordzelik and a high school friend, Rick, built an Estes *Saturn V* kit and launched it in the front yard of the Gordzelik house.

>This was the big Estes *Saturn V*—about three feet tall. It launched, and then it arced over there on Buffalo Trial—the street we lived on—and we lost sight of it. We ran after it and soon found it. Two streets west of us lived an attorney and his wife in a two-story house. The *Saturn V* had hit the roof of that house. It must have made a heck of a noise because the lawyer's wife was out there, and as we were picking up pieces of the rocket, she was screaming and hollering and shaking her fists at us. She said that she knew who we were. She also said that we were worthless and we would never amount to anything.

After high school, Gordzelik and his friend, Rick Husband, kept in touch. Husband joined the United States Air Force and became a test pilot and then a NASA astronaut. He was the pilot of the Space Shuttle *Discovery* during a successful ten-day launch in 1999. Four years later, Husband was the commander of the shuttle *Columbia*, which near the end of another successful mission broke apart on re-entry into the atmosphere on February 1, 2003. All seven astronauts were lost in the skies over Texas. It was the first shuttle accident since the *Challenger* disaster in 1986.

After the *Columbia* tragedy, the Amarillo airport was renamed in honor of Rick Husband. Gordzelik, who was then running his own successful business, was at the ribbon cutting ceremony at the airport. A slight, elderly lady approached him. It was the same woman who had yelled at Husband and Gordzelik when their Estes *Saturn V* had accidentally struck the roof of her house in the early 1970s.

"Years ago, I told you and Mr. Husband that you would never amount to anything," she said wistfully to Gordzelik. "And I have never been so wrong about anything in all my life."

After high school, Pat Gordzelik earned a college degree in business administration, married his high school sweetheart, and had twin girls.

His first marriage ended in divorce in 1980. Then he met his wife, Lauretta, on a blind date. They married in 1981.

By the early 1980s, Gordzelik had been away from hobby rocketry for years. In 1982, his younger sister, Myra, called and asked Pat if he would be interested in helping her sons get their rocketry badges in the Boy Scouts. "I had a good time doing that," he said. "But I didn't really get back into it yet. It just planted the seed and got me thinking about rocketry again."

Meanwhile, Gordzelik took a job in the fastening business, selling all types of fasteners. Eventually, he borrowed money and opened his own fastener company. It was called PGP, Inc. His first store was in Amarillo. Later, he branched out to several other locations in Texas and New Mexico. "We expanded to power tools, construction supplies—just about anything that anyone needed to put something together," he said.

With his new business, Gordzelik had little time for hobbies, or much else outside work. Then in 1985, he came across a book called *Amateur Rocket Motor Construction* by David Sleeter. "I ran across it by accident at the Hobby House in Amarillo," said Gordzelik. "Way in the back of the store by a bunch of old kits and stuff, there was a stack of magazines, and underneath them, I saw this white book, and it had a picture of a rocket motor on the cover. I asked the guy how much he wanted for the book, and he said it had been there so long I could have it for free. I took it home and started looking at it, and then I told Lauretta about how I had burned my grandmother's basement down. I said I think I can do a better job this time."

Gordzelik read Sleeter's book, and he was hooked. "Within a month, I had dropped about $1,500 in tooling to make black powder motors," he said. "I got charcoal, I got medical-grade potassium nitrate—Lauretta was a pharmacist—and I'm soon making motors up to G size." Gordzelik loved making rocket motors, even more than building rockets. "Rockets became a necessary evil," he said. He only wanted rockets to launch and test his home-built motors.

Gordzelik honed his motor-making skills over the next several years. He created black powder motors up through J impulse and experimented with ammonium nitrate motors. "I mixed it all by hand in a stainless steel bowl with a wooden spoon," he said. "There was a lot of trial and error, which means boom, boom, and boom!"

He launched rockets at his eighteen-acre ranch in canyon country on the edge of Amarillo. Eventually, he went to the FAA for an altitude waiver. "It was hard to do because the Amarillo airport was only 9 miles away, and when the winds shifted, the planes would fly right over our house," said Gordzelik. But that was only half the problem. No one had ever asked the local FAA for a rocketry waiver before.

"They thought I was nuts," recalled Gordzelik. "They had never granted a rocketry waiver, and this was the first time they had a request. 'We don't know if it is legal or not,' they told me. Then they got ahold of the regional office in Albuquerque, and I was eventually able to get a 5,000-foot waiver for my property."

By the early 1990s, Gordzelik was quietly enjoying rocketry at his ranch. He had discovered APCP motors at a local hobby store—AeroTech G motors—but he was not part of any formal rocketry club. He had no knowledge of Tripoli or high-power rocketry. He never heard of LDRS. Then while visiting one of his fastener stores in Lubbock one day, a customer walked in with a magazine that he thought Pat might find interesting. It was Bruce Kelly's *High Power Rocketry*. "I was fascinated, and I went completely through that magazine. It was like Nirvana for me. I was not alone! And that's how I discovered the Kloudbusters," said Gordzelik.

Gordzelik and his wife attended their first high-power rocket launch at a Kloudbusters event in Kansas in latter half of the 1990s. "I thought a 29 or 38 mm motor was big back then," he said. "Then I saw Darren Owens launch a *Mosquito* with an M motor!"

Pat and Lauretta both joined Tripoli, and Gordzelik expanded into making AP motors after attending a Thunderflame class hosted by Jim Mitchell and Sonny Thompson. In 1999, he found two more hobbyists in Amarillo who were also Tripoli members. The three of them joined forces to create Tripoli's newest prefecture, the Panhandle of Texas Rocketry Society, also known as POTROCS.

When POTROCS formed in 1999, they had a decent rocketry range outside Amarillo with an impressive FAA waiver of 19,500 feet. Unfortunately, the club lost that field in 2000 after a rocketry-related fire attracted the attention of the local authorities. A total of thirteen acres of grassland burned during that fire. After the fire was put out, the property owner decided Tripoli flyers should fly their rockets somewhere else.

That winter, when it was slow at the office, Gordzelik began searching for a new field.

"I would get in my truck, and I'd drive a half a day every week just knocking on doors," he said. When a door was answered, Gordzelik introduced himself and did his best to sell the property owner on the merits of sport rocketry. "We have a hobby that's fascinating," he would explain. "It probably got lots of astronauts involved at NASA flying model rockets, and we get waivers from the FAA, and we have insurance, and we're looking for a field.

"'Oh yeah,' they would say to me with a smile. 'We know you rocketry people—you're the guys who burned down Brian's place, aren't you?'"

That was how it went—at least for a while. "I must have knocked on a hundred doors," said Gordzelik. "I even put a packet together with rocketry photos and went to various chambers of commerce looking for launch sites—but no luck." Eventually, his persistence paid off.

"One day a man with whom I had done business for years was in the store," recalled Gordzelik. "His name was Wylie A. Byrd, and he owned one of the biggest companies in the town of Tulia, Texas—not far from Amarillo. He was a very successful man, yet you wouldn't know it. He drove a ten-year-old pickup truck. While he's in the store, I said to him, 'Wylie, don't you own some land out there in Tulia?' Of course, I knew he did, and I had already looked around there for a field."

Gordzelik explained his rocketry hobby to Byrd and told him he was looking for a suitable range to launch high-power rockets. "I do have some land you can see," Wylie replied. "Why don't you meet me over there the day after tomorrow?"

Before his next visit with Byrd, Gordzelik checked with the FAA and learned he might be able to secure a 25,000-foot waiver in the vicinity of Wylie Byrd's land. He then got out his aviation maps—Gordzelik had a private pilot's license—and he found a perfect spot for rocketry on the map near Tulia.

He took out a pen and marked the spot on his map.

The following day, Gordzelik met Wylie for their drive. While driving around in the old pickup truck, Wylie passed the exact spot Gordzelik marked his map the day before. It was a great location in person too, thought Gordzelik. He asked Wylie to stop the truck right there.

"I asked him if he knew who owned this land right here," said Gordzelik.

"Well, how much land do you need?"

"I just need about a section of land."

A single section of land is equal to 640 acres.

"Well," said Wylie, waiving his hand nonchalantly in the air, "these *eight sections* are mine."

Gordzelik was stunned. It was early 2001. No one knew it yet, but he had just found the location for LDRS-21. Byrd would eventually allow POTROCS the use of seven full sections of his land—more than four thousand acres of flat, wide open space to launch their high-power rockets into the stratosphere.

Yet Byrd offered the Texas flyers even more, said Gordzelik. "So we have the site, and we need to get a few launches under our belt before we can ask for an LDRS, and we are having launches every month. We have a large portable generator out there for the launch, and one day Wylie comes out to visit and to see our rockets, and while he is there, we were shouting over the noise of the generator, and he asks, 'What do you need this electricity for?' I explain to him that we need a PA system and we need to charge batteries, et cetera, and he says okay, and that's all. Then he drove off."

The next month as Gordzelik drove up to the range head, he noticed there was something different about the site. Looking around, he finally figured out what it was: There were new power lines at the field leading to a new power pole with an electrical outlet at the base of the pole. There was electricity at the launch site—they would not need a generator.

"Wylie shows up at the launch later that day, and he says, 'I didn't know if you needed 220, so I put in 110 and 220, and you are welcome to it.'"

POTROCS now had all the power they needed at the range head. At another launch later, Byrd dropped by the range again and asked Gordzelik why the club needed fifty-five gallon drums of water with buckets beside them all along the range.

"We already lost one field to a fire," Gordzelik told the rancher. "And we don't want to lose this one." Byrd's generosity revealed itself yet again. "The very next month we come out to the range and there is a new well and a cattle tank filled with water!" recalled Gordzelik. "Wylie told us, 'You know, I need water for my cattle too, so it's for both of us.'"

LARGE AND DANGEROUS ROCKET SHIPS

"So now we had it all," said Gordzelik. "We had the land, the space, a great waiver, electricity, and even water. Our membership had grown from eighteen to forty in six months. We decided it was time to host LDRS." The Tripoli Board of Directors readily agreed: LDRS-21 would be held in Texas. And by the time the launch came around in the summer of 2002, POTROCS had 110 Tripoli members.

In the days and weeks after September 11, 2001, launches around the country were canceled because of the FAA's reevaluation of all airspace restrictions. The annual BALLS launch was one of many events canceled and not rescheduled. Airspace restrictions related to rocketry were eventually lifted, thanks in no small part to the lobbying work of Tripoli Board member Dick Embry. Rocket launches resumed in most places in late 2001.

Then in October 2001, barely a month after September 11, a commercial high-power motor shortage began that would last well into 2002. That shortage was caused by a terrible fire at the manufacturing headquarters of America's most well-known motor builder.

At a little after noon on October 15, a fire broke out at AeroTech's manufacturing facility located on the outskirts of Las Vegas, Nevada. The fire resulted in the tragic death of one employee and the destruction of both AeroTech's facility and the surrounding businesses. Several lawsuits arose out of the fire, some taking years to run their full course.

According to a lawsuit filed by AeroTech, there were actually two fires that broke out in their plant on October 15. The first fire started when an employee inadvertently left a slotter machine plugged in during the cleaning process. While cleaning the area, the employee accidentally started the machine, which caused a metal-on-metal spark that, in turn, ignited propellant cuttings on the slotter table. The fire quickly jumped to a nearby container filled with APCP propellant shavings, resulting in either a flash fire or an explosion that killed Avelino Corpuz and seriously injured another worker. Firefighters from the Clark County Fire Department arrived within minutes, quickly containing the blaze. However, during the operation to put out this first fire, a large drum of powdered magnesium was partially filled with water. A combination of water and powdered magnesium can lead to excessive heat, hydrogen gas, or even an explosion. So although the primary fire was reportedly extinguished prior to 1:00 p.m., a hazmat team was dispatched to deal with the magnesium drum, which was now "foaming" and spilling its contents over the top lip of the drum.

Over the next few hours, the fire department allowed AeroTech and other tenants to reenter separate portions of the premises while the hazmat team dealt with the smoldering magnesium container.

Then at 3:48 p.m., fire officials attempted to remove the magnesium drum from the building. Suddenly, the simmering magnesium, which had been growing hotter with time, flashed into a full-fledged fire. Fire officials, along with everyone else, fled the building as this second fire grew in intensity, ultimately spreading throughout the facility.

According to AeroTech's lawsuit papers, Clark County firefighters then proceeded to direct "large caliber master streams of water" into the building to contain the fire, causing the magnesium blaze to spread further and grow in size. Then there was "a violent explosion that could be seen and heard for miles."

Ultimately, the fire department was forced to let the facility burn. Hundreds of homes in the immediate vicinity were reportedly evacuated. The second fire went out on its own accord the next morning, October 16. By then, the entire building used by AeroTech was destroyed. Several nearby businesses were seriously damaged. Damages were in the millions of dollars.

In 2001, AeroTech was the largest high-power rocketry motor manufacturer in the world. News of the fire resulted in high-power hobbyists rushing to purchase motor inventory wherever they could find it. By early 2002, there was a sense that high-power motors were scarce. Manufacturers like Animal Motor Works and Cesaroni rushed in to help fill the void. But no one could instantly fill the shoes of rocketry's largest motor maker.

In the meantime, Gary Rosenfield moved forward to reestablish his manufacturing ability, ultimately picking a new location for his facility in Cedar City, Utah.

"I would personally like to thank all of you who have shown your moral support to [me] and especially to the employees so deeply immersed in the recovery efforts evolving from the October tragedy," Rosenfield said in an open letter to the high-power community in early 2002. "Your phone calls, letters, and e-mails have been a tremendous encouragement to all of us."

There were several legal actions that arose after the fire, including a criminal investigation of AeroTech by the Clark County District Attorney's Office. That investigation was abandoned in 2003. AeroTech was also a defendant in at least two civil cases. AeroTech sued the Clark County Fire Department, among others, in federal court, alleging that the second fire was caused by the negligent actions of county fire officials. That case was dismissed in late 2003. AeroTech would eventually reopen and resume manufacture of rocket motors. But legal fees related to the fire were more than a million dollars.

The twenty-first annual LDRS was hosted by POTROCS near Amarillo, Texas, on July 11–16, 2002. It was one of the largest LDRS events ever held, with nearly 1,300 flights taking to the air.

Flying on day 1 started off well with blue skies and plenty of sunshine in the Texas Panhandle. The range opened just before 8:00 a.m., and activities commenced with the landing of an emergency medical services helicopter on the field. The chopper would be present for the entire duration of LDRS—a first at any LDRS launch.

Over the next several hours, 220 rockets were launched, including at least 9 hybrid flights and plenty of large APCP projects through N power. Flyers from Illinois, Indiana, Oklahoma, Michigan, and Texas launched multiple M projects. There were successes, and there were failures, including Nebraska flyer Kevin Trojanowski's ill-fated *Lusty Corn Maiden* rocket—built to look like a 7.5-foot tall replica of a giant ear of corn—which was seriously damaged on takeoff when his ejection charges fired halfway up the rail. The rocket, sans the nose cone, still streaked up more than 2,000 feet on hybrid power.

Late that afternoon, a tornado touched down several miles south of the range head.

"At the launch site, the clouds darkened and within just a few minutes the sky went from a few large drops of water to dumping rain in torrents," wrote *Extreme Rocketry*'s Brent McNeely. "Unlike ordinary rainstorms, this rain was propelled sideways at almost ninety degrees to the ground. Everyone outside of a vehicle was soaked instantly, as though they had jumped into a pool."

"I shut down the range even with rockets out on the pads right before the downpour," said Pat Gordzelik, who was the launch director for the event. "The grumbling from some who were tasked with turning off electronics on racked rockets turned to 'thank-yous' when a microburst hit the site head-on. What a mess! Vendors had their wares out in the open, and everyone and everything was drenched from pouring rain coming from all directions." A large number of tents and awnings were crumpled by the fast-moving storm, and the grassland rocketry range was transformed to several inches of mud, trapping many vehicles. The following morning, cars returning to the range were rerouted to avoid the muddy and difficult roads to the west.

The ground dries up quickly after a thunderstorm in a Texas summer, and over the next two days, more than 650 additional rockets were launched. There were 19 Ms and 4 Ns on Friday alone, with more flyers arriving from places like Toronto, New York, California, Florida, and Kansas. The number of Level 3-type motors was even bigger on Saturday. M-powered rockets awaiting flight were "stacked up like cordwood," said Gordzelik, who, at one point, counted 18 such rockets ready to fly. It was nearly impossible to keep track of all the unusual or mega-sized rockets. They seemed to be everywhere.

On Friday, Bill Wagner of Texas impressed the crowd with his flight of a 10-foot-tall *Mercury Redstone* on an AeroTech M1939. Fellow Texan Dave Schaeffer and his team sent skyward a replica of a beautiful remote-controlled *X-30* hypersonic vehicle on L power. The scratch-built rocket was 18 inches in diameter and nearly 10 feet long. After the ride skyward on an L850, Schaeffer took over the remote controls to bring the rocket back. "The *X-30* circled the range several times," reported one observer. "But unfortunately, on its final approach, it stalled. Dave almost pulled it out. Thankfully, the damage was minimal."

One of the most powerful rockets of the weekend was an upscaled Estes *Big Daddy*, created by Rick Waters and Chris Standish of Florida. At 110 pounds loaded, the 18-inch-diameter rocket held three 75 mm M1315 motors. The rocket had a great liftoff and carried almost a mile high into the air. Then the nose cone separated and came in ballistic, while the rest of the *Big Daddy* landed gently under chute. The nose was repaired, and the rocket would fly again Monday on an experimental O motor.

Kimberly Harms, rapidly becoming known as the leading woman experimental flyer in the hobby, and her team of Washington State flyers launched a hundred-pound rocket, *Community 8*, to 10,000 feet on a cluster of M and K motors. Tripoli treasurer Bruce Lee launched his *Super Mario* rocket on a Hypertek M1015 to nearly as high, and Texan

Dan Stroud thrilled spectators with his *Zeus, Ruler of the Sky* rocket that achieved almost 20,000 feet on an M2500 motor. It was Stroud's Level 3 certification flight.

One of the most striking projects to fly on Saturday was the incredible *Ariane 4* hauled to Texas by Swiss flyers Jurg Thuring, Mathias Gloor, Herbert Gort, Christoph Graf, and Daniel Flury. The magnificent 1:13 scale *Ariane* had made its debut at LDRS in Lucerne in 2001 but did not fly because of high winds and other issues. A year later, it was ready again. The rocket was more than just a showpiece: The 15-foot-tall ship held a central M1939 surrounded by four more J and K motors. At ignition, the rocket rushed skyward on all five motors. After motor burnout, the four outboard booster pods separated cleanly to fall free on their own under chute. It was a great ride up and back, marred only by some damage to the booster when it experienced parachute trouble on the way down.

Tripoli's annual banquet and members' meeting was held at the Radisson Inn and Hotel in Amarillo on Saturday night. Guest speakers include author William Gurstelle (*Backyard Ballistics*) and also Quentin Wilson of *October Sky* fame.

The last day of commercial flying at LDRS-21 was July14. Dave Zupan started the day early with his cluster of seven J570 motors in his *Seven-Toe-Dawg* to more than 13,000 feet. Fellow Illinois flyer Tim Lehr, a rising star in the high-power world, launched his N-powered modified *Ultimate Endeavor* to 16,000 feet. Hal Ellis of Oklahoma launched an *Honest John* replica on a K550 with a twist: As the LCO punched the launch button, four Estes C6 black powder motors ignited sideways on the rocket's upper airframe, causing the entire vehicle to spin with vectored thrust to 3,000 feet. (In a couple of years, this feat would be repeated by another team of flyers in a much larger *Honest John* at the Black Rock Desert in 2004 at BALLS).

One of the most anticipated flights of the day belonged to Wedge Oldham, who returned to LDRS with his repaired *Nike Hercules* launched at LDRS-20 in Lucerne. This year, Oldham and his crew upped the motor power for the project: The booster held four M1315s that staged to an N2000 in the sustainer. The rocket was ready on Saturday but ran into a snag. "While arming the altimeter for the nose cone, the electronics were inadvertently activated, and they fired the ejection charges," reported Brent McNeely. The nose shot off the rocket. No one was hurt, but the flight was delay to work out the bugs. On Sunday, the 265-pound rocket roared off a custom-built trailer launcher with all four M motors lighting on cue. The White Lightning motors left a massive trail of thick white smoke as it climbed perfectly into blue skies. Separation of the second stage went as planned, and the N motor lofted the upper stage to more than 12,000 feet above the Texas grasslands.

The high-power rocketry phenomenon known as Gates Brothers Rocketry, first seen nationally in 2001 at LDRS-20 in Lucerne, continued in Amarillo, Texas, in 2002. In five days of launching unparalleled in the history of LDRS, brothers Dirk and Erik Gates fired off a spectacular array of high-power motors, including the following:

Three AeroTech J570s

Six AeroTech K1050s

Five AeroTech M1315s

Six Animal Motor Works M1850s

One Animal Motor Works M5100

Four AeroTech N2000s

The two Southern California Tripoli members began their ascent into LDRS history on Thursday, July 11, launching their 70-pound rocket *Aramis II* on an N-2000 to more than 14,000 feet. The rocket recovered perfectly, landing without a scratch. On Friday, the Gates Brothers returned with *Porthos*, flown on a central N2000 surrounded by six K1050s. The 220-pound, 16-foot-tall rocket carried JVC camcorders and multiple electronics. It was another textbook flight and recovery.

On Saturday afternoon and with the assistance of Gary Rosenfield, Dirk and Eric launched an upscaled version of an AeroTech *Sumo* rocket. This vehicle tipped the scales at 220 pounds with a central N2000 and four M1315s. It held two bowling balls in the nose, an 8-pound ball and a 12-pounder, for stability. At ignition, the rocket screamed off the pad on the N and two of the four Ms. Several seconds into the flight, onboard electronics airstarted the last two M motors, taking the rocket to an altitude of 12,211 feet. Although the *Sumo* was damaged on recovery, it was another showstopping flight.

The Gates Brothers' LDRS-21 show was still not over.

On Sunday, they launched an upscaled version of a Public Missiles *Bullpup* rocket. It sported a gleaming white automotive paint finish and was meticulously detailed. The *Bullpup* was wheeled out to the away cell astride a custom-made carriage pushed by several crewmembers. The rocket transporter had six wheels and looked as though it belonged in a NASA satellite lab, not some dusty Texas prairie. The *Bullpup* weighed 125 pounds on the pad. It was 11.5 inches in diameter and stood 12 feet tall. This was the fourth day in a row the Gates Brothers were launching a mega rocket, and they pulled it off without a hitch. The N2000 in the *Bullpup* took the missile to 6,300 feet, before it returned to land softly under a 26-foot-diameter parachute.

On Monday, July 15, 2002, Dirk and Erik returned to the field with their biggest and most ambitious rocket yet at LDRS-21—the *Athos II*.

The *Athos II* was a two-stage rocket. From the aft end of the booster to the tip of the sustainer's nose cone, the rocket was 22 feet long. It started out 12 inches in diameter at the base and slimmed down to just 7.5 inches at the upper stage. The rocket weighed 265 pounds on the pad, holding seven AMW M-class motors in the booster (a single M5100 surrounded by six M1850s) and four AeroTech motors in the sustainer (M1315 and three J570s). This motor combination likely made *Athos II* one of the single most powerful high-power rockets flown on commercial motors in the history of LDRS—with a combined total of 55,580 Newton seconds.

The standing waiver for LDRS-21 was 20,000 feet. A higher altitude window was available on request by calling the FAA's Albuquerque, New Mexico, station ahead of time. The FAA's window for *Athos II*, which was projected to clear more than 25,000 feet, was from 1:00 to 3:00 p.m.

In the months leading up to LDRS, a 2-hour launch window seemed like plenty of time for *Athos II*. But after nearly a week of showstopping flights and recoveries every day, things looked a little different to Erik and Dirk Gates on Monday. "We were exhausted from the pace of the prior four days," said Erik later. "We toyed with the idea of simply taking Monday off and launching the *Athos* two-stage on Tuesday. But after chatting with more than a few folks, it became apparent that a number of people had extended their stay to Monday just to see this project."

The Gates Brothers and their crew worked feverishly all day long on Monday to prepare for the flight, to no avail. By late afternoon, it was clear the seven-motor mega rocket would not leave the pad on Monday: "When we finally had the entire vehicle assembled, all motors installed, it quickly became apparent that we would have to break the vehicle in two to transport it to the pad," said Erik. "During our tests at home, we had centered the completed vehicle on the [transporter] with a good portion of it hanging off the rear. That works until you install seven M motors in the booster. Hanging all those motors off the rear of the [transporter] didn't seem like a good thing to do. We broke the vehicle in two just above the sustainer booster and strapped the upper two sections to the side of the [transporter] for rollout. It was now 3:00 p.m. We had burned through the first hour of our high-altitude waiver and hadn't even made it to the pad. We had 1 hour left to get it in the air."

Even though they scrubbed the flight, the Gates Brothers and their crew spent another 3 hours on Monday going over flight procedures. They also practiced the use of a rented automotive cherry picker at the pad. This would allow them to reach the upper portions of the rocket to arm the electronics before flight. "We repositioned and leveled the unit three times before we were satisfied we could maneuver it up and down the length of the rocket and [then] set it down in a safe position for launch."

On Tuesday, the team arrived at the field at 10:00 a.m. Their FAA launch waiver to high altitude was from 2:00 to 4:00 p.m. "The standdown call the day before had been the right one," said Erik. "We had spent another 6 hours prepping the flight since that call. Fortunately, we were able to do it at a leisurely pace."

But they were not out of the woods yet. At just after noon, Launch Director Gordzelik alerted everyone to approaching cloud cover that looked ominous. Was a repeat of last Thursday's ferocious afternoon storm on the way? Gordzelik was not taking any chances. "I approached Erik and Dirk at 12:30 p.m. and asked them how long it would take to ready *Athos*. They replied, 'One hour,'" said Gordzelik. "I got on the phone with Albuquerque FAA and asked them if we could push up our window slot to 1:30 p.m. The reply was 'We'll check and call you back.' Five minutes later, Erik, Dirk, and their ground support crew stood next to me when my cell phone rang. 'Albuquerque FAA here. You are cleared to 30,000 at [1:30]. Call 5 minutes before launch. This is a 1-shot, 10-minute slot. Cleared for go!'"

The *Athos II* crew finished preparations at 1:25 p.m. Gordzelik called Albuquerque back. It was time to go. At 1:30 p.m., the go button was pushed by Animal Motor Works owner, Paul Robinson, who had designed the green propellant in the seven-motor booster. With a blinding green light, all seven of the booster motors came to life. Moments later and while the rocket was still in one piece, the M1315 in the sustainer fired. "The booster had a little AP residue on it, but otherwise, there was no damage from the sustainer lighting while still being attached," said the brothers later.

Several seconds after the sustainer's M motor fired, the three remaining AeroTech J570s kicked in as well, and soon, the upper stage of *Athos* was out of sight. Meanwhile, the booster was already beginning its descent under a military-style parachute. At apogee, the sustainer suffered a broken U-bolt, which resulted in the lower half of the rocket plummeting back to Earth with no chute. But the upper half of the sustainer returned under parachute, recording an altitude not as high as they hoped but still over 20,000 feet.

This Gates Brothers' LDRS-21 high-power rocketry show was over. There has never been another LDRS performance like it, either before or since.

LDRS-21 was a very successful launch—even with the bad weather of day 1. And with two days of experimental flying, this was the first six-day-long LDRS in history. POTROCS carried it off well.

There were 69 flights during the research portion of LDRS. In addition to the Gates Brothers' *Athos II* flight, there were rockets launched on every type of motor, G through O power. This included community space program's flawless launch and recovery of their 200-pound rocket on an O motor and six Ks, Dan Stroud's *Zeus*, *Ruler of the Sky* to more than 18,000 feet, and Walt Stafford's *Red Devil* to over 13,000 feet on a homemade L motor.

As with every research event of its day, there were plenty of CATOs, skywriters, loop-d-loops, cruise missiles, and other mishaps. But with every failed experimental flight, new lessons were learned, making this part of the hobby stronger. All told, there were 1,190 flights during the first four days of LDRS-21. With the two days of experimental flying, that number climbed to, making the total flight count 1,259. It remains one of the largest LDRS events ever held.

31

Passing the Baton

In the summer of 2002, and after serving nearly eight years as leader of the board of directors, Bruce Kelly stepped down as president of the Tripoli Rocketry Association.

Kelly had been a board member since 1989. He was an integral part of the reshaping of high-power rocketry from its modest beginnings in the mid-1980s to its modern international form. Kelly was just a two-year board member when he created *High Power Rocketry* magazine in 1991. Tripoli's income that year was $47,624.10. By 1994, when he was elected president, the paid membership of the organization had more than doubled. Much of that increase was because of Kelly's personal efforts and nationwide distribution of *High Power Rocketry*. The magazine brought many new members to the hobby, to the great benefit of Tripoli. Between 1995 and 2002, these numbers continued to climb. By 2002, and while Kelly was still president, there were more than three thousand active Tripoli members. The annual income for the corporation that year exceeded $300,000.00. This was a far cry from the organization's modest finances when Kelly discovered high power at Lucerne in 1987.

The Tripoli Rocketry Association during Kelly's tenure grew into a large and at times unwieldy organization. Yet the accomplishments of the association during this time were numerous: the obsolete confirmation program was scrapped, replaced by the modern three-tier certification system still in place today; the Tripoli Advisory Panel (TAP) was created; hybrids were integrated into the hobby; and experimental motors were recognized as a legitimate aspect of high-power rocketry. The basement bomber was laid to rest. While Kelly was president, relations between Tripoli and the NAR also improved. The ratification of NFPA 1127—the Code for High Power Rocketry—occurred during Kelly's presidency, and where possible, he worked with government regulators to ease regulatory burdens on the hobby. And when negotiations with the ATF over federal explosive regulations reached a dead end in the late 1990s, Kelly was an outspoken proponent of taking on the federal government in court.

Kelly faced opposition at times, and like any leader, he made some political enemies during his long run as president. Critics chided Kelly for lack of communication with the membership, or for his stance against the ATF, or a perceived tendency to micromanage some of the issues that faced the board. But throughout his presidency, Kelly remained an immensely popular figure.

Between 1995 and 2001, he won every election he faced, usually by a landslide. Kelly faced several electoral challenges from well-known, popular figures in high-power rocketry. The results were always the same: no candidate came close to beating him. In the popular or "advisory" vote for president cast by Tripoli members each year, the numbers in Kelly's favor were always lopsided. In his last election for president in 2001, he received 658 popular votes for president; the next closest candidate received 141. It was that way almost every year.

For the many Tripoli flyers who discovered high power during the 1990s, Kelly was the most recognized person in the hobby, second only to AeroTech's Gary Rosenfield.

———————————

The success of the *High Power Rocketry* magazine in the 1990s was not without setbacks. When he created the publication in 1991, Kelly was determined to use the magazine as a means to increase Tripoli's membership and spread the word of high power nationwide. He was successful on both counts. In the early 1990s, the magazine's production runs rose to almost nine thousand copies per issue. In 1994, that number jumped briefly to twenty-three thousand copies per issue. This included wholesale nationwide distribution to not only hobby stores but also booksellers, grocery outlets, and other retail sellers.

Unfortunately, the actual sales of the magazine did not mirror the rise in production to twenty-three thousand. When printed in these higher numbers, the sales were between 9 and 13 percent. The rest of the issues were returned—at significant expense to Kelly. So although the hobby was growing, *High Power Rocketry* eventually returned to its production runs of nine thousand per issue. Still, this was a phenomenal number in the history of hobby rocketry.

When Kelly became Tripoli president in 1995, he was publishing seven to nine issues of *High Power Rocketry* annually. This schedule continued to at least 1999, even in the face of Kelly's increasing duties as Tripoli president. This was the second time in Tripoli's short history that the club's president was also responsible for the organization's primary publication. In the late 1980s, Tom Blazanin tried to fulfill duties as both Tripoli president (and later as a board member) and publisher of the *Tripolitan*. It was a nearly impossible task. By 1995, Kelly was in a similar position—with a much larger Tripoli Rocketry Association—and a much bigger magazine. The result was not difficult to predict.

The first signs of trouble with *High Power Rocketry* began in 1998, when personal illness and the press of other business forced Kelly to drop the issue count to only six issues that year. In 1999, Kelly was at full speed again, publishing nine issues. Then in 2000, the magazine production dropped off again to five issues in the calendar year. That July, Kelly assured readers that he would catch up to nine issues in 2000 while also making quality changes in printing and binding of the magazine. Unfortunately, despite good intentions, Kelly was unable to achieve that production goal, and in 2001, the problem got worse. Only four issues of *High Power Rocketry* were produced with a 2001 calendar date, and the launch coverage that year was thin and outdated. The coverage by *High Power Rocketry* for LDRS-20 at Lucerne was brief and did not appear in the magazine until almost a year after the event.

Veteran Tripoli members took the sporadic production of the magazine in stride. They knew Kelly created the publication from scratch and that his hard work was in the best interests of Tripoli. Newer members, however, were somewhat less patient, and some were resentful of paying for a publication and receiving no magazine. It was not long before complaints regarding the magazine reached the Tripoli Board of Directors.

Kelly acknowledged the problem of missing issues in early 2002, admitting he was seriously behind in publishing issues of the magazine. He attributed his tardy production to a "heavy regulatory workload," that is, the time-consuming problems Tripoli was facing in dealing with government regulators and the lawsuit against the ATF. This was undeniably true. Kelly's presidential workload increased tremendously over the prior few years. His board duties were taking a toll on not only his personal life but also his ability to publish the magazine. Still, he was optimistic he would catch up.

"Those who have subscribed direct through the publisher will get every issue paid for," he promised. "Those who subscribed through [Tripoli] will have an opportunity to have their memberships extended to make up for missing issues."

A few months later, and just prior to LDRS-21 in Texas, Kelly resigned as president of Tripoli. It was time to choose between his two competing jobs in high power. "The job of managing Tripoli's affairs became too great to continue, along with the duties of publishing this magazine," explained Kelly at the time. "The volunteer presidential 'part time' job gradually became full time. I had to make a choice between the two jobs. I could no longer do both. Therefore, I stepped down as [president], and I'm now back to editing the magazine full time. You can look forward to see HPR get back on a more regular schedule."

Although he resigned as president, Kelly ran for another term on the board in 2002. However, publication difficulties with *High Power Rocketry* were now an election issue. Kelly lost his bid to keep a board seat by only 21 votes out of 931 cast. (He was replaced by a chemistry professor from Kentucky, Terry McCreary.) Despite his election defeat, Kelly remained at the helm of *High Power Rocketry* and would maintain his position as one of the high-power's most experienced and influential voices. But for the first time since 1989, Bruce Kelly was not on the board. In his place, Board Member Dick Embry of Kansas became the fifth president of the Tripoli Rocketry Association.

Dick Embry was born in Fayetteville, North Carolina, in 1953. He fell in love with rocketry at an early age. "My first experience was when my dad was stationed at McCoy Air Force Base in Florida. I was there when the first *Mercury* series went up," Embry said during an interview in 2002. "It was funny because you could look on TV and see the rockets go up or step out our back door and see them go up. We were so close— it was only 40 miles away—and you could see them come up over the next hill. I was actually able to go to Cape Kennedy and watch it in real time. That lit a fire that always stayed there."

Embry wanted to be a fighter jet pilot, and after high school, he attended Auburn University, where he earned his bachelor of science degree in engineering. After graduation, he enlisted in the United States Air Force. The Vietnam War was winding down, and Embry was assigned his first duty as a weapons controller in Alaska. From this humble start, he worked his way to flight training and was eventually selected to fly one of America's premier fighter jets of the era, the General Dynamics *F-16*. Over the next two decades, Embry flew many types of air force aircraft, including combat and combat support missions in the Gulf War in 1990–1991.

Embry launched his first model rocket as a teenager.

"I was about thirteen at the time," he said. "It was a three-stage model rocket where the stages were the same length as the motor. It was an awesome flight. I believe it was beginner's luck." Embry was unable to launch many rockets after that because he lived on military bases while growing up. But in the late 1980s, he found the hobby again, this time through an advertisement in an Oklahoma City hobby store. "My first high-power rocket was launched on a Vulcan I motor," said Embry. "That was really an experience, especially given how long it had been since I flew my last one."

Embry became Tripoli number 961 in 1990. He earned his confirmation for single-use and then high-power reloadable motors in 1990. President Chuck Rogers signed Embry's confirmation card.

By the early 1990s—interrupted by calls to active duty overseas—Embry was a local Tripoli prefect. He ran for a board seat in 1993 and was elected and began his first term in 1994. He was on the board that oversaw the creation of Tripoli's certification and TAP programs as well as final adoption of NFPA 1127. He was an early supporter of Tripoli's experimental/research rocketry rules, and he was Tripoli's insurance coverage liaison for several years.

One of Embry's personal projects as a board member was interacting with the federal government and the Commercial Space Transportation Office—later renamed the AST. Embry was Tripoli's representative with AST in the mid-1990s, where he lobbied to exempt high-power rocketry from certain federal regulations pertaining to amateur rocketry. Some of these regulations required FAA approval for rocket motors with a burn time greater than 15 seconds. Fortunately for Tripoli, Embry had plenty of prior experience with the FAA. At one time, he was the United States Air Force airspace manager for the Southwestern United States. "I had worked with the FAA extensively while in the military. In a nutshell, I was responsible for twenty-six low-level routes, five warning areas off the coast, and a smattering of restricted areas. The dynamics of managing, modifying, and monitoring all that airspace required daily conferences and coordination with all of the FAA regions concerned."

Through these experiences, said Embry, he earned the trust of the office of the FAA, and that trust would one day benefit rocketry. With his aviation background, Embry worked well with federal officials, such as Randy Repcheck at AST, to raise burn times and ease regulations that applied to high power. "I think it opened the door to both commercial manufacturers and the then-embryonic research venues to start thinking about bigger motors and pushing the envelope, which, if you look at the basic tenet of Tripoli, was the reason we started in the first place," said Embry. "It also lit a fire under AST to revise the federal regulations. Randy Repcheck's first love has always been amateur rocketry. He said several times that he and his office 'had to get on with the rewrite of the regulations.'"

By 1996, Embry's work with Repcheck resulted in an increase in motor burn times from fifteen to sixty seconds, and he continued to work with federal authorities to renew and obtain additional exemptions over the next few years. This included increases in altitude waivers and, ultimately, the creation of the Tripoli Class 3 Committee. This committee was delegated the authority by the AST/FAA to internally approve extreme high-altitude flights. Today it governs flights by Tripoli members in excess of 50,000 feet. Without it, and the work by Embry, there would be no Tripoli flights to the edge of space today. Embry's experience with federal authorities was also instrumental in re-obtaining FAA waivers for high power after the events of September 11, 2001. "After 911, there was a time when all flights of aircraft and rockets were banned by the government," said Embry. "They shut everything down. It was an understandable knee-jerk reaction." Embry, along with Wyoming senator Mike Enzi, worked hard to get the waivers reactivated within several weeks. "Our relationship with the FAA/AST paid massive dividends when a four-star air force general stopped all amateur rocketry flights," said Embry. "Obviously, we had the restriction lifted but with a dire warning from the general who made it clear that if he heard of any rocket being fired at any aircraft, he would turn off rocketry *permanently*."

Embry's first task when he became Tripoli president in 2002 was to reorganize the way the board functioned on a day-to day basis. He wanted individual board members to take greater responsibility in running the organization. For Embry, the process of delegating authority came naturally.

"I learned very early on in the air force and the deployments pre- and post-Desert Storm that my job was to manage and think," he explained. "It's impossible to run a deployment involving hundreds of people and billions of dollars of equipment by yourself. It's a recipe for disaster."

Embry's vice president, Ken Good, agreed with this philosophy. "It became quite clear in the later years of Bruce Kelly's presidency that running Tripoli had become something of an all-time-consuming monster," said Good. "Our organization had grown tremendously in size and complexity during the Kelly years . . . There was simply no way our organization's president, a volunteer position that must coexist with a person's real full-time working job, can possibly do everything on his own."

With a new leader at the helm and a change in philosophy at the board level, the Tripoli Rocketry Association continued to move forward into 2003, still fighting the federal government over explosives rules but on the verge of a publicity bonanza that would spread the word of high power more than ever before.

High-power rocketry was about to get its own television show.

32

The Rocket Challenge

In the late 1990s, there was a television show in Great Britain called the *Scrapheap Challenge*. In a typical episode, four-person teams were let loose in a scrapyard to see which team would be the first to create a specific mechanical device—the device changed from show to show—from materials scattered about the yard. The program was so popular in England the Discovery Channel decided to create an American version of the show. They came up with a new title using the colloquial American term for scrap.

They called the program *Junkyard Wars*.

On January 24, 2001, Tripoli members Ky Michaelson and Bruce Lee appeared in an episode of *Junkyard Wars* in which two teams were tasked with building a rocket that could carry aloft and then return safely to the ground an ostrich egg. The teams had 10 hours to accomplish their mission, and their rocket vehicle could be made only from parts they could find in a junkyard outside London. Rocket motors would be provided to the teams at the conclusion of their builds. Recalled Bruce Lee, "It was really cool because it was a real junkyard on the river. They chopped a corner of it and put a fence around it. Actually, I think it's one of the most dangerous places I've ever been. All the stuff in there is sticking out, sharp and pointy. There was broken glass everywhere. Once we got there, we went to the wardrobe area. After that, they let us go through the junkyard briefly to see what is was like. But they wouldn't let us stay long or look because they didn't want us to know what was there. Once the contest started, as team leaders, we had to stay in the build area, and only two other guys were allowed to go into the junkyard to find things."

In the end, Lee's team prevailed, returning the ostrich egg safely to the ground following a flight powered by three 38 mm I-impulse motors.

Six months later, Lee received a telephone call from a Discovery Channel production company interested in filming a rocketry-related television show. The company was called First TV, creators of a contemporary show called *Battlebots*, which was then airing on Comedy Central. In *Battlebots*, competitor-built robots battled each other to destruction in a small arena. First TV told Lee they were interested in doing another program like *Battlebots*, only this time the show's competitors would be engaged in high-power rocketry.

"They are looking for a fresh new idea with which they can make a show," Lee reported to the Tripoli board at the time. "They are working on ideas for competition with individuals and teams to compete in head-to-head events in such venues as rocket drag racing, spot landing, and altitude and speed records."

"Bruce brought it to the board before LDRS-20 in Lucerne," recalled Ken Good, who was a board member at the time. Afterward, President Kelly authorized Lee and Good to travel to Los Angeles during LDRS-20 to meet with First TV and see what they had in mind.

"We drove there and met with the producer and his team," recalls Good. "They were, of course, interested in shows that could be sold and not necessarily a one-off. Their most recent success was *Battlebots*, so they had some sort of competition in mind. They were really interested in the general concept of our big, noisy, fast-moving rockets but had few specific ideas in mind as to how that sort of action should be captured and packaged into a sellable show. They had a few suggestions, such as firing rockets at targets, which we quickly dissuaded them from considering. We really had to educate them on what high-power rocketry is about, our safety codes, et cetera. In essence, we told them that anything on which the show would be based would have to fall within one of our two safety codes if it were to be part of any official Tripoli event. Actually, I think we came away from that meeting with more ideas that could not be supported than ones that could and in which they were interested."

After their meeting, Good and Lee gave the production crew a tour of the rocketry activities at LDRS-20 at the Lucerne Valley. "The TV folks were quite amazed at what they were seeing," said Good. "We introduced them to flyers, explained what was going on, took them out to the pads, helped set up some takeoff shots, and ultimately, we were the subject of short interviews with them. At one point, we even introduced them to Frank Kosdon . . . Suffice to say that Frank was not at his most photogenic that day, but he did closely resemble the caricature of him [torn T-shirt, dirty gym shorts, and mud-caked knees] . . . Frank was also a bit less than congenial, mixing and sipping lemonade from a wide plastic container as he gave rather curt answers to questions. Indeed, First TV had met one of rocketry's 'colorful' figures."

The television crew watched launching all day on Saturday, July 20, 2001—the busiest day of LDRS. Hundreds of rockets of all shapes and sizes were flown, including Andy Woerner's 285-pound *V-2*, the John Coker-inspired Soviet *N1*, and multiple M-powered flights. The really big and unusual projects, and a number of flight mishaps earned the full attention of the television observers, said Good:

> While conducting our media friends around, I could not help but notice that there seemed to be a high incidence of "things going awry" with the launches that were taking place . . . Even more concerning was the large number of prangs and free-falling bowling balls, some surprisingly, from experienced flyers. One of the [scariest] of these was the flight of a huge one-third-scale *V-2*, which struggled on its not-all-lit engine cluster to leave the pads, immediately arcing over the crowd and crashing, fortunately in a clear area beyond the spectators.

In fact, perhaps thanks to the excitement of such disasters, the television representatives were very enthusiastic about what they saw that day at LDRS-20. Shortly thereafter, Tripoli and First TV entered into talks regarding the possibility of a primetime television show dedicated to high-power rocketry.

"Bruce Lee remained our point person for further contact," said Good. "After some time, the ideas that were forming were based on an upcoming LDRS and special competitions that would be conducted within that context. There was a great deal of back and forth between the board and First TV on some of the ideas."

"First TV needs our help to make it all work," Lee reported to Tripoli several months later, in the spring of 2002. Lee believed high power would benefit from a rare chance at national media exposure. "We have the knowledge of how to organize and run launches and work with the FAA and other government agencies," said Lee. "We also represent an existing base of competitors who can compete in high-power rocketry events. Not only that, there will be cash prizes for competitors, money for the prefecture hosting an event, and pay for some people in a support role. We are also proposing that all prefectures benefit with some portion of the profits. [The] Tripoli Rocketry Association will also be acknowledged on the show."

Lee also suggested some of the monies generated from a rocketry television show could be used to purchase private property on which future rocket-related events could be held without the worry of needing special permission from private property owners.

In late 2002, the Tripoli board voted unanimously to accept a contract with First TV for a high-power show based on an LDRS launch. By the time the contracts were signed, LDRS-22 in Argonia was the first available LDRS that could host such an event. The tiny launch event that started on a rural farm in Medina, Ohio, in 1982 was about to get its own national television show.

LDRS-22 was held on July 17–22, 2003, in Argonia. There were 1,080 flights during the first four days of commercial flying at LDRS-22, followed by another 117 launches over two experimental days. As with every other contemporary LDRS, mega projects flew almost every day. First TV was there to record all of it.

Team Gator returned to LDRS from Florida for a repeat flight of the prior year's *Big Daddy* rocket that thrilled the crowd at LDRS-21 in Texas. This time the 18-inch-diameter yellow-and-black rocket ripped off the pads on three Animal Motor Works M1850 Green Gorilla motors for a great flight and nice recovery. Jeff Purvis launched his giant yellow rocket on an M1939, and four doctors from Kansas City, Missouri, launched the world's biggest bottle rocket from a carefully crafted two-story tall green soda pop bottle aptly named *Thrust*. There were *Saturn V*s, *SR-71*s, and a K550-powered vehicle called *Alien with an Attitude*.

"I was a child of the Space Age," explained flyer Randy Braye to a writer from *Air & Space* magazine, which was also on hand for LDRS-22. "I flew my first rocket in 1968, built three *Saturn V*s, had a National Geographic map of the moon on my wall so I could plot where every [*Apollo*] mission landed. But then I found out about girls and cars, and I forgot about rocketry in 1971. I didn't get back into them until 1997. I wanted to buy my son a birthday present, and when I started snooping around the Internet, I couldn't believe how much rockets had changed. I thought to myself, 'Wow, they've got some really big stuff now.'"

Braye and others like him are sometimes referred to as Born-Again Rocketeers.

On Saturday night at the annual members' meeting at the Hyatt Regency in nearby Wichita, space shuttle astronaut Charles Donald "Sam" Gemar was the featured speaker. Gemar had flown on three shuttle missions, completing hundreds of Earth orbits and nearly 600 hours in space. He regaled Tripoli members with NASA and shuttle stories and joy of flying in outer space. It was the first time an astronaut addressed Tripoli members at an LDRS—another coup for the Kloudbusters.

The biggest project of LDRS-22 was a 4-fined, 12-inch-diameter rocket from Maryland painted like a cow. The rocket was called *Udder Madness*. The rocket was to fly on a 55,000 Newton second P12,000 motor created by Darren Wright. Upon arriving at Argonia, the rocket team discovered that one of the grains in their 6-inch-diameter P motor was defective. So Wright and his crew gathered what they could from fellow flyers, including a 10-quart commercial mixer another flyer brought to the launch, and recast a new APCP grain.

"At this point, Darren had his work cut out for him," said team leader Neil McGilvray. "Some of the Kansas Tripoli members sent Darren over to a small shop in Argonia, and he began the mixing process. While all of the elements of the original formulation were not going to be included in the replacement grain, the confidence level was still high. Darren and Eric Hall worked for the better part of Sunday afternoon mixing and packing the 12-pound grain."

The *Udder Madness* crew was ready for launch on Sunday. It was the most anticipated flight of the weekend. The television crew was all over the rocket as it was raised on the pad. They installed an onboard video camera for use in the television show later. After a ten-count, the launch button was pushed. There was a flicker of flame beneath the bottom of the rocket as the igniter inside the motor fired, and the motor came up to pressure. McGilvray described what happened next:

> [T]he first blast of thrust from the powerful motor kicked the rich Kansas soil 20 feet into the air as the rocket slowly began to move up the tower. Within the next split second, all hell broke less with a resounding *thud*, followed by an eerie silence—then the hissing of free burning propellant

and the painful thuds of the heavier pieces of the rocket falling back to the ground. The motor had over-pressurized and CATO'd! The 60 pounds of flaming propellant was now back at atmospheric pressure and was making a quick exit from the motor casing. There was a huge cloud of Kansas topsoil being kicked up from the pressure release and the cow was quickly being turned into ground beef. Fins were seen distinctly flying off in four different directions. The parachute/payload section was thrust straight up the tower as in a normal launch but without the bottom half of the rocket. The ARTS system that was still functioning was recording as the nose cone was sent 275 feet into the air on a ballistic trajectory. The 28-foot military main parachute caught its share of hot gases being released from the motor and was now a useless rag. The XXL Sky Angle [parachute] that was to bring the nose cone section back was in the same condition. Both [parachutes] flailed helplessly as the airborne sections [of the rocket] dropped painfully back to the ground. About half of the propellant remained lit and burned where it lay in the dirt field. The other half was extinguished as a result of the extreme pressure drop that was experienced from the peak of over-pressurization to the low of atmospheric pressure . . .

"Rockets are just another name for trouble," said McGilvray afterward, repeating a rocketry quote he once heard from the *Viking* program director at White Sands in New Mexico: "Either you just had trouble, you are having trouble, or you are going to have trouble."

In the early 2000s, Public Missiles Systems cofounder Frank Uroda was concerned about the future of high-power rocketry. A national economic downturn in the early 2000s hurt rocketry manufacturers. Uroda and other suppliers to the hobby were feeling the pinch—sales were down, and for a brief time, so were the membership numbers of Tripoli and the NAR. As Uroda watched smaller vendors close up shop, he worried that rocketry's future might be in jeopardy. Then at LDRS-22 in Kansas, he came up with an idea.

As Uroda watched the Discovery Channel cameras and crews scrambling all over the field, it dawned on him that the hobby was being presented with a unique opportunity to recruit new members to the ranks. High-power rocketry needs more than just a television show about what we do at LDRS, he thought. High power needed to make its own television commercial to air for the show. Following the end of LDRS and in late September 2003, Uroda started a campaign to create rocketry commercials to air during the upcoming Discovery Channel program in November. He called his campaign Save Rocketry Now.

Frank Uroda was born in Detroit in 1959. He discovered rocketry as a child in the back pages of a comic book. "There was an Estes ad with the title 'Model Rockets You Can Really Fly,'" said Uroda. "I only half-believed the title because I had bought several other things, like Sea Monkeys, which were very disappointing when I got them."

There was no disappointment when Uroda's first model rockets arrived. "It was the coolest thing I had ever seen," he said. He launched rockets off and on through high school, and then he moved on to other pursuits.

Uroda was good with his hands, and in the late 1980s, he created a cabinet-making business manufacturing custom office furniture. Through work, he also made friends with a rocketry enthusiast. Soon, he was flying D through F motors in model rockets he built on his own. "Because of the F motor, we began looking for bigger and stronger components to build from," said Uroda. "We tried everything we could find, including plastic tubes, Formica, and other stuff." Uroda was also looking for larger motors. "There were rumors going around about bigger motors. But I hadn't been able to find any," he said. "One day I picked up a rocketry magazine and began writing to different companies asking if there was anything else out there. A guy named Steve Buck sent me back a handwritten letter telling me to buy a LOC *Magnum* and to fly it on an AeroTech J motor. I remember thinking, 'You're kidding! That stuff really exists?'"

Uroda joined Tripoli, and soon, he was flying high-power rockets. He and his friend Gerald Kolb were both members of the NAR. They helped found the first Tripoli prefecture in Michigan. Shortly thereafter, the club planned a larger-than-life rocket. "I made a proposal to build what I called the M Project," said Uroda. "The primary purpose of the project was to help us advance our knowledge of rocketry. It was a

6-inch-diameter rocket about 8 feet tall, and it was supposed to have the latest and greatest things we could get in a rocket."

During the construction of the M Project, Uroda searched for a new type of airframe for their big rocket. "I didn't have any experience with fiberglass at the time. I started hunting around for better tubing. I made hundreds of phone calls to different companies. I found lots of tubes, but they all had drawbacks—this one was too heavy, this one cost a fortune, this one was not the right size, et cetera. I ordered lots of samples until we finally got the right stuff. It was called *phenolic* tubing."

Uroda believed he had discovered a perfect material for high-power airframes. And with this new product, Uroda and Kolb went on to found one of the longest-running commercial rocketry companies in hobby rocketry: Public Missiles Systems. Phenolic tubing was the company's first product.

Uroda marketed his new airframe tubes with small advertisements he purchased in rocketry magazines. The response to his ads was good. Soon, his company added another exotic material to their inventory: G-10 fiberglass. He and Kolb formed a corporation and released their first complete rocketry kits: the *Io*, the *Calisto*, and the *Phobos*. All three rockets were based on phenolic tubing airframes; all three had recovery systems that deployed parachutes by way of a small piston that forced the recovery gear out of the rocket tube at apogee. These were the original kits sold by Public Missiles Systems in the early 1990s.

"Because they were so unique, a lot of people were reluctant at first to try the new kits," said Uroda. "But after they did, they loved them. It got to the point where people got to know our kits in general and the quality of our products [and], then it became easier and easier to introduce more new products."

Uroda continued his woodworking business through the early 1990s. But as Public Missiles generated more income, he had less time for his day job. "There comes a point where your side business, like rocketry, is too much work to do on the side but not enough money to quit your first job," said Uroda, who slowly transitioned from his woodworking career to high-power rocketry. By 1995, he was fully immersed. Public Missiles was his full-time work; as high-power rocketry expanded in the 1990s, so too did Uroda's business. In 2000, he and Kolb had several employees, and the company was now marketing a full line rocket kits, rocketry electronics, and recovery gear. The company also sold components for nearly every aspect of the airframe. Public Missiles Systems rockets were seen at every local and regional rocket launch in America, and Uroda was among the most respected commercial vendors in the hobby.

When Frank Uroda conceived the idea for a national high-power television commercial, he had no experience in television advertising. He also did not have much time.

LDRS ended on July 22, 2003, and the Discovery Channel show was scheduled to air in November. Uroda had only a few months to get things moving. If the Save Rocketry Now campaign was to be effective, there was no time to waste. The first thing he did was contact Discovery Channel in New York to secure advertising space. He was told there were eighteen commercial spots still available for the November television show, each spot 15 seconds in length.

"I gave them my proposal for a commercial," recalled Uroda. "They said it would cost $90,000."

Uroda was used to dealing with advertising salespeople. Public Missiles Systems had run ads for years in magazines like *Popular Science*. "We went back and forth for a while and negotiated the number down to $63,000." He had saved some money. That was the good news. The bad news was the deadline. Uroda had only two weeks to generate all the money needed to secure ad space for the show.

"I came up with the idea that I would ask rocketry vendors everywhere to donate prizes, and we would then sell raffle tickets, at $100 each, for a chance to win a prize. All of the money collected from the tickets would pay for the ads," he said.

Uroda took his proposal and the Save Rocketry Now idea to the Tripoli and NAR boards for help. Their initial response was not what he expected. "Nobody is going to pay $100 a piece for a ticket, they told me. They thought the $100 number was ridiculous."

He moved forward anyway. High-power flyer Greg Deputy created a rocketry website called Flyrockets.com. The purpose of the website was to collect the raffle prizes from the vendors and checks from consumers. The site would also serve as contact point for viewers of the commercial that would air in November during the LDRS-22 special.

"Flyrockets.com was the whole point of the ad," said Uroda. "We wanted everyone to have easy access to rocketry through the web page, and Flyrockets.com was easy to remember." The idea was for television viewers to see the commercial for the web page during the show, and when they logged on afterward, they would be redirected to the web pages of local and national Internet sites for Tripoli and the NAR.

"We have a once-in-a-lifetime opportunity with the Discovery Channel," Uroda told the rocketry boards and commercial vendors. "They are producing a 3-hour special on rocketry and LDRS, and the show will reach six million viewers. This show is the equivalent to millions of dollars of advertising."

In a lengthy e-mail circulated throughout the rocketry community, Uroda explained his vision for a commercial, the manner in which money could be raised by raffle, and proposed additional advertising in print media that should be conducted after the show was aired. The response to Uroda's plea was extraordinary. In less than two weeks, his Save Rocketry Now campaign generated the $63,000 needed for the commercial air time—and then some.

"Many in the community, including myself, were doubtful it could be done," wrote *Extreme Rocketry* publisher Brent McNeely in early November, just days before the television show was to air. "However, Frank persisted. Despite naysayers he pushed forward with determination and vision [and] not only did Frank generate the funds needed in less than two weeks, he produced over $83,000 for promoting rocketry."

To create the commercial, Uroda turned to Pat Gordzelik, who, with the help of Kloudbusters' member Bob Brown, put together a professional ad featuring one of the space shuttle astronauts, Sam Gemar. "We cut the deal, and Bob got a film company, and we hired them, and Sam did the commercial at the Kansas Cosmosphere and Space Center," said Gordzelik, who also helped arrange for the domain name of flyrockets.com. "There was a guy in Texas who already had that page. I called him and told him what we were doing and asked if we could have the domain name, and he said yes. He just gave it to us at no charge."

With the commercial paid for and ready to go, and the Discovery Channel special only a couple of weeks away, Uroda discovered one last hurdle—the Internet web pages of rocketry clubs around the country needed to be updated immediately. "I started to realize that almost all of the club information on every website was way outdated," he said, "and that included the websites of Tripoli and the NAR. Some of these websites were also not very user-friendly. This would make it harder for new people to join these clubs. In the interim, I started pushing for everyone to get their websites up to par. We had Greg Deputy help us send out guidelines, and there was a good response to that. There must have been a hundred websites that were reworked in the last two weeks."

Uroda was finished. In a matter of weeks, he had gathered everything together for the rocketry commercials that would eventually air on national television during the LDRS show. All that was left was the broadcast. "It was like a shooting star," he said. "It all happened so fast. It was just nonstop."

High-power rocketry made its primetime television debut on Sunday, November 9, 2003, at 8:00 p.m., on the Discovery Channel in the United States. The 3-hour program was broken into three 1-hour episodes. It was narrated by the Emmy Award-winning television personality Kerry McNally.

"We're here in the middle of nowhere, which is just where you have to be to safely launch some of the biggest monsters you are ever going to see!" shouted McNally from a wide-open range in Kansas as the show began. "Not long ago, this entire area was a wheat field. Today, however, we're going to be harvesting a different kind of crop: rockets—big rockets!"

With a jazzed-up version of the song "Somewhere over the Rainbow" playing in the background, Kerry continued. "We're in sweltering Argonia, Kansas, for the twenty-second annual LDRS launch, which sounds like a very scientific acronym. But it actually stands for Large Dangerous Rocket Ships, which is precisely what you are going to see dueling head to head in a spectacle that we call *The Rocket Challenge*."

High-power rocketry, with all its engineering, slang, and quirky personalities, was opened up to the world for 3 hours. There were spectacular flights and flaming disasters: thirty-nine rocket mass drag races and two-story-tall bottle rockets; rockets that flew low and slow, and missiles that raced into the stratosphere. Television being television, there was also plenty of hyperbole: Newton seconds were converted into horsepower, Level 3 flyers became rocketry's PhDs, quotes that probably should have been left on the editing room floor made their way to the forefront of the show. "We're shoving 160 pounds as high as an airliner can fly!" said one excited flyer, whose offhand remark was among the first statements in the show's opening segment.

"A lot of guys close their eyes and see women," said another flyer. "I see *rockets*."

In another portion of the program, Derek Deville scrambled for cover as ejection charges on a full-scale *Nike Smoke* inadvertently exploded right behind him.

It certainly looked dangerous.

The Rocket Challenge revolved around multiple contests, most of which were created just for the show. There was the target landing contest, where a group of flyers launched rockets to see who could come closest to a target not far from the pad, and the Supersonic Showdown, to see who had the fastest rocket of the launch. There was also a contest called From the Ground Up, in which eight teams of several flyers each were set loose in an impromptu rocketry supply store under a massive tent at the field. The plan was to grab rocketry components from boxes loaded on a table and then scratch-build an 8-foot-tall rocket as fast as possible. Amazingly, the first rocket was assembled, glued, and ready to launch in less than 1 hour, and when all eight rockets were launched, every single rocket was recovered safely under chute.

The show also revealed a contest First TV did not have to make up—the "Bowling Ball Loft," where flyers built rockets propelled by a K motor to see who could lift a 16-pound bowling ball to the highest altitude and return it to the ground safely under a parachute. That sounds safe.

The Bowling Ball Loft had been around since the 1990s. It required serious engineering to place a heavy ball within the confines of a working rocket. It was also uniquely dangerous, especially since balls ball would, on occasion, separate from their parachutes and return to Earth like some sixteenth-century cannonball. "I met my first goal," one flyer told the television audience after his ball cleared more than 3,600 feet and then landed under chute: "I didn't kill anyone."

The First TV crew captured it all and delivered it to the living rooms of America. "They had personnel everywhere with movie cameras," said launch director Gordzelik. "They even had a remote-controlled helicopter with an onboard video camera. LDRS-22 was definitely not a good place for Jimmy Hoffa to hide out. But after a short period of time, the film crews with their trailers, equipment, and golf carts merely blended in with the rest of us weirdos."

Prior to LDRS-22, the production company invested hundreds of hours traveling to communities all over America for interviews of flyers and their teams to help explain to the television audience why rocketeers do what they do. It was a nuts-and-bolts segment of high-power rocketry, covering each project from its inception in someone's garage or barn to the ultimate launch in Kansas.

For example, First TV traveled to Arizona for a look at Hillbilly Rocketry's several-hundred-pound *Gila Monster* and then matched the preflight interviews with the actual launch of the *Gila Monster* in Argonia. The massive rocket performed flawlessly, powered by six M motors and reaching more than 6,500 feet. First TV was also in Florida watching Team Gator put together their *Big Daddy* rocket, and they traveled to Texas to meet Dan Stroud and his team working on a rocket called *Aurora*. The *Aurora* project began with a 6-inch-diameter diameter P motor built by Pat Gordzelik. The motor was 5 feet long and weighed 92 pounds. Gordzelik called his homemade fuel Polish Rojo. The motor would generate 2,000 pounds of thrust and burn for nearly 8 seconds, said Stroud. The rocket's 20-foot-long airframe was made primarily from carbon fiber. The total weight of *Aurora* was 175 pounds at the pad.

Aurora was crammed with state-of-the-art high-power electronics and camera equipment and sophisticated tracking gear. *Aurora* also became the first successful P motor launch in the history of LDRS. Nearly 30 minutes after it left the pad, *Aurora* touched down gently under parachute in the middle of the launching range. The flight was a total success. "After going over 29,000 feet, it came down soft and gentle," said Stroud. "Not a scratch and only 1,200 feet from the pad. We had a thousand dollars' worth of tracking equipment in the nose, and it comes right down in front of us."

It was great rocketry television. *The Rocket Challenge* producers also created state-of-the-art computer graphics to show viewers how a high-power rocket worked in flight and recovery; introduced the audience to the world of reloadable and scratch-built motors, altimeters, and carbon fiber rockets; explained the meaning of insider terms like "certification," "land sharks," "shreds," and "core samples." "In rocketry, your rocket never explodes or blows up," exclaimed McNally. "You CATO!"

There were many successes aired during the show. There were also plenty of failures. Disasters are an inherent part of high-power rocketry. It is how flyers learn. There are no feel-good awards for participation, no phony ribbons or trophies, no social counseling when a rocket that took two years to build disintegrates on the pad in a microsecond. You suck it up and move on in rocketry, or you get out. If a rocketeer hasn't destroyed at least one rocket yet, they haven't been flying very long.

"Right before they push the button and you're waiting, you're wondering why they haven't pushed the button yet. That's when I get nervous," whispered one flyer to the national television audience.

Tim Lehr's *The Beast* had a great flight up. But it ended life bouncing off the ground, destroyed on recovery. "If it worked right every time, it would be too easy—and no fun," Lehr told the audience.

The on-camera destruction of the biggest rocket of LDRS-22, the P-powered *Udder Madness* rocket, was captured in full-color glory. It was a spectacle on television. For 3 hours, high power received the kind of attention most hobbies can only dream of: a national audience supported by advertisers like General Electric and Tide. For many flyers of the future, *The Rocket Challenge* was their first introduction to the hobby.

33

A Space Shot

By the fall of 1996, Ky Michaelson had made a name for himself in high-power rocketry. In the hobby only two years, he had already created his own line of rockets, parachute equipment, launch videos, and related high-power paraphernalia. Outspoken and confident, Michaelson was an advocate for, among other things, a massive increase in Tripoli's membership, purchase of private property for launch sites, and a monthly magazine for the hobby.

Michaelson did more than just talk the talk; he walked the walk. He funded and launched several ambitious projects of his own. In less than three years, he destroyed at least two extravagant high-power rockets in Nevada and another huge rocket at the Orangeburg LDRS in South Carolina. One of his bigger mishaps was an O-powered vehicle that came in ballistic in Nevada from a purported apogee of 50,000 feet; another was a P-powered rocket at Black Rock that was destroyed by a massive CATO at the pad.

Michaelson remained undeterred and enthusiastic. And success finally found him at the BALLS launch in the fall of 1996, when he sent a Joe Boxer Corporation–financed P-powered rocket to an estimated altitude of 50,000 feet, with a perfect recovery. "I'm out of the altitude business for now," he declared after that flight. "I lived my dream! Now it is time to build a rocket to go into space!"

To help fulfill his ambitious goal, Michaelson created the Civilian Space eXploration Team or CSXT. Beginning in late 1996, Michaelson and his team started planning what he hoped would be the first amateur rocket to reach the boundary of outer space, defined by many observers as the Kármán line, an altitude of 100 kilometers above the earth or 328,000 feet.

Michaelson's quest for space progressed slowly at first. Among other things, he was overwhelmed by bureaucratic requirements to obtain permission from the government to launch a private rocket into space. Such requirements were set by the FAA and the Commercial Office of Space Transportation (AST). For a time, it looked as though Michaelson's efforts with government agencies were heading nowhere. His space launch idea might be a hopeless pipe dream. "Then I received a phone call from one Jerry Larson," recalled Michaelson. "And his call literally changed the entire direction of my space program forever."

Larson was a real-life rocket scientist who worked for Lockheed Martin and was a flight planner and ran flight simulations for the *Athena* launch vehicle program, including the NASA's Lunar Prospector mission, which had orbited the moon.

"I couldn't quite figure out why this guy was calling me," said Michaelson. "But he quickly told me that he's actually heard about all the trouble I was having trying to get my program licensed. Then he dropped the most exciting news on me since I started my quest for space. He told me he had worked with the office of Commercial Space Transportation [AST] on a regular basis and that he actually helped file the fight plan for the *Athena* rocket. I didn't even know what to say. He seemed to sense my apprehension and immediately assured me that he could answer any technical questions the AST office threw at me. Here I had been as close to a standstill with my space launches possible, and just like that, along came Jerry."

After several conversations with Larson, Michaelson was convinced it was possible to turn his space exploration dream into reality. "Jerry very quickly informed me that an effort such as this would require team players and lots of teamwork too," said Michaelson. "He immediately came on board as a very active important team member himself and became our project manager . . . In no time at all, he had completely turned things around for us and jump-started the whole program."

By early 2000, there were at least two more members on the CSXT team: Larson and longtime Tripoli member Bruce Lee. Over the next few years, the team attracted dozen more volunteers, gaining help and expertise with motor and airframe design, onboard avionics, flight simulations, and logistical ground support. The team also picked up much-needed financial assistance through corporate sponsorships. It was a long shot, but Michaelson and his crew might make it to space.

The first significant rocket fielded by CSXT was called *Space Shot 2000* and was launched at the Black Rock Desert on September 29, 2000. The rocket was 9 inches in diameter and nearly 15 feet long. It carried an R18,000 solid fuel APCP motor with a fifteen-second burn time. The four-finned vehicle weighed 437 pounds on the pad; 228 pounds was propellant. The motor would generate thousands of pounds of thrust.

"The sun came up, and we were blessed with a picture-perfect day and excellent launch conditions—very sunny and the sky was as clear as it could be," recalled Michaelson of the launch day. The CSXT team was required by the FAA/AST to send up weather balloons to measure wind speeds at various altitudes along the rocket's anticipated trajectory. The balloons went up, and by 8:40 a.m., everything was ready.

For electronics, the rocket carried two flight recorders, each having its own GPS receiver and antennas to verify altitude. "The U.S. government imposes a 1,000-knot [1,152 mph] restriction on the usable velocity of an off-the-shelf commercial GPS unit," said Jerry Larson. "So it cannot be used in the manufacture of ballistic missiles. Because of this restriction, the GPS receivers on board will shut down during the rapid ascent phase." At or near apogee, said Larson, and when the rocket has slowed to only hundreds of miles per hour, the GPS units will reactivate to record the rocket's peak altitude—and to determine whether they had, in fact, reached space.

The plan for the *Space Shot 2000* flight was for 15 seconds of powered flight followed by approximately 90 seconds of "coast" to the edge of space. At apogee, the rocket would separate into two pieces, the nose cone and the booster, and each component would return to Earth via their own parachutes. The total mission time, from liftoff of the rocket at Black Rock to space and back again, would be only 5 minutes.

The aluminum-based rocket roared out of its launch tower into the air. "The motor lit, and the rocket looked like it was flying perfectly," said Michaelson. "At about 15 seconds, the motor cut out, and the rocket was traveling at over 3,000 miles per hour when something happened.

"There was stress on the rocket, one of the fins broke loose, and it fell apart like a two-dollar suitcase," said Michaelson. "Everything came apart and went into a flat spin. It all landed within our range, and we got it all back except for a few fins."

Postflight analysis revealed that the rocket broke apart at approximately 40,000 feet. "I was in a state of disbelief," wrote new team member Eric Knight, who had recently written a $10,000 check to advertise his company's logo on the side of the rocket. "I intellectually knew this mission was a long shot. But after seeing and touching the magnificent vehicle, space never felt so close, so entirely possible to reach. After watching our rocket roar into the sky just a few seconds before, the machine seemed unstoppable."

It was back to the drawing board for CSXT.

Between 2000 and 2002, the CSXT grew to at least twenty-six official members. The team built another high-dollar, technology-loaded rocket while picking up an important new corporate sponsor along the way—Primera Technology, Inc., a Minnesota-based manufacturer of CD/DVD disc duplication and printing equipment.

The *Primera* vehicle, as it came to be known, was the same diameter as the *Space Shot 2000* rocket—approximately 9 inches but was almost 4 feet taller. Most of the extra length was for additional fuel. *Primera* was 18 feet long and was topped off by a 5:1 conical nose cone spun out of a solid block of 6060-T6 aluminum. On the launch pad, the rocket weighed more than 500 pounds. Of that amount, 290 pounds was solid fuel propellant. The 11.5-foot-long motor was in the S-class impulse range.

"At the bottom of the rocket were four sharp-edged fins," said Knight, who was now the leader of a new electronics platform for the rocket. "A keen eye at the launch pad would have noticed that the fins were not absolutely straight on the rocket but angled ever so slightly—about a half a degree from vertical. This subtle cant of the fin gave the rocket an intentional bullet-like spin." This spin, according to Jerry Larson, would be approximately six cycles per second and would improve trajectory and landing accuracy.

The four-day launch window for *Primera* was on September 16–19, 2002, at the Black Rock Desert.

"The whole crew arrived on Monday, September 16, and it was like a family gathering," recalled Michaelson. "The mood was extremely upbeat and optimistic as we all got right to work. A four-day window can go by very quickly if everyone doesn't pull together and stay focused, or if unforeseen problems occur, so we wasted no time. The avionics team changed the batteries in all the equipment, while we spent the day assembling the rocket and getting it placed in the launch tower."

By the end of the day, everything was in place for a high-altitude rocket launch on Tuesday. "I made a solemn vow to myself right then and there that whatever was about to happen wouldn't be for a lack of effort," wrote Michaelson afterward. "We had all given 110 percent, and if this attempt failed, I had no more to give and would just walk away."

Unfortunately, the weather conditions on Tuesday were not conducive to flight. The launch was scrubbed because of cloud cover and high winds. The inclement weather continued into Wednesday morning but then improved later that afternoon. It was too late to fly that day, perhaps tomorrow. On the last day of the window—Thursday, September 19—the team arrived at the pad at four o'clock in the morning to find a "clear, dark blue sky above."

The weather was fine. The launch was a go. The precise FAA window for Thursday's attempt was 9:00–9:20 a.m. In the last couple of hours before launch, the team grappled with weather balloon problems as well as an early Union Pacific train that agreed to wait at a station outside the launch and recovery area for an additional 30 minutes.

At 9:15 a.m., with the rocket loaded and ready to fire, the firing switch for the S-impulse motor was pushed. A few seconds passed, nothing happened.

"The motor failed to ignite. We were absolutely stunned. How could this possibly happen when we'd come this far?" said Michaelson.

Team members John Gormley and Chet Bacon jumped into a car and sped across the desert floor to the launch pad a quarter of a mile away, a tall rooster-tail column of fine dust trailing their vehicle the whole way. "They had less than 5 minutes to resolve whatever electrical problem had caused the igniter to fail," recalled one team member. If they could not do it in time, the FAA waiver would expire, and the launch would be scrubbed.

After 3 minutes, they returned; the countdown resumed. This time, *Primera* roared to life. The rocket raced into the sky ahead of the screaming S motor. But not for long. Seconds into flight, it all disintegrated, scattering parts and pieces all over the desert floor. "We all ran for cover and stood by helplessly," said Michaelson as the fragmented remains of the rocket hit the ground like so much burning confetti.

"I saw the smoke long before the boom reached my ears," said Steven McMacken. "Like [some] stupendous Fourth of July fireworks, the rocket had separated into a thousand pieces. Each piece trailed a white streamer as it plunged back to Earth."

The *Primera* flight was a total loss. Investigation after the flight revealed the likely cause of the disaster was a catastrophic motor failure a few thousand feet into the flight. "From what we could determine, the initial explosion originated near the nozzle, just above the tail fin assembly," said McMacken. "The force of the explosion blew the nose cone completely off the body. It landed a quarter mile west of the launch tower. The severely damaged avionics section remained intact, however, and remarkably was sending back data up to the moment of impact with the ground. Everything else was pulverized by the explosion. We were only able to find small 1- and 2-inch pieces of what was once 18 feet of high strength aluminum alloy."

Later, it was determined the motor sustained a burn through; the superheated propellant combustion gases melted through the aluminum motor case in flight, leading to the rocket's destruction. Team member Knight, like most of the CXST personnel, was crestfallen. "In 2000, our rocket flew to over 40,000 feet," said Knight. "Now with two more years of personal and financial sacrifices, it only got to 3,000 feet—not much higher than a $20 hobby rocket would fly. The agony and disbelief were beyond any possible words," he said.

Patience is a virtue, especially in rocketry. Author David A. Clary in his 2003 biography of rocket pioneer Robert Goddard recounted that Goddard's persistence—even outright stubbornness in the face of repeated disasters—is what set him apart from those around him. "We could go out on a trial and the rocket would explode," said one of Goddard's friends."Some of us standing around would say, 'Gee, this is terrible,' and Goddard would say, 'We learned something today. We won't make this mistake again. We'll correct it.'"

Following the *Primera* loss, the team regrouped and vowed to try again, perhaps in 2003 or 2004. It was ultimately determined *Primera* only had a 60 percent chance of reaching space. To increase the likelihood of success, any CXST rocket needed to shed some weight and be propelled by a much more powerful motor. It was back to the drawing board—again.

Between 2002 and 2004, the CSXT went through several more changes in the creation of a launch vehicle to reach the limits of space.

Two of the most important changes involved corporate sponsorship and motor design.

For sponsorship, Primera was replaced by the Go Fast Sports & Beverage Company and an Internet web design firm called Fuscient. Go Fast and Fuscient became the lead financial contributors for the rocket. The new sponsorships provided a fresh infusion of financial assistance to keep the CSXT moving.

The other significant development involved the rocket's motor design. Rick Loehr, who had built the solid fuel motors for the 2000 and 2002 space shot attempts, stepped down from the project. In Loehr's absence, the team considered hybrid motor technology as an alternative to solid fuel. To help analyze the possibilities, Michaelson enlisted longtime high-power enthusiast and friend, Korey Kline. At the time, Kline was working in Florida as one of the founders of Environmental Aeroscience Corporation, a hybrid rocket motor research and development company. Together with motor maker and Tripoli member Derek Deville, Kline ran a number of simulations for the proposed CSXT flight, ultimately selecting a hybrid motor profile that could do the job.

But there was still a problem. Flight simulations revealed that although the thrust-to-weight ratio with the hybrid motor was high enough to be safe at liftoff, there wasn't enough instant acceleration to prevent weather cocking off the pad.

"Our previous rockets were solid fuel rockets and accelerated in a blink past the speed of sound," explained Eric Knight. "They reached top speed very quickly—blasting right through low-level winds. But our newly proposed hybrid-powered rocket was lumbering by comparison."

"The slow takeoff was allowing the winds to play too great a role in the trajectory," said Deville, who reported that the possible downrange landing zone with the hybrid-based computer simulations were simply too wide for even the vast expanse of the Black Rock Desert to support. "We were faced with a difficult decision," explained Deville. "Either we continue with the hybrid design and find a new place to launch from or change course altogether and develop a solid motor with a higher thrust that would overcome the wind sensitivity."

"I didn't think we had to reinvent the wheel to make this rocket work," said Michaelson. "But it was starting to look that way. I called Jerry [Larson] and told him we needed to stop beating our heads against the wall and go back to our ammonium perchlorate motor." Larson then asked Deville and Kline whether they could build a large enough solid fuel motor for the vehicle, which was by now named the *GoFast* rocket. Deville was not sure. "We both had solid fuel experience, but this was a whole new ball game. This was going to be the largest solid fuel motor we had ever built, and if successful, it would be the largest successful amateur motor ever. With this in mind, we replied to Jerry with a timid yes, and so the design process began anew."

It was February 2004. The next window for another space shot flight, which was picked not only for the benefit of the team but also to ensure that the rocket had a chance to be the first amateur vehicle to space, was in May. There was no time to waste. "The first order of business was vehicle sizing," said Deville. "I began by reviewing my existing propellant formulations, taking into account the average impulse density and specific impulse. From this, I worked with Jerry to determine a preliminary mass estimate." The estimate Deville and Larson came up with suggested at least 425 pounds of propellant in a rocket with a gross weight of 700 pounds. With these numbers, which would make the *GoFast* vehicle much larger and heavier than *Primera*, Larson set to work, leading the team building the new rocket, while Deville took over with motor testing and design.

The solid motors for the two prior vehicles had utilized separate fuel grains of APCP, much like the grains in the typical high-power reloadable rocket motor, only bigger. Deville, however, opted for a different approach this time. Instead of several fuel grains stacked together inside the motor case, he would cast one super grain that would be the entire length of the motor. He believed the single-grain motor, bonded directly to the motor case, offered performance advantages over multiple grains. "We needed the motor to yield a neutral thrust trace," said Deville. "This can be accomplished in a single continuous grain motor by using port geometries." He continued:

> In a round port [fuel grain], as the propellant burns away the port circle will grow larger in diameter and will expose more propellant, increasing the chamber pressure and thrust to possible uncontrollable levels. We needed to have as nearly a constant as possible amount of propellant surface exposed. This can be done with an irregular-shaped initial port that exposes a larger area, such as a star-shaped port. If done correctly, the change in area over the duration of the burn can be minimized. Examining the choices of port

shapes that would work in this case, we narrowed in on what is known as "fin-o-cyl" geometry. This is essentially a round port or cylinder that has rectangular or nearly rectangular fins around it . . . This geometry can be easily fine-tuned to the desired thrust trace and has the additional benefit of being fairly straightforward to fabricate.

Despite success at narrowing down the optimal motor configuration for reaching space, Deville faced an additional hurdle that he could not resolve on his own. "The aspect of the motor design that had the greatest uncertainty for me was erosive burning. I had done significant research into the issue but had not yet developed any models or methods for predicting or accounting for the phenomenon," said Deville. "I refer to it as a phenomenon because it does seem to be a somewhat mysterious effect, with the kind of black magic feel to the lore that surrounds it."

Enter former Tripoli president, another real-life rocket scientist and then-current Tripoli Board member Charles "Chuck" E. Rogers, who had done research on erosive burning as far back as the 1980s. "I relayed my situation to Chuck and told him of the design work that had been done to date and what my anticipated problem with erosive burning was," recalled Deville. "And let me tell you, I found the right guy!"

"Erosive burning is in the core of the motor," explained Rogers. "As the hot combustion gases are coming down the core of the motor toward the nozzle, the high-velocity flow of the combustion gases over the burning propellant surface causes an increase in the local propellant burn rate, known as erosive burning. The highest velocities, and thus the highest Mach numbers in the motor core, are at the end near the nozzle, where the Mach numbers can reach Mach 0.7 or higher compared with Mach 1 at the nozzle throat. As a result of the erosive burning, the propellant burn rate in the local areas of the motor core can increase by up to 20–30 percent."

"To get the performance they needed to make it into space—that would be 100 kilometers—they needed to load the motor with as much propellant as they could," said Rogers. "They had to make the core of the motor as small as possible to get as much propellant in as possible. Also, they had to make the motor long and skinny to lower the overall drag of the rocket, but both of these things increase erosive burning in the core."

Rogers was quickly recruited to the CSXT project to assist in the erosive burning calculations Deville needed for the rocket's S-class motor. Rogers developed erosive burning design criteria that helped the team design the core of the motor. He and Deville also came up with a plan for the subsequent erosive burning testing. "We didn't have the time to do a whole bunch of erosive burning tests," recalled Rogers, "but we did have time to do a few tests, and I worked with Derek to figure out which erosive burning tests we wanted to perform."

Deville's plan for the *GoFast* motor was to build a couple of smaller motors first before he moved to the final product. "The first few series of tests were performed with 3-inch-diameter dynamic propulsion systems hardware," he reported. "This data was analyzed by Chuck, Jerry, and me and was very useful in determining the propellant burn rate . . . and the increase in burn rate due to erosive burning. As a side benefit, Chuck was able to use the data in his nozzle performance study, which was also used to optimize the nozzle for the CSXT motor."

The team fired a 6-inch-diameter test motor before working their way up to a full-size version of the *GoFast* rocket motor, which was ready for test-firing on Environmental Aeroscience's thrust stand in Florida on April 12, 2004.

The full-scale motor was massive. It was 10 inches in diameter and more than 15 feet long. The aluminum case and propellant weighed more than 600 pounds. For the test-firing, the motor was placed in a horizontal position parallel to the ground. It was secured to a massive I-beam that ran the entire length of the case. "With the cameras rolling, the moment of truth had finally come," wrote Deville afterward.

"The igniter popped, and flames grew from the nozzle and WHAM! In only a fraction of a second, the motor came up to thrust and then the [forward] end closure let go."

The end opposite of the nozzle side of the motor was blown free of the S50,000 motor—with nearly catastrophic results. With an open end, the propellant rushed out of the rocket spitting fire and smoke in every direction. Ky Michaelson, listening on the telephone to the test-firing from his home in Minnesota, knew something awful had gone wrong. "I heard Jerry give the countdown, [and] seconds later, we heard this huge roar that lasted for less than 1 second and then quit." The S motor not only lost the forward closure, but it also broke completely loose of the huge I-beam and launched itself across the property, zooming like an unguided cruise missile into a nearby lagoon, where it quickly sank, the thrust pouring out the open end of the motor case until it disappeared under water.

Through the phone, Michaelson could hear Jerry Larson shouting. "It's under water! It's under water, and there's water and fire shooting 30 feet in the air!"

Fortunately, no one was hurt.

"We need to make two more motors," Deville told the team after the accident. "One to static test and one to fly. We've got less than a month before our first scheduled launch day."

After yet another disaster, Michaelson began to lose confidence in his timeline:

> The fact of the matter was . . . that the entire project had not been easy on anyone, and we'd been snakebitten from the start. Unfortunately, the pattern continued, and the bites got harder each time. We ended our conversation in total agreement that we were going for it one last time, and each of us had a list of things to do. Derek and Korey ordered the chemicals, and Jerry ordered the aluminum. From Minnesota, I started back at square 1 in hopes of finding someone to help us build the two motors. I was lucky enough to find a machinist who needed some extra cash and was willing to work for us at night when he got off his regular job. Out on the East Coast, fellow team member Eric [Knight] and his crew were finishing up the avionics and getting them ready to ship out to Jerry for testing. To make things even more hectic, several other groups with multimillion-dollar budgets and the same ambitions we had were nipping at our heels to get into space first. The ultimate low blow came . . . when we learned that one of those teams heard we had been talking to the [Bureau of Land Management] about launching a rocket on May 24, so they blocked out that date for themselves to ensure we didn't get it.

When Michaelson got the news about losing the May 24 launch date, he called Jerry Larson. "I told him we simply had no choice but to bypass testing another motor," said Michaelson. "Time was no longer on our side, and we just needed to fly."

Larson and the rest of the team agreed, and in late April, it was decided there would be no more testing. The next motor they built would be taken directly to the Black Rock Desert for flight.

"We understood that such a decision meant going into the flight with increased risk," wrote Deville later. "But we felt that sufficient testing had been done to give us enough confidence to lay it all on the line. Indeed, that was how it was, because Ky had previously let us know that if this attempt failed, he was done. Without him as a motivating force behind the project, this team was not likely to ever attempt another space shot." The scheduled launch window was now May 17–21, 2004—less than four weeks away.

The first thing to be done after the April 12 motor failure in Florida was to figure out what happened. Scuba divers were hired to retrieve the S-motor case from the bottom of the lagoon. When they dragged the motor tube out of the water, the cause of the mishap was obvious: The threads on the forward closure were too fine to hold the closure in place. When the motor came up to full pressure on the test stand, the threads did not hold. The closure was simply forced off. The remaining fuel in the motor rushed out of the front of the 15-foot-long tube—opposite the nozzle end—sending the motor flying.

The engineering fix was not terribly difficult. The threads on the forward closure were redesigned, and after the closure was screwed in to the new motor tube, two rings of hardened bolts were installed around the outer circumference of the tube and into the closure, securing the closure in place for good. A double ring of bolts was also applied to the nozzle end of the case—just in case. This bolt design had been used before by Korey Kline on some of his larger high-power rocket motors. He called them Frank-en-bolts.

With the motor case issues addressed, the propellant for the final motor was cast and ready by May 7. Three days later, Michaelson arrived in Florida from Minnesota with truck and trailer to haul the S motor to the Black Rock Desert in Nevada. Meanwhile, other team members hurriedly finished their respective tasks and hit the road to get the rocket and all the remaining equipment to the desert on time. On Saturday, May 15, 2004, Michaelson arrived with the S motor at Brunos Country Club and Casino in Gerlach. All the remaining rocket gear had arrived too.

"The plan was for all of us to meet out in the [desert] on Sunday to assemble the rocket," said Michaelson, "leaving Monday open for our first launch attempt." By the end of Sunday, the rocket was on the pad in vertical position.

This was the third high-altitude attempt by Michaelson and CXST since 2000, the previous two rockets having been destroyed. This was also the largest and most sophisticated of the three—and by far the most expensive. In its custom-made launch tower, the rocket stood 21-feet tall and weighed 724 pounds. The S50,150 motor was likely the most powerful amateur motor ever built up to that time. The motor weight alone made up for almost 84 percent of the rocket's mass—nearly 607 pounds of propellant and case. Tens of thousands of dollars were invested in the project, along with thousands of hours of design, construction, and testing.

"There was no sleeping that night," said Michaelson, "only lying awake, watching the minutes tick by on the clock, as if time had suddenly stood still. We were all out on the lake bed before the morning's sunrise." By late morning, the rocket was ready to go.

"The final 15 minutes before launch were like no pressure I have ever felt," said Deville afterward. "I looked around the desert and saw dozens of people and tons of equipment that represented years of efforts, and it was all riding on the motor that we had built. Without the [second] full-scale static test, I didn't have 100 percent confidence that everything was going to work flawlessly. Had I mixed the propellant right? Was the propellant strong enough? Did we account for erosive burning properly? These and hundreds of other questions constantly raced through my mind. I couldn't sit down. I paced the desert, running through scenarios in my mind, making myself sick. The butterflies that I normally enjoy before a launch had changed into carnivorous beasts from the netherworld."

At 11:12 a.m. on Monday, May 17, 2004, the launch button for the *GoFast* rocket was pressed. "Nothing happened for a moment, but suddenly, there was an enormous roar," said Michaelson as his third space shot rocket came to life: "Flames spewed from the huge 15-foot motor, and the rocket was pulling twenty-three Gs as it lifted from the tower, leaving the desert floor behind in a gigantic whirlwind of sand and dust."

In less than 10 seconds, the *GoFast* rocket cleared 40,000 feet, traveling an estimated 3,420 miles per hour—above Mach 5. The motor burned nearly 14 seconds. After burnout, the entire vehicle disappeared from sight high in the sky. There was no motor CATO, no vehicle disintegration in flight. The rocket remained intact, heading for an estimated altitude of more than 62 miles. Now it was up to the tracking team to locate the rocket—if it survived and deployed its parachutes as designed.

Several minutes went by.

Then the team received a radio signal indicating onboard ejection charges had fired. "Because of the high speed at which they deployed, there was a chance they could have torn right off the rocket's booster or payload section," said Michaelson.

"Everyone, please be quiet!" shouted Jerry Larson. "If the parachutes didn't open properly, there's a good chance we'll hear two sonic booms."

Moments later, and just as Larson predicted, two distinct sonic booms thundered across the desert floor. "We didn't know what to make of them," said Knight. "If the [two] rocket sections were floating down on parachutes, what would be moving fast enough to create sonic booms? It didn't make sense."

One possibility for the sonic booms was that the parachutes had not deployed properly, and the parts were hurtling back to the ground at high speed. "We all concurred that the rocket had most likely hit the ground at supersonic speeds," said Michaelson, "and had plowed so deep into the earth that we'd never successfully recover the precious payload section, meaning we had no way of proving how high it had flown."

However, the team's initial reaction to the sonic booms was premature. One of the trackers suddenly yelled out. He had a signal.

"I've located it! I've located the payload! It's falling too slow!" he announced. "It has to be under parachute!"

And so it was. The following day, and after an exhaustive search involving a helicopter and multiple search teams, the payload section of *GoFast* was found on a remote mountainside approximately 20 miles from the launch site. The rocket section was resting at an elevation of 6,129 feet above sea level.

"[T]he spacecraft impaled itself a few feet deep into the rocky landscape," said Knight. "The exterior was scorched. Most of the decals and paint were melted off by the air friction of its Mach 5 flight into space. But overall, it was in terrific shape." The search for the *GoFast* booster section was not as successful. The search went on for days and then weeks and then months. By the fall of 2004, there were no signs of the rest of the rocket. It was presumed lost.

LARGE AND DANGEROUS ROCKET SHIPS

On November 10, 2004, a BLM surveying crew was conducting an aerial population count of wild horses when they spotted a tubular object that seemed out of place on a barren mountain slope not far from the Black Rock Desert. The BLM contacted the CXST, and searchers arrived by foot at the location of the object ten days later, on November 20. It was the missing section of the *GoFast* rocket. "The booster was in sad shape," said Knight, who jumped on a plane from his home in Connecticut to participate in the search. "There were pieces strewn about the impact site. It had been a violent collision with the earth, that's for sure. But it was not buried underground. And there wasn't a parachute to be found. Not a stitch of it."

Members of the team speculated the booster's parachutes may have worked for some portion of the descent but were ripped off the vehicle as it headed back to the ground. However, the parachutes were never found, and so what really happened during the booster's descent will never be known. The team used metal detectors to recover smaller broken pieces of the rocket at the impact site, and when they measured the booster, they were astounded to learn the impact presumably crushed the airframe such that it was now more than 2.5 feet *shorter* than it was before flight. The impact also caused the entire fin canister to slide to the forward end of the rocket. The booster was located a little more than 4 miles southwest of where the team found the payload bay in March.

In a press release by CSXT published one year after the flight, in March 2005, along with separate books subsequently written by Michaelson and Knight, it was claimed the *GoFast* vehicle reached space. According to team members, the rocket generated a peak thrust of 16,000 pounds and a top speed of 3,420 miles per hour. Motor burnout was at 49,000 feet, with the rocket spinning at eight revolutions per second at that point in flight. The rocket achieved an altitude of 379,900 feet or 72 miles, well beyond the boundary of the beginning of space, defined by some as the Kármán line at 100 kilometers (approximately 62 miles) or 328,000 feet.

Unfortunately, there was no radar tracking of the *GoFast* rocket in flight. Also, the onboard video system in the rocket failed, so there is no video recording of the rocket in flight from extreme altitude or any altitude. Not all the onboard electronics functioned as planned either.

However, onboard electronics did yield productive data that was analyzed by both Jerry Larson and Chuck Rogers. Rogers explained during an interview in 2014,

> The onboard data that were recorded during the flight were three-axis accelerometer data from a Crossbow CXL25L3 unit and attitude measurements using a three-axis magnetometer, which made measurements relative to the earth's magnetic field. The Crossbow had a 25 G axial acceleration limit, which the *GoFast* rocket came close to but did not exceed, during the initial acceleration right after liftoff. The rocket also had a Global Positioning System [GPS] unit, but after losing lock during the high initial acceleration right after liftoff, the GPS unit did not get a lock for the rest of the flight, including no lock through apogee. The GPS unit regained lock just prior to the payload section landing on the ground. The unit accurately recorded the landing time [840 seconds after liftoff] and the final location of the payload section, but no other flight data was recorded. There was also a problem with the three-axis magnetometer attitude measurements as the rocket's launch tower, which was wrapped around the *GoFast* rocket on the ground, interfered with the initial attitude measurements from the magnetometer. The launch tower also had a structural failure during launch, and while tower elevation and azimuth were measured prior to launch and postlaunch after the structural failure, there was no way to know the precise elevation and azimuth when the rocket had cleared the tower. From the flight data, the *GoFast* rocket clearly had a nearly straight-up flight, but we were unable to determine an accurate "zero" for the attitude data at the moment of launch. So the rocket's attitude data in flight could not be accurately reported.

Using the three-axis accelerometer data, primarily the axial acceleration, and the three-axis magnetometer attitude measurements, Larson modified the inputs to the preflight trajectory simulation he developed until it matched the accelerometer and attitude data, in particular, matching the axial acceleration, and he also matched the landing location of the payload section of the rocket. Using this trajectory reconstruction technique, Larson determined the *GoFast* rocket reached an altitude of 379,900 feet or 72 miles.

A second postflight analysis was then performed by Rogers, and using the axial acceleration from the three-axis accelerometer data, he was able to confirm the altitude for the rocket. "I took the axial acceleration data from the *GoFast* rocket, and I assumed a straight-up flight, and I came up with a certain attitude," continued Rogers. He added,

> The question with the *GoFast* rocket remained: what was the angle of the trajectory? You take the onboard axial acceleration data, and you go straight up, and you get one altitude, but the actual altitude was based on some trajectory. I knew the payload section was recovered 20 miles from the launch site. What I did was I took the flight profile to burnout from the axial acceleration, and I assumed it was a straight-up flight. Using this assumed straight-up flight to burnout, I had the burnout velocity and burnout altitude as initial conditions for running a series of trajectories starting with different flight path angles at burnout. Using a rotating oblate earth trajectory simulation program I coauthored at the Air Force Flight Test Center, which included the earth's rotation and the oblateness of the earth, which you have to include for rockets reaching Mach 5 and nearly 400,000 feet, I ran a series of trajectories using the burnout velocity and burnout altitude and the drag coefficients measured in flight during the early part of the coast phase after burnout. I adjusted the flight path angle at burnout until the rocket landed 5 miles downrange, and then 10 miles downrange, and then 20 miles and so on. I even ran the case where the rocket would land 40 miles downrange, and at that downrange distance, the apogee altitude of the flight was still above 100 kilometers. But at 20 miles downrange, I confirmed the altitude of 72 miles.

Rogers later calculated that the *GoFast* rocket reached the incredible speed of 3,420 miles per hour based on axial acceleration data. The maximum Mach number was 5.18, he said, adding, "It was the first hypersonic amateur rocket." Rogers rejected any claims that there had been a burn through the case at the end of the burn.

"There was no burn through of the motor case, but 5 seconds into the approximately 13.5-second burn (and after about 60 percent of the total impulse had been expended), a portion of the nozzle exit cone had ablated/eroded to the point where it got too thin, and a symmetric segment of the nozzle exit cone broke off and was ejected," explained Rogers. "You can see the point where part of the nozzle exit cone came off in the thrust curve based on the flight data published in a tech article on the 2004 CXST *GoFast* S motor in the April 2012 issue of *Rockets Magazine*. The loss in thrust when a portion of the nozzle exit cone broke off was 8 percent. The nozzle flow expanded from where a portion of the exit cone broke off into the inner portion of the fin can, with the fin can acting as a poorly shaped final portion of the nozzle exit cone for the rest of the burn."

The CSXT team and Tripoli member Ky Michaelson had achieved their dream of reaching space—a dream that for amateur rocketry stretched back as far as the first few decades of the twentieth century.

"The amateur pioneers in the early days of rocketry like Werner von Braun, Robert Goddard, and those working in Russia all sought to reach space for the love of science and exploration," said Jerry Larson in 2018. "Yet they never achieved that goal of reaching space as amateurs. The early rocketry societies around the world were instead swallowed up by governments for military and political purposes. The CXST team set out to finish the journey of reaching space as an amateur rocket society. It took us only six years, but we achieved the goal of putting an amateur rocket into space."

34

Rocketry vs. the U.S. Government

Part 1

The Lawsuit

In March 2004, and while the *GoFast* rocket was being prepared for its historic launch, Tripoli and the NAR were awaiting word on the outcome of a battle they had been fighting for more than four years. The struggle was being conducted pursuant to the Federal Rules of Civil Procedure in a courtroom in Washington, D.C. That spring, a federal judge was tasked to answer a question that affected every high-power motor user in America.

Was APCP an explosive or not?

The ATF said yes, subjecting rocketry to the bureau's authority and control. Tripoli and the NAR said no. In rocketry's view, the regulation of the hobby under the guise of federal explosives laws was an example of overreaching by the national government. Ultimately, rocketry leaders would have to decide whether to forego common sense and simply accept the government's authority. That decision would come in the first year of the new millennium.

By 1999, rocketry leaders had negotiated with ATF for five years over whether federal explosives laws granted the agency jurisdiction to regulate the hobby. In ATF's view, APCP was an explosive because it was placed on the federal explosives list by the agency's predecessor in 1971. ATF officials did not seem to care that the government never actually tested APCP as an explosive or that the agency had ignored hobby rocketry motors from 1971 to the mid-1990s. Nor did ATF accept any longer—as they stated in prior correspondence—that rocket motors were exempt from explosive regulations because they were "propellant-actuated devices."

Although ATF told high-power leaders in the 1990s that it was willing to work with them to minimize regulation of the hobby, there was never any real change in the government's attitude. In ATF's opinion, once a substance was on the explosives list, the agency was duty-bound to apply federal explosives laws to rocketry. This meant people who wanted to fly commercial high-power motors must comply with the same rules and regulations that applied to other explosives on the federal list, such as dynamite, blasting caps, or nitroglycerin. To comply with federal law, high-power flyers must obtain a low explosives users permit.

The LEUP required the filling out of a federal application and an initial fee and then a renewal fee every third year. The application also required fingerprints and included an agreement signed by the flyer that stated, among other things, that he or she was waiving their Fourth Amendment rights against unreasonable search and seizure. Failure to have an LEUP exposed flyers in possession of high-power motors to potential fines up to $10,000 and up to ten years in prison.

In 1999, Tripoli and the NAR hired lawyers well-versed in dealing with the government to try and convince ATF that APCP was not an explosive. These lawyers reminded ATF that traditionally, rocket motors were exempt from explosives rules because they were PADs. The attorneys for the agency met with the rocketry lawyers, and the two sides later exchanged letters on the subject that fall. Yet nothing changed.

In a final letter written on December 22, 1999, ATF rejected all arguments put forth by rocketry and its attorneys. APCP was on the explosives list to stay, said ATF. There would be no exemptions for high power, and the agency was not obligated to hold public hearings when it changed its position and began regulating hobby rocketry in the mid-1990s. If high-power rocketry wanted any more consideration on the subject, they would have to sue ATF.

LARGE AND DANGEROUS ROCKET SHIPS

On February 11, 2000, a lawsuit was filed by hobby rocketry at the clerk's office in the United States District Court in Washington, D.C. The plaintiffs—that is, the parties who were suing somebody—were the Tripoli Rocketry Association and the NAR. The party they were suing—the defendant—was ATF. (That same afternoon and several hundred miles to the South, Space Shuttle *Endeavor* took off from Pad 39a at Cape Kennedy to begin an eleven-day journey in the earth's orbit. It was the ninety-seventh flight of the shuttle since the first orbital flight in the spring of 1981. *Endeavor* would cover more than 4,000,000 miles on this trip before landing safely at the Cape on February 22.)

The complaint was several pages long and filled with legalese. The pleading alleged ATF was trying to "regulate high-powered sport rocketry out of existence." Specifically, the complaint said ATF was abusing its power in at least three ways:

First, the agency had inappropriately placed APCP on the federal explosives list. Second, the complaint alleged that regardless of the type of fuel used in a rocket, whether it was APCP or some other propellant, rocket motors were exempt from ATF regulation because Congress had specifically created an exception for "propellant-actuated devices" from the explosives list, and rocketry qualified for that exemption. Finally, the complaint claimed ATF's actions were the result of a new interpretation of federal explosives laws by the agency in the 1990s, and ATF had failed to provide the public with notice and opportunity to comment on such changes; therefore, ATF's newer view on high-power fuel was invalid.

Once a plaintiff files a complaint and delivers a copy to the defendant, the defendant must respond in writing within a specified period. The usual reply to a complaint is called the answer. An answer addresses the allegations in the complaint and either agrees with them or denies them. In an answer, the defendants typically agree with some things in a complaint but deny the critical matters. In the alternative, a defendant may file a motion to dismiss the complaint. A motion to dismiss challenges the sufficiency of the complaint. It asks the judge to dispose of a case now because the claims set forth have no merit. In the case of rocketry's lawsuit against the defendant ATF, the government responded to the complaint by filing a motion to dismiss. Government lawyers asked the judge to throw the case out without any further action. In support of their motion, ATF advanced several arguments.

The first and most formidable argument was based on the statute of limitations. Statutes of limitations are bright-line deadlines. They describe the time in which a plaintiff must file a lawsuit or lose their right to sue. The law requires a party to take action in a reasonable amount of time; otherwise, the case is simply dismissed, regardless of the merits of a claim. For example, in some states, people who are injured in an automobile accident have two years from the date of the injury in which to file their complaint against the driver who caused the accident. If they fail to file the complaint within two years, and this usually means two years to the day, then a court will dismiss the case at the request of the defendant regardless of the merits of the case.

ATF argued that since APCP was placed on the federal explosives list in 1971, Tripoli and the NAR had six years—the applicable statute of limitations when suing the U.S. government—to challenge that designation. They claimed anyone who wanted to challenge APCP as an explosive had until 1977 to do so; after that, they were barred by the six-year statute of limitations. It was a potent argument, and if the judge agreed with ATF, it would mean a quick end of the lawsuit filed by rocketry.

In its motion to dismiss, ATF also claimed the agency was not required to hold public hearings or provide public notice or opportunity to comment when, in the mid-1990s, it began enforcing its rules with regard to APCP in quantities that exceeded 62.5 grams of propellant. In other words, even though ATF ignored the use of APCP in rocketry for more than two decades, it was free to implement new interpretations of its rules without any further discussion with anyone. If these new interpretations harmed businesses in the hobby, said ATF, well, so what? The agency was not required to take any further action.

ATF's motion to dismiss was filed on May 11, 2000. Counsel for rocketry filed a written opposition, repeating in detail the reasons why ATF was wrong in seeking a dismissal and asking the judge to allow the case to continue forward. The next step for the parties was to obtain a hearing date so the judge could review the papers, hear oral arguments, and render a decision. However, before that happened, rocketry leaders and their lawyers held out an olive branch to the agency. They wanted to sit down again with their ATF counterparts to see if a settlement could be reached that would be acceptable to both sides. Rocketry wanted to retain some control of the case and its potential outcome before it was placed in the hands of a federal judge.

In the summer of 2000, and while Orangeburg played host to the largest LDRS in history, rocketry representatives had a face-to-face meeting with ATF bureaucrats in Washington, D.C., as well as lengthy telephone conferences between both sides. The lawyers for both sides asked the judge to temporarily "stay" the lawsuit while settlement negotiations ensued. A stay would mean the lawsuit would not move forward, albeit temporarily. The judge readily agreed—the parties were free to negotiate without any further action by the court. Unfortunately, by the end of that summer, there was no progress. ATF would not budge.

In a joint statement issued by Bruce Kelly and Mark Bundick, the memberships of each club were updated on the failure of these negotiations.

"I am disappointed that we have to go through the courts to solve [the] problem," said Bundick shortly afterward. "My previous experience of dealing with regulatory issues was much more productive. For example, our relationship with the FAA is at a very professional level. We don't always agree, but with the FAA, there is an understanding of how business is to be conducted, something that is not present with the ATF. I really never could have predicted we would have to spend all this time and money in the courts for a legal resolution."

The stay was lifted by the judge in December 2000. The future of high-power rocketry was now in the hands of a federal court judge.

When a lawsuit is filed in federal court, it is randomly assigned to one of several judges at the courthouse where the complaint is lodged. Each judge has their own courtroom and their own calendar. They take a new case and make it part of their calendar, usually on a first-come, first-served basis. New cases go all the way to the back of the line.

The lawsuit filed by rocketry against ATF in early 2000 was initially assigned to Federal District Court Judge Richard W. Roberts. Roberts had previously worked as an assistant U.S. attorney and had been chief of the Criminal Section of the Civil Rights Division of the Department of Justice. He was appointed to the bench in 1998. It was Judge Roberts who granted and then dissolved the stay in late 2000, and it was Roberts who ordered the dismissal hearing back on calendar in early 2001.

That was when the case stalled. Judge Roberts had too many other cases on his calendar, most of which were older than the lawsuit brought by Tripoli and the NAR. For the rest of the year, nothing significant happened in the case. Then in late 2001, the court administrator decided that this case, along with about one hundred other cases, be reassigned to another judge. That judge was recently appointed by Pres. George W. Bush. His name was Reginald Bennett Walton.

Judge Walton was born in February 1949. His father worked in the steel town of Donora, Pennsylvania. Walton was accepted to West Virginia State College on a football scholarship and earned his undergraduate degree in 1971. He earned a law degree from the American University, Washington College of the Law, in 1974. After practicing as a lawyer for several years, he was appointed by Pres. Ronald Reagan to the superior court in Washington, D.C., where he served for nearly twenty years. He was appointed by President Bush to the federal bench in Washington, D.C., on October 29, 2001.

Over the years, Walton earned a reputation as a no-nonsense judge who was tough on crime. This tough, no-nonsense temperament occasionally revealed itself outside the courtroom too. One night, after he had been appointed as a federal judge, Walton was driving his wife and teenage daughter to the airport for a family vacation, when he noticed someone attacking a cab driver on the side of the road. Walton immediately pulled his car over to the curb, jumped out of his car, tackled the assailant, and held him down on the ground until the police arrived.

In addition to the sincere thanks of the grateful taxi driver, a police spokesperson said of the incident, "God bless Judge Walton. I surely wouldn't want to mess with him."

In the spring of 2002, the attorneys for the government and high-power rocketry squared off against each other in a courtroom for the first time, in two hearings, before Judge Walton. The first hearing addressed rocketry's request for a court order directing ATF to immediately cease its regulation of high-power rocketry. This emergency relief was called an injunction. It would not be an end to the case; Tripoli and the NAR were simply asking the judge to rule that until the lawsuit was decided on its merits, ATF needed to back off its regulation of hobby rocketry.

Attorneys Joseph "Joe" R. Egan and Martin "Marty" Malsch appeared in person on behalf of rocketry. Malsch and fellow lawyer

John Lawrence had prepared the written briefs that were submitted to the judge prior to the hearing. Egan would be the voice of rocketry in court. He would argue the merits of the injunction. Egan, forty-seven, graduated from the Massachusetts Institute of Technology with degrees in physics, nuclear engineering and technology, and policy. He attended Columbia Law School after which he practiced nuclear law with firms in New York and Washington, D.C. In 1994, he started his own law firm devoted exclusively to nuclear-related law issues.

Egan was also a lifelong member of the NAR. In the early 1990s, he helped the NAR traverse the mine field of administrative law working on issues involving the FAA and rocketry. In 1999, Tripoli and the NAR hired Egan, Malsch, and Lawrence to represent hobby rocketry in their negotiations with ATF. Their firm would now represent rocketry in federal court.

The scope of the preliminary injunction hearing in 2002 was very narrow. It was just to determine whether rocketry was entitled to emergency relief, requiring ATF to ease off in its regulations while the suit was pending. Joe Egan took this first opportunity to introduce Judge Walton to the underlying merits of the entire case:

> Your honor, my name is Joe Egan. I am a physicist and an engineer by background and a lawyer representing the NAR and the Tripoli Rocketry Association. I would like to introduce my clients. The presidents of both organizations are here—Mr. Mark Bundick, the president of the NAR, and Mr. Bruce Kelly, the president of [the] Tripoli Rocketry Association.
>
> Rocketeers number tens of millions in this country. [These] organizations are basically umbrella organizations that represent the serious hobbyists that also propagate the hobby through the school system and the educational system in this country. Rocketry has been a very important scientific tool in this country since 1957, and many of the astronauts and engineers in this country, especially aerospace engineers—myself included—started out as avid hobbyists. I have been a member of the NAR since I was nine years old. I fly these things with my kids today. My kids are learning about science from these rockets.

Egan then gave Judge Walton a brief history on the events that led to the suit, including ATF's actions in the 1990s when the agency began to extend its regulations to rocketry. He explained the error made by the government when it classified APCP as an explosive in 1971 and how ATF had for years impliedly agreed that hobby rockets were propellant-actuated devices and were therefore exempt from explosives regulations anyway. Egan also challenged the legitimacy of ATF's most recent actions, requiring hobbyists to obtain a LEUP if they wanted to purchase reload packages with more than 62.5 grams of propellant for their rocket:

> [W]hat we have here is a situation where the ATF, I think, has a genuine concern about rocketry. I think they are concerned that somebody would pull up to the White House and shoot one of these things at the White House or similar things that they have actually said to us. And what they are really trying to do here is *regulate rockets*. And they don't have any statutory jurisdiction to regulate rockets . . . Rockets are regulated by many other agencies . . . But ATF, in their concern about rockets, would love to regulate rockets, and so they are trying to bootstrap and jawbone their way into regulating them by regulating the motors. We say that's akin to regulating the size of a sport utility vehicle by telling anyone if they buy more than two gallons of gas, they need a permit. That is essentially what the ATF is trying to do with these rockets.

Rocketry was already heavily regulated, continued Egan, by agencies such as the FAA, DOT, and CPSC. This was in addition to rocketry safety rules approved by the NFPA, he added. Egan argued that allowing the ATF to continue its actions against rocket users was causing irreparable daily harm to the hobby, resulting not only in a decline in membership in Tripoli and the NAR but killing sales for motor vendors and other commercial dealers too. Surely, said Egan to Judge Walton, allowing flyers to continue launching rockets without permits from the ATF, which had been the case in rocketry for many years, would not pose any danger to the public during the pendency of the lawsuit.

The counsel for ATF was Assistant United States Attorney Jane Lyons. "The regulation of explosives is a business of line drawing," argued Lyons. "And the agency is in the best position to do that. They regulate materials to guard against their misuse. The ATF does not have, obviously, as much concern with Mr. Egan and his children learning about science as they do about the possibility of obtaining these materials for improper use. And that's why they draw the lines that they do. The bottom line, with respect to their request for injunctive relief, is there is no emergency here. Nothing has really changed."

Lyons told Judge Walton that ATF's requirement of an LEUP applied only to persons who were traveling with high-power motors between states in interstate commerce. Flyers were free to launch high-power rockets and possess motors in their own jurisdiction, she said.

In rebuttal, Egan argued that the distinction between interstate and local commerce was a false one since dealers traveled among states, and this made sales in almost every state an interstate transaction, especially since APCP was only manufactured in Texas and Nevada.

Although he let both sides get into some discussion of the merits of the case, Walton pulled them back to the only issue that was to be decided that day. Since the ATF's regulation of the hobby had not changed since the mid-1990s, why should he issue an order for an injunction to alter things right now? Where was the irreparable harm to rocketry, right now, if the status quo was simply maintained? Clearly, Judge Walton was not convinced by the anecdotal evidence of dropping motor sales or declining membership numbers. He would need to see more before emergency relief in the form of an injunction was warranted.

A lawsuit is in some respects a war between opposing parties. And in every war, some battles are won and some are lost. In this first battle between ATF and rocketry, the government emerged the winner. Judge Walton denied rocketry's request for an injunction. The status quo would remain, he ruled. Until the lawsuit was over, APCP would remain on the explosives list; rocketry members must comply with ATF rules for LEUPs and storage requirements, or they could refuse to comply and take their chances. It was still as simple as that.

For Joe Egan and his legal team, and for Tripoli and the NAR, it was time to move on to the next battle, and that contest was right around the corner.

———◆•◉•◆———

With the request for their injunction denied, rocketry's legal team shifted to the next pressing issue in the case: surviving ATF's motion to dismiss the entire case. The hearing on the injunction was on Tuesday, April 30. The next day, Wednesday, May 1, 2002, everyone was back in court for the dismissal hearing. This time ATF was on the offensive. ATF's primary argument in its motion to dismiss was rocketry enthusiasts had waited too long to challenge the inclusion of APCP on the federal explosives list. APCP had first appeared in the 1971 list, Attorney Lyons told the judge, and since the statute of limitations was six years, the complaint filed by rocketry in February 2000 was many years too late. Judge Walton appeared to agree that the six-year statute of limitations applied. "Since this action was initiated over thirty years after APCP was first listed in the *Federal Register* as an explosive, the timeliness of the filing of this action is obviously suspect," he said.

The lawyers for rocketry were prepared. They had another way of looking at the statute; they agreed the six-year statute applied to their case. But ATF's trigger date for that statute to start was wrong, they said.

Citing a line of court cases that supported him in this setting, Attorney Egan argued the clock involving rocketry did not begin to run in 1971. Rather, the six-year time began in either 1994 or 1999, when ATF indicated, for the first time ever, that it was going to apply its explosives rules to high-power rocketry. These actions, argued Egan, "made a sufficiently significant extension of [ATF's] explosives regulations by concluding that they were applicable to the high-powered sport rocketry hobby, thus triggering a new statutory review period for judicial review of the interpretation."

In other words, the 1971 date was not the important date for the six-year statute. The critical dates were when ATF advised high-power rocketry that it intended to apply its explosive list to APCP used in the hobby. Judge Walton agreed with Egan, holding that the plaintiffs were entitled to the alternative, more recent trigger date based on when ATF first applied that list to rocketry. In so ruling, Judge Walton handed rocketry its first victory in their case against the federal government:

> ATF wrote two letters to AeroTech in 1994 and another to plaintiffs' counsel on September 7, 1999, in response to inquiries regarding its regulation of APCP and certain high-powered sports rockets that use APCP as their fuel source . . . In those letters, ATF reiterated that APCP was an explosive and stated that particular rocket motors that used APCP would also be regulated by the ATF . . . Plaintiffs alleged that the 1994 and 1999 statements were the first indications that ATF intended to regulate high-powered sport rocketry. As a result of ATF's statements in its September 7, 1999, letter,

plaintiffs directly challenged ATF's listing of APCP as an explosive and [ATF's] indication that it intended to regulate high-powered sports rockets that used motors fueled by 62.5 or more grams of APCP. In response, on November 24, 1999, ATF's representatives orally informed plaintiffs that ATF rejected plaintiffs' contentions that it could not regulate APCP and the specified rocket motors that are propelled by APCP. Then on December 22, 2000, ATF reiterated its positions in a letter that was sent to plaintiffs' counsel.

In count 1 of the amended complaint, plaintiffs plainly alleged that on several occasions since 1994, ATF has exceeded its statutory authority by indicating that it will regulate high-powered sports [rockets] that use rocket motors that contain 62.5 grams or greater of propellant. Accordingly, within six years of when plaintiffs commenced this action, ATF, for the first time, indicated its intent to apply its regulation of APCP to plaintiffs. Therefore, this action having been initiated within six years of those statements, count 1 is not subject to dismissal on timeliness grounds.

This important ruling meant the lawsuit brought by Tripoli and the NAR would survive. Rocketry would get its chance to try and prove APCP was not an explosive. Judge Walton also sided with rocketry by finding that ATF's new actions vis-à-vis rocketry beginning in 1994 constituted "rulemaking," which required the agency to have a public comment period before any new rules could be given effect. Said the judge, "Because §847 does not exempt certain types of rules from its notice and comment command, these unqualified procedural prerequisites had to be employed before the rule was adopted. There being no claim that notice and opportunity for comment were afforded, ATF's motion to dismiss count 4 of the amended complaint must be denied."

During the summer of 2002, and while LDRS-21 went forward in the Texas Panhandle, rocketry's legal team went back to work, marshalling their evidence in preparation for a showdown on the merits of their case against the government.

Rocketry survived the first phase of the lawsuit. It was time to move into the next phase of the legal case, also known as "discovery."

Discovery is a term that describes a period of many months—even years in complex cases—where each side investigates the merits of the other side's claims. Experts are hired to review documents and express expert opinions. Affidavits or depositions are taken of witnesses. The parties exchange information regarding the specifics of their claims or their defenses. The underlying purpose of discovery is twofold: it helps parties prepare for trial, and it prevents undue surprise. When discovery is over, the parties have a pretty good idea where they stand. They know what the evidence will be, and they can estimate what their chances are at trial. They can then make reasoned decisions as to whether a case should be settled or taken to trial. Modern legal discovery practice removes the old-fashioned "trial by ambush."

In some cases, parties to a lawsuit don't want to wait for trial. They believe that the facts of the case are so clear that any judge presented with those facts can reach but one decision: to rule in their favor. For these parties, the legal system has invented the "motion for summary judgment." A motion for summary judgment is a legal tool used to shortcut a case. It is often called a minitrial in the sense that it requires a party to present its best evidence to the judge, and then in reply, the other side has to present information that disputes such evidence. If a party fails to refute critical evidence offered by the other side, then the court grants summary judgment. A trial is not necessary.

On July 22, 2002, Judge Walton ordered the parties to prepare for summary judgment on the main issues in the case. This meant that he expected both sides to present their best evidence and legal briefs to the court for his consideration. He directed both sides to have all their motions, briefs, evidence, and opposition filed between August 20 and October 16, 2002.

The issues for summary judgment were simple for both rocketry and ATF.

For Tripoli and the NAR, it was vital that APCP be removed from the federal explosives list. They would have to provide admissible evidence to convince Walton APCP was not an explosive. If rocketry could present evidence that established this as an *undisputed fact*, then Judge Walton would grant summary judgment, and APCP would be off the list. If rocketry lost this issue, then a secondary argument would be that rocket motors were exempt from the explosives list because they qualified as PADs. This was still a good argument, but it was rapidly losing strength as ATF had recently decided to embark on new rulemaking—with a public comment period—to establish that PADs were, in fact, much more limited and did not cover high-power rocket motors.

ATF's course was also clear: they filed a motion for summary judgment to establish that APCP was an explosive, that it was properly placed on the explosives list in 1971, and that high-power rocket motor users were therefore subject to ATF regulation in the form of an LEUP.

To support its position, ATF argued that when Congress delegated its authority to create the federal explosives list, it defined "explosives" in at least two ways. One definition was if a chemical compound detonates—if it clearly explodes—then it can be placed on the explosives list. The other way to get on the list, argued lawyers for the ATF, was if a substance "deflagrates," that is, it burns very rapidly but does not detonate. APCP deflagrates, they said, citing several explosives publications that *could* be interpreted to support the decision to include APCP on the list. Since the ATF was the primary government agency responsible for maintaining the list, said counsel, ATF's decision on this issue should be given deference over claims made by high-power rocketry.

The lawyers for Tripoli and the NAR had a very different view of the explosives definitions used by ATF. First, they challenged the definition of "explosive" that was being used by the agency. The proper definition was found elsewhere in the law enacted by Congress, they said. An explosive was defined as "any chemical compound, mixture, or device, the primary or common purpose of which is to function by explosion." The primary or common purpose of APCP is not to function by explosion, argued counsel for rocketry. Therefore, it should not be on the list.

In the alternative, rocketry told Judge Walton even if ATF's definition of deflagration was used, there was no evidence that APCP does, in fact, deflagrate. "Although ATF points to various documents in the administrative record in an attempt to support its conclusion, infra, ATF's conclusion that APCP deflagrates when ignited is not supported by any tests or other forms of analysis presented in the administrative record. Rather, a detailed review of the documents relied upon by ATF reveals that, in fact, APCP does not deflagrate as suggested by ATF but instead merely 'burns' when ignited [i.e., a much slower process]. Thus, ATF should not have classified APCP as an explosive."

The lawyers for rocketry also urged the court to reject any claim that ATF's decision on APCP should be given special deference since the agency had not made a proper determination in the first place.

By the end of 2002, all the briefing papers and evidence had been submitted to Judge Walton. It had been nearly three years already since Tripoli and the NAR had filed suit. It was time for the court to reach a decision. Then the rest of Judge Walton's courtroom calendar caught up to him. Weeks passed and then months with no decision from the court. And the case of rocketry versus the U.S. government returned to limbo once again.

35

Rocketry vs. the U.S. Government

Part 2

SB-724 and Other Setbacks

By mid-2003, there was still no decision from Judge Walton on the motions for summary judgment filed by rocketry and ATF. Another LDRS would soon come and go—LDRS-22 in Argonia covered by the Discovery Channel—and Tripoli would elect its first new president in almost a decade, Dick Embry. Yet there was no news from federal court in Washington, D.C. Around the country, many Tripoli and NAR members were filling out applications for their LEUP—just in case—and were ordering ATF's orange book, which outlined the responsibilities that went along with holding a low explosives users permit. For many in the hobby, regulation by ATF seemed a foregone conclusion. The lawsuit was just delaying the inevitable.

However, some flyers were taking their chances. Neither Tripoli nor the NAR was enforcing rules that applied to an LEUP. That was ATF's job, said the boards of both organizations. Some flyers ignored the federal rules, preferring to stay off the government's radar. They bought their motors from dealers who did not require an LEUP.

Meanwhile, Tripoli and NAR leaders decided to try another approach with ATF. That spring, high-power leaders would look to the legislature for help. Perhaps it was time for an end run around the courts.

On April 1, 2003, Republican senator John Enzi of Wyoming stood before the U.S. Senate asking for support for a bill he was sponsoring to ease regulations imposed on rocketry by ATF. Enzi said to his congressional colleagues in the Senate chambers,

> Start counting backwards from 10 to 0: 10, 9, 8, 7—and depending on the context, people will be instantly reminded of their youth, sitting in front of a dimly lit television, watching a rocket take flight as we began the study of space flight and space travel. We were much younger then, and all around me kids from all over the state and all around the country were excited and fascinated by the new age of rocketry and, later, space travel.
>
> When Russia launched its *Sputnik*, it created a sensation, and their success, spurred on by the climate of the Cold War, challenged us in the United States to reach for the skies.
>
> Wyoming isn't called the pioneer state for nothing, and so my classmates and I were determined we would do everything we could to learn about this new branch of science and involve ourselves in the race for space. It was not too long after that President Kennedy issued a challenge to the nation to land a man on the moon and return him safely to earth.
>
> What seemed to be against all odds soon became a reality when Neil Armstrong walked on the moon, taking a small step for man and a giant leap for mankind.
>
> Even today, those of us who saw those events firsthand on the television will never forget what a miracle it was. It fired our imaginations as it taught the nation a powerful lesson: if we can make this impossible dream come true for the nation, of what more are we capable if we dare to try? Perhaps that lesson is what made our nation what it is

today and why we have continued to defy the odds of what is possible for us as a nation and even for each of us as individuals.

Then came September 11, and we, as a nation, faced another challenge. The call for increased security that resulted from those cowardly and cruel attacks has had some unforeseen consequences, however.

Enzi explained that over the past several years—and now even more so with proposed provisions of the Homeland Security Act—ATF was overregulating hobby rocketry through extensive licensing for the purchase and use of APCP. Enzi asked lawmakers to join him in passing Senate Bill 724 which would provide rocketry an exemption from explosives laws for APCP, much like the exemptions created for users of black powder and antique firearms. Enzi continued. "My concern about the impact of these regulations, and the process necessary to obtain permits, and the bureaucracy that would be necessary to do that, and to fulfill the requirements for background checks is that it will certainly slow the participation of our young adults in studying rockets and pursuing their dreams of space travel."

"John Wickman kicked this off," recalled then-Tripoli vice president Ken Good. "He was friends with Senator Enzi, and he thought we could resolve all of these problems with the ATF through Senator Enzi in the United States Senate."

Wickman was well-known as the founder of CP Technologies. Since the mid-1990s, he had been a proponent of homemade rocket motors and was the author of one of high-power's early books on the subject titled *How to Make Amateur Rockets*. In his book and related videotapes, Wickman not only provided education on building rockets, but he also detailed formulas and advice for ammonium nitrate and AP-based propellant for high-power rocket motors. Indeed, education was one of the touchstones of Senator Enzi's speech on the Senate floor:

> As I learned from my own experience—and I was one of those rocket people back at the time of *Sputnik*—the study of rockets had a ripple effect throughout my own education. It taught me a lot about math, when we had to calculate the amount of fuel we needed and the rate at which the rocket would travel at speed-calculating heights, figuring trajectories,

figuring the amount of Gs that would be on the passenger. It taught us about the study of weather as we would examine reports about our own launch date and temperature and cloud cover that would affect our ability to observe the launch and weather balloons for measuring the winds aloft to better tell where it would go and to make calculations about how high we were able to fly on any particular day.

Together with NAR president Bundick, Tripoli leaders threw their support behind Senator Enzi's bill as a real alternative to the court battle in Washington, a battle that remained in limbo. One of the first things they did was hire John Kyte, a Washington, D.C., lobbyist, to help enlist more sponsors for SB-724. "John was working directly to try and get more senators behind this effort," explained Good. "We got some sponsors, it got out of committee, and we made some real progress. We did a lot of letter writing to other senators too."

As Enzi's bill picked up momentum in the Senate, on June 10, 2003, the Department of Justice sent a letter to the Senate Judiciary Committee, critiquing Senator Enzi's bill and rejecting the notion of offering any exemption for APCP to rocketry. In that letter, a Justice Department representative called Enzi's bill well-intentioned but misguided—and even dangerous. "In formulating the 62.4 gram exemption, ATF determined that this threshold afforded a reasonable balance between the need to prevent terrorists and other criminals from acquiring explosives and the legitimate desire of hobbyists to have easy access to explosives for lawful use," they said.

What happened next was comedy.

On July 30, 2003, and as SB-724 continued to gain support, two senators from the Northeast called a press conference to denounce Enzi's bill: Democrat Charles "Chuck" E. Schumer of New York and fellow Democrat Frank R. Lautenberg of New Jersey. In a made-to-scare-the-press photo opportunity, both politicians stood grim-faced alongside a 7-foot-tall high-power rocket. The senators told reporters that Enzi's bill would allow terrorists to gather explosive materials for use in bombings.

"Why anyone in the post-9/11 world would think making it easier to get bomb-making material is a good idea is beyond me," whined Schumer, who accepted without question ATF's definition that APCP was an explosive. "In essence, the bill would create a new loophole which would allow terrorists to stockpile explosives."

"Model rocketry is a hobby enjoyed by many," echoed Lautenberg, who, along with Schumer, likely never touched a hobby rocket in his life. "But it's not worth compromising our safety. Inconvenience is something that we all have to put up with to ensure our safety."

Schumer went even further, suggesting hobby rockets could be used as vehicles to deliver nuclear and biological weapons. "The bottom line is that some of these people are building not rockets. They are missiles," he said.

President Bundick rejected Schumer's claims, telling the *New York Times*, "These rockets can't be used as weapons. They don't have guidance systems, they don't have enough payload, [and] they don't have enough range, so to suggest that these materials could be used by terrorists is just untrue. And you cannot turn this stuff into a bomb."

Senator Enzi also responded to the press attack by Schumer and Lautenberg, telling anyone who would listen that the two democrats were trying "to squash efforts to preserve a constructive, educational, and important hobby enjoyed by millions of Americans." The opposition to derail the proposed bill "doesn't make Americans that much safer," argued Enzi. "But it does make us more fearful and less free."

Unfortunately, the damage was done. The anti-high-power hyperbole by the two democrats and their press show would not be undone.

"Senator Enzi was mad as a wet hen when the Justice Department letter went out," recalled Ken Good, who said Enzi and other senators were already working on a compromise bill after the Justice Department had privately voiced its concerns. "And when Schumer and Lautenberg saw an opportunity to embarrass Republicans with this dangerous thing they were doing and they had this photo-op with this big rocket, saying, 'Look, these Republicans want to deregulate this dangerous thing,' well, the bill died right then and there. It was over. Even Senator Enzi realized it was done."

Rocketry's attempt at a legislative solution to overregulation was dead in the water. Several more months would pass before there was any action by the court in Washington.

Judge Walton announced his decision regarding the competing motions for summary judgment on March 19, 2004. He ruled in favor of ATF. APCP belonged on the federal explosives list, he said.

Walton acknowledged APCP did not detonate or explode. The important inquiry to him was whether Congress intended the explosives list to include chemicals that "deflagrate" and whether APCP did, in fact, deflagrate. In both regards, he found the arguments put forth by the government lawyers convincing.

"A court should review scientific judgments of an agency not as the chemist, biologist or statistician that we are qualified neither by training nor experience to be," he wrote, "but as a reviewing court exercising our narrowly defined duty of holding agencies to certain minimal standards of rationality." Walton said that based on the evidence before him, he was satisfied the definition of explosives urged by ATF was the correct one and that the agency had made a determination, supported by expert opinions in the record that APCP did, in fact, deflagrate.

It was a serious setback for rocketry, and in some ways, it looked like the end of the lawsuit filed by rocketry in 2000. Yet there was a small silver lining in the judge's twenty-three-page written decision. He disagreed with ATF's argument that rocket motors were not propellant-actuated devices. He also said that since ATF's decision on PADs was made without notice-and-comment rules as required by law, it was invalid. So the PAD exemption for rocketry remained alive—at least for the time being. Judge Walton observed ATF was now engaged in rulemaking and a comment period to evaluate the exempt status of motors containing more than 62.5 grams of propellant, so he was delaying final judgment on that issue.

This meant rocketry's case against the government was still alive but just barely. Tripoli and NAR leaders put on their best face following Walton's decision in favor of ATF. Presidents Embry and Bundick reported they were encouraged by discussions with their lawyers who were offering "strategic directions" on what to do next. Tripoli vice president Good agreed. "So what does all this mean, and where does this place us right now? Obviously, viewed in a general way, we won a count, we lost a count, and two counts are still 'on hold for now.' All members clearly want to know what our 'net position' is as a result of this and under what conditions they should expect to pursue high-power rocketry as we move ahead. While it is tempting to try to immediately set forth a more significant and detailed interpretation of the case rulings, at this moment, it would be both premature and tactically unwise to do so."

Tripoli and the NAR had already spent three and half years and over $200,000 in legal fees in their fight against ATF. Yet in the wake of Judge

Walton's decision that spring, was all this money simply delaying the inevitable? Was it really a good idea to take on the federal government simply because ATF wanted flyers to obtain an LEUP for their rocket motors? Was it time for the leadership of Tripoli and the NAR to concede defeat and move on? What were the alternatives?

"This is the close of one inning," wrote Good in an editorial in the *Tripoli Report* after the judge's ruling. "Everyone needs to understand that we are not at the end. Maybe Churchill's quote about the victory at El Alamein is most apt: 'It is not the end, or even the beginning of the end, but it is the end of the beginning.'"

Several weeks after Judge Walton's ruling against rocketry, and two days before the historic flight of the *GoFast* rocket at Black Rock, a team of East Coast flyers launched the then-heaviest high-power rocket in history from a farm on the Maryland seashore.

The *Liberty Project* was put together by a team of nearly eighty people, most of whom were members of the Maryland Delaware Rocketry Association, also known as MDRA. The rocket replaced Chuck Sackett's *Project 463* as the biggest amateur rocket ever launched. *Project 463* was a two-stage vehicle that stood 43 feet tall and tipped the scales at 1,200 pounds. In flight, the rocket's recovery equipment failed to work, leading to the destruction of *Project 463* when the rocket smashed into the desert floor at Black Rock in 1995.

The *Liberty Project* began in the summer of 2000 and proceeded slowly over the next three years. As the vehicle grew in size and complexity, additional team members were recruited, and various motor formulas were reviewed and analyzed, including fuel profiles created by Darren Wright, Jeff Taylor, and Jim Mitchell. After a detailed analysis, five motors were chosen to power the massive rocket—a central P10,500, two O8200s, and two more N5400s—giving the rocket a total impulse of 112,000 Newton seconds, enough high power to launch this 1,366-pound rocket into high-power rocketry lore.

"I considered going with a single big motor for the project but knew that there would be little or no future use for such a motor after the *Liberty* flew," explained Neil McGilvray, one of the team leaders for the rocket. "The decision was then made to make use of the previous Darren Wright and Jeff Taylor successes with the 150 mm P and 114 mm O motors." Each motor would be a bates-grain type, including the two N motors that were based on Jim Mitchell designs.

The recovery and electronics gear for the big rocket was complex, starting with the 105-foot-diameter military surplus parachute that would return the payload section back to Earth. "The parachute came packed in a 24-inch square box and weighed 114 pounds," said McGilvray. "It was a beast to even try to move, let alone what it might be to pack. At 114 pounds, it would require the power of an M2500 motor just to launch the parachute!"

The plan was to split the rocket in half at apogee—estimated to be around 4,000 feet—and then fire off the nose cone to release the main parachute at 2,600 feet. The rocket would hold seven parachutes and carry multiple onboard altimeters and remote control units—built by ARTS, black sky, and Missile Works—to activate more than 280 grams of black powder ejection charges designed to separate the rocket into several different sections for recovery.

One of the most important aspects of the *Liberty* launch would be to make sure that every motor in the rocket worked when the ignition button was pushed. To help light everything on time, Darren Wright introduced the project team to thermite igniters, a relatively new development in high power but one that would become commonplace in the years ahead.

"Initial tests on these igniters were discouraging," said Wright, who worked on the concept with Jeff Taylor. "But subsequent discussions with Gary Rosenfield at LDRS in 2003 led me to make some significant changes in the size of the igniter charge." Wright also spoke to a chemist, John Gustavsen, who suggested that the composition of the thermite igniters be changed from iron to copper. "These changes resulted in a much more successful igniter, which has proven itself countless times since," said Wright. Indeed, within a few years, copper thermite igniters would become the standard technique in large rocket cluster ignition.

"The discovery and subsequent sharing of the knowledge of Darren Wright was probably one of the most important aspects of this project," said McGilvray, who emphasized the vital importance of all five motors lighting simultaneously on the pad. A failure of any of these motors to fire on time could lead to an unstable, underpowered rocket, and disaster. "The thermite mixture used in these igniters not only ensured a quick light. They guaranteed one."

LARGE AND DANGEROUS ROCKET SHIPS

Prior to the flight of the *Liberty Project*, the largest high-power rocket attempt east of the Mississippi was likely Mark Lloyd and Kelly Mercer's impressive *Bigger Dawg*, a P-powered single-stage rocket that weighed nearly 650 pounds on the pad. On an April morning in 2002 at the Whitakers launch field in North Carolina, *Bigger Dawg* sustained a massive motor CATO shortly after takeoff. Much of the rocket was destroyed.

The launch of the *Liberty Project* two years later was scheduled for Saturday, May 15, 2004, at the Higgs Farm in Maryland. A custom-built 30-foot-tall launch gantry was built by MDRA flyer Fred Schumacher, who also lent his professional crane operator services to help lift the 24-foot-tall rocket in place on the launch pad. With the rocket standing vertical and the motors and igniters all in place, the countdown for launch was ready to commence.

"A project like this requires hundreds of hours of dedication and thousands of dollars to pull off," noted McGilvray. "Typically, this would be untouchable for most clubs for lots of reasons. We are very fortunate that the MDRA was able to put all of the pieces of the puzzle together and make the dream a reality. Now the question was, 'would it work?' We had reached that point in every project where there is only one way to find out—you have to push the button."

At ignition, the 1,366-pound rocket roared into the sky, firing all five motors. "The rocket left an enormous cloud of smoke and dust at the pad," said one observer. "The flame was over 40 feet long and changed color three times until it disappeared in a huge trail of gray-brown smoke."

As the rocket continued its climb, the N and O motors burned out, while the P motor continued to push the vehicle to an apogee of around 3,000 feet. The ejection charges started firing, including four black powder charges totaling 150 grams that blasted away simultaneously—an unplanned event. The 105-foot-diameter parachute was out of the payload bay, but the large booster section of the rocket was falling free without any parachute at all.

What happened? "We would have to wait to find out as we watched the 550 pounds of booster come crashing back into the field," said McGilvray. "With a sickening hollow thud, two of the fins were instantaneously smashed, and the body tube was broken in half."

Postflight inspection revealed a 12,000-pound rated nylon webbing line snapped because of heat from the ejection charges, setting the booster free of its recovery gear. The nose cone and payload bay, which were supposed to come down under separate parachutes, drifted away together under the 105-foot mega chute, landing in dense forest a mile away. It took several days and many hours of work to retrieve the parts and pieces dangling from the tall trees near the Higgs Farm.

The launch and recovery of the *Liberty Project* was far from a total success. But it would reign for the next five years as the heaviest high-power rocket ever flown. And although recovery of the vehicle from apogee was plagued with problems, the lessons learned in launching this massive rocket would prove invaluable later. Five years from now, another flyer would enlist MDRA's help to launch an even larger rocket at the Higgs Farm—a one-tenth scale *Saturn V*—and the experience gained through the launch of the *Liberty Project* would help turn that flight into a near-perfect success.

36

Geneseo

While Tripoli and NAR leadership weighed the possibility of appealing Judge Walton's summary judgment ruling against rocketry, LDRS-23 was held on July 1–6, 2004, in the small farming community of Geneseo, New York.

The flying range at LDRS-23 was one of the most picturesque in LDRS history, a four-square-mile hay farm resting in a shallow bowl-shaped river valley that was lush and welcoming. Geneseo is home to the State University of New York, and the campus was easy to see from the range—its low buildings set against a small hill a couple of miles east of the range head. To the west, and immediately adjacent to the farm, was a rural airfield, home of the 1941 Historical Aircraft Group. A grass runway ran alongside the range, and throughout LDRS, aircraft arrived and departed from a 1940s-era hangar at one end of the field. These planes were flying museum pieces, like the single-engine T-6 trainer that buzzed the field all weekend long or the four-engine Boeing B-17 Flying Fortress—one of the great relics of the World War II—that took to the air periodically during the launch. Painted on the forward fuselage of theB-17 were the words *Memphis Belle*. One afternoon that weekend, both the T-6 and B-17 passed together over the rocket range, barely a hundred feet above the ground.

This was the first of several LDRS events held in New York. It was also the first time in four years Tripoli's national launch was back east again, the last time being at Orangeburg in 2000. The launch was hosted by the Buffalo Rocketry Society. Day-to-day range responsibility was handled by several prefectures, including Tripoli Pittsburgh, the Maryland Delaware Rocketry Association, Tripoli Southern Ontario, and the Long Island Advance Rocketry Society.

Flyers began arriving on Wednesday and Thursday while farm tractors and balers were still harvesting hay on the recovery field.

"Believe it or not, on Monday, the hay was this tall," said launch director Lloyd Wood as he held his arm out shoulder high. It was touch-and-go for a while as to whether the field would be ready by the first official day of the launch. "I was thinking of getting my family into a car and heading to Canada," he said jokingly. But everything was in order when the first day of flying rolled around on Thursday.

The waiver for LDRS-23 was lower than most contemporary LDRS launches, at 8,000 feet, with windows for slightly higher altitudes from time to time. A few flyers grumbled about the low waiver. Yet the waiver did little to deter more than one thousand enthusiastic East Coast and Canadian flyers from launching hundreds of rockets under ideal weather conditions for six days. Before it was over, more than 1,200 rockets would take to the air. There were no mega projects at LDRS-23. But there were plenty of big-ticket flights.

One of the first flights was by Dennis Lappert with his 57-pound Public Missiles *Intruder* on an Animal Motor Works M1350. Lappert's rocket had a good flight and recovery and was among dozens of L- and M-powered projects using a fully mechanized trailer pad provided by the Maryland Delaware Rocketry Association. The big blue trailer carried a unistrut rail mounted to a tubular radio tower that lay flat on the trailer for easy loading of heavier projects. Once a rocket was firmly in place on the rail, the push of a button raised the entire assembly, rocket and all, into vertical position for launch.

Phillip Hathaway made the trek to LDRS from Maine to achieve Level 3 with his gold-painted *Phoenix* on an M1419, while John Russo had success with his *Strato Cobra* that cleared 7,000 feet on an AMW M1850. Rick Dunseith was part of a large contingent of Canadian-based Tripoli flyers who launched rockets all week at LDRS. Dunseith launched his 80-pound scratch-built *Paralyzer* on a Cesaroni N2500 motor—the first flight of this soon-to-be-popular motor anywhere in the United States—for a successful Level 3 certification. Maryland flyer Dan Michael launched a fine-looking 7.5-inch-diameter *Patriot Missile* on an M1315.

There was a plethora of *Nike Smokes* at Geneseo; they came in all sizes, from model rockets to large scratch-built projects. Blair DuPont achieved Level 3 using a Smokin' Rockets 7.5-inch *Smoke* on an M1315. Dan Lord launched a *Nike Smoke* that was likely the most powerful rocket at LDRS-23. The three-quarter-scale rocket held a pair of L777 motors, two more M1350s, and a central N4000. The rocket was dedicated to the lost crew of the Space Shuttle *Columbia* and was signed by the *Columbia's* very first pilot, Jack Lousma, who was aboard the shuttle for its maiden voyage in 1981. The inscription read, "To Dan: Wishing You Blue Skies and Happy Landings! Jack Lousma, Cmdr. Columbia III." There was a ten-count for the rocket's launch, and when the countdown reached zero, the crowd held its breath in anticipation. Nothing happened. Then one of the L motors came to life. Yet it was too little thrust to budge the heavy rocket. The L burned away right there on the pad. There was no damage to the rocket. Moments later, the second L came to life, with exactly the same result: lots of fire and smoke but no liftoff.

Lord and his team waited several minutes, and when nothing else came alive, they cautiously approached the pad, pulled out the two unburned M1350 motors, and placed a new igniter in the central N motor. At ignition, the big N motor came to life instantly, carrying the *Smoke* high into the air for a perfect flight and recovery.

Another big *Nike* was Jerry O'Sullivan's 11.75-inch-diameter rocket powered by a trio of Loki Research M2000 demonstrator motors. The rocket carried video cameras and all sorts of electronics and weighed 150 pounds fully loaded. O'Sullivan's *Smoke* ripped off the pad on Friday morning, reaching 6,000 feet above the hayfield. At apogee, the rocket unexpectedly deployed its large, 24-foot-diameter main parachute. The missile then hung in the air for at least 10 minutes, drifting slowly off the rocketry range, over another farm, past a local highway, and then up over a hill into dense forest 2 miles away. The rocket seemed recoverable—at least to all eyes on the field—but when O'Sullivan and his crew made their way to the forest, the *Smoke* was nowhere to be found. Over the next few days, repeated searches failed to turn up the expensive rocket, which was presumed lost.

The Discovery Channel did not return to LDRS in 2004. The many contests that dominated Argonia the year before were gone. But the annual Bowling Ball Loft, hosted by the AHPRA, returned for plenty of exciting flights, all aiming for a $1,000 cash prize offered by the reigning bowling ball champion, Geoff Elders. "Win Geoff Elders' Money" was the theme of the contest, and plenty of people tried. The goal of the contest was to attain the highest altitude possible using an I-impulse motor propelling a rocket holding an 8-pound bowling ball.

The bowling ball contests in high-power rocketry likely originated with members of AHPRA in the mid-1990s. "The Bowling Ball Loft was first proposed in 1996 by other AHPRA members and [me]," said member Mark Clark. "We were trying to come up with a standard payload for a large high-power rocket that would be available everywhere and [would be] inexpensive. Other objects were suggested, but the 16-pound bowling ball was the final choice. The original rules were for parachute duration using any motor. Parachute duration was selected at that time over altitude as it [was] easier to time the flights, and there [was] no argument over altimeter accuracy. [The contest] was announced at LDRS XV to a roar of laughter."

Between 1998 and 2000, the contest was a regular event at the BALLS launch at the Black Rock Desert. Multiple 16-pound bowling balls were launched on O and P motors at BALLS in 1998. Some of these bowling balls, held in 9-inch-diameter rockets, reached altitudes as high as 30,000 feet, said Clark. In 2000, the contest was also held at LDRS-19 at Orangeburg. But the rules were altered to reflect safety concerns at the much smaller field, said Clark. "Parachute duration was changed to altitude to keep the rockets out of the trees and corn. The weight of the ball was reduced to 8 pounds. With the new rules in hand, we contacted [LDRS officials] and received their approval."

Bowling ball rockets typically took on one of two popular designs. Most flyers placed the ball in the nose cone area of the rocket; others located the ball somewhere along the airframe. The highest bowling ball flight at Orangeburg was Mark Ketchum's *Bowled Over*, which attained 4,833 feet. Other contenders included Jeff Taylor's *Eleventh Frame* (4,275), Rev. Brad Wilson's *Scarlet to White* (3,629 feet), and Derrick Deville's *Cat's Eye* (3,869 feet).

Bowling ball rocket designers were some of the most sophisticated flyers on the field. As it turns out, to place an 8-pound ball in a rocket that weighs only ounces—and to keep it stable during flight—takes some real rocket science. Bowling ball flyers spend their spare time discussing fluid dynamics, computational physics, stability rings, and launch tubes. And they all have a bit of Captain Ahab in them, weight reduction being their white whale. The lighter the rocket's airframe, the higher the flight and chance to win the contest.

"I dumped my first altimeter because it weighed 3 ounces," said one contestant. "The new one weighs only half an ounce. That will do."

"I took the terminal blocks off my altimeter and soldered the wires directly to the board to save more weight," declared another. One flyer even shaved some of the paint off his bowling ball to ensure that the ball weighed 8 pounds and not 1 gram more.

One of the first bowling ball rockets at Geneseo rose perfectly to more than 2,000 feet. At apogee, the rocket turned over and, to the dismay of flyers and photographers on the field, lost its bowling ball *without* the parachute. All eyes remained fixed on the tiny ball as it grew larger and larger on its return, whistling first and then screaming its way in before impacting with a heavy thud in the soft farm soil.

Disqualified.

The next rocket was owned by reigning champ Geoff Elders, who was putting up the $1,000 first prize for the contest. His rocket was painted bright yellow and decorated with cartoon figures. And it did not launch from a conventional pad. Instead, it was shot out of a 10-foot-long medium green PVC tube pointed straight up into the open sky. "This gives it a little cannonball effect," explained Elders. At ignition, the rocket shot up and out of the tube to nearly 2,800 feet. Recovery was perfect.

Richard Hagensick, who one day soon would be launching rockets at Black Rock to 50,000 feet and higher, watched the bowling ball contest for a couple of years before diving in as a contestant at Argonia in 2003. "I didn't like it at first," said Hagensick. "But somehow I became intrigued about how light you could build the rocket and have it survive the flight."

"Someone told me I couldn't do it," said contestant Drake Damerau. "And that's all it took—science rules!" In the end, Geoff Elders kept his money. He won the contest for the second year in a row, with an altitude of 2,755 feet, followed closely by Damerau and Hagensick at 2,713 and 2,487 feet, respectively.

One of the truly unique rockets at LDRS-23 was a Chinese Long March rocket scratch-built by NAR/Tripoli member Rock Boyette of Florida. The 10-foot-tall rocket held five motors: a central L952 and four I205s. The smoke and fire at ignition made it look like the real thing as it climbed into the sky above Geneseo; everything looking perfect. Unfortunately, the rocket shredded a few thousand feet in the air, parts and pieces falling all over the range. "The number 2 booster came apart at the coupler joint between the motor mount and a section of tube forward of the fins," explained Boyette. "Apparently, I never properly glued the tube to the coupler."

New England flyer Robert Dehate had better luck with his N-powered *Orient Express*, which had a fine flight. Young flyer and future NASA astronaut Woody Hoburg delighted the crowd with his upscaled *Mosquito*, which boosted on two M1315s but came in hard for a rough landing, and Ed Miller treated everyone to an M-powered orange flying saucer powered by an M2000 Redline motor.

The annual Tripoli members' meeting and banquet was Saturday evening at the RIT Inn and Conference Center in Rochester, an hour away from Geneseo. Two lifetime memberships were bestowed—one to Guy Soucy, who had run Tripoli's annual election process for many years, the other to Mark Clark for his contributions to high power and tenure as Tripoli Motor Testing chair, which he was now relinquishing to Paul Holmes, who would handle the task for many more years. A new Tripoli honor was also established at LDRS-23: the President's Award, bestowed by the Tripoli president for significant one-time achievements in any given year. The first two President's Awards went to Frank Uroda for his Save Rocketry Now campaign of 2003 and Ky Michaelson for his role in the CXST *GoFast* rocket that spring. (At the time of the award, the booster for the rocket was still missing. It would not be found until November.)

Commercial flying at LDRS-23 ended on Sunday, July 4, a day with fewer flights because of high winds that developed late in the afternoon. It was the only inclement weather of the week. The search for Jerry O'Sullivan's missing *Nike Smoke* was now into its third day. "I was out there again last night," said the tired flyer, who was now offering a $500 reward for return of the rocket. "As I got farther and farther away from the road, the forest got real quiet and real dark," he said. "Then I stepped into the huge paw print of a bear track, and I knew it was time to quit."

Research flying started on Monday, July 5. There were L, M, and N experimental flights, including Neil McGilvray's *Cats in the Cradle* flying pyramid on an N4500. But the winds returned again, so several anticipated flights were forced to wait another day. On Tuesday, there were more flights, as well as some memorable CATOs, including the at-the-pad destruction of Robert Utley's *Sky in My Eye*, as the sixth day of LDRS-23 came to a close.

Meanwhile, there was good news for Jerry O'Sullivan's missing *Nike Smoke*. On Monday, O'Sullivan hired a student pilot to fly over the forest where his rocket disappeared a few days earlier. From the air, the pilot spotted what appeared to be a large parachute well into the forest, and O'Sullivan and his crew returned for another look on foot. On Tuesday morning, O'Sullivan was all smiles. The rocket was located and was in perfect condition after resting undisturbed for four days in the wilderness. "It was the longest-winded recovery in rocketry history," said a smiling O'Sullivan as he and his team packed the big *Nike* into the car for the return trip home.

Well, almost. Another flyer had to wait nearly a year to recover his LDRS-23 rocket. His name was Shannon Rollins.

Shannon Rollins traveled to Upstate New York from Alabama for his NAR Level 3 certification at LDRS-23. His scratch-built rocket was named *Daedalus III*. It was simple in design—four fins and a nose cone—and built to last. Rollins wrapped the entire 7.5-inch-diameter airframe and fins in several layers of fiberglass and carbon fiber. The payload bay housed altimeters by Olsen and Missile Works and carried a Panasonic DVD camcorder. The rocket also held a black sky automatic release mechanism to keep a tight rein on the main parachute until it descended to a low altitude.

Rollins placed his rocket on the pad at Geneseo on Saturday afternoon. The bright blue vehicle was almost 15 feet tall and weighed 70 pounds holding a Hypertek M1000 hybrid motor. With NAR Level 3 Certification Committee member Rich Pitzeruse standing by, Rollins backed away from the pad and gave a thumbs-up for launch. At ignition, *Daedalus III* left its away cell with the typical high-pitched whine of a hybrid motor. "It was spectacular to see the rocket I had spent six months building finally ride a column of light-colored hybrid smoke high into the New York sky," said Rollins. "The motor was supposed to burn for 9.1 seconds, but it actually burned for 13!"

The rocket nosed into the wind on ascent, and after what appeared to be a good deployment more than a mile high, it began drifting off the main recovery area. But it was still easy to see, and when the main parachute—a Rocketman R18C—deployed at 1,500 feet, the crowd applauded what looked like a textbook Level 3 certification flight.

"I was ecstatic that it had all worked as planned," said Rollins. "But my relief was short-lived as the rocket disappeared beyond a tree line more than a mile away." Still, he wasn't too worried. "Hey, how can you lose a fourteen-and-a-half-foot-tall rocket with a big red-and-yellow parachute?" he thought.

It was a long way to the tree line, which was on private property. Rollins and his recovery team secured permission from the owner of the adjacent farm to search for the rocket. For the rest of Saturday, they found nothing. A couple of miles in the opposite direction, Jerry O'Sullivan was still looking for his missing *Nike*. On Sunday morning, Rollins and his girlfriend, Erika, continued where they had left off in their search. "We looked all day long," he said. "We searched the fields, the woods, and the riverbanks. All we found were some strange tracks in the mud. They were bear tracks."

There was no trace of the *Deadalus III*. No nose cone, no main airframe, no parachutes, nothing. This seemingly perfect Level 3 flight had vanished. Without recovery of the rocket, Rollins was denied his Level 3 certification. It was a long drive from Upstate New York back to Alabama.

"It was hard to leave without the rocket," said Rollins. "Somewhere in Geneseo, New York, about 1,000 miles from my home, my rocket was lying in the woods or dangling from a tree or who knows what. It was a sickening feeling to know that all that work was lost." By the end of 2004, Rollins started to think about another Level 3 attempt, once again starting from scratch.

Enter a farmer by the name of Will Wadsworth.

Wadsworth owned a small farm near the Geneseo airfield. On a spring day in 2005—eight months after LDRS-23—a friend offered Wadsworth an airplane ride above Geneseo. Wadsworth said sure, and soon, the two were flying above the family farm. It was an exceptionally clear day, and as Wadsworth was surveying his property from the air, he noticed a strange silhouette in the river adjacent to his land.

There was something under the water, and it wasn't natural, he thought, and when he returned home, he decided to investigate a little further. He invited his daughter for a canoe ride down the river, and the two of them set out for the spot where he thought he saw an object from the air. But when he arrived there, he found nothing. The view from the air was quite different from what they could see at ground level. He decided to turn the canoe around and head home.

Suddenly, the little boat bumped into something. Wadsworth and his daughter peered over the side of their craft, and there, seemingly flying upstream in the middle of the river, was a shape that appeared to be a rocket. It was the airframe of *Daedalus III*.

The rocket had been under the water since LDRS-23. It rested flat on the shallow river bottom with its parachute still deployed. One of the parachute shroud lines was snagged across a submerged tree stump. Wadsworth and his daughter grabbed the shroud line and yanked it to shore, where together, they pulled the entire rocket out of the river. They loaded the rocket into their canoe and headed back home.

Moments later, something else caught their eye. There was another strange object floating beneath the surface of the river. It was the nose cone of the *Daedalus III*, and it too was held in place by parachute lines snagged on another stump. They hauled the cone out of the river and took the entire rocket home. The Wadsworth family was not sure what to do with the rocket, and for the time being, they let it dry out in the barn.

Several weeks later and during a local rocket launch at Geneseo, another flyer wandered on to the Wadsworth property, looking for another downed rocket. The family didn't have that flyer's rocket, but they told him about a big blue rocket they found in the river that was now sitting in a barn. Rich Pitzeruse, who was a NAR L3CC member for the Rollins Level 3 flight, and who lived in nearby Syracuse, heard the news of the mysterious blue rocket. He drove over to the farm to investigate.

When he arrived, Pitzeruse could not believe his eyes. There was Rollins's *Daedalus III*, missing for nearly a year. It was complete and intact. Pitzeruse got on the phone to Rollins. Shannon and Erika, who was now his wife, drove back to Upstate New York from Alabama. The rocket was still at the farm, and Wadsworth was waiting for them.

"He walked me to the rocket, and to my surprise, the *Daedalus III* looked great, considering it had been swimming in the river for a very long time," said Rollins. "Just as we were getting ready to leave and go home, I got another surprise," said Rollins. "Rich produced a signed Level 3 certification form!"

Pitzeruse had inspected the rocket, and with the recovery of the airframe and the nose cone in good condition, he signed off on the certification flight—nearly a year after the rocket left the pad. Rollins had a successful flight and recovery at LDRS-23 after all, said Pitzeruse; he just had to wait a year to prove it.

"My Level 3 certification was a long strange trip, but it was worth it," said Rollins, who credited the Wadsworth family and Pitzeruse for returning the rocket to him. "And knowing that my membership card with the new Level 3 stamp on it would be on the way made the trip back to Alabama with my rocket even sweeter."

After serving as association leader for nearly two years, Dick Embry stepped down as president of Tripoli at LDRS-23. In Embry's place, Ken Good was elected sixth president of the Tripoli Rocketry Association.

Good was a member of the original Tripoli Federation founded in Pittsburgh in the mid-1960s, and he had attended the first LDRS in Medina, Ohio, in 1981. He then left rocketry, nearly forever, a couple of years after LDRS-2. Yet he found his way back to the hobby in the late 1990s and quickly assumed leadership roles, first with his local Tripoli club in Pittsburgh, then on the Tripoli Board of Directors, and now as president. Good would remain at the helm of Tripoli for the next six years, completing the second-longest term of any person to hold the office as Tripoli president.

Ken Good was born in July 1952 in Greensburg, Pennsylvania, and grew up in nearby Larimer, a small mining town east of Pittsburgh. He was the youngest of five children of English-Irish descent. "The house we lived in was a small two-bedroom home. My sisters will tell you that I was spoiled, but it was a rather humble upbringing. We didn't eat very well, but it was a more or less happy childhood. My mother was more religious than my father, who was a driver most of his life, and she sang in the choir and taught at Sunday school."

His paternal grandmother, Sarah Fallon Whittle, who was widowed before Ken was born, was a big influence on his life:

> My mother and father separated when I was nine, and we lived just down the street from my grandmother. I grew up in that house. It was very interesting as it was very much a link to books because my grandfather had a vast library, and there were many artifacts and pictures from England. My grandmother would tell me lots of stories about her life and her family. She came to America in 1905 aboard the White

Star Lines RMS *Baltic*, captained by E.J. Smith, who would later become captain of the *Titanic*. This was his assignment just before *Titanic*, she said, and she recalled actually meeting him on the voyage. She came over in second class because her family was too proud to have her travel third class.

"My grandmother was ultimately very happy with her life in America," said Good. "But when she first came to Pennsylvania, there was nothing but little coal mining communities that did not even have sidewalks. She said she could have wept. She was used to the civilized part of England, and she did not expect America to be so rough and backward at the time."

Good discovered rocketry at fourteen. "It was about the same period of time that the NASA space program was moving right along," he said. "In our parents' generation, the big upcoming things were air travel, radio, and television. Yet in many respects, these things made the world seem smaller and more understandable. When space travel came along in my generation, there was suddenly this whole new vista. We had all of these new horizons to explore. I remember at one time looking at a NASA chart and reading—this was way back in the 1960s—that we would be on Mars in 1986. I remember very vividly that we were going to be on Mars. It was a real disappointment when that didn't happen."

Good's first model rocket was an Estes *X-ray*, which included a clear plastic payload section. "We launched a live chameleon in it, which today would be a violation of the Tripoli Safety Code. But back then, I didn't know any better."

Good was not a founding member of the Tripoli Federation. He joined in 1967, and over the next several years, he and other club members launched rockets of all types, including the *Gloria Mundi* and later a rocket design of his own creation.

One of my early efforts was to get into massive staging of rockets," recalled Good during an interview in 2014. "Back in the 1960s, if you had a rocket with more than three stages that flew successfully, you were extremely lucky. I came up with a design called the *Rack Rocket*, which is basically an open framework rocket where the motor is ejected out of the framework. The framework has to be heat resistant for this to work. My first rack rocket was called the *Achilles*. This model rocket attained an altitude of about 1 mile on six stages of black powder B motors. This was my early claim to club fame. It was projects like that which showed we were thinking beyond typical model rockets. We were thinking big about where civilian rocketry could go back in those days. The multi-staging of an unusual nature is something I later revived in the 2000s when I became reengaged with my 'motor feed' staging concept."

Good considered careers in archeology or perhaps teaching history. But like most teenagers, he was low on money and uncertain as to the best career path. "I worked briefly making little travel trailers at a fabrication plant," he said. Later, he landed a job at U.S. Steel in one of their plants near Braddock. "I worked in open hearth labor. That means you do a lot of dirty work, and you shovel soot and cinders. After two months of that, I decided I better get my butt back to school. Still, the job paid well, and when I was thinking about quitting, one of my friends said I was nuts to leave, but I wanted to do more. It may have gotten better with time but not for me. Within a decade, the industry took a downturn anyway."

Good enrolled at Penn State in 1972 and attended classes on and off for a couple of years. He also got married, had two children, and took jobs doing machine shop work and turning a wrench on sports cars. From 1976 to 1983, he sold auto parts at a foreign car store.

"I ended up managing people at the auto parts store, but I also realized that I did not want to do this for the rest of my life," he said. "Then my Dad asked me if I was ever going to go back to college, so I enrolled at Pitt in the information sciences program and attended school while I was working for the next few years."

Good dropped in and out of organized rocketry in the early 1980s. He was one of the few flyers to attend LDRS-1, with his friend Curt Hughes, in 1982.

"Curt heard about it at KentCon in 1981, which was a conference of like-minded rocketry people that was hosted by Kent State University," remembered Good. "We both attended KentCon, and we drove there and met a lot of different people, including Chris Pearson and the Ohio guys who were really pushing beyond model rocketry in those days. We were really excited about what these people were doing, and we realized that there were other people out there who were also thinking that we needed to get beyond model rockets."

Unfortunately, the press of everyday life would soon take Good away from the emerging high-power rocketry scene. He was working and attending school full time. There was little time left over for anything else. "I don't think I slept for about three years during this time," he said. He made it back to LDRS-2 in 1983 and a few more launches after that. But by 1984, he was out of rocketry. He was not around to see his local Tripoli club transformed into a national high-power association, nor did he witness the evolution of rocketry technology from the mid-1980s through most of the 1990s.

In the meantime, he earned a degree from the University of Pittsburgh in information sciences, and in 1986, he went to work as a procedures analyst for the Federal Reserve Bank in Pittsburgh. It was a job that would change his life. For the next twenty-two years, he rose through the ranks at the Federal Reserve, moving into management of the bank's information systems and ultimately supervising more than two hundred employees. It was a role that would later help Good when he accepted the job as Tripoli's leader in 2004.

"I learned to understand what motivates people and what gives them joy and also the type of stuff that just turns people off. You really need to learn the right way to work with people. The experience at the Fed where there were some people who were not getting paid very well translated directly to the volunteer system at Tripoli, where people are not getting paid at all. Nobody is getting paid for anything at Tripoli! And so at Tripoli, you really need to find out what gives people happiness."

Good's return to rocketry began with a telephone call in 1996 from friend Art Bower. "He invited me to fly ultralight airplanes with him, Francis Graham, Curt Hughes, and a few others at Joe Galando's farm," recalled Good. "When we finished flying for the day, Joe and Curt showed me a publication called *Rocket Mail* that was created by Francis. This was the way Francis tried to keep the original Tripoli Pittsburgh prefecture together. It was a great publication but of a very cut-and-paste quality, which was then photocopied."

Good recommended to Graham that they revise the newsletter production process with a desktop publishing program. Soon, Good was using a modern program to publish the newsletter himself. The club did not have a field at the time; Good helped locate one. One thing led to another, and he was soon elected as the club's prefect.

"In the late 1990s, around the time I became the club prefect, Bruce Kelly made a trip back here to visit our launch," said Good. "Several of us met for dinner afterwards, and we all chatted, and I found that I really liked Bruce. He had a great handle on the organization, and he was super engaged. I also thought his *High Power Rocketry* was a great publication, and I was really impressed by it. After that, in the 1999 election, Bruce contacted me and really encouraged me to run for an open board seat that was being vacated by another director. He said, 'I really think you should be on the board.' Actually, it was an easy sell by that time. I had been engaged with Tripoli back in the early days, and I thought, sure, why not? So I ran for the board seat and won it in a very close race with Derek Deville."

Good became Tripoli's secretary and took over as the editor of the *Tripoli Report*. "I really enjoyed the fact that I could now meet a lot more people in the hobby and could become more engaged in it. I wanted to be helpful in taking the organization somewhere." One of Good's first acts after being elected president in 2004 was to declare he was going to follow Dick Embry's practice of delegating authority to fellow board members.

"It's not always easy to see behind the scenes, but Dick invariably took the trouble to involve me as his vice president and the entire board of directors in a great deal of business," recalled Good. "This ensured that he took advantage of the talent on the board by delegating tasks to appropriate members. I have always felt this was the best way to have our board function. Apart from being an effective use of resources, it also ensures that board members are engaged in matters that will help the next president step up to the plate well prepared and ready to go."

Good also instituted monthly board of director teleconference meetings and the increased use of e-mail as the preferred method of communication among board members. "I wasn't going to individually call eight other directors to discuss an issue," explained Good, "especially when there were often questions, comments, and proposed changes. It could run into dozens of calls on only one issue."

Good's larger plan for the future of Tripoli was a broad one. He wanted the organization to focus on pushing the limits of putting rockets into space. "Tripoli had the reputation early on as being cowboy rocketry. But I never thought of us as cowboy rocketry. I saw us as cutting-edge civilian rocketry. That, to me, is the mission of Tripoli and will continue to be the mission of Tripoli," he said.

He also agreed with Bruce Kelly's vision of increasing rocketry's presence as much as possible. "I believe we need to promote rocketry and make more people aware of it. I pushed for the television exposure of the LDRS event, and I believe it was a success. Many new people joined and told us they did so because of seeing *The Rocket Challenge* television show." Like many high-power flyers, Good was drawn to the hobby because he found nothing else in life quite like it. "I really believe rocketry is a creative outlet. It's an artistic endeavor with a strong science undercurrent. When you build a rocket that has a specific design, with an individual paint color, to be flown on a specific motor, you are blending together many different aspects of creativity and science."

Good did not pay much attention to rocketry's simmering feud with the ATF in the late 1990s. But as a new board member in late 1999, he had little choice but to get involved in Tripoli's early litigation struggles with the government. "My first board meeting with Tripoli was where they announced they were going to join with NAR in the lawsuit against the ATF. I thought this could be long shot, but I did believe we were correct on legal grounds, at least from what I could understand at the time. The ATF had put APCP on their explosives list in a very arbitrary fashion, and they never provided a substantiated reason as to why it should be there."

One of the issues he and his fellow board members faced in the early days of the lawsuit was member opposition. "Some people thought it was a waste of time and money and that we might not win," he said. "Could we really prevail in a lawsuit against the federal government, they asked? Chances are we would not. I mean, who beats city hall? That's the cliché.

"I talked to a member at an Orangeburg launch, and he let me know that he thought the lawsuit was all wrong," said Good. "I had only been on the board about a year, and he came up to me and said we should not be doing this. We are just putting ourselves on the federal government's radar when we should be flying under their radar, he said. I told him I understood his position, but I believed that the days of being 'under the radar' of the government were long over. The feds know what we are doing, I said. The FAA knows what we're doing. These were no longer the days when, as in 1969, a few of us launched the *Gloria Mundi* near a housing development, and no one knew anything about it. High power was now on everyone's radar."

Good believed that ATF's requirement that flyers obtain an LEUP was more than just an inconvenience. It was, in fact, a hurdle, especially for younger flyers. "For example, if I were eighteen and I had to tell my mother that I needed to get an LEUP and that ATF agents were going to come to my home and inspect the place, she would have pleaded with me to find something else to do," said Good. "She wouldn't want federal agents on her property. Who would? There was just a certain amount of stigma and a degree of difficulty that was added to people who were just coming into the hobby. Now you have to get a permit and have federal agents around, and the fire marshal may get involved, and it just looks like a cascade of intrusion that could inhibit people from joining the hobby."

When he became vice president of Tripoli in 2002, Good was assigned the task of point man for Tripoli in the lawsuit. After SB-724 failed and while still waiting for Judge Walton's decision regarding summary judgment, Good and other rocketry leaders met with ATF representatives to try and settle their dispute out of court. "Pat Gordzelik, Mark Bundick, and I went to Washington to try and pitch some sort of compromise, something that, if nothing else, would establish a higher threshold for an LEUP." They suggested ATF create a new class of LEUP that would be specific to rocket motor users but would not have the same rules and requirements as actual low explosives, a sort of middle ground that might be acceptable to both sides.

Recalled Good, "We went in there for a meeting with Lou Raden and others at ATF, and it was a polite meeting, and they politely said we will listen to whatever you have to say, and they politely heard us out. But Raden then told us, in his own way, that unless there was something in it for them, they were concerned about getting their butts kicked in this post-911 environment by any public perception that they were relaxing controls over anything that they considered to be a risk involving dangerous substances. They heard us out, but it was pretty clear that they were not going to work with us on a compromise."

There would be no compromise with ATF, the Senate Bill soon failed, and in 2004, Judge Walton ruled that APCP was properly on the explosives list. Rocketry was beaten. For new Tripoli president Good and his board of directors—and their counterparts at the NAR—it was time to choose between two unpleasant alternatives: spend tens of thousands of additional rocketry dollars to appeal Judge Walton's decision against the hobby—with the outcome of such an appeal far from certain—or accept the inevitable and give in to the federal bureaucracy on the allegedly explosive nature of APCP.

It was now the early fall of 2004. The lawsuit was about to enter its fifth year.

37

The International Launch

The enthusiasm for model rocketry that swept through the United Sates after *Sputnik* was not immediately duplicated in Canada. One likely reason for the slower movement of the hobby in Canada was the Canadian Explosives Act, which banned all rocketry in Canada, except by professionals. In 1964, the Canadian Aeronautics and Space Institute approached the Royal Canadian Flying Clubs Association to undertake the task of obtaining an exemption to the act and organizing model rocketry in Canada. In 1965, these groups created the Canadian Association of Rocketry—also known as CAR.

In 1966, model rocketry was legalized in Canada. Yet restrictions continued to choke the hobby's growth. These restrictions included, among other things, altitude limits of 1,200 feet and the requirement that launch sites be on military ranges or other areas specifically approved for rocketry by government. The CAR, along with regional model rocketry clubs, lobbied the government to relax some of these regulations and finally, in 1989, restrictions on the import of composite motors were lifted. High-power rocket motors began to legally make their way into the country. Soon thereafter, Tripoli prefectures began to appear in Canada. In 1993, a federal ministry approved the first set of high-power rules and regulations in Canada. That fall, the first modern high-power launch in Canada was held at the Sullivan dry lake bed in Alberta.

The Sullivan Lake launch, billed as "High Power Rocket One," was a milestone for Canadian rocketry. It was held on October 23–24, 1993, and featured a waiver to 12,900 feet, with motors through K power. Tripoli Canada prefect Dale March, along with fellow Tripoli member

Carl Benson, obtained special permits to import AeroTech high-power motors for the event. They obtained the motors and then traveled all night from their homes in British Columbia, arriving at Sullivan Lake barely 3 hours before flying was scheduled to begin. "Had they not made it, the launch would have been scrubbed as they were transporting the entire motor order from AeroTech," said flyer Garth Illerbrun. "Their Herculean efforts did not go unappreciated."

AeroTech representative Denise Savoie met with government officials at Sullivan Lake to discuss future import of high-power products. At the launch event hotel, Savoie addressed flyers regarding AeroTech's plans to gain access to the Canadian commercial motor market. March also brought Canadian flyers up to date on the status of high-power rocketry in the United States.

High-power rocketry spread across Canada after Sullivan Lake. Although the CAR became the predominant force in the hobby, Tripoli's presence in Canada grew too. The number of prefectures in Canada grew to six by the early 2000s.

Tripoli Toronto, organized in 1998, became Tripoli's 110th prefecture. Its leader was a man unknown in high-power circles. One day the products bearing his name would be ubiquitous in the hobby—in Canada, the Unites States, and all over the world. And rockets powered by his products would breach the upper limits of the earth's atmosphere. His name was Anthony Cesaroni.

The idea for the first international LDRS likely began in a garage in Lethbridge, Canada. It was there that Tripoli members Max Baines and Tim Rempel conceived of an LDRS held outside the United States.

By 2002, the ROC Lake launch site was Canada's largest high-power rocket launch, recalled Baines. Flyers believed the site was big enough to host an LDRS. Baines flew to LDRS-21 in Amarillo in 2002 to discuss the possibility with the Tripoli board. "Surprisingly, the board did not think the idea was crazy," said Baines. "I asked for a date down the road in order to allow me the time to meet with Canada's explosives branch and other regulatory branches. This would ensure that motors and rockets could be moved across the border smoothly. Mike Dennet, then with Cesaroni Technology, attended with me and helped with the proposal."

LARGE AND DANGEROUS ROCKET SHIPS

On July 21, 2002, the Tripoli board unanimously approved Tripoli Alberta to host LDRS-24 in 2005. It was the only time in LDRS history that the location for Tripoli's annual event was picked three years in advance.

The city of Lethbridge takes its name from William Lethbridge, president of the North Western Coal and Navigation Company in the latter half of the nineteenth century. The town was incorporated in 1906 and began as a center for agricultural and coal. Today the city is a commercial, transportation, financial, and industrial center. It is the largest urban area in Southern Alberta, with more than ninety thousand residents. It is also the site of the Lethbridge Viaduct, a railway bridge more than 300 feet tall, and the largest structure of its type in the world.

The launch site for LDRS-24—ROC Lake—is 45 minutes southeast of Lethbridge and is surrounded by miles of agricultural land. The flying range site is about an hour over the Canadian border with the United States.

"The original ROC Lake site is actually some 25 miles north, located on what was then our family farm," said Baines. "It was right next door to the Rock Lake Hutterite Colony and the real Rock Lake." When the family farm was sold a few years later, the launch site was moved to the location where LDRS-24 would be held, he added. "The site is owned by my wife's aunt and uncle, and when we moved the site to their gravel pit, we renamed it ROC Lake."

"It does not have a single rocket-eating tree in sight," the local flyers said. "There is, however, a rocket-swallowing stream to the south that enjoys the occasional rocket snack, especially electronics bays."

"It is one of the few permanent amateur rocket launch sites in the world, with a standing waiver of 20,000 feet and windows to 50,000 feet that can be activated by a NOTAM and seven days' notice," said Baines.

By the early 2000s, ROC Lake rocketry events were attracting flyers from both sides of the border. Tripoli flyers from the Midwest and Pacific Northwest would join forces with CAR members to launch all sizes of rockets, A through O power, at the site.

One of the concerns for Americans traveling to a Canadian LDRS event was the international border crossing into Canada.

"Black powder and certain igniter types are going to make things a little sticky for you at the border," reported Canadian flyers Ian Stephens and Kyle Baines in a *High Power Rocketry* article prior to LDRS-24. "To resolve this, we will be offering all our flyers their black powder for free. That's right, for all your black powder needs, you can take what you need at no charge." The Canadians told American flyers to leave their igniters at home too, promising to sell igniters on site at LDRS for their cost or $1 each. This would also help ease questions along the border crossing, they promised.

But what about high-power rocket motors? How would they be transported across the border? Baines and his fellow flyers had a good working relationship with government regulators in Canada, including explosives and transport officials. "In the end, these regulatory branches did everything possible to allow us to not only have the event but to also have foreign flyers bring in their own motors," said Baines. "They also allowed American suppliers, such as Tim Lehr of Wildman Rocketry, to import and sell motors at the site."

To bring commercial motors into Canada from the United States, flyers had to fill out a government form and pay a fee of $30. The form was available at the official LDRS-24 website. To help offset this fee, launch organizers provided a $30 discount to flyers who registered early. The government fee would be a wash for those who registered early. In the alternative, Americans could simply drive to Canada without motors, which could then be purchased on site at LDRS-24 from vendors like Lehr and others. For the return trip home, however, flyers with H-impulse motors and above would need an LEUP to bring them back into the United States. This encouraged people to use all their motors at the launch.

LDRS-24 had four days of commercial and two days of experimental flying, July 14–19, 2005. To prepare the field, organizers relied on their own efforts, as well as equipment donations, from rocketry clubs in Canada and the Canadian Association of Rocketry. They were fifty launch pads, twenty-eight built by the Canadians just for LDRS. Scott Bartel at black sky donated a number of launch pads and rails too. The rail donations were important as Canadian high-power rules required rails—no launch rods—for flight. "We knew this rule would catch many flyers off guard," said Baines. "We had literally hundreds of rail buttons available free of charge to allow all rockets to be Canada legal."

Tripoli's UROC club in Utah also volunteered to lend their electronic launch control system for LDRS-24. The set up of the field began on Saturday, July 9, five days before launch. To encourage people to lend a hand early, Baines promised an extra day of flying to LDRS, before the official start of the launch.

"Yes, we are looking for all the help we can get, and we would love to see as many people as possible take a few extra days off from work and join us in setting up the range. Any hands will make light labor," announced Baines. "It is our hope that by Wednesday morning, we will have everything set up and ready for testing. To that end, I have requested a launch authorization for Wednesday, the day before LDRS actually starts. This extra day of launching will be available to everyone who has helped set up the event . . . to ensure that those who will be the most active in the actual running of LDRS . . . would at least get one day to fly, and it also helps work as a test session for all the equipment and the angle of the rails. In this way, we hope that come Thursday morning when the event actually opens, we will have a fully operational and tested system, along with some happy volunteers who have at least had a chance to get in a flight or two."

"Unlike most LDRS launch sites, the ROC Lake site had the advantage of extremely high waivers," observed Tripoli member Rick Clapp of Oregon, who wrote the lead magazine article on LDRS-24 after the launch was over. "It didn't take long to see that flyers were planning on making serious use of the tall airspace. Several two-stage projects were planned, and a [Cesaroni] product demo of a full O motor would awe spectators and flyers alike."

One of the more ambitious multistage rockets was the ATHA (Access to High Altitude) Aerospace *Altus 40*, built by David Buhler, Wayne Gallinger, Ian Stephens, and Scott West, all of Canada. The *Altus 40* was designed around a 4.5-inch-diameter two-stage airframe more than 17 feet tall. It was powered by a CTI N2500 in the booster and a CTI N1100 in the sustainer. At ignition, the *Altus 40* raced up the rail nicely but almost immediately ran into trouble. "A large flame shot out of the side of the rocket near the booster's forward closure, which caused the whole rocket to do a slow cartwheel until stresses broke up the airframe," said one observer. And then the second stage ignited.

The sustainer, powered by the long-burning Cesaroni N motor, raced across the sky like an unguided cruise missile. As the rocket dipped toward the ground, the unlikely target appeared to be a herd of cows grazing in the distance. "The rocket plowed into the ground near the cows, which scattered and ran for their lives," said Clapp. "The bellowing sounds of protesting 'mad cows' reached all the way back to the flight line, but no animals were injured."

Canadian rocketry rules required a large unoccupied splash zone be set aside as an area where rockets can impact if the recovery system fails and a rocket comes in ballistic. "They were also very strict about launch rail angles, noted one American flyer. "The rails were required to be angled toward the splash zone. The pad managers were responsible for setting the rail angles, and the flyers had no real say about it. You don't like it, you don't fly. The explanation was that rails provided a much more predictable ballistic trajectory into the splash zone."

Also, all rocket crashes were individually reviewed by the RSO. "It took a little while to learn the [new] rules," said another U.S. flyer. "But soon, everyone adjusted, and all launches proceeded smoothly."

Kristofer "Kip" Daugirdas, a sophomore at the University of Michigan studying aerospace engineering, put his skills to the test with a two-stage rocket that tipped the scales at 135 pounds. It was powered by two Cesaroni motors: an N2500 in the booster and M1400 in the sustainer. The rocket roared off the pad, separated cleanly, and the sustainer lit on cue, taking the upper stage to almost 16,000 feet. Recovery was perfect. (One day in the distant future, Daugirdas would build rockets that would push the altitude envelope at Black Rock.)

Another fine flight was in a rocket powered by a demonstrator motor from Mike Dennett of Cesaroni: a 6-inch-diameter O6800. The motor was fit into an all-metal rocket built by Scott Bartel. The O motor held 40 pounds of APCP and had a burn time of almost 6 seconds. The rocket, which weighed 117 pounds on the pad, screamed off the rail and reached Mach 2.4 on its way to more than 44,000 feet above the range. Recovery was flawless.

"After recovery, Scott put the nose cone on display at the registration tent so everyone could see what high-Mach [flight] does to the paint job," wrote Clapp. The flight marked the first time that a commercial O motor had been flown at any LDRS. And Dennett, who was also a licensed pyrotechnics operator, treated everyone on Friday night to a spectacular firework show.

Tripoli's annual members' meeting and banquet was held Saturday night at Steve Erickson's Restaurant, located a block from the official hotel, the Ramada Inn in Lethbridge. Among other things, European Tripoli member Rolf Orell was presented with a Tripoli Lifetime Membership for his tireless work in support of Tripoli and high-power rocketry in Europe. It was the first time the award was given to a non-U.S. Tripoli member.

After the usual festivities and Tripoli business were conducted that evening, a special guest speaker arrived, courtesy of some string-pulling by CAR president Ian Stephens. Through the Canadian Space Agency, Stephens arranged for Astronaut Dr. Robert Thirsk to address the rocketry community at LDRS. Dr. Thirsk had flown on the space shuttle (STS-78) in 1996, and his speech was the highlight of the weekend.

"Dr. Thirsk gave an excellent presentation about the state of the space program, and where it is going in the near and distant future," wrote Clapp, who said that Thirsk also described his space shuttle mission and the many experiments he performed while on board. "During the presentation, Dr. Thirsk reminded everyone of the many contributions made by Canadians to the space program, both in people and in high-tech equipment. Canadian astronauts and space hardware like the well-known Canadian robotic arms have flown extensively on the space shuttle and the International Space Station." After LDRS-24, Thirsk would return to space aboard a Russian *Soyuz* spacecraft. He would one day set the Canadian record for the most time in outer space, at 204 days.

Weather issues came and went at ROC Lake over the weekend, delaying some of the flying and pushing commercial flights into Monday, which had previously been designated an EX day. Flyer Woody Hoburg entertained the crowd with his stars-and-stripes-painted 200-pound rocket on multiple motors. "It was a beautiful, slow liftoff," said Rick Dunseith of Toronto, "and the K's air started beautifully. The last M didn't light on cue, though, and the rocket arced over and deployed its drogue, and then the M lit, spinning the booster section around and around."

Vern Knowles and several Tripoli members from Idaho launched numerous rockets at LDRS-24, including one Knowles flight to over 16,000 feet. "A group of five of us rented an RV together and traveled the 1,750 miles from Boise, Idaho, to the launch site and back," said Knowles. "It was a thoroughly enjoyable time! We got to see some old friends as well as meet a lot of new people and fly our rockets at one of the best launch sites anywhere." Other American flyers included future Tripoli president Steve Shannon of Washington and Mike Worthen with his M-powered *Highway 101* rocket.

When it was all over, approximately 250 registered flyers attended the ROC Lake launch, sending into the sky several hundred rockets. It was not a Tripoli flight record. Yet there was no doubt the Canadians did a great job of hosting the first-ever international LDRS.

"Anyone that elected not to attend LDRS missed a very fun time with some really great hosts in Canada," said Knowles, who described ROC Lake as a great venue. "The lack of attendance was probably partly due to the long distance for some people to travel but probably also due to the perceived difficulty of getting motors across the border. However, we had no problems at all. I imported quite a few commercial motors for myself and some friends I was traveling with. The LDRS organizers made it very easy to get the proper paperwork filled out ahead of time. The Canadian border official asked to see our paperwork and was very satisfied with it. Our reentry into the U.S. was a nonevent. The topic of rocket motors never even came up."

"The event itself was first class, extremely well-run and a lot of fun," echoed Tripoli president Ken Good: "The ROC Lake site is one of the finest high-power launch sites anywhere, and LDRS attendees appreciated the wide-open spaces with hardly a tree in sight. Another consideration I hadn't thought about ahead of time was the long summer days at the higher latitudes of this part of the world. It was a novel experience for me personally to witness daylight until after 10:00 p.m. and to be able to see some exceptional flying conditions from 7:00 to 9:00 p.m. A hearty 'well done' to our Canadian colleagues."

At the annual Tripoli members' meeting at LDRS-24, the board of directors was faced with a difficult decision: should the corporation continue its relationship with the magazine that had helped build high-power rocketry?

Bruce Kelly's *High Power Rocketry* was the leading voice of the high-power movement from 1990 through the early 2000s. Kelly's vision of a magazine that would spread the word of high power to every hobby store in America was fully realized. And as the magazine's star rose in the 1990s, so too did the success of the Tripoli Rocketry Association. In the first ten years of *High Power Rocketry*, Tripoli was transformed from a small organization of several hundred flyers into an international membership in the thousands. Many of the future leaders in the hobby obtained their first look at high power via Kelly's magazine. The thought that the magazine might go out of existence was too painful to consider. Yet something had to be done, and at LDRS-24 at Lethbridge, the board took action.

Kelly stepped down as president of the Tripoli Rocketry Association in 2002 to devote more time to *High Power Rocketry*. He knew he had to turn the magazine around quickly. The publication was in arrears and was losing readers and advertisers. "I regret that the problem went as far it did," Kelly told the board that summer. "I probably should have stepped down earlier as president and focused attention on the magazine."

Kelly knew that Tripoli leaders were taking heat from disgruntled members regarding the status of the magazine, which had dropped to just a few issues per year. Many members paid Tripoli directly for their subscriptions to *High Power Rocketry* when they paid their annual dues. And the board was concerned—rightfully so—that money was being accepted for a product that was not being delivered as promised. People were paying for nine issues per year and receiving just three or four, if not less. Longtime Tripoli members, who knew of the significant contributions *High Power Rocketry* made to the hobby over the years, hoped for the best. Some newer members, however, were confused. From their point of view, they paid for a magazine and received little in return. Something had to change.

Tripoli vice president Good, speaking in 2002 just after Kelly resigned as president, summarized the problem with *High Power Rocketry* and what needed to be done immediately:

> As I see it, what we need to accomplish is this: to get *HPR* back up to nine issues per year, to ensure members have had their outstanding issues resolved to their satisfaction, [and] to put into place some means of ensuring we have

accountability for *HPR* meeting its production agreement from this point forward. I also believe that the production schedule should be back on line by the end of the year, and if not, we will need to consider the option of severing the relationship between *HPR* and [Tripoli].

Kelly assured the board he would get things back in order. After all, prior to becoming Tripoli president in 1994, Kelly created *High Power Rocketry* from almost nothing and was very successful. Now that he was out of office and could focus his energies on the magazine again, he expected to repeat his successes of the 1990s.

However, the magazine landscape in 2002 was very different from what it had been when Kelly entered the hobby fifteen years earlier. Kelly's practice of giving away free advertising for local and regional launches, of not paying writers for stories, and the sporadic printing schedule of *High Power Rocketry* between 2000 and 2002 made his renewed efforts in 2002 more difficult. The steady rise of the Internet was also cutting into revenues for all magazines everywhere.

Something else was different in 2002 as well. *High Power Rocketry* was no longer the only high-power publication in town. There was a new rocketry magazine created by Tripoli member Brent McNeely. His magazine was called *Extreme Rocketry*.

Brent McNeely was born in Denver, Colorado, in 1965. His father was a mining engineer who traveled the country doing the legwork needed to set up mines before they could be put into production. "I moved twenty-seven times before graduating from high school," said McNeely in an interview in 2004. "Most of the moves came because we would move to where a mine needed to be prepared for production."

With every new mine, McNeely's father would secure temporary housing for a few months. If the economy and the mining were good, the family would eventually purchase a new home. But they were never in any one location for too long, recalled McNeely. At some point, the local economy would decline, or his father would be asked to help start another mine, and they would be on the road again. "Just counting the states, not moves within each state, I lived in Colorado, Utah, New Mexico, Arizona, Florida, Idaho, and Nevada," said McNeely.

LARGE AND DANGEROUS ROCKET SHIPS

McNeely was an avid reader, and he had a talent for art, inspired perhaps by his mother, who ran an engraving business. After high school, he considered majoring in art at college but was concerned he could not make a living out of it. He studied several other subjects instead, including computer science and religion. He graduated from Brigham Young University in 1990 with a bachelor of arts degree in Ancient Near Eastern studies.

"During my time at BYU, I did research for five different professors in the Department of Religion," he said. Some of this research involved ancient languages and texts, such as the Dead Sea Scrolls, the *Book of the Dead*, and other ancient documents. "It was at this time that I started doing editing for the professors on books they were writing," he said. McNeely also became involved in Mormon scholarship, eventually editing several volumes of work by a well-known Mormon scholar. "At BYU, I also worked on the staff of two different college magazines, including being the editor in chief for an eastern studies publication."

After college, McNeely was hired to work at BYU's Jerusalem Center for Near Eastern Studies in Israel. In anticipation of his new job, he moved to Israel and took a temporary position at a kibbutz, where he learned to bake bread while working on his Hebrew language skills. Several months later, and before his work at the Jerusalem Center began, Saddam Hussein invaded Kuwait, throwing the Middle East into political turmoil and war. McNeely returned home. He became disillusioned with his religious studies, and eventually, he settled in Las Vegas, Nevada.

"I decided to look at what I could do well, which was computers and art," recalled McNeely. "I got a job with a small graphics design firm using the skills I learned working on books and magazines while at BYU. I worked there for a short period of time before they went out of business. A few of their customers approached me and asked if I would be willing to continue doing work for them, so in 1991, I started my own graphic design business."

McNeely's interest in rocketry—like the experiences of so many other high-power pioneers—began early in his life. He built his first model rocket, an Estes *Alpha III*, when he was in the fifth grade in Utah. When the family moved to Wyoming, he continued building and launching model rockets, eventually becoming president of a local rocketry club. He left model rocketry for other pursuits when he got older. Then in 1994, McNeely rediscovered the hobby. He purchased a few model rockets at a hobby store in Las Vegas, and he launched them with his nieces and nephews.

"I then decided to see if there was a local rocketry club and did a search on the Internet," said McNeely. "I discovered Tripoli Vegas and called the phone number provided, which put me in touch with Tom Blazanin." The former Tripoli president was living in Las Vegas now, and he invited McNeely out to a launch at the lake bed near Jean. "I found eight or nine cars there with a dozen or so people," said McNeely, "and as I was pulling up, I noticed three or four guys in the back of a pickup with what appeared to me to be a huge rocket as compared to the rockets I was flying. It was probably 6 inches in diameter and 10 feet tall. Three of them hoisted it onto their shoulders to carry it out to the pad, and I was just awestruck."

It wasn't long before McNeely was a Tripoli member. With his writing and editing skills, he started writing feature stories for *High Power Rocketry*. In 1999, he achieved his Level 3 certification with one of Gary Rosenfield's first 75 mm M1315 motors. Afterward, McNeely and his rocket appeared in a widely circulated full-page color advertisement for AeroTech's popular new motor.

Brent McNeely's path to creating his own rocketry magazine began with mentor Bruce Kelly. Kelly published several stories by McNeely in *High Power Rocketry*, and in the late 1990s, when Kelly ran into problems getting the magazine out on time, McNeely offered to assist him using new design programs he had with his own company, *Rocketeer Media*.

"Bruce declined my offer for assistance," said McNeely. "He told me he enjoys doing the layout and design, and he preferred to do it himself." Indeed, one of Kelly's favorite aspects of creating *High Power Rocketry* was designing the magazine. McNeely kept to his graphics business and continued to enjoy rocketry as a hobby. He also started designing model rocket-powered cars that he and other hobbyists would race in the desert.

He wanted to do more. In late 1999, McNeely turned to Darrell Mobley of the Internet rocketry site *Rocketry Online* to create a survey asking people if there was any interest in a new rocketry magazine. Mobley, an Internet wiz who created his website from scratch in the mid-1990s and then developed it into the premier rocketry website in the world, reported his survey revealed widespread support for a new magazine.

LARGE AND DANGEROUS ROCKET SHIPS

McNeely decided to test the waters with a trial publication, available on the Internet only. Using his talent as a graphics designer, he put together a trial issue of a new magazine in early 2000. Before running his first edition, he asked friends for ideas for a name for the publication. Several were suggested, but McNeely went with the moniker proposed by Gary Rosenfield. It was a catchy phrase, thought McNeely: *Extreme Rocketry*.

"I was able to start the magazine with $30 down to Network Solutions to register the domain name," he said, "[and] I was able to produce the initial magazine for free with donated articles and photographs." McNeely's online trial run was an immediate success.

"I had previously contacted every possible advertiser I could think of and offered them an 80 percent discount off of standard advertising rates to be included in the electronic version. I then decided that if I received at least five hundred subscriptions, I would take the magazine to press, and then the advertisers would owe me the remaining 80 percent towards advertising in the print edition. The advertisers liked the idea."

McNeely's strategy was the right one at the time. In March 2000, he gathered enough advertisers and subscriptions to physically print the first issue of *Extreme Rocketry* magazine. It was a revolutionary product from the start, with professional design features never seen in any rocketry publication before.

In his first issue, McNeely premiered several features for which the magazine became well-known. The most important of these was the *Extreme Rocketry* interview, a personal discussion, usually led by McNeely, with a prominent rocketry personality—in model or high-power rocketry—in a question-and-answer format that introduced flyers to some of the most prominent personalities in the rocketry community. Quite fittingly, his first interview was with Gary Rosenfield. In the years to come, McNeely would publish interviews with people such as Frank Kosdon, Bruce Kelly, Vern Estes, Frank Uroda, Ken Good, Trip Barber, and scores of others. With these interviews, McNeely created an unprecedented historical record for the hobby.

The magazine also featured a regular letters section, publisher and guest opinion columns, launch coverage and technical articles, and a centerfold photo contest, in full color, related to some aspect of rocketry. Winning photos submitted by readers would receive a $100 check for their efforts. Then in July 2000, McNeely did something else unprecedented in the hobby. He announced he would pay writers for stories that appeared in his magazine. It was the first time any high-power or model rocketry-related publication had made such an offer, and it was long overdue.

With *High Power Rocketry* struggling in the early 2000s, *Extreme Rocketry* rose quickly, gathering advertisers desperate to reach consumers on a regular basis. McNeely's magazine had growing pains, as all publications do, but it was printed on a regular basis, nine issues per year every year, throughout the 2000s. By mid-2002, many of the largest advertisers in the hobby were using *Extreme Rocketry* in addition to, and in some cases in lieu of, *High Power Rocketry* as the place to market their rocketry products to consumers.

AeroTech was one of the very first advertisers in the early issues of *High Power Rocketry* in the 1990s; indeed, its reloadable motor ads financed the *Tripolitan's* first color cover—created by Bruce Kelly—and remained a mainstay of *High Power Rocketry* for years. Yet by late 2002, AeroTech too had made the jump to *Extreme Rocketry*—with consequences that would prove devastating to *High Power Rocketry* in the years to come.

"I'm so far behind that I'm basically starting over," said Bruce Kelly in late 2002. "I need to build subscriptions. I'm removing a lot of fluff from the magazine, and I'm cutting back on advertising. I might have to cut back on my distribution for a few issues. I just want to get it out to the people who have paid for it already and to get out the number of issues we promised in a year. That is not going to happen this year. There is no way there are going to be nine issues this year. But there are going to be four more."

Kelly's decision to cut back on advertisers—the lifeblood of every magazine—was, in retrospect, not the best course for *High Power Rocketry*. For most magazines, advertisers are the only way to stay in business. If advertising revenue is lost and not immediately replaced, it is only a matter of time before a publication goes out of business. And by the fall of 2002, too many advertisers were leaving *High Power Rocketry*.

The last full-format size edition of *High Power Rocketry* was printed in September 2002. It ran about thirty-eight pages, cover to cover, and the lead story was about a launch that had taken place two years earlier—the BALLS launch in 2000. Then in the next published issue (October/November 2002), Kelly made a dramatic change in the physical appearance of the magazine. To lower printing costs, he reduced the size of *High Power Rocketry* from its customary 8¾ by 11-inch format to a digest size of 6¾ by 9½ inches.

"We've had to take some extraordinary measures to keep the publication in the black," explained Kelly. "The new size format is one example. It allows us to print more pages per press pass. This change was inspired by a similar change made to Tripoli's newsletter last week. So we'll do this for a couple of issues and see how we like it."

As it turned out, the new format was easy to read, and between late 2002 and early 2005, Kelly produced another eighteen issues of the magazine on a sporadic schedule. The magazine's coverage of LDRS-21 and LDRS-22 was excellent—some of the best LDRS coverage to ever appear in any magazine—and there were plenty of technical articles that appeared during this time too. However, regional launch coverage dropped dramatically, there was no coverage at all of LDRS-23, and advertising space in the magazine dropped to new lows. In some issues, there were barely half a dozen ads in the magazine. Many advertisers were turning to either *Extreme Rocketry* or the Internet to reach their rocketry audience.

"When I brought the magazine back on to a regular publishing schedule last July, I purposely kept advertising at a bare minimum," explained Kelly in 2003. "After being out of production for a while, it was important for me to get some articles out quickly. Since then, I have let advertisers know they are welcome to advertise. However, as the number of paid ads continues to increase, we will be limiting advertising to only eight pages per issue . . ."

The announcement limiting ads in the face of increasing competition would only make matters worse. Advertisers need ad space and regularly scheduled issues to get their message out. Now they might not be getting either with Tripoli's flagship magazine. Kelly also announced a return to his prior policy of providing Tripoli, LDRS, and regional launch organizers with free ads for their activities. Kelly was a proponent of free advertising for launch organizers, including LDRS. It was a purely altruistic act and was in accord with Kelly's love for the hobby and his desire to spread the word of high power far and wide. "It was not a demand of Tripoli for me to run free launch ads," said Kelly at the time. "It was my own choice."

By some accounts, Kelly had handed out nearly $80,000 in free advertising space over the years. This was a boon to the growing high-power hobby in the 1990s. Yet by the early 2000s, this policy made little sense. By this time, Tripoli was running budgets in excess of $300,000 per year. The corporation could well afford to pay for ads and for prefecture listings in *High Power Rocketry*. Free space to LDRS promoters was also unnecessary. LDRS was no longer the modest venture it had been when Kelly entered the hobby in the 1980s. The national launch entertained thousands of flyers and spectators annually and grossed many tens of thousands of dollars every year. Organizers of LDRS and similar large regional events could have, and should have, paid for their advertisements in *High Power Rocketry*. But Kelly did not require it, and the magazine's bottom line suffered as a result.

In late 2003, Kelly and the Tripoli board agreed to move to a twelve-issue-per-year schedule for *High Power Rocketry*. There were also changes made to the accounting procedures regarding subscription funds Tripoli collected for the magazine. These changes were purportedly made to ensure that subscription monies were transmitted more quickly to Kelly to keep the magazine running. In mid-2004, Tripoli board member Derek Deville expressed his hope the magazine would soon be back on track.

"The [magazine] situation has been very difficult for all of those involved. I have not always been satisfied with the delivery or the content of the magazine, and I was particularly displeased with the mixed accounting of [the magazine and Tripoli]. In this regard, I helped to orchestrate and strongly supported a separation plan that took effect just after LDRS in 2003. As a result, the funds for [*High Power Rocketry*] are now passed through TRA, without any holding period, straight to [the magazine]. Many people are still unsatisfied with the state of [the magazine]. However, I do believe that Bruce is now doing everything possible to get things back on track. In these efforts, he has had reasonable success, and I believe the situation will continue to improve."

Other Tripoli board members expressed similar sentiments. "I thought that Bruce had done so much for Tripoli, and he had contributed such a wonderful publication that if we could just give him as much leeway as possible, then maybe he could get back on track," recalled Good.

Unfortunately, the problems with the magazine did not improve. By September 2004, only four issues had been produced that year. And the Tripoli board was feeling more heat from members who were paying for a magazine and not receiving a product. In the board's view, they had to act. In a teleconference on October 5, 2004, the option of severing Tripoli's relationship with *High Power Rocketry* was openly discussed and then rejected by the board. Although it was clear Kelly would not be able to produce twelve issues in 2004, a nine-issue schedule still seemed feasible. The board agreed to nine issues while also requiring, for the first time ever, that Kelly execute a formal written contract with Tripoli. The contract would set forth the rights and responsibilities of the parties with regard to *High Power Rocketry*. Among the responsibilities for Kelly was to produce nine issues of the magazine postmarked in 2004, followed by nine more in 2005, including at least four postmarked by July 1, 2005. Failure to meet this schedule would be grounds for termination of the relationship between Tripoli and the magazine.

As it turned out, even with the contract, the publication schedule would never fully resume. The last issue of *High Power Rocketry* bore the date of January 2005. There was no mention in the magazine that this would be the last issue. But there would be no more.

By the time LDRS-24 rolled around in the summer of 2005, there had been no issues of *High Power Rocketry* produced in months. At the board of directors meeting in Lethbridge, President Good formally notified Kelly, who was in attendance, that *High Power Rocketry* was in default on the written contract with the association. It was decided that Tripoli would send Kelly a formal default notice with the message that the contract would be terminated in ninety days.

"When, after years of lacking one, a contract was written between Tripoli and [the *High Power Rocketry*] magazine to explicitly define responsibilities and expectations, and in less than a year [the] magazine was in default of contractual production terms, there was no longer any real choice," said Good later. Afterward, it was decided that any further monies sent to Tripoli for *High Power Rocketry* subscriptions be returned to the subscriber. The board also decided that the publications committee—of which Kelly was a member—be reformulated. Kelly was out, and *High Power Rocketry*, which for years was the very symbol of the Tripoli Rocketry Association and the high-power rocketry movement, was no more. "This was a very difficult decision in many respects," Good reported after LDRS. "[The *High Power Rocketry*] magazine at its best is a fine publication and has played a key role in many current rocketeers becoming interested in Tripoli and high-power rocketry. However, the problems of production over the past several years have not gone away . . . and the magazine issue has become difficult and divisive, and it was becoming more and more clear that the majority of the membership favored a separation."

In the end, the demise of high-power's most iconic rocketry publication was likely due not so much to Bruce Kelly's management style but to the immutable vicissitudes of the publishing business. The fact of the matter was this: the printed word was dying in 2005. And nowhere was it dying so fast than in the newspaper and magazine industry.

Publishing a magazine has always been a tough business. But in the early 2000s, magazines were disappearing in droves. *U.S. News & World Report*, *PC Magazine*, *Disney Magazine*, *Home*, and *American Heritage* are but a few examples of the big-league publications that closed their doors in the 2000s. One of the primary reasons for their disappearance was the rapid rise of the Internet, which was steadily draining advertising dollars from nearly every form of traditional media. That *High Power Rocketry* was a publication with a small and transient readership made the situation for Kelly and others like him even more tenuous. Publications with decades of experience and hundreds of thousands of subscribers were feeling the pinch; for smaller magazines, the pain was even more acute.

Yes, Kelly could have printed the magazine on a more regular basis between 2000 and 2004; yes, he could have considered paying writers for stories; yes, the infrequent publishing schedule caused him to lose advertisers; yes, he could have done this; or yes, he could have done that. But no—and despite claims to the contrary—he was never going to make much money doing it. The fact of the matter is that all these things and more have been attempted by magazine publishers in high-power rocketry over the last forty years, and not a single one of them—or any combination of them—has yet to result in a viable business model that over the course of time allows its owners to make money, let alone break even.

Extreme Rocketry's Brent McNeely published valiantly for nine years, from 2000 to 2009, and was forced to close up shop because the numbers just didn't add up. Bob Utley and Neil McGilvray at *Rockets Magazine* did the same thing for almost eight years and were then forced to change from a printed magazine to an e-magazine in 2014. There was simply not enough money to keep printing hard copies of *Rockets*. In

2018, *Rockets Magazine* called it quits too. "No matter how many corners are cut or pennies are saved, the magazine isn't viable in this current media environment," wrote McGilvray when *Rockets* published its last electronic issue in late 2018. "In a world of Tweets, Instagrams, Live Streaming, Facebook, and other evolving technologies, the standard of a periodically produced magazine format was heading the way of the dodo bird."

Even *Launch Magazine*, a very slick and attractive Madison Avenue creation with professional writers, accountants, and a fully paid staff, could not survive more than two years in the hobby in the mid-2000s. Indeed, the only magazine that has continued to be printed over the many years since the early 1990s is *Sport Rocketry*, which is underwritten by the membership of the NAR and survives in large part, thanks to the tireless and voluntary efforts of its editors.

In retrospect, and considering the history of magazine publishing in this hobby, in many ways, it is a wonder why Kelly did not give up on *High Power Rocketry* sooner rather than later. Kelly could have walked away in 1998 when he was ill or in 2000 when the responsibilities of being president of Tripoli became too great. He also could have left in 2002, when he stepped down from the Tripoli board. But Kelly was never a quitter in his life. He kept trying, producing eighteen more issues of the magazine over the next three years. He continued to follow his passion for high power on a determined path, a path that had a modest beginning at the Lucerne range in the 1980s, to becoming president of Tripoli and publisher of *High Power Rocketry* in the 1990s to an unfortunate and disappointing end with the magazine in 2005.

Yet in his wake, Kelly left a legacy of service and commitment almost unparalleled in the hobby. Between 1990 and 2005, Bruce Kelly—and very few others like him—was high-power rocketry *personified*.

38

Rocketry vs. the U.S. Government

Part 3

The Appeal

The failure of SB-724 in 2003 and Judge Walton's subsequent order granting summary judgment to ATF in 2004 were significant defeats in rocketry's battle with the government. Tripoli and the NAR were beaten in the legislature and in federal district court—and beaten soundly. Rocketry leaders had spent four years and hundreds of thousands of membership dollars taking on the government, and they had little to show for it. By the middle of 2004, there was no reason to believe APCP was coming off the federal explosives list anytime soon—if ever.

This was the legal landscape facing Ken Good when he became president of Tripoli that summer. And one of the first decisions he had to make was whether to accept ATF's authority over the rocketry world. Any other course of action, such as an appeal of Judge Walton's decision, was not only a long shot but would also take more years and hundreds of thousands of additional dollars without any guarantee of a different result.

───◆◆◆───

Since ATF's victory in 2004, there were signs the agency was not paying any attention to Judge Walton's other ruling that PAD exemptions applied to rocketry—at least until ATF went through a public notice-and-comment period in support of new regulations to remove PADs.

In the first few months after the summary judgment ruling, rocketry's lead attorney, Joe Egan, wrote to the lawyers for ATF repeatedly, warning his adversaries that around the country, ATF field agents were ignoring Judge Walton's decision on PADs. Egan also challenged ATF with other letters, demanding the agency cease implementation of arcane rules on federally required locking mechanisms for rocket motor storage magazines. In Egan's view, ATF was revising its storage rules at will without any input from anyone. This cycle of revision to the rules was wreaking havoc with dealers and flyers who were trying to comply with constantly changing federal rules.

"Over two thousand magazines have been sold to hobby rocketeers," said Egan. "The dealers were told by [ATF] officials—repeatedly—over these many years that those very robust magazines, which have a double-lock system, were perfectly acceptable under [ATF's] rules. Then last December, with no notice, [ATF] changed its interpretation of the types of locks that should be required on these magazines. The dealers and manufacturers scrambled and marketed a modification kit that could convert the locks to acceptable form. [ATF] told the dealers, and the hobbyists, that these retro kits were acceptable. Then last week, [ATF] again changed its interpretation, again with no notice. [ATF's] latest about-face apparently involves its continuing uncertainty over the meaning of a 'mortise lock.' [ATF's] new interpretation apparently now requires that those locks be internally mounted within the storage magazine wall, something that is impossible to retrofit and which essentially will put all the hobbyists (and dealers) out of business, at least until they can write off their original investment, induce a manufacturer to make new magazines, and purchase those new ones."

Egan also reported incidents that amounted to outright harassment of individuals. In Florida, for example, one member in Lea County was forced to give up his LEUP. "[ATF] agents twice visited his home to inspect his magazine," reported Egan. "First, they told him last week that it was acceptable. Then they returned this week to tell him it was not acceptable. As a result, he is being forced today to surrender his license to the agency." The ATF took similar action against a well-known and longtime flyer, continued Egan, and trouble was reported by Ohio rocketry dealers too.

"Obviously, this is most disturbing to the rocket organizations since it smacks of trying to obtain indirectly what the court refused to give the [ATF] directly," said Egan. "[ATF's] actions this time are likewise unlawful. First, to the extent these magazines contained fully assembled rocket motors (many of them contain only such motors), they are propellant-actuated devices which are currently exempt from [ATF] regulation and, technically, do not even need to be in a double-locked heavy-duty magazine. Second, I would have hoped that the [ATF] would know by now that changing its rules with no notice, and no opportunity to comment, is not something that federal administrative law can countenance. This latest round is particularly onerous since it will force the hobby rocketeers to surrender their licenses and abandon their considerable investments in their already double-locked, and recently retrofitted, magazines. If this is not arbitrary, nothing is."

ATF officials were apparently emboldened by Walton's summary judgment ruling in their favor. Actions by agents around the country were inconsistent, but there was a trend toward increased inspections, quibbling over minor issues, and even outright harassment of LEUP holders, confirming for many high-power flyers they were better off simply ignoring the LEUP process entirely. Why get an LEUP, asked these flyers? It was better to stay off the federal government's list, they reasoned.

"It is abundantly clear that some elements of the ATF do not feel that they need to follow legally promulgated rules," remarked Good at the time, who urged all flyers to donate to the legal fund that was keeping rocketry's counsel on board. "TRA and NAR leaders are resolved not to stand by and let our members and dealers who supported us to be bullied. This outrageous behavior must be answered in the strongest manner. Help us fight the fight for rocketry and freedom and donate now!"

As hope for a negotiated settlement faded away, rocketry leaders were faced with another important decision: Should they appeal Walton's decision that APCP was an explosive? Or should they just accept defeat?

"The summary judgment decision against us was certainly deflating," said Good. "But we remained in very close communication with Joe Egan and Marty Malsch all of the time, and we all believed that the main reason Judge Walton ruled the way he did was that he granted too much deference to the opinions of ATF. We believed ATF did not deserve that deference because they had not demonstrated any particular expertise with APCP, so we were intent on filing an appeal."

Yet what about the cost? Were Tripoli and NAR members willing to spend even more money after being defeated the first time around? Good knew it was a difficult question to answer for the board.

"We knew it was going to cost at least $25,000–$100,000 just for the appeal," he said. "And that was a really tough sell. But as Tripoli and NAR leaders, we felt we were 'in for a penny in for a pound.' We believed that if we could just get this case to an appellate court—where there would be three judges involved—there would be a better chance of getting one or more of those judges to see the arguments that we felt Walton had just let go."

Good faulted no one for reluctance to go any further with the lawsuit. After all, the Tripoli leader could not guarantee an appeal would be successful. If rocketry moved forward, the only certainty was that the lawyers for Tripoli and the NAR would have to be paid. Should rocketry cut its losses and call it a day, or spend more money that may not change the result? Why not quit now?

"Having taken things so far and having spent this much time, money, and effort, I wouldn't feel that I was doing the responsible thing for Tripoli and the membership if we had just let it all go," said Good. His view was also a broader one, influenced perhaps by the simple idea that might did not make right.

"I became rather determined that an arrogant enforcement agency with patently false arguments was not going to prevail if I could do anything to prevent it," he said. "Mark Bundick shared this sentiment with me."

In late 2004, Tripoli and the NAR decided to appeal Judge Walton's decision. "I think at the time, we had so much invested that we couldn't back away from it," recalled then Tripoli vice president Patrick Gordzelik. "Ken Good, Dick Embry, and I had just finished off a meeting with ATF to see if we could find some common ground. The assistant director treated us like dirt. We had no common ground. We were dug in so deep and had a mind-set that we were not going to quit, and they were not going to beat us, and we didn't have any decision other than to go ahead and move forward."

In the federal court system, there are three primary courts: district courts, appellate (or circuit) courts, and the United States Supreme Court. The district courts are trial courts, where people with federal disputes have their matters settled by a judge or a jury. There are ninety-four federal district courts in the United States. Most cases filed in the federal district courts begin and end right there. The judge (or jury) renders a decision, and the matter is concluded.

Sometimes there are grounds for a losing party to appeal the judgment of the district court. A party may claim evidence was admitted that should not have been allowed, or the judge's decision was not legally correct based on the facts presented in court. These and a host of other reasons can be grounds for an appeal by the losing party to the federal circuit courts of appeal. The losing party sends the appeals court an application, called a brief, and the court decides whether it will accept the appeal.

In rocketry's lawsuit against ATF, any appeal regarding Judge Walton's summary judgment would go to the United States Court of Appeals, D.C. Circuit. This court reviews Washington D.C. District Court decisions and other cases involving the rulemaking of federal agencies of the government such as ATF. For cases on appeal, three of the judges on this court—called justices at this level—will hear a party's appeal and then render a decision.

If a party loses its appeal at the federal appellate court level, it might appeal to the United States Supreme Court. However, the Supreme Court does not take cases absent special circumstances or where there is a disagreement among the different federal circuit courts on an important federal issue. In the lawsuit brought by Tripoli and the NAR against ATF, if the circuit court justices ruled against high-power rocketry, the chances the Supreme Court would accept another appeal would be slim to none. So for all intents and purposes, the D.C. Circuit Court of Appeal was rocketry's last chance. If ATF prevailed here, the case would be over.

Tripoli and the NAR filed their notice of appeal on December 23, 2004. The appeal will cost a minimum of $25,000, reported Good and Mark Bundick to their respective members. "We estimate the total costs of remaining legal work on both the litigation and appeal will take approximately $100,000 over the next two years."

Both presidents urged their members to donate whatever they could to rocketry's legal fund, and both organizations continued to split costs for the litigation evenly. If this cooperation between Tripoli and the NAR had not taken place, the action against ATF would never have gone forward. It was just too expensive.

The appeal was not only going to cost money, but it was also going to take time. Rocketry would file an opening brief and the evidence in support of their appeal by August 2005. ATF would then file a reply brief. All evidence and briefs were to be filed by October 14, 2005. The appellate justices would then review the entire administrative record and make their own determination as to whether ATF acted correctly when it classified APCP as an explosive.

Two weeks after the last flight at LDRS-24 in Lethbridge, Canada, and on August 5, 2005, rocketry's lawyers filed their opening brief with the appellate court. The issue they framed in their brief was a simple one:

> Did ATF exceed its authority when it ruled that the APCP used in hobby rocket motors is an "explosive" within the meaning of 18 U.S.C. § 841(d)?

According to rocketry attorney Marty Malsch, who authored the brief, the answer was yes, ATF had gone too far. "When ignited, an ignitable material will 'burn,' 'deflagrate,' or 'detonate,'" wrote Malsch, "depending on how fast the chemical oxidation reaction occurs, with burning being the slowest, detonation being the fastest, and deflagration being somewhere in between."

Using the record from the lower court, Malsch outlined in his brief the difference between deflagration and detonation. "In a detonation, the reaction is so fast that a supersonic shock wave occurs, producing a characteristic noisy explosion," he explained. Malsch also pointed out that the reaction rate in an explosive detonation is approximately 1,000 meters per second. In contrast, the burn rate for deflagration is much slower. In fact, according to ATF, it is 1 meter per second, he said. APCP, whose only purpose is to serve as rocket fuel, burns at a rate that is a fraction of this speed. It does not deflagrate, and it certainly does not explode, argued Malsch in his brief. "The very last thing any rocketeer wishes to see upon motor ignition is an explosion," he said.

"The administrative record here shows without contradiction that the APCP in a model rocket functions by burning in a controlled manner, with the burn front proceeding at a speed of from 3.81 to 101.6 millimeters per second," said Malsch. These measurements, he added, were not from evidence created by high-power rocketry but from a printed table in an explosives' encyclopedia relied upon by ATF:

> Thus, the administrative record relied on by [ATF] establishes without contradiction that the highest burn rate for APCP rocket fuel (101.6 millimeters per second) is a *factor of ten* below [ATF's] own burn rate threshold for deflagration . . . In sum, the APCP is appellants' members' model rocket motors does not function by deflagration under [ATF's] own definitions and data.

"This is not a case where a reviewing court should defer to an agency's interpretation of ambiguous data," argued Malsch. "This is a case where all of the data contradicts the agency!"

Not surprisingly, ATF had a much different view of things. ATF lawyers urged the court of appeals to uphold Judge Walton's opinion. ATF conceded the primary purpose of APCP was to function as rocketry propellant (and not to function by explosion), but it maintained the compound qualified as an explosive because it did, in fact, deflagrate. Since APCP deflagrated, it was properly included on the explosives list, they argued.

ATF did not provide the court with data that was any different from the data cited by rocketry. Instead, the government asked the justices to again defer to the scientific opinions that were offered to Judge Walton by the agency in district court. Yet those opinions did not come with any data as to how APCP, or any other substance for that matter, fit into any continuum that differentiated burning from deflagration from detonation. Precise measurements were not necessary, argued lawyers for ATF. All that was needed was for ATF "experts" to simply opine that a deflagration reaction is much faster than burning. Based on those expert opinions, APCP deflagrates and is an explosive, argued the government.

In a final written brief in the fall of 2005, Malsch outlined what he believed was at the heart of ATF's argument for control over high-power motors:

Ultimately, ATF is reduced to simply begging for deference. It urges the court to ignore the proverbial "elephant in the room" (the actual scientific data about how APCP-based rocket motors function) and to defer to a novel agency theory that actual scientific data are not necessary to describe and reach conclusions about natural processes because a few unsupported and highly generalized "opinions" about rocket propellants from authoritative sources (ATF refers to this as a consensus) are all that count. Thus, ATF concedes that "it certainly could have conducted experiments or otherwise researched burn rates specific to APCP used in model rocket motors to reach its conclusions . . . but says it was not required to do so . . ."

After all the briefs and papers were filed, the appellate court told both sides that it expected to hear oral arguments in the case in January 2006, before three justices of the circuit court. The court would grant each side only 10 minutes to argue the merits of their case.

The Federal Court of Appeals for the D.C. Circuit is one of the most important appellate courts in the United States. It is responsible for reviewing the rulemaking of many agencies of the federal government as well as hearing appeals from lower courts regarding federal agency actions. In a legislative system where unelected bureaucrats constantly seek to increase their power over the citizenry, the D.C. Circuit Court is often the final arbiter on the limits of that power. For this reason, this court has grown in stature over the years and is now a frequent supplier of judges nominated to the United States Supreme Court. Of the nine justices on the Supreme Court today, four of them—John Roberts, Brett Kavanaugh, Clarence Thomas, and Ruth Bader Ginsburg—were all chosen from the D.C. Circuit.

Physically, the courthouse is located in the E. Barrett Prettyman building on Constitution Avenue, an easy walk to either the U.S. Capitol or the White House. The structure is named after a former chief judge of the court and is listed on the National Register of Historic Places. Pres. Harry S. Truman laid the cornerstone for the eight-story building in June 1950. The courthouse opened in 1952.

There are eleven justices that sit on the D.C. Circuit. Three of them would handle the case of *Tripoli Rocketry Association, Inc. and National*

Association of Rocketry, Appellants v. Bureau of Alcohol, Tobacco, Firearms, and Explosives, Appellee. The three justices were Harry Edwards, Merrick Garland, and David Tatel.

Justice Harry Thomas Edwards was the senior member of the panel. Edwards graduated from law school with honors in 1965 and then landed a job with a Chicago law firm, where over the next several years, he specialized in labor relations and collective bargaining. He later returned to academia, taught law, authored several legal books, and was appointed to the D.C. Circuit Court in 1980.

Justice David S. Tatel earned his law degree and went into private practice while also serving as the director of several civil rights foundations. In the mid-1970s, he was appointed director of the Office of Civil Rights of the Department of Health Education and Welfare under President Carter. Tatel was later appointed to the D.C. Circuit in 1994, taking the seat vacated by Ruth Bader Ginsburg, who was then appointed to the U.S. Supreme Court. In a *Los Angeles Times* profile in the mid-1990s, Tatel was described as an avid Baltimore Orioles fan, outdoorsman, marathon runner, and a mountain climber. Tatel is also legally blind. He lost his sight after law school to retinitis pigmentosa.

Justice Merrick B. Garland was the youngest member of the panel. Garland, fifty-four, was born in Chicago and was a National Merit Scholar in high school. Following law school on, he worked as a law clerk for U.S. Supreme Court justice William Brennan. Between 1979 and 1997, Garland worked in private practice and was also as an attorney with the Department of Justice. He was appointed to the D.C. Circuit in 1997.

The hearing date for rocketry's appeal was Tuesday, January 10, 2006. Oral arguments began at 9:30 a.m. Each side had 10 minutes to present their arguments.

Lawyer Joe Egan presented the merits of the case for rocketry. He quickly went through the high points of Malsch's written brief, explaining the critical distinctions among burning, deflagration, and detonation and how APCP was not an explosive. He pointed out where the lower court made its error in granting too much deference to ATF's opinions, and he answered questions posed by the judges. On behalf of ATF, Atty. Jane Lyons urged the court to confirm Judge Walton's opinion. She argued it was important for the court to allow ATF to keep rocketry propellant on the federal explosives list.

"During oral arguments, both parties were questioned by multiple panel members regarding explosives definitions, criminal and civil penalties for explosive law violations, and case citations," reported Ken Good, who was also present in the courtroom. "Joe Egan did a superb job of presenting our best case in the face of very tough quizzing," said Good later. "Responding to repeated tough questions about burn rate data and related technical assertions that she was having difficulty answering, Ms. Lyons was reduced to openly asking for the ATF to be granted deference. Justice Edwards quickly interjected, saying, 'I thought that was where you were going and that just doesn't cut it for me.' Years later, Lyons would admit to me and Trip Barber that she was not eager to repeat that experience at the appellate court ever again."

The justices did not reach a decision at the hearing. In a joint statement to their respective members a week after the hearing, Good and NAR president Bundick asked for continued patience with the lawsuit, now in its sixth year. Although Good and Bundick were optimistic based on what they observed at the hearing, past experience taught them to be cautious. "We do not expect to receive a ruling until sometime in the summer of 2006," they warned their respective members.

But they were wrong. In a remarkably quick turnaround decision issued on February 10, 2006, Justices Edwards, Tatel, and Garland *overruled* Judge Walton and held that based on the administrative record in the court below, ATF's decision to place APCP on the explosives list was unsupported by the evidence.

"This court routinely defers to administrative agencies on matters relating to their areas of technical expertise," wrote Judge Edwards, who authored the written opinion of the court. "We do not, however, simply accept whatever conclusion an agency proffers merely because the conclusion reflects the agency's judgment. In order to survive judicial review in a case . . . an agency action must be supported by 'reasoned decision-making.'"

The problem with this case, explained the court in its fifteen-page written decision, was ATF's explanation for its determination that APCP deflagrates lacked any coherence whatsoever. There was no reason for Walton to have deferred to ATF's findings, said the court. "We therefore owe no deference to [ATF's] purported expertise because we cannot discern it. [ATF] has neither laid out a concrete standard for classifying materials along the burn-deflagrate-detonation continuum nor offered data specific to the burn speed of APCP when used for its 'common or primary purpose.'"

The court rejected ATF's assertion that a chemical compound such as APCP could be placed on the explosives list simply because it burned "much faster" than other materials. The much faster standard was simply too vague and ambiguous:

> The fatal shortcoming of [ATF's] position is that it never reveals how it determines that a material deflagrates ... We understand that it may be necessary for [ATF] to define a range flexibility to characterize a particular substance. But as a reviewing court, we require *some* metric for classifying materials not specifically enumerated in the statute, especially when, as here, the agency has not claimed that it is impossible to be more precise in revealing the basis upon which it made its determination. Yet in this case, [ATF] has provided virtually nothing to allow the court to determine whether its judgment reflected reasoned decision-making.[ATF's] unbounded relational definition—i.e., "the deflagration reaction is *much faster* than the reaction achieved by what is more commonly associated with burning"—does not suffice because it says nothing about what kind of differential makes one burn velocity "much faster" than another. Ten millimeters per second? A hundred? A thousand?

The court also faulted ATF for what it didn't say in its briefs—namely, the agency ignored rocketry's claim that evidence submitted by ATF actually gave credence to the proposition that APCP did not deflagrate. ATF had relied on the *Encyclopedia of Explosives* as an authoritative source in the case. Yet that publication, as pointed out by Tripoli and the NAR, supported a finding that APCP was not an explosive:

> [Tripoli and NAR] focus on the range of burn speeds illustrated in the *Encyclopedia of Explosives*, arguing that "the administrative record relied on by [ATF] establishes without contradiction that the highest burn rate for APCP rocket motors (101.6 millimeters per second) is a factor often below [ATF's] own burn rate threshold for deflagration (1,000 millimeters [or 1 meter] per second)." The agency's brief says virtually nothing in response to this ... Moreover, the burn rates that [ATF] attributes to detonation support [rocketry's] contention that detonation occurs at a speed representing a

different order of magnitude than the speeds reflected in the *Encyclopedia of Explosives*.

Turning to ATF's claims that it was in a better position than the courts to determine what was or was not an explosive, the court observed that although ATF admitted it could have conducted experiments specific to APCP, it chose not to. "Unsurprisingly then, rather than resting on concrete evidence to support its judgment, [ATF] simply points to evidence relating to the properties of 'rocket propellants' and claims deference on the basis of its presumed technical expertise and experience," said the court. "The purported evidence cited by the agency does not support its determination in this case, and the cry for deference is hollow."

The speedy judicial opinion was a stunning victory for rocketry. It meant Judge Walton was wrong when he ruled that ATF had proved that APCP was an explosive. Yet the ruling did not end the case. Although the three-judge panel could have reversed the lower court entirely and simply ordered the removal of APCP from the federal explosives list, they chose a more conservative approach. The judges reversed and remanded the case back to Judge Walton to start over again—only this time he was to apply the correct legal standard as articulated in the appellate court's written opinion. ATF would have to prove—scientifically and based on real discernible standards—that APCP was an explosive. If they could not, APCP would come off the list.

The tables had turned. ATF's team now had the important legal decision to make. The appellate court made it clear that based on the evidence—both from ATF and high-power rocketry—APCP might not be an explosive. Where was the evidence that APCP deflagrates? asked the court. So far, ATF had presented none, they said.

Given the tenor of the appellate court decision, one option for ATF was to concede that rocketry propellant did not deflagrate and delist APCP from the explosives list. By doing so, they could end a legal battle that had cost the government hundreds, if not thousands, of hours of legal time and expense. Alternatively, ATF could sit down with rocketry leaders to try and negotiate a settlement that kept the chemical on the list but left open an exemption for hobby rocketry—similar to the PAD exemption that had been in place for other items on the explosives list for many years.

A third option was to fight, to return to the district court and stonewall as long as possible or perhaps conduct scientific studies that demonstrated APCP did deflagrate, that it was an explosive. In the meantime, perhaps hobby rocketry would give up or maybe even run out of money. Herein lies the inherent injustice of any fight against a government bureaucracy in this country: unlimited resources and unelected officials who answer to no one. Rocketry, in the role of David, had dealt the government bureaucracy a surprising blow. And ATF, in the role of Goliath, chose to fight on.

In the spring and summer of 2006, both sides returned to Judge Walton's courtroom. ATF told the judge it would prove APCP belonged on the list, that it did more than just burn, and there was no reason to remove it from the explosives list. Judge Walton created a new scheduling order, giving ATF until the early fall to present its new evidence. He also gave rocketry time to do the same. The court then set up a schedule for another round of summary judgment briefs, which were due in early 2007. At least another year would pass before there would be any resolution of the case.

On October 17, 2006, Joe Egan received a letter from ATF assistant director Lewis Raden. The letter was written in response to another inquiry from rocketry's legal team asking ATF to reconsider its position on APCP. In his letter, Raden told Egan that ATF had now conducted testing of APCP that proved—not surprisingly—that hobby rocketry propellant should remain on the explosives list. Attached to the letter was an internal ATF memorandum summarizing a study that summer conducted at the United States Air Force Research Laboratory at Tyndall Air Force Base in Florida. This testing was ATF's alleged proof that APCP was an explosive, said Raden. The letter, testing protocols, test results, and a new administrative record were then sent to Judge Walton. The judge gave both sides until January 31, 2007, to file legal briefs in support of summary judgment—this time based on the new administrative record and the instructions set forth by the court of appeal.

Rocketry's lawsuit against the federal government was about to enter its eighth year.

39

The March of Time

LDRS-25 was hosted by POTROCS on June 29–July 4, 2006, near Wayside, Texas. It was the second time in five years the national launch was held in the Panhandle. And in terms of total Newton seconds burned, this may have been the biggest LDRS in history. In six days of flying, more than 100 M motors or larger were burned. More than seventy such motors were fired during the commercial launch alone. There were multiple O motors, at least four P-powered flights, and the first Q motor to ever launch at LDRS. More than 2,000,000 Newton seconds were expended by 238 flyers who logged more than 900 flights, making this another LDRS mega-launch.

Flying started on Thursday, June 29. Launch director and Tripoli vice president Pat Gordzelik and his team set up rows upon rows of pads on a range located in the middle of nowhere. Big sky and empty land for as far as the eye could see. It looked like the set of the movie *Giant* from the 1950s. This was also the first LDRS to have a wireless launch control system, which worked fine for the entire week.

The record-setting pace for big motor flights started on day 1, with M- and N-powered flights by people like Mac and Steve Heller of Connecticut who launched their rocket with an M1480 or Mike Abbott of Texas with his *Regulus* on a Loki M1882. Texan Jim Parker flew two M rockets on Thursday—his *Concept 75* on an M1060, followed by *Little Blue Pill* on an M1939. There were low and slow flights with heavy rockets as well as minimum-diameter vehicles headed for the stars, taking advantage of the 50,000-foot waiver for the launch.

One of the most interesting flights at LDRS-25 was Art Upton's *Booster Bruiser* launched on a Cesaroni N1100. Upton was the owner of Booster Vision, a company specializing in rocketry video equipment. His modified LOC *Bruiser* was loaded with multiple electronics, GPS, and Booster Vision video equipment. It tore off the away cell on the long-burning N motor for what looked like a perfect flight. Several seconds into flight, which Upton hoped to reach 17,000 feet, the rocket suddenly veered off course, made a series of violent turns, and then totally shredded at high altitude. The cause of the mishap was anyone's guess. The rocket was too high in the sky when it disintegrated. But Upton recovered his onboard video equipment. And his cameras had a story to tell. In one of the more remarkable film sequences in high-power history, with Upton's Booster Vision camera looking down the side of the airframe at two of the fins, the cause of the disaster was clear. The rocket had three oversize black-painted fins. As the rocket increased in velocity, two of the fins began to rock back and forth. As the speed of the rocket increased, so did the movement of the fins. They were increasing in tempo quickly—but in slow motion on the video—until they were in full fin flutter looking like the movement of the wings of a bird instead of solid rocket fins. A fraction of a second later, the fins were ripped off the rocket, leading to the in-flight disaster. It was a tough break for Upton. But his video captured precisely how fin flutter develops and how it looks in flight right up to the moment of disaster. Upton released the video for all to see, and it became part of the *Rockets Magazine* coverage of LDRS-25.

Upton was not alone in less-than-optimal flights at LDRS.

Over the weekend, there were several spectacular mishaps, including an O-powered hybrid that sustained a CATO seconds into flight as well as a two-stage M-powered rocket that lifted off and then quickly turned around to plow itself into the ground under full power. The rocket pierced the hard Texas ground as if it were soft putty.

There were several mega projects over the weekend, each with an incredible amount of firepower. With these rockets and their high-dollar propulsion systems, one could point to the aft end of the vehicle and simply say, "Right there—that's where the money comes out."

Woody Hoburg led the charge with his familiar American flag-themed rocket. The 16-inch-diameter, 131-pound rocket blasted off on an N4000. A few seconds into flight, two M1297 motors airstarted, followed a few seconds later by two more K555 Sparky motors. It was a perfectly executed flight and recovery. The Hillbilly Rocketry team from Arizona launched their famous *Gila Monster* propelled by four M motors and a central N that was going to be airstarted. The igniter for the N motor was toasted by the flaming exhaust surrounding the M motors, either during takeoff or on the ascent. There was no N airstart. Still, the 325-pound rocket had a great ride up and settled back to the ground beneath a huge military surplus parachute. Tim Lehr and friends successfully lofted a rocket called the *Megg* on a central N, three Ms, and three more L motors—all seven motors firing as planned on the pad.

As with every modern LDRS, there were plenty of special events. In addition to bowling ball contests, flying saucers, and even a flying table, there were raffles, special presentations, and a local helicopter company POTROCS had available for hire to take people on aerial tours of the area. On more than one occasion, the helicopter helped flyers locate downed rockets on the vast Wayside range. There was even an airplane performance. "Those of you who slept in on Saturday morning missed a special treat," said Gordzelik. "Before we opened the range, we were provided an air show at no charge by Don Johnson piloting a 1942 *Stearman* and Mark Britain piloting an acrobatic *Pitts*."

On Saturday night, the annual Tripoli members' meeting and banquet was held at the Ritz Hotel in Amarillo. There were speeches by Tripoli leaders and the announcement of the annual election results. Guest speakers included author and engineer Bill Gurstelle and Dr. John Chandler of Texas Tech University. More than three hundred flyers and friends had a great evening that ended with longtime Tripoli members Mark Clark and Robin Meredith entertaining all with their rocketry-related stand-up routine.

The research launching at LDRS-25 began on Monday. It was filled with numerous M-, N-, and O-powered flights, including the 365-pound *Event Horizon* from Colorado on three N4000 motors, Ron Rickwald's U.S. Navy *Standard Missile* on a O7148, and Robert Utley's *Black Eye* on a O5000 motor. On Monday alone, there were ninety flights that accounted for more than 845,000 Newton seconds of impulse—possibly a one-day LDRS record in terms of Newton seconds. Dan DeHart of Tripoli Houston—one of Tripoli's oldest prefectures—launched his *Spartacus* on an experimental O4800 and eight more Cesaroni L730 motors to airstart. The 17-foot-tall rocket was more than 11 inches in diameter and weighed over 200 pounds on the pad. "*Spartacus* left the pad on a huge plume of smoke and fire," wrote Neil McGilvray of *Rockets Magazine*. "The airstarts kicked in as planned leaving a sweet trail of smoke behind the rocket." As big as the project was, it was still hard to see *Spartacus* at its apogee near 16,000 feet, added McGilvray, where the unexpected happened. The apogee ejection charges failed to separate the rocket. The rocket was coming in ballistic, with only the main charges left to possibly save it.

"At the scheduled 2,000 feet altitude, the charges fired for main deployment," said McGilvray. "[But] the rocket was going way too fast . . . and the forces of the high-speed deployment tore the beautifully finished rocket to a million pieces."

There were other big projects waiting in the wings. For the first time at any LDRS, there were several P-powered rockets, including *Spinal Tap* built by Ed Rowe and Mike McBurnett, hoping for at least 20,000 feet on their P9811 motor. The rocket blasted off into beautiful blue skies for a perfect flight and recovery. Walt Stafford of Alabama launched his own specially formulated P18,000 in his *Red Devil* for a great flight. "The tall rocket was looking at over 30,000 feet on the long-burning P motor," wrote McGilvray. "The liftoff was everything you would expect out of a P motor—long flame, loud report and a rocket that went on forever, it seemed, until gravity said, 'You've had enough fun, time to come home.' The flight was textbook perfect."

Another anticipated flight was a *Delta III* scale rocket, powered by a central P8600 and 9 L1200s, all motors created by Gordzelik. The total impulse of the rocket was in the Q-motor range. The one-fifth scale *Delta* was created by a team of more than twenty Tripoli members who designed the rocket to include nine boosters, each holding one of the L motors that would separate from the main airframe during flight by explosive bolts—just like the actual NASA rocket. The flight of the *Delta III* was superb, with all motors lighting on the pad and the rocket shedding seven of the nine boosters perfectly during its ascent.

The Wayside LDRS also saw the first single-motor Q-powered rocket at any LDRS. This was the *Phoenix XL*, led by flyer Sterling Edmunds, who joined with Ed Rowe, Mark Lloyd, and Blaine Jefferies to create a massive 500-pound vehicle that was the heaviest rocket of the launch. The *Phoenix XL* was powered by Lloyd's 8-inch-diameter Q13,800 motor, which accounted for more than 100,000 Newton seconds of total impulse. The skin of the booster was completely removable in one piece. The airframe was bolted in place on the rocket's exoskeleton after the motor was installed. Spectators could see the inner structure of the booster while the team prepared the vehicle for flight. This was one of the biggest single-motor vehicles to fly at any LDRS, ever. The rocket had a spectacular liftoff and rose to an estimated 20,000 feet before deploying all recovery gear on cue and drifting perfectly back to the field.

The final moment of the landing was memorable too. When the rocket touched down, the airframe landed on one side of some power lines near the edge of the range. The parachute then lay down gently on the other side of the lines. The long recovery harness connecting the rocket and the parachute then settled right on the electrical lines, knocking out power to the entire town of Wayside. "Fortunately," noted *Extreme Rocketry* magazine, "the power company was able to respond quickly, removing the rocket from the power lines and restoring power to all affected."

Not a single complaint was received from the citizens of Wayside for the temporary interruption of their power to remove the downed rocket.

In the fall of 2005, the Tripoli Board of Directors solicited bids from anyone interested in creating a new magazine for the corporation.

"As we move ahead to forge a relationship with a new organizational magazine, distribution to potential members is an important goal," said Pres. Ken Good in late 2005. Good and other board members wanted a return to the days when a magazine such as *High Power Rocketry* could be found in hobby stores all over the country. "Those issues grabbed the interest of many current [Tripoli] members and was the medium through which quite a few people became engaged with Tripoli and high-power rocketry," said Good. "Additionally, we must not forget that those manufacturers and vendors who support our form of rocketry are very eager to have advertising access to a reliable publication that has a strong base of distribution."

There were two magazines providing some high-power coverage in America at this time. The first was Brent McNeely's *Extreme Rocketry*, now about to enter its sixth year of production. The other was the NAR's *Sport Rocketry* run by Thomas Beach. Although primarily devoted to model rocketry, *Sport Rocketry* increasingly covered high-power as more NAR members became certified high-power users every year.

In late 2005, the Tripoli board asked for bids for the creation of a new magazine. Among other things, the board said it was looking for a publisher to guarantee at least six issues per year—one every two months—at a minimum of forty-eight pages per issue. The publisher would also have to agree to provide Tripoli with at least six pages per issue (or alternatively, 10 percent of the page count) with what amounted to free advertising for Tripoli. This "advertising" could include, but was not limited to, no-cost ads in the form of prefecture listings, membership applications, or related Tripoli business information. The magazine was required to be available by both subscription and by individual purchase at hobby stores within the first year of publication. In return, Tripoli promised to process magazine subscriptions through the membership applications (at a cost of 10 percent to the publisher), to provide magazine owners with access to the Tripoli membership list for marketing purposes, and to include website links and front-page advertising for the magazine from Tripoli's official web pages on the Internet.

There were several responses to Tripoli's request for bids, including interest from both Brent McNeely at *Extreme Rocketry* and also from the NAR's primary publication, *Sport Rocketry*. In the end, however, the board chose a proposal submitted by a new rocketry company based in Maryland called *Liberty Launch Systems*—owned by Tripoli members Neil McGilvray and Robert "Bob" Utley. They called their proposed publication *Rockets Magazine*.

McGilvray and Utley discovered high power in the mid-1990s. By 2005, they were both in leadership positions with the Maryland Delaware Rocketry Association, arguably the most successful rocketry club in America. Utley and McGilvray were part of the team that launched *The Liberty Project*—the heaviest amateur rocket ever flown as of 2005—and both were former disciples of Jim Mitchell's Thunderflame homemade motor-making courses. They both graduated from Mitchell's classes and went on to make scores of their own motors up through P power. They were outspoken proponents of the increased use and availability of research motors at all Tripoli launches and were frequent participants at rocketry events all over the country, from Whitakers in North Carolina to the Black Rock Desert in Nevada. They were part of the team that created MDRA's semiannual and wildly successful Red Glare launches.

"Our goal is to keep the content positive and fun," they wrote in the first issue of *Rockets* in the spring of 2006. "Rocketry is a hobby after all. It is not our goal to right every wrong or to settle every score of the past. We are only concerned about what is in front of us and from our perspective the future is bright. All you have to do is look around. The rocket clubs from around the country and the world are still in operation, the big projects are still flying and most importantly of all, people are still showing up at launches—the passion is alive and well."

Their first big test for *Rockets* was at LDRS-25 at Amarillo in 2006. The issue covering the event hit the newsstands within ninety days of the launch, providing competition for *Extreme Rocketry*, which also produced its story on LDRS at or about the same time. For the first time in nearly twenty years, LDRS enthusiasts had two stories on the national launch from two different magazines.

"As Tripoli celebrates 25 years of LDRS, *Rockets Magazine* will be there to cover the action," they promised. "This year the event will get unprecedented coverage, with a cover-to-cover issue dedicated to every aspect of LDRS." Indeed, the entire September/October 2006 issue was devoted solely to LDRS-25. It was the beginning of a new tradition that *Rockets* carried forward to every LDRS it would cover in the future. Utley and McGilvray also created a new line of DVDs providing video coverage of LDRS and other launches around the country. Between 2006 and 2018, the pair would attend more regional launches and cover more rocketry events in person than anyone in the history of high-power rocketry.

From Wayside, Texas, LDRS moved to the Jean Dry Lake Bed near Las Vegas, Nevada, in 2007. LDRS-26 went forward on July 12–17. It was the second time in six years the event was hosted by Rocketry of California (ROC), which had hosted LDRS-20 at ROC's home field at Lucerne in 2001. This year, ROC teamed up with Tripoli Las Vegas to host the big event.

The town of Jean is a small commercial center of merchants and casino operators located about 30 minutes south of Las Vegas on Interstate 15, not far from the California-Nevada state line. The town was originally named Goodsprings Junction. In 1905, the local postmaster renamed it Jean to honor his wife. There are no official residents in Jean today, but people in nearby communities such as Primm and Sandy Valley list Jean as their address because that is where the main post office is located. The banquet for LDRS-26 would be held at the Primm Valley Resort and Casino, between Jean and the state border.

The rocketry field at Jean is similar to other Western desert ranges. It is flat dry lake bed, free of obstructions. High-power flyers used the lake bed and similar locations around Jean for many years. Its close proximity to the interstate highway and waivers of more than 15,000 feet made it ideal for rocketry, although the Mojave Desert heat made summer launches a physical challenge.

Although well-planned and coordinated by ROC and Tripoli Las Vegas, LDRS-26 faced a number of hurdles. First, in the fall of 2006, LDRS codirector and ROC leader Greg Lawson—a longtime Tripoli member and high-power flyer—succumbed to liver cancer. Then on the eve of LDRS, a series of unrelated automobile accidents tested the commitment of ROC and its vendors, leading one organizer to call LDRS-26 "the LDRS that shouldn't have happened."

The story begins on the freeway more than a thousand miles from Jean, along Interstate 80 in Illinois. On July 5, rocketry vendor Tim Lehr of Illinois started his trip to LDRS-26. He was driving a full-size RV that towed a large trailer crammed with rocketry supplies of all kinds, including hundreds of pounds of rocket motors. Lehr was one of the biggest rocket motor vendors in America; scores of people planning on flying at LDRS-26 had preordered their high-power motors from him.

Barely a half hour into his journey, Lehr glanced into the rearview mirror of his RV and was astonished to see a huge fireball following closely behind his rig. His trailer—loaded with APCP rocket propellant—was engulfed in flames. He pulled to the side of the highway and scrambled out of his motor home. He was safe. But it was too late for his rocketry supplies. Sparks from either the rear axle, or its bearings, caused overhearing that, in turn, set the wooden frame of his trailer on fire. From there, the fire spread quickly. Once he was off the road and out of his vehicle, Lehr could do little more than just watch as a raging blaze enveloped the entire trailer, sending huge columns of smoke into the air and destroying his entire stock of rocket motors and everything else inside. The intense heat of the fire shut down the interstate freeway, leading to a traffic jam 10 miles long.

"If APCP truly was an explosive, I wouldn't be here right now," said Lehr afterward.

Despite his losses, Lehr continued his trek to Nevada. Along the way, he arranged to pick up additional motors to help customers whose fuel was destroyed by the fire. Once he arrived at Jean, Andy Woerner of What's Up Hobbies organized a raffle to help Lehr cover some of the expenses associated with the fire as well as airfare home after the launch.

Lehr's loss was not the only setback at LDRS-26. On the other side of the country in California, ROC was about to have its own close call on another freeway. On Sunday, July 8, ROC member Wedge Oldham and friend Melinda Catalano were towing all of ROC's launch equipment from Southern California to the dry lake 175 miles away. The equipment housed in the trailer included dozens of launch pads, signs, electrical necessities, and everything else needed to run LDRS-26. Not far from the desert city of Barstow, California, and while on Interstate 15, Oldham sensed the big trailer starting to fishtail behind his Chevrolet Suburban. In a matter of seconds, the vehicle and trailer were headed for the center divider of the freeway.

"Everything went in slow motion," explained Catalano later. "We hit that center divider with such force that I could hear the sounds of breaking glass all around us and the sounds of crunching metal. Then we were being tossed around, and everything went upside down. I could see my cell phone tumbling around as if it [were] in a dryer."

When the vehicles finally came to rest, the Suburban was on its roof with Oldham and Catalano suspended upside down in their seat belts. The trailer somehow was still on its wheels but was seriously damaged. Oldham and Catalano unbuckled themselves and walked away virtually unscathed. But the Suburban was totaled, and the rocketry trailer was inoperable and sitting on a freeway, in the middle of nowhere, miles from LDRS.

Oldham had to find a way to immediately pick up the pieces and move forward. "Fortunately, almost all of the equipment survived the accident and was in good working order," said Catalano. "The only question was how to get it out to the launch site. The trailer was not totaled in the accident, but it was damaged to the point where it could not be used."

Once again, Andy Woerner stepped up to help, arranging for another trailer to transport the launch equipment from the scene of the accident to LDRS-26 in Jean. Incredibly, that replacement trailer was then involved in yet another accident, this one in San Diego County when the truck transporting it was T-boned on the way to assist Oldham. It seemed as though this year's national launch might not go forward after all. But ROC members were determined to find another way. They rented a U-Haul truck, drove to the scene of Oldham's accident, unloaded the contents of the damaged trailer, and got everything to LDRS-26 with time to spare. By Wednesday, July 11, nearly eighty pads were set up and ready to go. Never in the history of Tripoli's national launch had organizers and vendors persevered through as much personal loss, fire, and multiple accidents to get the event off the ground.

Yet there was still one more challenge left overcome: the weather. It is always hot in the summer in Jean. The average high temperature in July is 104. As it turned out, flyers at LDRS-26 would have been happy if they had been stuck with that kind of heat. Instead, for the first four days of LDRS, the temperatures on the dry lake bed soared to 115 degrees and higher, posing dangers of heat stroke for flyers and spectators alike.

"The heat was virtually unbearable at times," said Oldham, who credited launch organizers for imposing mandatory breaks for the entire launch crew during the day for rest and water. The temperatures also changed the habits of many flyers, who adjusted their schedules accordingly. "The campers [were] prepping their rockets throughout the night to take advantage of the cooler conditions. What a choice, prep in the dark of the night or melt in the heat of the day," observed *Rockets Magazine*. All flyers were grateful to the creators of the short-lived *Launch Magazine*, which erected a 12,000-square-foot enclosed tent—that's twice the size of an NBA basketball court, complete with air-conditioning—that remained open all week so everyone could briefly escape the heat and cool down during the day.

LDRS-26 would not be a record setter in terms of registered flyers or total flights. But it was still a successful launch. More than 900 flights were recorded, A through O power. At least sixty motors were M impulse or higher, and there were 150 flights were on research motors on Monday and Tuesday.

The rockets at LDRS this year were big and brutish, burning up thousands of dollars of APCP to get a few thousand feet into the air. Among the biggest flights of the weekend—and one of the most powerful rockets of LDRS-26—was the 350-pound *Gemini DC* built by the Upscale Rocketry Team from Arizona. The 30-foot-tall rocket had a main airframe more than 14 inches in diameter plus two outboard pods, 10 inches in diameter each. The rocket was powered by a single M1939 and four M1315s. All five motors were fired on the pad, and the heavy rocket lumbered up a few thousand feet, turned over at apogee, and deployed nearly twenty parachutes, big and small, from all over the airframe.

Another group of flyers launched an upscaled *Thor-X* on an Animal Motor Works N4000 and three M2200 Skidmark motors. The

335-pound *Thor* lifted quickly off the pad for a great—but not very high—flight and recovery. Hillbilly Rocketry of Arizona also fired up a rocket over 300 pounds. Their *Hillbilly One* was a 16-inch-diameter rocket powered by a central O-impulse motor surrounded by four M motors. The big rocket reached more than 9,000 feet in altitude and had a fine recovery under parachute. Additional memorable launches included Gary Byrum and Terry Drake's beautiful red *Bomarc*, which at more than 100 pounds had a terrific flight on two L1420 motors off the pad, followed by airstarts of two additional K550s, Team Rage's 20-foot-tall *Hawk Missile* that blasted off on an N4000, Jack Garibaldi's 160-pound upscaled Estes *Citation* on an AeroTech N2000 and three Cesaroni L730s, and a 13-foot-tall-scale *Mercury Redstone* on M power flown by Mark Hayes.

Darrell Burris launched his rocket, *Sweet T*, with a central M1315 and six K550s. The rocket lifted on the M motor, and seconds into the flight, three of the K550s airstarted for a great special effect, followed in another several seconds by the last three K motors. The rocket cleared 12,000 feet and had a fine recovery. Veteran flyer Vern Knowles of Idaho flew several different rockets on M power, all recovered successfully.

The Gates Brothers returned to LDRS after a two-year absence to fly their *Porthos* on Sunday, powered by a central M2500 and six K1100 motors. The 200-pound, 12-inch-diameter rocket looked fantastic on the pad as the crew and a group of photographers and videographers got their cameras ready, awaiting countdown. Then according to *Rockets Magazine* writer Neil McGilvray, the unexpected happened.

"*Porthos* was ready to launch just as a major dust devil was racing toward the away cell," said McGilvray. "It was decided to allow the cloud of sand to pass. As the last remnants of the storm passed the tower, and with everyone's cameras tucked away to avoid infiltration of dust in the sensitive lenses and electronics, a blue flash left the ground with a mighty roar. *Porthos* was soaring straight up on a huge trail of blue flame. Everyone was looking at each other, all asking the same question: 'What happened?' Some people used more flowery language."

There were several explanations for the premature launch. The best guess was that a final gust of wind from the dust devil knocked over a ladder that had been left standing near the rocket. The ladder fell on a relay that was part of the launch electrical system, and this unintended electrical contact fired the 200-pound *Porthos* into the sky.

"It was a beautiful sight to see the big blue rocket flying on seven Blue Thunder motors, and sad that nobody even raised their camera," noted Neil McGilvray. "Which begs the questions: 'If a rocket flies at LDRS and no one takes a picture of it, did it really fly?'"

Research launching on Monday was cut very short by yet another dust storm. Launch organizers took the downtime as an opportunity to introduce many people to the future of high-power rocketry's igniter systems. In a group discussion attended by as many as fifty flyers and led by Pat Gordzelik, Wedge Oldham, Rick Dickinson, Darren Wright, Doc Hanson, and others, the use of copper thermite was discussed as the latest preferred method to ignite multiple motors in large-scale rockets, especially multiple M, N, O, and larger rockets.

LDRS returned to Kansas in 2008. This was the fifth time Kloudbusters had hosted Tripoli's national launch in Argonia, and they ran the entire event like a machine. The launch was almost professional in nature. In the months leading up to the event, Kloudbusters secured the necessary waivers, made the appropriate hotel and banquet arrangements, set up and maintained an attractive dedicated web page, and attended to the hundreds of other details that make any LDRS worth traveling to.

And they made it look easy.

When dawn came on day 1 of the launch, the range at Rick Nafzinger's farm was meticulously laid out with more than eighty rocket pads. "We held four major organized work parties leading up to LDRS-27," said LDRS-27 codirector Lance Lickteig. "[W]e laid 540 feet of buried cable for power at the storage trailer and LCO position, cleaned and repaired signs, cables, and other infrastructure, performed general site cleanup, and assembled all the flyers' packets." Kloudbusters also secured volunteers for pad manager duties and parking enforcement, an often overlooked detail at national launch events. They made liaisons with the local authorities in Argonia, and they had representatives to deal with the launch hotel back in Wichita. Kloudbusters even had their own firefighting crew, just in case.

"It's been our history that we try to make people happy when they come out here," explained Bob Brown, who, together with Lickteig, codirected LDRS-27. "We treat everyone like friends and family, and we hope that comes across."

In six days, nearly 1,100 flights were logged, A through Q power. The total attendance at the launch was in the thousands. At one point on Saturday afternoon, there were nearly nine hundred cars, trucks, and other vehicles in the range parking area. Yet for the flyers arriving from thirty-eight states and several countries, it was all play with no side effects. There were no long lines, no shortage of RSOs, no issues with pad equipment, away cells, or range facilities.

"The people, camaraderie, and friendship are what drive an Argonia LDRS," explains Kloudbusters cofounder John Baumfalk, who organized the first national launch here in 1993. "Rockets are just something we do together. Rocket people are like family. And that's why they keep coming back here because we treat them like family."

There even seemed to be less CATOs in 2008, especially during the research days on Monday and Tuesday, suggesting the proficiency of the research arm of Tripoli was reaching a level unparalleled in high-power history. More people were building their own motors in the upper impulse range—L through Q power—and they were doing it without any problems. Their motors were working, and they were working well.

In fact, the most memorable piece of destruction at LDRS in 2008 was a research Q-motor project where the motor outlived the rocket. The result was a spectacular ride up for the 425-pound vehicle and an equally memorable dive back with the rocket in pieces because of airframe and altimeter problems. The motor worked perfectly.

Even the weather cooperated in 2008. To avoid a repeat of having to launch in the extreme temperatures at LDRS-26 in Nevada, the Kansas launch was moved from mid-summer to Labor Day weekend, August 28–September 2. It was warm every day, but a steady breeze—even some wind at times—kept the range temperatures well under one hundred degrees. Humidity was mild too, and for the majority of the launch, the skies were either blue or blue with rolling white clouds. This ideal flying weather allowed several flyers to take a crack at the FAA waiver of 50,000 feet at Argonia.

There were few mega projects at LDRS-27, but there were scores and scores of flights with L motors and above—more than 150 such motors by one count, including 87 M motors, 24 Ns, and 4 Ps burned. Several flyers, including Jim Parker of Texas, Andrew Grippo of Louisiana, and Bob Haas of Alabama, launched multiple M- and N-powered projects—almost a rocket every day. Haas reached a combined altitude on his multiple projects of more than 100,000 feet, including two flights at or above the 30,000-foot mark.

Vern Hoag launched a 1/25 scale *Saturn V* that was 16 inches in diameter and stood 15 feet tall. It was powered by an AeroTech M1850 and four Animal Motor Works L777 Skidmark motors. The 200-pound *Saturn V* lifted perfectly and reached 4,000 feet before turning over and returning via parachute for a flawless flight. For the time being, Hoag's rocket was the largest *Saturn V* in high-power rocketry history.

Another big project at LDRS-27 was the *Sky Maven* by Tripoli Minnesota. *Sky Maven* was powered by a Richard Hagensick P2700 motor. The 20-foot-tall rocket was adorned with World War II pinup artwork, and after ignition, it reached nearly 15,000 feet for a nice flight and uneventful recovery. Walt Stafford of Alabama had success with his P-impulse flight on Monday in his *Red Devil*, which reached 26,000 feet before setting back down perfectly a mile away. Ed Rowe and Mike McBurnett returned to another LDRS with their well-worn *Spinal Tap* on a P11,718 motor. Their rocket cleared 30,000 feet.

The scale versions of the *Patriot Missile* were everywhere during LDRS-27, a reminder of the Argonia events in the mid-1990s. Paul Nossman of Texas flew the biggest *Patriot*, a three-quarter scale, 180-pound vehicle that had a good launch on a central N motor and four Ks. Lee Brock of Alabama launched a *Patriot* on a J motor, Mike Tyson of Maryland sent up an unpainted *Patriot* on an M1939 for a successful Level 3 certification, Ken Herrick of Illinois flew a *Patriot* on N power, and Vern Hoag added an M-powered *Patriot* to his *Saturn V* flight.

Carl Hicks of Alabama sent up what may have been the most exquisite *Patriot* to ever grace any LDRS. Hicks's *Patriot* was a museum piece down to every last nut, bolt, and decal. It was too nice to fly really, but Hicks launched the rocket on an N2000 anyway. "It did everything I wanted it to do except it deployed the main at apogee," said Hicks. "The flame from the motor was longer than the rocket, which reached 7,431 feet. We recovered the rocket a half a mile away, and the only damage was a little bit of dirt and dust, which I polished up quickly."

There was no bowling ball loft at LDRS this year, but there were drag races, raffles, and a contest called Mach Madness, an event that presented a prize to the fastest rocket of the week. There were plenty of contenders, with the winning team reaching achieving a speed of almost Mach 3.

40

Rocketry vs. the U.S. Government

Part 4

Friday the Thirteenth

By the spring of 2007, rocketry and the ATF were ready to do battle again in the federal courts in Washington, D.C. The case of *Tripoli Rocketry Association, Inc. and National Association of Rocketry v. Bureau of Alcohol, Tobacco, Firearms, and Explosives* was sent back to Judge Reginald Walton in 2006 after rocketry's victory in the Circuit Court of Appeals. That victory reversed Walton's 2004 summary judgment decision where he ruled APCP was an explosive.

Now that he had the case back in his courtroom, Walton ordered the parties to again brief the issue of whether APCP should be on the federal explosives list. There was no hearing date scheduled for oral arguments. But opening, opposition, and reply briefs were filed by both ATF and Tripoli/NAR.

In the opening brief for rocketry, lawyer Marty Malsch respectfully reminded Judge Walton of the reason why the D.C. Circuit Court sent the case back to him for another look. "[ATF] lost on appeal because it had 'neither laid out a concrete standard for classifying materials along the burn-deflagrate-detonate continuum nor offered data specific to the burn speed of APCP when used for its common or primary purpose.'"

In other words, to deflagrate or not to deflagrate? That was the question ATF had to answer, said Malsch. "The fatal shortcoming of

[ATF's] position is that it never reveals how it determines that a material deflagrates," argued Malsch. "[T]he agency never defines a range of velocities within which materials will be considered to deflagrate."

In an attempt to correct the mistake they made first time around, ATF supplied Walton with a new administrative record that allegedly proved APCP deflagrates. Yet to rocketry supporters, the evidence presented by ATF was laced with subterfuge, sleight of hand, and misinformation.

In 2006, ATF commissioned a study by John Hawk and Robert J. Dinan at a United State Air Force Base in Florida. The Hawk-Dinan report tested five different AeroTech motors: F50-9T, G40-10W, G80-10T, H124-PFJ, and J350W-M. The purpose of these tests was to extract a burn rate for hobby motors when used for their "common or primary purpose." The higher the burn rate, the more likely it could be classified to deflagrate, believed ATF. The study also determined burn rates for candles and bond paper to have other items that burn to compare with rocket motors.

To establish the burn rate of each motor, Hawk and Dinan divided the *total length* of the motor by its burn time. Thus, the two experts—whose expertise was not disclosed in the record but was later called into question by rocketry—determined the burn time for APCP motors was in a broad range between 36 and 143 mm/second.

Surprisingly, these burn rates were lower than the deflagrate definitions ATF had previously presented to Judge Walton before the case went to the court of appeals. In 2004, and then at the court of appeals, ATF suggested a definition of deflagrate that was much faster than ordinary burning. It began at *meters per second*—not millimeters per second. And the low end for deflagration at that time was 1,000 millimeters per second. Using the 2004 standard set by ATF, the Hawk-Dinan findings that APCP burned in a range of 36 to 143 mm/second proved that APCP in hobby motors did not deflagrate.

This was a problem for ATF. But the agency thought it had found a way around this conundrum. The government would simply change the deflagrate definition. "Deflagrate" no longer began at 1,000 millimeters per second, argued ATF's new experts, ignoring the agency's prior position. Instead, they urged Judge Walton to look at the definition in an entirely new manner by looking at another substance that was placed on the explosives list and then comparing that substance's burn properties with APCP.

The substance ATF chose to use for comparison with APCP was safety fuse. On the burn-deflagrate-detonate continuum, safety fuse clearly did not detonate; it burned. Yet if Congress placed safety fuse on the explosives list, claimed ATF, it must deflagrate. (Materials that simply burn do not make it to the list, they argued.) ATF's experts took a length of safety fuse, measured its burn rate, and defined a new floor for deflagration. That floor was an incredibly slow 7.5 millimeters per second. Now their attorneys argued that if a substance burned at a rate equal to or faster than safety fuse, or 7.5 millimeters per second, the agency was justified in finding that said material deflagrates. Comparing the Hawk-Dinan AeroTech burn rates (36–143 mm/second) with the new deflagrates definition of 7.5 mm/second, APCP belonged on the federal explosives list, they said.

It was a clever argument. But to rocketry's legal team, it was all nonsense.

The attorneys for Tripoli and the NAR told Judge Walton, among other things, that it was absurd to use safety fuse, which burns at a low rate, as a benchmark for deflagration. "Safety fuses are licensed and regulated like explosives not because they deflagrate or explode themselves," argued Malsch, "but because their purpose is to initiate an explosion in something else." Malsch pointed out that in ATF's air force study, even ordinary bond paper burned at a rate that was as high as 55.8 mm/second, leading Malsch to point out that the government was adopting "the patently ridiculous position that APCP burns fast enough to 'deflagrate' like an explosive, even though it burns no faster than ordinary white office paper."

Perhaps bond paper should be on the federal explosives list, suggested Malsch.

There was something else wrong with the Hawk-Dinan report, noted rocketry's lawyers. They had not only used the wrong definition for deflagrate, but they had also used the wrong methodology to determine the burn rate. ATF's experts arrived at their burn rate numbers by dividing the length of each rocket motor by its burn time. But APCP in rocket motors does not burn in a linear fashion along the length of the motor; rather, the fuel burns from the inside out, starting from the central core and moving directly outward to the motor case. This is a *radial burn rate*, explained Tripoli vice president and rocketry expert Dr. Terry McCreary, an associate professor of chemistry at the University of Kentucky, who supplied an affidavit in support of rocketry's new summary judgment motion:

> The APCP rocket motor grains tested in the Hawk-Dinan report have hollow cores that are ignited almost instantaneously along the entire length of the propellant grain or grains. As a result, the APCP in the tested hobby rocket motors burns along the entire length from the inside surface (the surface of the hollow core) to the outside surface, a much shorter distance than the total length of the motor or the propellant grains. Calculating a scientifically valid burn rate for this hobby rocket, APCP would therefore entail dividing the measured burn rate durations by the difference between the inside and the outside diameter of the fuel grain or grains (thickness), not the length of the grain or grains (or motor). Since the Hawk-Dinan Report used the wrong dividend, its reported burn rates are scientifically invalid.

In other words, ATF was comparing apples with oranges. The essential measurement was not length of a motor—as urged by ATF's experts—but its width or, more technically, the web thickness. This was the distance between the inner surfaces of the core and the outer edge of the propellant grain. Since the web thickness of a hobby rocket motor is only a fraction of its length, the "corrected" burn rates for APCP were far less than those ATF had presented the court. Relying on AeroTech's published data on burn rates and diameters for its motors, McCreary determined the burn rate for APCP in hobby motors was not 36–145 mm/second as suggested by ATF but 4.25–7.33 mm/second—less than the benchmark definition of safety fuse (7.5 mm/sec) adopted by the ATF. So even if the court were to adopt the absurd deflagrate definition urged by ATF, that definition would still not include APCP since it burned slower than 7.5 millimeters per second.

ATF was in a pickle. To refute the new data put forth by rocketry, ATF's counsel again asked Judge Walton for deference to the agency's own experts. The counsel also questioned Dr. McCreary's conclusions regarding the burn rate of the AeroTech motors at issue. Using an affidavit from another alleged expert—David Shatzer, an employee of ATF—agency lawyers said Dr. McCreary had figured it wrong. His burn rate calculation using the web thickness of the motor instead of its length should simply be ignored, they urged.

In reply, and in a final affidavit filed with the court to counter ATF's newest assertions, Attorneys Joe Egan and Malsch turned for help to the father of high-power rocketry: Gary Rosenfield of AeroTech.

"To eliminate any lingering doubts about this," Malsch told the court, "plaintiffs asked Mr. Gary Rosenfield, the president of the nation's preeminent APCP vendor, and someone with decades of experience in designing and testing APCP [motors] to describe how the APCP [motors] tested by ATF actually burn and how burn rates are scientifically determined. It is important to note that the motors [ATF] tested were actually *his* motors."

Rosenfield's affidavit, filed with the Judge Walton's clerk on April 10, 2007, described the rocketry pioneer's education, training, and background in hobby rocketry and traced the development of AeroTech from its beginnings in 1982 to the present. There was no doubt Rosenfield was a bona fide expert the APCP used in high-power rocketry for more than thirty years.

"I have devoted most of my professional career to the science and study of APCP for hobby rocket motors and for rockets generally," testified Rosenfield before the court. "To the best of my knowledge, there is no living person who has manufactured, tested, and sold more APCP rocket motors than I have." AeroTech's founder then addressed the claims made by the government, summarily dismissing ATF's argument that propellant burns in a lengthwise manner:

> This is not an accurate description of how APCP actually burns in the hobby rocket motors tested by ATF. Nearly all APCP hobby rocket motors, including those tested by ATF, share the characteristics of having an open "core" of some shape and size running roughly through the middle of the propellant, sometimes all the way to the top of the motor. Mr. Shatzer appears to have failed to account for the significance of this core. Upon ignition, the entire inner core is ignited virtually instantaneously (within a small fraction of a second) and the propellant burns *radially* from the core outward until the propellant is consumed. This core-to-casing outward radial burn of the propellant takes place uniformly and virtually simultaneously along the entire radial length of the propellant.
>
> Thus, the true "linear distance" that is measured, if one wants to measure the burn rate of the propellant by observing the burn of an actual assembled motor, is the *radial distance* between the inner surface of the core to the inner surface of the cylindrical casing. So in fact, the overall length of the

motor has nothing whatsoever to do with measuring the burn rate of the propellant in an actual rocket motor, contrary to Mr. Shatzer's view.

Using this correct method to determine burn times, Rosenfield provided the judge with the accurate APCP burn rates for the five of his AeroTech motors that had been tested by ATF. The rates ranged from a low of 4.23 mm/second for the G40-10W to a high of only 7.31 mm/second for the G80-10T.

All were below ATF's arbitrarily low figure for safety fuse at 7.5 millimeters per second.

"[ATF's] peculiar testing methodology for the APCP motors it tested has no basis in the technical literature, is not used by anyone in the rocketry business, and is not endorsed by any professional or standard-setting body," concluded Rosenfield in his affidavit. "Indeed, I have never heard of anyone trying to measure burn rates in this manner."

By the summer of 2007, the court had all the documentary evidence it needed. All that was left was oral arguments by the lawyers. Yet a hearing was still not scheduled. The primary reason for this latest delay was Judge Walton had fallen behind in his calendar. Indeed, for much of the first half of 2007, Walton was presiding over what was at that moment the biggest jury trial in America: the criminal action filed by the government against I. Lewis "Scooter" Libby, a former high-ranking official in the Bush Administration. Libby's trial began in Walton's courtroom on January 16, 2007, and ended in early March with a conviction on some of the counts against him. However, posttrial motions and sentencing hearings continued the case all the way to June. (Ultimately, Walton sentenced Libby to thirty months in prison and assessed fines of $250,000.)

The Scooter Libby trial and other pressing issues meant delays for all other litigants in Walton's schedule for the rest of the year. This included rocketry's case against ATF. In the meantime, another Tripoli national launch, LDRS-26, went forward at Jean Lake in Nevada that July. This was the eighth LDRS held during the life of the lawsuit.

The end of the lawsuit seemed to be in sight now, but rocketry would have to wait for Walton to find the time to review his file and take the next step. As the 2008 New Year holiday came and went, the lawsuit entered its ninth year on the federal court docket.

———◆———

Joseph R. Egan, the NAR-member-turned-rocketry-lawyer, would not live to see the end of the battle he had fought so hard to win. He died of cancer at his home in Naples, Florida, on May 7, 2008.

The son of Minnesota turkey farmers Dick and Lucy Egan, Joe Egan joined the NAR when he was nine years old. Later, in response to a call for legal help in the early 1990s, he provided valuable services to the hobby as a lawyer. "Joe walked into the NAR board meeting in Phoenix in 1992 in response to a call for a volunteer lawyer for the NAR," recalled Mark Bundick. "He proclaimed, 'I think you have a problem with the FAA, and I think I can help you.' Working then for Shaw Pittman, he secured nearly a quarter million dollars of pro bono work that resulted in changes to Federal Air Regulation Part 101. The changes that he secured for all U.S. rocketeers now benefit our clubs and, more importantly, the high school students involved in the Team America Rocketry Challenge."

Egan's primary field of expertise was nuclear law. He earned degrees in physics and nuclear engineering, and at one time, he worked in the control room of a nuclear power plant. He earned his law degree from Columbia. After law school, he worked for private firms in New York and Washington, D.C., before starting his own law practice—Egan, Fitzpatrick, and Malsch—in 1994.

Egan represented national governments in disputes over nuclear waste as well as whistleblowers reporting illegal dumping of radioactive materials in the United States. He was a past president of the International Nuclear Law Association and played a role in nonproliferation agreements and the return to the United States of weapons-grade uranium from more than forty nations around the globe. Egan's biggest nuclear battle involved the government's proposed $77 billion Yucca Mountain nuclear waste dump in Nevada. Although he was a fervent supporter of nuclear energy, Egan took the case in 1996 after learning that the Energy Department discovered the mountain to be highly porous and would therefore leak contaminated materials far sooner than originally predicted. According to the *New York Times*, the legal work performed by Egan and his firm "set back the Energy Department's project at Yucca by years."

Egan's firm was hired by Tripoli and the NAR in 1999 in hopes that the lawyers might be able to reach an agreement with ATF to back off hobby rocketry. When settlement negotiations failed, he led the fight against the agency at both the district court and then the appellate court level. With his partner Martin Malsch on the briefs, Egan argued rocketry's case before Judge Walton and then before the D.C. Circuit Court of Appeals.

"He was a rocket enthusiast and champion for our cause," wrote Tripoli Pres. Ken Good. "Seeing him in action in court, I can attest that he invariably presented our case brilliantly and compellingly. I will forever remember his calm composure in front of Judges Tatel, Edwards, and Garland of the appellate court, answering their tough questions and conveying our arguments in a clear, effective, and irrefutable manner."

"[W]e knew him and enjoyed his company as a rocketeer first and a lawyer second," echoed NAR president Bundick. "Whenever we got together for dinner prior to a court hearing, before we settled into the business at hand, there was always time to talk about rockets. Having your lawyer share the knowledge of your hobby, to say nothing of the passion of its practice, is an uncommon, unique experience. It's those rocketry times, more than anything else, which I'll treasure from my time with Joe."

Egan looked forward to the day when he would get a second chance to orally argue rocketry's position on APCP with Judge Walton in court. Now the case would have to go forward without him.

"This has been the best two years of my life," said Egan as he battled cancer while at the same time enjoying the many passions of his life—boating, fishing, traveling, cooking, and collecting wine and coins, along with spending time with his family and working until the very last day of his life.

Joe Egan was fifty-three.

The lawsuit between rocketry and ATF seemed to go nowhere in 2008. The most significant thing that occurred was the passing of Joe Egan. Otherwise, the case remained in a kind of legal limbo. LDRS-27 came and went at Argonia over Labor Day. It was the ninth LDRS to take place since the lawsuit was filed in 2000. Still, there was no news from the courthouse.

"Where in the world is Judge Reginald B. Walton?" asked *Rocketry Planet* creator and online Internet moderator Darrell Mobley. "No one in the hobby knows. Just like no one knows why the judge still hasn't issued a ruling. But the latest speculation that the judge was awaiting the November presidential election seems to be the only one with any validity, or at least believability, to it. But once November comes and goes and we still have no ruling, well, what then? Your guess is as good as mine."

In October, and having heard nothing from the court in months, Ken Good urged Malsch to ask the judge for a status hearing on the long-pending motions for summary judgment filed by the parties in 2007. "We had been part of several status hearings before Judge Walton in the history of this case. Most of them didn't amount to much," recalled Good. "But our strategy was to suggest that the requested hearing would afford the court an opportunity for the parties to provide any information needed that may be deterring a ruling on the case." Judge Walton granted the request, and he set a routine status hearing for early 2009.

The hearing on the status of the cross-motions for summary judgment was held before Judge Walton in Washington, D.C., on Friday, March 13, 2009. Since this was only a status hearing, there was little chance Judge Walton would make any important decisions. But the hearing would give the parties an opportunity to talk to the judge and inquire as to anything further that needed to be done while awaiting his decision. To prepare for the hearing, Malsch, Good, and Trip Barber—the new president of the NAR—planned a strategy for helpful remarks and, if needed, explanatory arguments.

"We agreed that it was essential that if Judge Walton seemed unsure about anything, we would need to explain with graphics that a layman could understand how a core-burning APCP motor actually functions," said Good. "We came up with some good graphics that we were ready to present as court exhibits. We also crafted a condensed explanation of how the ATF had erred in their testing and what the true burn rates for the AeroTech motors were. These arguments were in our papers, but Marty got prepared for a clear and concise outline by way of an oral presentation."

On the day of the hearing, Malsch, Good, and Barber met in the courthouse cafeteria for a last-minute strategy session. Along the way, they ran into a local Tripoli member, Mike Tyson of nearby Maryland, who had decided to attend the proceedings simply out of curiosity.

"I was relatively new to high-power rocketry, and I was interested in the status of the lawsuit," said Tyson. "When I saw that there was going to be a status hearing, I asked Ken Good if it was okay for me to go. Of course, he said yes. I commute through Washington, D.C., so it was easy to get there, so all that added up to me going."

When the hearing convened on the morning of March 13, Judge Walton took the bench and immediately commented on how long the case had been pending before him. He said he wanted to bring it to a conclusion. The nature of the status conference quickly changed. Suddenly, the gathering turned into a full-blown hearing on the pending summary judgment motions filed in 2007.

Atty. Jane Lyons again appeared for ATF. She urged the court to deny rocketry's request to remove APCP from the explosives list and asserted that the new administrative record—supported by the Hawk-Dinan study and the David Shatzer affidavit—satisfied the requirements imposed by the court of appeals that ATF establish standards for deflagration. She also asked the court—yet again—to defer to the federal agency's findings.

Malsch took the place of Joe Egan arguing for rocketry. He urged Judge Walton to grant summary judgment in favor of rocketry on the grounds ATF's studies were erroneous, they had relied on the wrong methodology, and the agency failed to follow the order of the D.C. Circuit Court of Appeals. There was still no standard for deflagration, argued Malsch, other than what ATF already decided to do beforehand with APCP.

"I was initially slightly concerned that we did not have Joe Egan and that Marty would have to handle our oral presentation," said Good later. "I had always regarded Marty as the best author of our written papers, while Joe was the attorney with the courtroom skills. But I shouldn't have worried. Marty handled himself very well, fielded the judge's questions decisively and with clarity, and was able to lead Walton to understand the fundamental flaw of ATF's testing and conclusions."

As the hearing progressed, Judge Walton nodded in agreement with Malsch regarding an important objection to some of the evidence offered by ATF. The judge then ruled to exclude the burn times suggested by ATF employee David Shatzer on the grounds that the affidavit was not prepared by the agency during the course of their investigation into APCP but later after they had already decided to label the compound as an explosive. Walton also expressed doubt that ATF had satisfied the requirements that had previously been set for defining deflagration by the D.C. Court of Appeals.

As the questions, answers, and arguments continued to unfold in the courtroom, Mike Tyson, who had been sitting quietly in the gallery and who had never seen any of the proceedings before this one, turned quietly to Ken Good.

"This looks like it is going pretty well for us," he whispered. Good nodded in agreement. It was going well.

As the judge continued questioning Lyons, Walton distilled his concerns into one important question for ATF's attorney:

> Since rocketry had presented burn data evidence that seemed to be reasonable, scientific, and based on industry standards, and this evidence was not in agreement with the ATF's calculated test data, where was it substantiated in the agency's record that it had reviewed and considered this contrary evidence and set forth the rationale for not agreeing with it?

At this point, said Good, Attorney Lyons began a semi-frantic search through a large binder she had brought in to the courtroom, presumably holding the ATF's administrative record. Good turned to his left and whispered to NAR president Barber, "She can look all day through that material. She won't find it," said Good. "I know because I have looked at every page of it."

As Lyons combed through her files, the judge looked around the courtroom briefly, and then he announced it was time to take a recess. "I will permit counsel to continue to look for a few minutes while I take a short break," said Walton, who got up from the bench and left the courtroom through a private door in the back wall. It was quiet now; the only sound in the courtroom being ATF's attorney continuing her furtive search through the many binders, boxes, and files she brought with her to court.

Five minutes passed. Then ten.

Judge Walton slipped through the back door and into the courtroom again. He quietly resumed his position on the bench. He asked Lyons if she had found any pertinent citations in the record to answer his question.

No, she had not, she replied.

A few more moments went by. The judge looked around. He glanced down at the papers in front of him. He looked up again.

Then Judge Walton announced he had made up his mind. He was ruling in favor of rocketry. APCP was coming off the federal explosives list, he announced. The nine-year legal battle between Tripoli and the NAR against the U.S. government was over. Just like that.

"It seemed like a bolt from the blue!" exclaimed Tyson, who had come to court that morning to observe a routine hearing and ended up being witness to rocketry history. "We were totally unprepared for a ruling, and there it was. We won! No fanfare. It was just there. Boom! It's over."

"We were surprised—and of course, overjoyed—that he stated this flat out and without qualification," said Good, who immediately reported the judge's stunning decision to the Tripoli board. "I was ecstatic," recalled board member Pat Gordzelik. "I had just sold my company and was living in Texas in my motor home and pouring concrete when Ken Good called me. I was light-headed from the exhilaration of the victory. I didn't think the news was going to come back that quickly and be good. It was a pleasant blindside."

But Tripoli's president cautioned the board with a warning too. Because of Judge Walton's case backlog, it could take a few weeks before he sent his formal written decision to the parties. So there would be a little more waiting, explained Good.

Yet the judge's written opinion was issued the very next business day—Monday, March 16, 2009. In that decision, Judge Walton granted summary judgment to Tripoli and the NAR, finding that ATF failed to prove the rocketry fuel used by hobbyists all over America belonged on the federal explosives list. Wrote Judge Walton, "Specifically, [ATF] did not adequately explain why it came to the decision it did in light of contrary evidence in the administrative record submitted by the plaintiffs, which tended to show that APCP can burn at a rate lower than that which the defendant designated as a threshold, and which, if true . . . would require a change in the proposed rule."

In other words, and unlike the decision he made in 2004, this time Judge Walton was not going to defer to ATF's alleged expertise, particularly where the record showed there was plenty of evidence to the contrary and that evidence was simply ignored by the government. Having made his decision that ATF had improperly placed APCP on the explosives list, Walton had two choices for a remedy. On the one hand, he could have done just what the appellate court did—remand the case back to ATF for yet another chance to get it right and fix the administrative record. Instead, he took the alternate approach. APCP would be removed from the explosives list, he ordered. And since APCP was no longer on the list, any remaining issues in the lawsuit related to propellant-actuated devices (PADs), or allegedly inadequate public comment periods, or the 62.5 gram limit on motor size, were irrelevant.

"As we parted company that day after the hearing, we all agreed it was unfortunate Joe Egan didn't live to see the victory he had worked so hard to achieve," said Good. "Marty had done a superb job, and we told him so. Trip and I agreed that he, Mark Bundick, and I would meet again to have a victory toast in Joe's honor."

"We were truly fortunate to have Joe Egan and Marty Malsch representing us," said Ken Good in 2014, more than five years after the ATF's involvement in rocketry had ended. "Joe, in particular, understood rocketry, having been an enthusiast himself. But the challenge was still for Joe and Marty to fully appreciate the scientific nuances of APCP and how to translate important technical facts into language that a layperson could comprehend. Frankly, when I attended my first hearing with Judge Walton, it was clear to me that he found the science underlying the case a bit baffling. It was not surprising when he largely gave deference to ATF in his initial rulings."

Good was the third Tripoli president to deal with the lawsuit, Bruce Kelly and Dick Embry having presided over their portions of the case too, along with their counterparts at the NAR, Mark Bundick and Trip Barber.

Kelly received a telephone call from Good immediately after the hearing on March 13, 2009. "Going against the ATF was really a step into the unknown for us," recalled Kelly, who was Tripoli president when the lawsuit was filed in 2000. "Here we were just a bunch of hobby guys wanting to be left alone. And you've got this big bully stepping on the playground kicking your rockets around and telling you to go home. It was time for us as an organization to show some backbone. I felt vindicated by the end result."

Good credited the joint defense undertaken by the NAR and Tripoli as a key to their legal success. Both associations provided the scientific minds and the requisite financial support to keep the case alive through thick and thin. Both organizations defeated the bureaucracy. "I cannot imagine that our outcome would have been as successful if we had not pooled our collective talents and provided our legal team with the well-considered information they needed to make our case to the courts, which they did masterfully."

Not long after Judge Walton's decision, and with some additional prodding from Marty Malsch, ATF published a delisting of APCP from their explosives list. "Several ATF field agents were making noises about being out of the business of regulating rocketry motors for the moment," said Good. "I suppose some of them had to get past the immediate slap in the face to their pride."

ATF officials soon changed their website to remove APCP references and brought their field agents up to speed on the change in the law. It was finally over.

Joe Egan's role as lead attorney in rocketry's fight with ATF was not the longest legal battle of his career. That distinction rested with his firm's lawsuit against the government's proposal to build a radioactive waste dump at Yucca Mountain in Nevada. As of the beginning of 2018—ten years after Egan's death—the Yucca project remained in limbo, thanks, in large part, to the efforts of Egan and his law firm.

In an obituary he wrote weeks before his death, Joe Egan directed that his ashes be spread at Yucca Mountain, with the following epitaph:

> Radwaste buried here only over my dead body

41

The Saturn V

Steve Eves discovered rocketry in Springfield, Ohio, in the 1960s. "I was in the fourth grade, and one of the kids in class had an older brother who was into model rockets," said Eves. "One day he put on a demonstration behind the school and launched some rockets on black powder A and B motors. That experience tripped a trigger in me." Soon, Eves was building his own model rockets while also keeping track of America's race to the moon.

"We only had one television at the time," recalls Eves, "a 13-inch black-and-white set that was always kept in the basement." When the moon landing occurred on July 20, 1969, the television was carried upstairs. "Sitting on the living room floor watching this historic event, with the words on the bottom of the screen that said, 'LIVE FROM THE MOON,' was an incredible experience," said Eves. "We had all followed the space program, and it was a mad dash to beat the Russians to the moon—just another part of the Cold War we had grown up with—and now we had won the race."

That night, Eves and his father, Donald, an engineer working for the Goodyear Tire and Rubber Company, stood outside in the warm humid air, staring at the moon. "Of course, we could not see anything new even with our telescope," said Eves. "But we knew that they were up there."

Eves lost track of rocketry when *Apollo* ended in the early 1970s. The pioneering spirit that had driven America's space program for two decades was discarded. The quest for space was replaced by endless personal comfort seekers and shortsighted, weak-kneed politicians. Hardship as the indispensable tool to build character was discarded; the era of the grief counselor arrived. Such changes in national attitude did not go unnoticed by America's youth.

In 1978, and after graduating from high school, Eves traveled to Grand Island, Nebraska, for a course on auto body framing. He liked working with his hands, and he loved working on cars. The course promised experience in both. "As soon as I arrived in Grand Island, the town was hit by this huge hailstorm—baseball size and larger hail—that damaged all of the cars in the area," recalled Eves. "The body shops there were booming, and I was just twenty years old and looking for some adventure, so I took a job at a local shop and went right to work. It was two years before the hail damage work finally slacked off."

Eves stayed in Nebraska for a decade. He married, had three children, and divorced. Eventually, he decided to return to Ohio. He settled in the small community of Lake Township, just outside Akron. Now a skillful body and paint expert, he worked in the car business and raised his children.

Not long after his return to Ohio, Eves rediscovered rocketry in a small hobby shop in Canton. It was there that he spied the first G motor he had ever seen in his life. "In all the years I had fooled around with model rocket motors, the biggest thing available was the D motor," said Eves. "And when as kids we launched a rocket with a D motor, you usually never saw it again. Suddenly, I was looking at a whole basket of F and G motors—the old AeroTech stuff, I think—and I decided to buy it all. There might have been fifteen motors in there, and I didn't even own a rocket, but I bought them all."

Eves joined Tripoli. Soon, he was launching rockets in Illinois, Ohio, and Virginia. "In those days, you flew by the seat of your pants," he said. "There was very little by way of electronics, and what was available was expensive. We relied on motor ejection for recovery."

Eves spent a lot of time launching H- and I-powered rockets, with the typical learning curve of damaged, destroyed, and lost rockets. He honed his skills, and in 1994, he decided to build something bigger.

"A friend came by one day, and he had these four huge dry chemical drum containers," recalled Eves. "They were 16 inches in diameter, and we both thought that we could make a rocket out of the stuff." Eves and two friends, Mark Rogers and Jamie Stark, converted the drum containers into a 135-pound, 16-foot-tall rocket, an upscaled version of the LOC *IROC*, a popular kit at the time.

The *IROC* was first launched at Danville, Illinois, in 1994. At the time, Danville was one of the premiere launch sites in the Midwest. The big project was a showstopper, launched on a central M motor and four K550s. "It was a really slow takeoff, but it did stay stable," said Eves. "But there was not enough time for deployment, and the parachute systems failed to fully open. The rocket was damaged on landing." Eves and his friends repaired the *IROC* and launched it again at Culpepper, Virginia. This time everything worked as planned. The rocket cleared 4,500 feet before settling back gently under parachute. The team was elated. Their hobby was now turning into an obsession. "There were some guys on the West Coast that were building and launching some big projects back then," said Eves. "For a while, I think our *IROC* was one of the biggest rockets launched east of the Mississippi."

It was about this time that Steve Eves began thinking more about his rocketry experiences as a child and the Space Race of the 1960s. One of his favorite model rockets was a *Saturn V* he had built and flown as a boy. "The *Saturn V* had always stuck in my mind as the most awesome rocket ever built," said Eves, who was able to obtain actual plans for the rocket from old NASA drawings that were available on the Internet. "I began to seriously consider the possibility of building a truly large *Saturn V*."

The *Saturn V* remains the largest and most powerful rocket ever built. The rocket was 365 feet tall—the height of a thirty-six-story building—and was 33 feet in diameter. It was a multistage rocket with the first stage capable of developing more than 7.5 million pounds of thrust. The vehicle could lift a payload of 100,000 pounds. On the pad, the entire rocket weighed 6.7 million pounds.

Eves believed that with his automotive skills, he could construct a scale version of the *Saturn V*. After doing some calculations, he settled on building a one-tenth version, which translated into a 36-foot-tall rocket four stories tall. The weight of the finished project would be over 1,000 pounds. "I was intimidated by the weight problems, and the commercial motors that were available back at that time just weren't enough," said Eves. "I wanted to build the rocket. But how was I going to power it?"

As Eves began early work on the issues related such a big rocket, a series of events transpired that would ultimately halt his *Saturn V*, along with the rest of his interest in high-power rocketry. First, one of his rocketry friends, Mark Rogers, was diagnosed with an incurable disease. "Mark's days were numbered," said Eves. "We planned a trip to the Black Rock Desert, loaded up a couple of vehicles and the *IROC* rocket, and set off for the West."

Near Green River, Wyoming, the team lost control of the truck and trailer they were towing along Interstate 80. The trailer separated from the truck, careened into oncoming traffic, collided with a van, and exploded. It was a horrific accident. Luckily, there were no fatalities. But several people were hurt, including Rogers, who was airlifted to a hospital in Salt Lake City. There would be no launch at Black Rock. Not long afterward, Rogers passed away.

After his friend's death, Eves decided to take the *IROC*, which was not damaged in the freeway accident, to LDRS 15 in Orangeburg, South Carolina, in 1996. He thought the flight of the rocket at LDRS would be a fitting tribute to his friend. Unfortunately, the flight was a disaster. Powered by a central M, four K, and eight I motors, the big rocket struggled at takeoff. The motors did not all light on time. The rocket veered off course, turned into a cruise missile, and crashed into dense forest. The *IROC* was destroyed. The 6-foot-long nose cone was never found.

The Interstate 80 accident, the loss of his friend Mark Rogers, and the destruction of the rocket that the two of them helped build together were enough for Eves. He sold off all his rocket motors, parts, and kits. And then he left high-power rocketry seemingly forever.

Steve Eves remained unconnected to hobby rocketry for almost a decade. He was unaware of the *OuR Project's* flight to nearly 94,000 feet in 1997 or Ky Michaelson's repeated space shot attempts afterward. He did not see the continued expansion of high power around the country; the improvements in electronics, airframe design, and recovery in rocketry; the steady increase of commercial motor power; or the rise of the basement bomber. He never heard of the lawsuit against ATF.

Then one afternoon in early 2007, Eves received a telephone call that would change his life. An acquaintance involved in radio-controlled airplanes called to tell him that earlier that day, he was in a small warehouse in Ohio that was "chock full of rocket parts for sale." His friend knew Eves launched rockets long ago, and the parts in the warehouse were for sale. "Do you want to take a look at this stuff?"

It had almost been ten years since Eves left the hobby. It wouldn't hurt to look, he thought. "I could not believe what was there. There were more than 230 cases of launch equipment and rocket kits. And the best part was that the owner only wanted $1,000 for everything inside."

Eves bought it all. It took two trips that filled a 24-foot-long trailer. "I kept some of the parts," said Eves. "But I sold the majority of the equipment to a rocket manufacturer. This allowed me to make some extra money, and I decided to get involved in rocketry again."

Eves still had his old *Saturn V* drawings and plans, and he dug them out of storage. He was ready to build the *Apollo* rocket. And he was going to make it the one-tenth version he envisioned more than a decade earlier. He gave no thought to propellant yet or where or how he would launch such a massive rocket. He just forged ahead. Construction began in May 2007.

The booster section of the *Saturn V* by Steve Eves was built around eleven separate centering rings, each at least half an inch thick. Running between every pair of centering rings were sixteen one-inch wide wooden stringers that made up the airframe to which the outer skin of the rocket would be affixed. The wood used in the airframe was seven-ply aircraft-grade plywood imported from Sweden.

The outer skin of the rocket was made from several materials, starting off with one-eighth thick Luan plywood wrapped around the rocket and secured with nails. "I did not use a nail gun because the head of the nail you would use with a gun would not catch on the Luan wood," said Eves. "The nails in the gun are made to go into trim and be invisible; you do not see the nail. I did not want that. I wanted the head of the nail to catch on the Luan skin to hold it down firm." Eves used a hammer and ten thousand nails for the 300 square feet of Luan that served as the base of the outer skin of the rocket. He then applied two layers of fiberglass cloth to the Luan, using 14 gallons of resin and 45 square yards of material. When completed, the three-piece airframe was both strong and inflexible.

As the rocket took shape, Eves realized the weight of the *Saturn V* would require many special safety precautions. One of his biggest challenges was determining how to ensure that the rocket's 20-foot-long booster section—weighing more than 600 pounds—could be separated at apogee from the 400-pound upper stage. A clean separation of these two heavy components was essential. They had to come apart to release all the parachutes on board. If there was a snag or malfunction, the launch crew and any spectators would be subject to the risk of more than 1,000 pounds of free-falling rocket parts heading their way.

In the average high-power rocket, the upper and lower airframes are connected by a simple slip fit. The upper airframe has a coupler attached to one end that is slightly smaller in diameter than the booster. The coupler allows the two sections of the rocket to mate perfectly, with the upper airframe sliding into the booster.

"I was concerned that the conventional slip fit might fail," said Eves. "The forces that are involved in a rocket this big might cause the two sections to bind on the coupler, resulting in no deployment at all."

To address this potential problem, Eves fabricated a flat cylindrical metal plate that attached flush to the aft end of the upper airframe of the *Saturn V*. The plate could be made to rotate like a wheel ever so slightly, and it had eight small slots that accepted the same number of metal rods, or pins, that protruded from the booster section of the rocket. By putting the booster and the upper airframe together, and rotating the wheel slightly, the metal rods locked in place, holding the two sections firmly together during flight. To unlock the plate at apogee, Eves relied on the same pyrotechnic charges used to deploy automobile airbags. The charges were wired to onboard altimeters that would fire at apogee and rotate the wheel in an opposite direction to unlock and separate the rocket into two pieces, freeing the parachutes tucked away inside the booster.

Once finished with the rough assembly work on the rocket, Eves set out to make the entire vehicle as authentic-looking as possible. "The detailing was daunting," he said. "There were over one thousand wooden ribs that had to be attached to the outside of the upper and lower airframes. These ribs run vertically one next to another down the outside of the airframe, as only a quarter inch wide, and they had to be glued and nailed into place, each and every one of them. It got very monotonous. The ribs had to be cut in several different styles and lengths and then each had to be sanded, predrilled, and installed with nails and epoxy. I would run into bumps and get frustrated. But I never thought about quitting. I might go out into the garage at night after work and spend 3 hours on it and only end up attaching attach twenty-five ribs. I could only laugh to myself. 'Hey, only eight hundred more to go.'"

Eves finished basic construction in mid-2008. He sanded everything down and then sprayed the exterior with a flawless black-and-white paint scheme. He added *Apollo 11*–style lettering and insignias, making the rocket look like a museum piece. The rocket was stood up to its full height at an air show in Cleveland over Labor Day weekend in September 2008. It was almost four stories high. A commercial-grade lifting crane was needed to raise the 1,600-pound rocket into vertical position.

Onlookers were amazed. The *Saturn V* looked absolutely authentic. One young onlooker had a question for Eves. "Where does the astronaut sit?"

With the construction nearly completed, Eves faced his next big challenges: first, how would he power the big rocket, and second, where could he launch it?

Eves remembered a conversation he had with *Rocket Magazine*'s Neil McGilvray at a regional launch back in 2007. The two had met by happenstance, and Eves showed the magazine editor a photo album of his half-finished *Saturn V*. Not really believing that the rocket would ever be finished, McGilvray offered Eves the possibility of flying the rocket at MDRA's field in Maryland. Now in the summer of 2008 and with the rocket almost finished, Eves called McGilvray back.

"Were you really serious about your offer to help me get the *Saturn V* off the ground?"

After preliminary discussions with his fellow board members at the Maryland Delaware Rocketry Association, McGilvray—now joined by *Rocket*'s copublisher, Bob Utley—agreed to help Eves find a launch field. They also suggested Eves accept more help for his massive project. McGilvray and Utley were instrumental in the final motor selection for the project.

"Bob and I considered making one P-impulse motor for the project," recalls McGilvray. "We thought the initial dry weight of the rocket would be 600 pounds. Once the nearly complete weights came in, it was clear 'we needed a bigger boat!' There was only one way to launch this rocket, and that was with whatever impulse was required to give it a minimum 5:1 thrust-to-weight ratio. Bob and I quickly came to the realization that the right way to go with this was with a commercial manufacturer who was familiar with P-plus impulse motors."

McGilvray and Utley then pitched the idea for a power plant to commercial high-power motor maker Jeff Taylor at Loki Research. The three flyers had previously collaborated on other projects, including the *Liberty Project* in 2004. Taylor was excited to get involved with the *Saturn V*, and *Rockets Magazine* then took the lead in helping Eves secure the financing he would need to pay for propellant costs of what would become the most expensive fuel in high-power history. McGilvray and Utley also went to work, lobbying their board and the Higgs Farm landowner to allow the *Saturn V* to be launched at MDRA's home field on the eastern shore of Maryland.

By the fall of 2008, *Rockets Magazine* was running both print and Internet ads, soliciting sponsors for the nine high-power rocket motors needed to achieve a safe thrust-to-weight ratio for the three-quarter-ton *Saturn V*, a central six-inch-diameter P motor and eight four-inch-diameter N motors.

"The choice of motors was largely driven by the desire to use existing reliable motor designs rather than develop something new," said Taylor. "I had an 80,000 Newton second P motor that was well proven, and we simply added N motors until we had enough impulse for a safe liftoff. "The final impulse of the nine motors would be more than 163,000 Newton seconds, making the power plant of the *Saturn V* the equivalent of an R motor.

The total cost of the propellant, even with discounts from motor maker Loki Research, and motor cases donated by machine shop expert Tim Wilsey of MDRA, was more than $13,000. For $1,000 each, sponsors could purchase one of the eight N motors needed for the launch, and after the flight, they would get to keep the four-inch-diameter motor case. The price to sponsor the single P motor was $2,000.

By early 2009, and thanks to the national ad campaign by *Rockets*, financing for all nine motors was secured. "The combination of eight N motors and a central P motor will dwarf the power plant of the *Liberty Project*," predicted McGilvray. "This will be the most impressive display of raw power ever witnessed in an amateur rocket. The visceral impact to the senses will be overwhelming."

The final preparation for launch of the *Saturn V* began in early 2009. The Higgs Farm agreed to host the launch site. It was the home of

MDRA's Red Glare event and was spacious enough for a flight to several thousand feet.

The final takeoff weight of the rocket was more than 1,600 pounds. This would necessitate a large launch tower that would have to be built from scratch. In February, two dozen MDRA club members began gathering steel for the fabrication of a 2,400-pound gantry to secure the *Saturn V* on launch day. The tower was built at the Higgs Farm. It was painted deep red and was lifted into vertical position on the farm with the help of a 60-ton, eight-wheeled Terex hydraulic crane from A-Crane Service of Maryland. The same crane would later be used to hoist and match the *Saturn V* into position with the gantry on launch day.

The launch was scheduled for Saturday, April 25, 2009, just shy of the fortieth anniversary of the *Apollo* 11 flight to the moon in July 1969.

The final assembly of the individual components of the *Saturn V* took place at the field on Friday, April 24. The rocket was carried to the farm in a lengthy trailer that held three large subassemblies: the booster, the upper airframe, and the capsule/escape tower. The capsule had recently been displayed at a science exhibition in Huntsville, Alabama. Several astronauts, including Alan Bean of *Apollo 12,* signed the capsule. *Rocket Boys* author Homer Hickam, whose story was adapted to the movie *October Sky,* also signed the capsule:

> Keep aiming high, but don't blow yourself up!

Connecting the subassemblies of the *Saturn V* took all day on Friday. The rocket's separate pieces rested horizontal on work horses as technicians moved to and from each piece, checking for details of the fit before raising the vehicle in one piece. It required dozens of volunteers to maneuver the separate pieces of the rocket into position.

At just after 7:00 p.m., with an orange-tinted sun sinking low on the horizon, the fully assembled rocket was lifted into vertical off the ground. MDRA member Fred Schumacher—an experienced crane operator—took over the directing of the big Terex crane now being operated by James Bland. The telescoping boom of the crane stretched up more than eight stories high, lifting the nearly 2-ton rocket off the ground and then directly above the four-story-high gantry. For several minutes, the entire *Saturn V* hung motionless in the sky directly above the gantry.

"The 1,600-pound rocket seemed weightless as it hung under the tip of the crane boom," said McGilvray, who was perched on top of the red gantry, ready to guide the rocket's rail guides as it was gently lowered to the gantry. "The first rail guide slid in perfectly, and we all breathed a sigh of relief as the second and third rail guides were engaged," said McGilvray. The rocket was now resting on the base of the tower.

That night, the entire assembly was illuminated by a ring of spotlights surrounding the gantry and its special rocket. The *Saturn V* could be seen for miles.

Tripoli member Ken Sparks traveled from Arizona to witness the upcoming flight. "I liked the idea of putting lights on the erect rocket at night—just like the original," said Sparks as he joined others who had gathered to gaze in awe at the rocket that night. "As a teenager in high school, I had gone to Florida to watch the *Apollo 17* launch, which was the last flight to the moon and the only launch at night. Steve did a fantastic modeling job, and the resemblance to the original is inspiring."

"It's perfect," said another man standing quietly in the dark just beyond the glow of the spotlights. "It's just perfect."

At just before 10:00 p.m., a middle-aged woman wearing only pajamas and a sweater appeared out of the darkness, crept up to the rocket, pulled an instant camera from her pocket, and snapped a picture of the *Saturn V.*

"I live down the road," she said quietly. "I could see the rocket from my house. I just had to get a closer look." She turned and, without saying another word, disappeared into the darkness.

On Saturday morning, April 25, 2009, the Higgs Farm was overrun by spectators. The rocket was scheduled to take off at one o'clock in the afternoon. In the meantime, hundreds of people gathered around the base of the *Saturn V* to get a close-up view.

"It is absolutely incredible!" exclaimed Tripoli board member Erik Gates, who had boarded a plane from California at the last minute to see the launch.

"I wish Von Braun could be here to see it," said Scott Hungarter, who had driven down from Philadelphia for the flight.

Young James Bassette of New Hampshire made the nine-hour drive with a dozen Civil Air Patrol cadets from New England. As they walked up to the towering *Saturn V* that morning, they stopped dead in their tracks. "It was worth every traffic jam, car horn, gallon of gas, and hour spent in a van with no air-conditioning," said Bassette.

"When I first saw the rocket on the launch pad today, I thought it was a work of art," said Jackie Cullen of Maryland. "I was just amazed that this was a model rocket that someone had built in their garage and that it was being launched into the sky by a rocketry club and not by NASA."

Christian Brechbul, a teenager from Vermont, was too young to have seen the *Apollo* flights to the moon. He lingered for a while at the base of the *Saturn V*. "I knew it would be big, but not that big," said Brechbul. "I stood at the bottom of the rocket and imagined it being ten times larger, and I realized just how much courage it must have taken for the astronauts to sit in a tiny capsule on the tip of it all."

At a little before 11:00 a.m., all spectators and nonessential personnel were asked to move away from the rocket launch area to safe areas more than a quarter of a mile away. The final preparations for launch were now underway. Eves and friend Tom Erb climbed a long ladder to reach an access door on the upper airframe of the rocket. The door was opened, and multiple onboard electronics inside were activated. Then Tripoli member Dan Michael and his son Andrew were tasked with first testing and then installing the essential igniters that would fire all nine motors simultaneously at ignition.

"I used e-matches inserted into a thermite composition," explained Michael, who chose two matches per motor to ensure success. "In this manner, we would be firing eighteen e-matches at the moment of ignition, instantly and redundantly igniting the thermite."

The aft end of the *Saturn V* rested on metal crossbars at the base of the gantry. The bars were 3 feet off the ground. This extra space below the rocket allowed for access to the nine high-power motors in bottom of the rocket. It was in this small area that Michael and his son went to work installing the igniters. The first pair of igniters was installed by at 12:25 p.m.; the last igniters were in only 10 minutes later, at 12:35 p.m.

There were only a handful of people let around the pad now. *Rockets Magazine* cofounder Utley and LDRS founder Chris Pearson made final adjustments to onboard video cameras in the lower end of the rocket. Then everyone cleared the area near the pad. The rocket stood alone now, towering above a green field on a little farm in Maryland. Meanwhile, in a long line that stretched out more than 2 miles, an excited crowd of five thousand people stood in the distance, waiting for the flight of the biggest amateur rocket ever flown. People were on their feet and on one another's shoulders; some were standing on top of their cars or in the beds of their pickup trucks. Moments before the final countdown began, they surged forward in unison, held in check by the natural barrier of an irrigation canal that ran for much of the perimeter around the Higgs Farm.

At just before 1:00 p.m., a sounding rocket built by Ben Russell was launched to check the winds aloft. The winds were just fine. The sun was shining. Everything was a go.

LCO McGilvray began the countdown over a multiplex of public address speakers: "Ten, nine, eight, seven, six, five, four, three, two, one."

At zero, Steve Eves, who had given up on high-power rocketry in 1997, pushed the ignition button for his 1,600-pound commemoration of man's first flight to the moon. Instantly, there was a blinding flash at the base of the rocket.

All nine Loki Research motors came up to pressure at once. A column of dense white smoke kicked up around the base of the rocket, completely obscuring the four-story-tall gantry, as the *Saturn V* raced up the rail, leaving the Higgs Farm in a fury of flame, smoke, and noise. The force of the nine motors lighting simultaneously shoved the 2,400-pound launch tower 13 feet across the ground.

In seconds, the rocket was high in the blue Maryland sky, pulling seven Gs, powered by 11,500 pounds of thrust, and making a fiery trail in the air.

The crowd was stunned. The massive *Saturn V* left the pad so quickly it took many by surprise. It looked like 1969, the rocket on the leading edge of an incandescent, bright white light as it reached speeds of more than 440 feet per second on its way to an apogee of just under 4,500 feet.

"The sound of the rocket taking off reminded me of the launches that I had seen on television as part of the U.S. space program," said spectator Cullen. "I know the real *Saturn V* rocket was bigger, but this rocket was really big, no matter how you look at it. And to be at the Higgs Farm to witness this launch was exciting!"

"It was incredible. As the rocket was climbing, I was in total awe," echoed Michael Williams of Maryland.

The up part of the flight was flawless. Now the *Saturn V* relied on multiple onboard altimeters to detect apogee and fire the ejection charges that would turn the metal flywheel connecting the booster and upper airframes together. The turn of the wheel would separate these pieces of the rocket, deploying the parachutes. As a backup plan, Eves also placed a remote control receiver in the rocket so he could fire another ejection charge if the onboard altimeters failed to perform. It was his last resort—an emergency panic button of sorts.

The *Saturn V* slowed as it neared apogee. It seemed to linger for a moment, still moving higher, and then it turned horizontal, directly above the field and the crowd. It appeared to stop in midair.

"Push the button! Push the button!" yelled several people standing next to Eves, who was holding the remote control activator in his hand. He paused.

Suddenly, the rocket came apart, and within another few seconds, the sky was filled with multiple parachutes—some as large as 28 feet in diameter—as the booster and upper airframes fully deployed their recovery gear and floated down independently of one another. The cheering turned to a roar as the rocket got closer to the ground.

Then the unexpected happened. The main booster stuck the landing, standing straight up on all four fins half a mile from the launch tower. The parachutes gently dropped around it. A hundred yards away, the upper section of the rocket landed gently in the grass.

"That was the most amazing launch I have ever seen," declared Jeff Williams. "I think everybody will agree that the recovery just made the launch ten times better. Nobody could believe that the rocket landed straight up!"

Within minutes, scores of spectators, fellow flyers, and news media converged on the landing site, joined by an exhausted Eves and his family. The flight was a success. The rocket had sustained a small zipper and some minor cosmetic damage that was easily repaired.

Teenager Christopher Kondi made the 10-hour drive with his family from New Hampshire to see this flight. Like all the spectators at the Higgs Farm on April 25, he was not disappointed. "As the rocket climbed high into the sky, I discovered how a dream can really come true," he said. "And for Steve Eves, it really did. There was no part of the day that was not exciting, between getting up close to the rocket and the absolutely fantastic landing, when it stood back up, as if to say let's go again!"

A few months after its flight, the *Saturn V* was on display again, this time in Upstate New York at LDRS-28. For the second time in five years, the national launch was held in the Empire State. But this time it was not at Geneseo. It would go forward at an alternate location that had been in the works for nearly two years.

In 2007, Tripoli member Deb Koloms, a practicing ophthalmologist in New York, received a phone call from McGilvray, who wanted to discuss the possibility of another LDRS on the East Coast, perhaps as soon as 2009. He called Koloms because she had been part of the Buffalo Rocketry Society, the cohosts of LDRS-23 in Geneseo in 2004. Do you think we could hold LDRS at Geneseo again, asked McGilvray?

"While I was interested in doing LDRS-28 in New York, I did not think that Geneseo would be the best place for it," said Koloms. "The waiver is only 8,000 feet. It is an active airport, and the 1941 Historic Aircraft Group Museum had too many irreplaceable vintage aircraft on the field."

Indeed, these same issues had almost led to the cancellation of the Geneseo LDRS in 2004, recalled Ken Good. "The board of directors considered not having LDRS-23 at Geneseo," said Good. "But a potential alternate bid fell through."

For 2009, Koloms had an alternate suggestion for McGilvray. Located a little more than an hour from Geneseo and near the town of Potter, New York, was the rocketry range at Torrey Farms, run by the Potter Youth Rocketry Organization (PYRO), a chapter of the NAR. With thousands of acres of farmland and FAA waivers up to 23,000 feet, Torrey Farms was an ideal place for rocketry. Koloms was already talking to the PYRO leaders about the possibility of adding a Tripoli prefecture to their field.

"The PYRO Board thought this was a great idea," said Koloms, "and Tripoli Western New York moved to Potter and decided to bid on LDRS-28."

LDRS was coming back to New York, and it would be held on a NAR field. To spread the work around, it would be hosted by not only Tripoli Western New York but also MDRA, METRA, and PYRO. All four Tripoli organizations would work together—with *Rockets Magazine* as event sponsor—to put together the 2009 launch that would run five days, July 2–6.

The field for LDRS-28 was on a family-run farming operation dating back to the eighteenth century. The farm encompasses several properties at different locations and is one of the largest vegetable crop farms in New York.

Torrey Farms was still fairly new to rocketry in 2009, having been discovered in 2004 by Mike Dutch and some friends who were looking for an alternate field to Geneseo to fly high power in New York. Dutch, a member of both the NAR and Tripoli, had seen several large fields around Potter and was told by an aviation acquaintance that from the air, one could see several more. One rainy weekend, Dutch and the other cofounders of PYRO drove around Central New York, looking for a new rocketry launch site.

"We did find a few fields that would work for low-power ranges, but I kept telling the guys that we had to see this field in Potter before we made any decision," said Dutch. Eventually, the team found themselves at one end of Potter field.

"We came to Route 364 and found the south end of the field and the barns," said Dutch. "We drove into the drive, and as luck would have it, we came across the farm manager working on one of the plow rigs at the barn. We all jumped out and talked to the manager and explained that we were a nonprofit club that worked with all types of youth groups teaching them the safe hobby of rocketry and that we were looking for a new field that we might use to launch rockets on. The manager thought that was a great idea and said he would talk to the owner to see if he would give us permission to use his Potter fields to fly on."

Two months later, the permission to launch rockets there was granted. With vast open spaces and the supportive enthusiasm of the landowners, Potter soon became one of the best fields for high-power rocketry in the Northeastern United States. The range was on a 1,200-acre parcel of Torrey Farms near Potter and was surrounded by nearly 15,000 additional acres of farmland. There are trees here and there on the farm and also just off the flying range—typical for many Eastern rocketry fields—but it was a good wide-open space.

The waiver for LDRS-28 would be 15,000 feet, with windows up to 23,000 feet. Flying started on Thursday, July 2, 2009.

Well, sort of.

During the previous few days, a slow-moving low-pressure system began dumping rain throughout Central New York. Wednesday night brought lightning and thunder accompanied by heavy downpours. On

Thursday morning, the field was usable for launching rockets. But the problem was getting to the range head, which was almost impossible because of a thick river of sticky mud on the roadway. Observed one flyer, "Entering the long entrance road on the first day of LDRS-28 was challenging, even with the help of four-wheel drive. Depending on how adept you were with your vehicle and also how determined you were, you either got a white-knuckle ride into the field or a frustrating exercise in wheel spinning and mud flinging. Many of the fully loaded trucks with trailers, trucks without trailers, and TVs simply bogged down and were buried up to their axles."

Torrey Farms farm manager Rick Hall came to the rescue. Hall dispatched a convoy of several heavy farm tractors to pull the stuck vehicles out of the mud and tow them either off the property or even to their designated parking spots on the range.

"It was striking how efficiently and quickly the farmhands moved from vehicle to vehicle and gently pulled them out with the massive tractors," wrote McGilvray in *Rockets Magazine*. For the owners of vehicles who didn't have a chance to navigate the slippery road, Hall came up with yet another day 1 solution. "Rick directed his men to hook up two big hay wagons to transport the folks who had been exiled to the pouter parking lot into the field to participate and enjoy the event," said McGilvray. The hayrides were so popular that they were continued all day on Friday and Saturday too—long after they were necessary.

Launch activities on Thursday were slowed because of the inclement weather. Still, almost 100 rockets took to the air. The weather improved each day thereafter. Using more than fifty pads arranged neatly on the field, almost 1,200 flights took to the air at LDRS-28, making this among the biggest LDRS launches ever. Nearly 120 of those flights were with research motors. There were no single-motor P flights at LDRS in 2009, but there were plenty of high-power flights through O power. There were scores of M and N flights—more than 40 on Sunday alone. Several rockets cleared more than 15,000 feet. Several more had one form or another of a spectacular demise, including an N-powered rocket that sustained a CATO halfway up the rail, spewing out another four K motors into the air, all of them on fire.

The crowd loved it.

Although there were many rockets that weighed over 100 pounds, there was a definite trend away from mega projects this year. There were few rockets that weighed more than 200 pounds and only one that weighed more than 300. The heavyweight was Dan Michael's 310-pound *Patriot Missile*. Michael's rocket was the heaviest rocket of LDRS-28, and it had a flawless launch and recovery, powered by a central AeroTech N2000 and four M1315 motors, all lit on the pad.

"I have never been that close to that many motors being lit at the same time," said Gary Rosenfield, who helped Michael and others get the rocket on the pad. "This was an awesome launch!" The 15-foot-tall *Patriot* climbed to 9,000 feet before settling down gently under chute in a potato field.

Other big flights over the weekend included the Gates Brothers with their *Porthos* on multiple motors as well as their 150-pound *Sumo* on five M-impulse motors. Both launches had good flights and recovery.

Saturn V replica rockets were quite popular at LDRS-28, perhaps coinciding with the fortieth anniversary of the launch of *Apollo 11* in 1969. There were scale models of the big rocket of every size. The largest to fly was Vern Hoag's 15-foot-tall version, which thrilled all in attendance with a fiery launch on a central M and four L motors. The rocket emitted a huge tail of flame and smoke as it climbed high into the sky and then landed without incident after clearing more than 4,000 feet.

At the annual banquet on Saturday night, July 4, Steve Eves was presented by the Tripoli Board of Directors with the President's Award for his historic flight that year. Bruce Lee was also honored with a lifetime membership for his many contributions to high-power rocketry and to Tripoli. There was also another historic moment at the annual meeting. That evening, the results of the annual board of directors elections were announced. When all the votes were tallied, Deborah Koloms became the first woman in Tripoli history to obtain a seat on the board.

Koloms was not the first woman to run for a board seat. That distinction belongs to Terri LaMothe, who ran for the board in 1993 and 1994. LaMothe was unsuccessful in her bids, and it would be twelve years before another woman ran for a seat. That woman was Deb Koloms.

Koloms was born in 1958 in Illinois, and she enjoyed model rocketry as a child. "I think I built most every rocket Estes made back then and can still remember the fun I had," she said during her first run for a board seat in 2006. "As high school came along and then college, rocketry took a back seat while more pressing things such as getting a degree, starting a career and family became a priority."

Koloms earned a degree in electrical engineering from Washington University in St. Louis in 1980 and worked for a year with Motorola as a design engineer. She returned to school and earned her medical degree from Loyola Stritch School of Medicine in 1985. She completed her residency in ophthalmology in 1989. She practiced in the Chicago area until 1997 and then relocated to Upstate New York, where she started her practice in the city of Watertown.

Kolom's move to New York was not only good for her medical career, but it was also a boon to high-power rocketry. In only a few years, she was serving in rocketry leadership positions both locally and nationwide. "Around 2000, I caught the rocketry bug again after seeing a web page featuring high-power rockets," said Koloms. "I became a 'born-again rocketeer' and have not looked back since. I bought my first high-power kit—a LOC *IROC*—and built and flew it for my L1 and L2 certifications in 2001. In 2002, I flew my L3 flight on a large tetrahedral rocket."

Koloms was a board member of the Buffalo Rocket Society, where she also served as Tripoli prefect for the local club. In 2004, she was named to the Tripoli Advisory Panel. Her path to the board was not an easy one, but she proved that perseverance prevails. She first ran for election in 2006 and then again in 2007. She was defeated both times. She was still fairly unknown nationwide. Yet as word of her qualifications slowly spread, she nearly doubled the votes she received by 2008. She won easily in 2009—her fourth run—by defeating several highly qualified candidates.

"Gentlemen, move over. There is a lady in the house," announced Tripoli president Ken Good as Koloms took her seat at the annual meeting at LDRS-28. "Congratulations, Debra. Your persistence and hard work has paid off, and a bit of Tripoli history has been made. Debra is the first female board of director candidate elected by the membership, and as the numbers show, Debra's support from the membership was exceptionally strong."

One day in the not-too-distant future, Koloms would become the president of the Tripoli Rocketry Association.

The victory over the government in the federal lawsuit, the launch of the Steve Eves's *Saturn V*, and another successful LDRS were all bright moments for the high-power faithful in 2009. But there were some bad news in 2009 too. Most of it came late in the year.

On October 10, 2009, Animal Motor Works cofounder Paul Robinson died after a long struggle with pancreatic cancer.

Robinson first gained notoriety in high-power rocketry in the mid-1990s. He was one of the five members of the *OuR Project* rocket, launched at Black Rock Desert on August 16, 1996, powered by an R17,542 motor. The team was aiming for an altitude of 100,000 feet—an unheard-of goal for high power at the time. The vehicle reached an estimated apogee of 94,000 feet. Onboard video relayed live images to the ground during flight. "What was seen was a clear outline of the entire Black Rock Desert and the surrounding terrain, the curvature of the earth, and the great blackness of space," reported Robinson. It was an inspirational flight.

Robinson went on to run the commercial motor endeavor Kosdon East, followed by the even more successful Animal Motor Works (AMW), which he cofounded with flyer and fellow motor builder Jim Rosson, another member of the *OuR Project* team, in 2000. In the early to mid-2000s, Animal Motor Works was one of the most popular commercial motor lines in America—second only to AeroTech. Robinson and Rosson were big fans of manufacturing propellant that emitted different-colored flames, a technique Rosson helped perfect in the 1990s. A little barium made one flame green, strontium made it red, and magnesium for white, and tiny titanium chips created sparks. To market these colors, AMW sold many of his motors with nicknames, like Red Rhino, Green Gorilla, or Blue Baboon, each color depicted by the name. Robinson was called the zookeeper.

"Paul was a 'get it done' kind of guy," said Rosson. "Forget the paperwork, ignore the protocols, and just get it finished as best as you can. He always thought a 4-hour job took 15 minutes, and he shaved time wherever possible to prove he was right."

"The last time I saw Paul was at LDRS-28," recalled friend Neil McGilvray. "Paul knew his days were numbered. Yet he spoke openly of his prognosis, and he remained optimistic. Paul wanted to spend his last days among rockets and the flyers he considered his brothers and sisters."

Paul Robinson was fifty-eight.

Several weeks after the death of Paul Robinson, high-power rocketry suffered another loss. On Sunday afternoon, December 20, 2009, Tripoli board member Erik Gates, an electrician by trade, died after accidentally falling 30 feet through a skylight while working on the roof of a commercial building in Southern California.

In 2000, Erik and his brother Dirk were unknown Tripoli flyers who had been in high power for barely a year. Within a short period, the phenomenon known as Gates Brothers Rocketry swept over the hobby like a tsunami. At LDRS-20 in 2001, the Gates Brothers electrified the crowd and all of high power, successfully launching several mega projects that instantly established the pair among the top rocketry flyers in the United States. Over the next few years and at launches in Southern California, Black Rock, and several LDRS events, the Gates Brothers put on shows that attracted huge crowds—and many new flyers—to high-power rocketry. There has never been anything quite like them in high power, before or since.

In 2003, and thanks to his notoriety in high power, Erik was asked to consult on a new television show where a pair of oddball scientists used elements of the scientific method to test the validity of rumors, movie scenes, and other legends of fact and fiction. The show was called *Mythbusters*—one of the most popular television programs ever broadcast on the Discovery Channel.

"Erik Gates was with us on *Mythbusters* in our very first episode and was responsible for the rockets that we installed on the JATO rocket car," recalled star Jamie Hyneman of the program. "In my opinion, it is still one of our best episodes and the first of many episodes centered [on] unusual uses of rockets. I don't think we have done a single episode that used rockets the way they were intended, and Erik, to his credit, took on all these screwball rocket projects with enthusiasm—or I might say even glee."

"I've known Erik for quite a few years," said Gary Rosenfield. "AeroTech designed and produced motors for a number of his projects, including the *Porthos* rocket and several *Mythbusters* episodes. His craftsmanship was impeccable, and I always thought he was one of the best people in rocketry."

"Erik was very much aware of his position in rocketry," echoed flyer John van Norman. "He used that influence to become, as it has been said, an ambassador for the sport."

In 2007, Gates was elected to a three-year term with the Tripoli Board of Directors, where he took on an active role in the leadership of the hobby. "His passion for rocketry was one of the first things I noticed about him" recalled *Rockets Magazine* cofounder Bob Utley. "As a Tripoli board member, Erik worked with members to help bring good ideas to Tripoli. He worked tirelessly with others who had different opinions about policy to see if there was common ground to build on."

Gates's enthusiasm for high-power rocketry and the remarkable success he achieved in the hobby earned him a special respect that often motivated others into action. When Erik Gates asked somebody to help do something for the hobby, the answer was usually yes.

For example, Tripoli failed to attract a bidding prefecture to host LDRS-28 in 2009. The board was in a jam. So as the summer of the launch was approaching, Board Member Gates picked up the telephone and called flyer Neil McGilvray in Maryland.

"Tripoli doesn't have a bidder for LDRS-28," Gates told McGilvray.

There was a pause on the line.

"Well, it sucks to be you" was McGilvray's response.

"Well, I'm calling because I want *you* to organize LDRS-28 on the East Coast," replied Gates.

There was another pause on the line.

"Well, it sucks to be me," said McGilvray, who, together with Gates and several other prefecture leaders, went on to organize the successful event in Upstate New York—just six months prior to Gates's death.

"If it were not for that one phone call, who knows where LDRS-28 would have been held?" said McGilvray later. "Erik always found a way to get things done."

Pat Gordzelik, who had left the board as vice president in 2009, agreed to return to temporarily fill the empty seat left by Gates. "I loved Erik's can-do attitude and his 'you got it' answer any time I needed his help. But a true friend is one that never shies away from telling you when you mess up either. His legacy was his never-ending enthusiasm, his courage, and his good heart," said Gordzelik.

Erik Gates was forty-seven when he died, survived by his wife and three children. In 2011, the Tripoli Rocketry Association established an annual scholarship in his name.

42

LDRS Turns Thirty

In 2010, and after serving as a board member for more than a decade, Ken Good announced he was stepping down as president of the Tripoli Rocketry Association.

Good presided over some hard times within the organization. But nothing was as difficult as maintaining the seemingly never-ending battle with ATF that had just begun when Good started his first term as a board member in 1999. "Little did I imagine at the time how long we would be engaged with [that] case and that the leadership of the Tripoli half of this effort would be passed from Bruce Kelly to Dick Embry and then to me. I could not foresee that we would experience an ebb and flow of interim successes and setbacks, followed in 2006 by a dramatic appellate court overturn of our initial loss at the district court level.

"I view myself as the trench-warfare president," said Good, looking back over the long struggle with the federal government. He was also proud of how he and others also transformed the way the board operated. "I believe I was able to do at least two philosophical things beyond the lawsuit," he said. "Motivate the board and utilize everybody's talents in a way that reinforced their individual contributions. We created a board that was more of a professional, business-oriented model."

During his tenure as Tripoli's leader, Good had presided over, among other things, the commingling of research and commercial launches; the establishment of the Class 3 Committee, paving the way for ultrahigh-altitude attempts; and the creation of Tripoli's first Internet members' forum.

"I felt my presidency was a success, but it was time to step down. Why screw around when there are others who are ready to contribute? It was time for me to just get out of the way and let new people take Tripoli somewhere useful."

LDRS-29 went forward on June 11–16, 2010, at Lucerne. This was the second time in ten years that high-power's oldest venue—the Lucerne Valley—hosted the event, and it would become one of the biggest shows in LDRS history. At least 1,349 rockets took to the sky, including 32 M flights, multiple Ns, a few Os, and a P-powered project. The total flight count made LDRS-29 the second largest LDRS in history—second only to LDRS-19 in Orangeburg in 2000. There were no rockets over 300 pounds this year, but there were several bigger-than-life vehicles, including Tim Lehr's *Dark Star* and *Project Odyssey* flown by Doug Gerrard, Art Hoag, and James Russell. The largest motor was a P6300 in an upscaled 23-foot-long *Talon* by Chris Williams, Joe Cox, and Charlie Cox.

National television returned to LDRS in 2010. This time the Science Channel was on hand for the entire weekend. This was the first national rocketry event to be televised since the Discovery Channel's *Rocket Challenge* in 2003. And this would be the first of three annual shows in a row as the Science Channel contracted with Tripoli to cover LDRS through 2012. The Science Channel spent hundreds of thousands of dollars on LDRS-29. It generated more than 1,000 hours of live action video that was ultimately compressed into a 1-hour program that was first aired less than thirty days after LDRS-29 on July 5, 2010, at 9:00 p.m.

The television special was called *Large Dangerous Rocket Ships*. The program was hosted by the engaging and enthusiastic female scientist of *Mythbusters* fame, Kari Byron.

"This is the most important event of the year in high-power rocketry," Byron told the television audience at the start of the program. "Thousands of wild, obsessed rocket builders from around the country pack up their trucks, RVs, and trailers and descend on the Lucerne Valley in California to launch rockets that they've spent months, and even is some cases years, building. They're finally ready to blast off. Today the sky's the limit!"

Large Dangerous Rocket Ships took some liberties with its coverage and the facts related to rocketry—as all television programs do. They billed LDRS as a competitive event, which it can be, but made it look like that was really all it is. And the narration was filled with the usual hyperbole that only a television reality show can get away with. But it was all entertaining, and unless you were a humorless, hard-core rocket fanatic, it was filled with plenty of fun.

The program revolved around a few drag races, some low-altitude odd rockets, and a series of personal interviews of flyers from all over America.

For example, Science Channel followed Tripoli flyer Mark Hayes from his garage in Colorado, where he built an absolutely stunning upscaled *Aries Space Plane*, all the way to Lucerne for the flight. The rocket was powered by five motors, including an L995 Red Lightning, and featured separate deployment events calculated to bring the rocket back in three sections under multiple chutes.

"The heart and soul of the rocket is the motor," Byron explained during one segment of the show. She explored what a high-power motor was—and what the letters mean—in general terms. She also defined the quirky language of the hobby, with examples of each, including the CATO, land shark, skywriter, and lake stake, to name but a few.

The program also explored a little bit of LDRS history, going all the way back to the first launch in 1982. The show's producers—with the help of some high-power flyers—got some of that history wrong, but they got the main point just right. Explained Byron, "Nearly thirty years after the first LDRS, advances in technology, lighter building materials, and more powerful rocket motors make LDRS the event to see the cutting edge in high-power rocketry because LDRS opens the field to the biggest and most innovative rockets in the land."

During all four days of the launch in June, the film crew was everywhere. There were dozens of independent contractors, producers, screenwriters, and technical people at the launch. They had RVs and ATVs and were scrambling from one place on the field to another all day long. They had a central-located tent city-like headquarters as well as an elevated metal platform—at least two stories high—to capture the entire rocketry landscape at Lucerne.

One of the show's main features was the fastest-to-1000-feet drag race contest, where flyers launched 6-inch-diameter rockets powered by J motors only. The rocket that reached an altitude of 1,000 feet faster than any other would be the winner.

Another television event was the odd roc competition. "The goal is to take something that was never meant to fly and to launch it as low as possible," said Byron. The closest to but not below 100 feet would take home the odd rocket trophy. This was definitely not a typical LDRS competition, to be sure, but it was just a slight spin of a familiar sight at any LDRS going back to the 1980s—weird rockets built by flyers with lots of imagination. Among other flights, there was an airborne Christmas tree, a beer bottle rocket, and a flying coffin of death.

And on the nontelevision side of things, at the annual members' meeting, Ken Good received a Tripoli Lifetime Membership Award for his years of hard work and effective leadership as a board member and president of Tripoli.

The Science Channel coverage of LDRS-29 was a significant part of the national launch in 2010. This was the first national television coverage of high-power rocketry since Argonia in 2003. This show, and the two that would follow it over the next couple of years, would bring high power into millions of homes around the country—and the world.

Yet LDRS-29 is significant for another reason too. The most serious accident to ever occur at any national launch occurred on Saturday afternoon at LDRS-29, with several people sustaining injury—one of them a serious injury. The accident served as a wake-up call to all high power. Despite three decades of success and relative safety, this hobby remains a dangerous pastime.

At approximately five o'clock on Saturday afternoon, most of the LDRS crowd was gathered at the western edge of the flight line, awaiting final launch in the fastest-to-1000-feet contest. There were just three rockets left to fly. Each rocket was loaded and vertical on their respective pads. Kari Byron was there at the launch pad area, all the television camera crews were there, and hundreds of spectators were waiting for the last big flight of the day to decide the winner of the event. The countdown for launch was just about to begin.

Meanwhile, several hundred yards away at one of the faraway cells, a team of flyers was busy preparing one of the most ambitious flights of the weekend: the *Delta II Project*.

The *Delta II* was a meticulously designed one-ninth-scale replica of the rocket originally conceived by McDonnell Douglas, which had lofted satellites into orbit since the late 1980s. The two-stage rocket was nearly 14 feet tall. The main airframe was 10.75 inches in diameter with four strap-on boosters that were each 4 inches in diameter. There was an additional L-impulse motor in the upper stage. The rocket was built by experienced Tripoli and NAR members associated with Tripoli San Diego.

The *Delta II* rocket was flown in November 2009 at the Plaster Blaster regional launch east of San Diego. During that flight, the rocket lifted off on an AeroTech M1939 and four K550s in the booster and a single Cesaroni L730 motor in the sustainer. The flight was near perfect. After the burnout of the main motors, all four boosters were ejected away from the rocket, each returning under its own parachute. The L motor in the sustainer was fired as planned several seconds into flight. Although the rocket had some minor stability issues at high altitude, the flight and recovery were successful. It was an impressive project.

The *Delta II* team planned a similar flight profile for LDRS-29. Late Saturday afternoon on June 13, they hauled the big rocket, which weighed nearly 150 pounds, out to the most distant launcher on the range. At this away cell, and while the rocket was being prepared for launch, a small crowd of people gathered—some of whom were unnecessary to flight preparations. In addition to crew members, there were family members of the crew and even children near the pad, all milling about as the *Delta II* was readied for flight.

Shortly before launch, and after the rocket was raised to vertical on the launch pad, an all-terrain vehicle (ATV) pulled up to within 50 feet or so of the pad. On the ATV were two people curious about what was going on. The driver of the ATV was Graham Orr. His passenger was Kairee Goodin. Neither Orr nor Goodin was a member of the *Delta II* crew. They were also not members of Tripoli or any other rocketry organization. They were just two civilians spending their day on an ATV in the desert, which was open to the public. They saw the rocketry activity, and they decided to get a closer look.

After the ATV came to a stop, Orr got off the vehicle and walked over to the *Delta II* crew to talk. The rocket was already vertical. Moments later, the team decided they needed to lower the stop on the aft end of the launch rail. To do that, they decided to lower the entire rail and rocket back to a horizontal position to make their adjustments.

During this time and while her boyfriend was adjacent to the pad, talking to the crew, Goodin remained behind, sitting on the ATV, which was parked just a few degrees off the nose cone of the rocket. In other words, after the rocket was lowered, the pointy end of the *Delta II* was aimed in the general direction of Goodin and the ATV.

Apparently, before the rocket was lowered from the vertical to horizontal position, not all the onboard electronics were disarmed. Some of these electronics were supposed to airstart the upper stage of the rocket when it was high in the air. Seconds after lowering the rocket, and as the team slid the rocket further back to a newly adjusted rail stop, the upper stage L motor ignited without warning.

The sustainer immediately broke free of the booster, roared off the pad, hit the ground, and proceeded to careen all over the vicinity of the launcher, sending flames, smoke, and huge clouds of dust, dirt, and debris into the air. From a distance, the entire launcher and everyone around it was obscured by dense smoke. All activity on the range came to a halt, including the final countdown of the drag race contest. Hundreds of people stared into the distance of the away cell. None of them knew that the most serious accident in LDRS history had just occurred.

Several people were directly injured by the fire and heat from the powerful rocket motor. The most serious injury was to Goodin, who was severely burned while sitting on the ATV. Goodin and another person were airlifted off the range by rescue helicopters, which transported them to Arrowhead Regional Medical Center, nearly 75 miles away. At the hospital, Goodin was treated for first- and second-degree burns as well as shrapnel-related lacerations sustained during the accident.

Fortunately, no one was killed. But they could have been, and it was lucky the injuries were not even more severe than they were.

Goodin filed suit in San Diego County on June 11, 2012, almost two years to the day after her injury. There were three named defendants: Tripoli San Diego, Rocketry of California (ROC), and the Tripoli Rocketry Association. Goodin's attorneys alleged their client sustained permanent and disfiguring scars as a result of her burns. They were seeking compensatory and punitive damages from the defendants. The twenty-page complaint alleged, among other things, the defendants had engaged in the ultrahazardous activity of high-power rocketry without taking adequate safety precautions to protect spectators. The complaint also alleged that the first-aid facilities at LDRS-29 were woefully inadequate for the dangers present at the launch and that there were numerous and lengthy delays at the field before Goodin was finally treated and then transported to the hospital.

There were no depositions taken in the case, and there was little local or national publicity. Tripoli immediately clamped a lid on the whole incident—likely at the request of their own attorneys—leaving members with little to go on as to the actual facts that led to the accident. There were few news stories, although there was some Internet coverage of the accident two years later when the lawsuit was filed. The entire matter was then settled by Tripoli and the lawyers representing Ms. Goodin. The case was dismissed less than ninety days after the complaint was filed. The lawsuit was over quickly and quietly.

A confidentiality clause in the settlement agreement prevented any of the participants in the case, including the Tripoli board, from disclosing the settlement amount paid to Ms. Goodin. It seems likely it would have taken a number well into the six figures to end the case. It was undoubtedly one of the largest settlements ever paid on an insurance claim for Tripoli.

The June 13, 2010, incident involving the *Delta II* was the most serious range accident in the history of LDRS. When it was finally over, there was discussion of changes that needed to be made to help prevent a similar occurrence in the future.

"Rocket preparation safety should have been better managed at LDRS-29 and absolutely must be better managed as we proceed to fly dangerous rocket projects," wrote Ken Good after the event. The incident should be a high-power case study on how not to manage flight preparation of a high-power rocket, he said, adding,

> How the heck were two totally non-authorized people permitted not only to be out on the range, but in close proximity to a large rocket? The answer is that they were not noticed on their approach to the pad, and they were not challenged by the rocket flight crew. The LDRS-29 organizers personally pointed out to me the difficulty of securing the complete perimeter of a large BLM-controlled site, since many like Lucerne are not administered via a "granted sole use, permit." And cordoning off an entire dry lake range area is not practicable.
>
> But frankly, this is no excuse.
>
> We have to think like NASA does. They, too, have a large launch area at Cape Canaveral, and they do not

permit interlopers there. Yes, they have permanent barriers, surveillance and security personnel that we cannot have. But we can adopt some form of effective range security. We can and must have sufficient range safety officers actively scanning the range for unauthorized people. All authorized flyers and range personnel should be equipped with badges/name tags that identify them as authorized. Any/all "questionable" people must be accosted and credentials verified, while ALL pad activity is on hold until this is done. Last and not least is the responsibility of the rocket flight crew, as a "last line of defense." If someone shows up at the launch area who is clearly not part of the crew or launch organization, the crew needs to immediately halt activity and ask such people who they are and why they are there. The range safety officers can escort any interlopers off the range.

On the technical side of things, Good reminded members of the dangers of becoming too complacent on the range—a danger that, strangely enough, might be more serious among experienced Level 3 flyers than among novices. "There is indeed an unspoken code in which we tend to let people we regard as 'experts' proceed with what they are doing," observed Good. "But that is an error, and we must be brave, mouthy, rude, [and] outspoken enough to speak up to stop something that looks like it's not the correct thing to do." Continued Good:

> The LDRS 29 accident revealed a startling truth. The actual ground ignition of the rocket motors was not caused by an explicit breach of either of the TRA Safety Codes. But it was caused by not following accepted practices that are (or should be) well known and are often described in the documentation supplied by rocket electronic manufacturers. In brief, once a rocket's electronics are armed, especially those connected to motors and/or pyro charges, the rocket must not be moved without disarming those electronics. No, you won't find that in anyone's safety code. You also won't find any verbiage that says you shouldn't light a rocket motor in your garage, or shove igniters in your mouth either. It comes down to using sense and understanding the nature of the materials/components that are being used. Never proceed with any operation with rocket motors, pyro charges, igniters, electronic devices, etc., until you understand the procedures that must be followed for their safe use . . .

In the summer of 2012, the Tripoli Board of Director clarified its rules regarding launch crews and spectators at all high-power events. New guidelines were implemented to reduce the number of nonessential personnel—and wholly unnecessary spectators—at the launch pad area. The new rules called for specifically designated areas for spectators who, unless qualified under some other provision, are not to be allowed at any launch area at any Tripoli-sanctioned event. "Unless all high-power flyers move our collective risk management to the levels required by the rockets we are flying, we are living on luck and the law of averages," concluded Good in an opinion column he wrote for a Tripoli newsletter. "Perhaps that LDRS 29 incident was the sobering example we needed to see, to help us commit to the actions that will permit high-power rocketry to continue onward, as it has done since it was begun so many years ago."

Following LDRS-29, Vice Pres. Terry W. McCreary became the seventh president of the Tripoli Rocketry Association. McCreary was born in 1955, and he grew up in the tiny town of East Freedom, Pennsylvania. His parents, Harry and Charlotta, were from small towns in Central Pennsylvania.

"My dad drove a truck. He started out with the railroad when he was young and then went on to big trucks," said McCreary. "He drove locally in Pennsylvania for thirty years and more than 2,000,000 miles without an accident. My mother was a homemaker, and she was always pregnant or nursing a child. I was the fifth of fourteen children. It was a lot more fun in retrospect than it was at the time. You learned to eat the food whenever it was there and to wait your turn for the bathroom. We only had one bathroom, and an outhouse."

McCreary's path to high power is a familiar one. "I remember very vividly in the fifth grade there was a student, an only child, who had money, his own television set, which was unheard of, and model rockets. He brought an Estes catalog and a rocket to school one day. He launched the rocket during recess. I was hooked. Of course, this was in the early 1960s, and the space program was everything. We watched every mission. I still recall the last *Mercury* flight, and in school, every time there was a launch, they would stop everything and roll a television into the classroom so we could watch."

"Rocketry started at ten and continued until I was about sixteen," said McCreary. "I remember one rocket in particular—an Estes *Vega* model I had. The reason I got it was because I had appendicitis, and I was home from school for two weeks with very little to do, and my parents got together some money and bought the rocket. I built it and painted it and put the decals on, and it was beautiful. I put a B motor in it and pressed the launch button, and it went up fine, but the parachute didn't come out, and it came in and core-sampled. It broke my heart. But I was getting interested in cars and girls anyway. I soon got out of rocketry."

Even though he lost his passion for rockets, McCreary remained involved in science, especially chemistry. "I just loved to mix things and watch them foam up. I did not have a chemistry set growing up, but back then, you could go into a hobby store and buy little bottles of chemicals and equipment to mix them. When I graduated and started college, my main interest was chemistry and in making things too. I liked being active rather than passive. I loved the idea that I could make hydrogen and burn it, do electrolysis of water, bend glass tubing and make it into shapes."

McCreary attended St. Francis College in Loretto, Pennsylvania. He excelled in analytical chemistry, which is the study of the separation, identification, and quantification of the chemical compounds of materials. He earned his bachelor's degree from St. Francis in 1977—he and his brother Jim were the first in their family to earn college degrees—and enrolled at the University of Georgia in Athens.

"I was able to get an appointment as a teaching assistant and work on my advanced degree in analytical chemistry," said McCreary, who discovered he enjoyed teaching chemistry to students even more than doing research. After obtaining his masters, he packed up again and took a teaching position at Cumberland College in Eastern Kentucky. He taught chemistry there for six years, from 1979 to 1985. He was appointed as an assistant professor of Chemistry in 1984. McCreary decided that to advance further, he needed a PhD in his field. He left Cumberland for Virginia Polytechnic Institute in Blacksburg, where he received his doctorate in analytical chemistry in 1989. He accepted a teaching position at Murray State University in Kentucky, where he taught chemistry for the next thirty years.

McCreary's reintroduction to hobby rockets came in the early 1990s, when he saw an advertisement in *Popular Science* for a book by David Sleeter on how to make black powder motors. "Unfortunately, when I ordered the book, I got a note back from the author who said it was now out of print. But he copied the old book on a copy machine at four pages per sheet for me. He said he was going to make a new edition that was much more detailed, and he told me that since I had already paid for the book when the new edition was done, he would send me a copy."

McCreary's early work with black powder motors was less than successful. "Out of the first ten motors that I made, about eight or nine of them CATO'd," he said. "All of these were Estes-sized motors. I hadn't figured out some of the chemical relationships yet and was just following the instructions in the book. It turns out that I had mixed my mixture much too thoroughly for the length of core that I had, and that's why the motors would CATO. But I didn't realize that at the time."

Up to this point, the biggest motors he had seen—the ones in Sleeter's book—were 38 millimeters in diameter and perhaps 10–12 inches in length. He had not yet heard of Tripoli, high-power rocketry, or LDRS. Then in 1993, two things happened that would change the chemistry professor's view of hobby rocketry.

First, he saw an ad in *Popular Mechanics* for rockets made by Public Missiles Systems. "In the ad, there was a picture of four guys with a rocket that was 4 inches in diameter and five feet tall, and the ad said you could shoot for altitudes to 50,000 feet. I thought, 'Holy crap!'"

Next, McCreary came across a copy of Bruce Kelly's *High Power Rocketry*. He liked what he saw, and he immediately bought a subscription to the magazine. "I saw the prices for motors in ads in the magazine and was amazed that people were spending $10 to $20 on a single rocket motor! I was still used to Estes motors at .95 cents each."

McCreary learned of a high-power launch just 4 hours from his home. It was the Spears 2 event in Manchester, Tennessee, and he finally got a chance to see what high-power was all about—in person. "About an hour shy of Manchester, I passed a Subaru station wagon," he recalled. "I saw a beautifully painted tube in the back with some fins on one end. Yep, I'm headed in the right direction."

As he arrived at the field, he noticed a large structure that he thought was a funny piece of farm equipment. It couldn't be a launch tower, he thought. Nobody would launch a hobby rocket that big. Yet it was a launch tower, and as the day wore on and he saw one rocket after another leave its pad, he had an epiphany: "It was the type of thing that just overwhelms you. It was a little bit hard to take it all in. I ended up taking lots of notes while I was there and writing an article for *High Power Rocketry*. I was really impressed by the people at my first launch. They were incredibly friendly and willing to talk about what they were doing."

McCreary picked a good first launch to attend. That weekend, he saw all forms of high-power rockets, H through N power. He met Bruce Kelly, Dennis LaMothe, Tim Eiszner, and several other high-power veterans. He saw hybrid-powered rockets and K-powered flights to 10,000 feet. He was flabbergasted when LaMothe launched an N-powered missile that sustained a CATO mid-flight. Another flight was Eisner's rocket powered by an AeroTech K458 "coffee can" motor. "It went to Max-Q and then just came apart," said McCreary. "That was memorable!"

"Occasionally, I remembered to write something down, but ye gods, I just want to watch! I learned that Black Jack means a lot of black smoke. Blue Thunder propellant is aptly named. And the time from apogee to ejection is measured in hours, not seconds. Okay, well, it seems that long."

That evening at the flyer's motel, McCreary bought his first high-power rocket, a LOC *Starburst*. He also joined the Tripoli Rocketry Association. It was the first step in journey that would one day lead him to the national rocketry stage.

As a chemistry expert, McCreary soon gravitated toward the experimental or research side of high power. He met homemade motor guru Jim Mitchell, who lived less than an hour away, and the two became fast friends. "Jim was very funny, and almost everything he said made you laugh," said McCreary. Yet what really intrigued him about Mitchell was the fact he was making his own high-power rocket motors.

"When I went to my very first launch, I looked at all these motor reloads, and I realized that this type of propellant was completely out of my experience," recalled McCreary in 2014. "As a chemist, I wanted to know more about it, and I picked up something called Blue Thunder, and I asked, what is this? And people said it is called Blue Thunder. And I said yes, but what is in it? And they said well, it makes a blue flame. And that was about as much as most people seemed to know about it. And I looked at it closely, and I saw specks of stuff in it, and I thought, well, what the devil is this? I knew that the space shuttle used propellant that was different than black powder. But I didn't realize that what I was now looking at was essentially the same kind of propellant used in the shuttle. I wanted to find out more. And Jim Mitchell played an enormous role in that."

McCreary was soon plowing through as many books as he could find on solid motor propellant. But there really wasn't much available yet on homemade motors. He came across a dozen different formulas for propellant, but some of the ingredients were not even necessary. He started making motors with Mitchell and his friends.

"We learned by trial and error," he said. "We added things here and there and wrote down the results. And then one day it clicked in me. Oh, we are looking at a polymerization reaction! And this stuff is what cures that stuff, and that stuff is what makes it flexible, and this stuff isn't needed at all. It started to make sense to me, and I was then able to draw on my chemistry experience."

McCreary's first motors were 29 and 38 mm composite motors. "I thought it would be neat if other rocketeers who were interested in this would not have to go through the same sort of process. There is a lot of reinventing the wheel in amateur rocketry."

In 1997, he began collecting all his notes and materials together to write a book on making APCP rocket motors. "As a teacher, I decided I was going to write something so people could understand what they were doing instead of simply following a recipe."

McCreary began his *Experimental Composite Propellant*, a 248-page work on the subject, in 1997. When it was completed in 2000, it was an instant hit with a new generation of research flyers desperate for more information on building their own rocket motors.

"The book begins explaining how a composite motor works and the differences between it and a typical black powder motor," said McCreary. "It also gets a bit into ballistic calculations of pressure inside the motor and the parts of the rocket motor. The book also tells you what materials you need to get and describes how to make a very simple composite propellant."

By 2002 and as a result of his book, McCreary was a celebrity in high power. He was also an outspoken advocate for rocketry's suit with the ATF over the classification of APCP as an explosive. "Obviously, the people who make APCP professionally do not consider that it 'functions by explosion,'" he wrote in an opinion column in *Extreme Rocketry*. "An explosion in a grain elevator may occur, but that is not the function of a grain elevator. An explosion in a rocket motor is not the function of

APCP in a rocket motor." Recalls McCreary of the lawsuit with the ATF, "From the very beginning, I was aware that the classification of the propellants as explosives was not scientific. I was offended as an academic that there was no dividing line between low explosives and non-explosives. I knew you could make many things blow up—like even dry ice—so the definition being used by the government was insufficient."

McCreary was elected to the board in 2002. He had large shoes to fill. The seat he obtained originally belonged to Bruce Kelly. In 2007, he used his chemical expertise for an affidavit that was lodged with the federal court, challenging the conclusions on burn rates reached by ATF's experts in the Hawk-Dinan Report. When Ken Good decided to step down as president in 2010, McCreary was already vice president of Tripoli. The next step seemed obvious.

"Ken Good wanted to stay on to see through the lawsuit," said McCreary later. "And when it finally ended, he was tired. I decided it was time for me to run for president, and people on the board wanted me to run, and I didn't think anyone else was willing to do it at the time. As an academic, I had a little more free time than other people, and I thought I could do the job."

Like the Tripoli presidents before him, McCreary was soon devoting a tremendous amount of time to his new volunteer position as president. Much of his effort was spent responding to an almost daily barrage of telephone calls and e-mails on every conceivable subject related to rocketry.

"I recall one call where somebody had launched a rocket and they had put 10 pounds of lead shot in the nosecone to make the rocket stable, and it turned out that when the parachute opened and the nose cone was out, they had not glued the lead shot in place, so all of this lead came raining down on cars at the launch," said McCreary. "I get the call about the fifteen or twenty cars, all of which have little dents in them, and what are we going to do about it? I don't remember how we handled it, but I think most of it went to someone's homeowner insurance, and the claims ended up being resolved amicably. I think people realized that you shouldn't wear dress pants when you shovel manure, and if you bring your car to a rocket launch, you are taking a chance that something might happen to it."

On April 11, 2011, Dr. Franklin "Frank" Kosdon died of natural causes at his home in Southern California.

Kosdon was born in East Los Angeles in November 1941. He spent his formative years staring into the night sky through telescopes at his family home in the Central Valley of California. After high school, he was accepted at the Massachusetts Institute of Technology, where he studied physics and aerospace engineering. While at MIT, Kosdon and another student authored a research paper called *Experimental Development of an Isocyanate Solid Propellant*, which earned the pair a national award from the American Rocket Society and recognition at a luncheon where Vice Pres. Lyndon B. Johnson was the guest speaker. The research paper served as part of a larger project to construct and launch a probe intended to carry an instrument package to the edges of space.

Kosdon was not part of high-power rocketry in its early days. Yet when he found the hobby in 1987, he jumped in with both feet. Over the next few years, he became one of the pioneers, along with Gary Rosenfield, of the reloadable high-power rocket motor. His motor company, Truly Recycled Motors, debuted in mid-1991, just months after AeroTech introduced its first reloadable motors.

Kosdon was never one for formality, and his rocketry business reflected his own personality. He never had a professional sales plan for his company, and he lacked the marketing and organizational abilities of Rosenfield. Nevertheless, he developed a reputation for building unique high-power motors. In the view of at least one commentator, Kosdon created high-power motors that "really shook the ground."

At LDRS-12 in Argonia in 1993, Kosdon launched his *Full Metal Jacket* with an O-10,000—probably an ineligible motor for an LDRS event at the time. Yet Kosdon rarely paid attention to rules, and when LDRS event organizers decided it was more trouble to take the rocket down than to fly it, they moved everyone back into the parking area and launched the rocket. Undoubtedly, the organizers were just as eager to see the rocket fly as everyone else. *Full Metal Jacket* roared off the pad to an estimated altitude of 35,000 feet, giving all in attendance a brief glimpse at the future of high-power rocketry.

In 1996 at BALLS-5, Kosdon built the 10-inch-diameter R17,000 motor for the *OuR Project*, which was launched to an altitude just shy of 100,000 feet. The 700-pound R-powered rocket, which was the group effort of Kosdon, Paul Robinson, Jim Rosson, John Dunbar, and Phil Prior, plummeted back to Earth undeployed and disappeared into the ground, never to be seen again. The team effort proved Tripoli flyers were capable of reaching extreme altitudes and was an inspiration to the entire association.

Kosdon rarely used a personal computer. He preferred to write out everything by hand or used a typewriter, tools of a prior generation. He disdained conformity; he hated authority.

"It was Dr. Kosdon's trademark to show up at any venue he attended wearing well-worn T-shirts and shorts along with a pair of dirty sneakers or flip-flops," wrote *Rocketry Planet* creator Darrell Mobley shortly after Kosdon's death. "His eccentric attire was just as much a part of who he was as the Ziploc plastic bag he used for a wallet that contained a pocket-sized copy of the Unite States Constitution. Still, no one can deny his expertise in building rocket motors."

Kosdon's mistrust of authority made him a natural proponent of going after the federal government regarding the definition of APCP on the federal explosives list. He also went after high-power leaders when he felt constrained by rules and regulations in the hobby. He ran afoul of the Tripoli board in the early 2000s and was briefly suspended from Tripoli for violations of the association's motor rules. In 2002, he ran for a seat on the board with a campaign that urged members to use their votes to throw the entire Tripoli board out of office. He was not elected, and eventually, the stormy relationship between him and the board was mended. Yet Kosdon always remained a rebel at heart.

Through the early 2000s, Kosdon was a regular figure at high-power launches all over the west, such as ROCstock, DairyAire, BALLS or even LDRS. He remained involved in the motor-making business, and out on the range, he was willing to talk about motor selection or propellant theories with anyone who needed help.

"I have one of his letters, given to me by a friend, and I treasure it," wrote Tripoli president McCreary when he heard of Kosdon's death. "He didn't need a computer. He was a natural and his motor designs were often the product of just a few moments' thinking. Frank was truly one of the pioneers in our hobby and his influence on rocketry will long be remembered."

Frank Kosdon was seventy.

The thirtieth anniversary LDRS went forward on September 1–6, 2011, at the location where it had been conducted six times since 1993—the pasture on Rick Nafzinger's farm near Argonia, Kansas. It was the first time LDRS had a starting date in September and was the latest LDRS ever held. But there was no shortage of flyers, spectators, or excitement. Despite a few days of oppressive heat accompanied by high winds, 936 flights took to the air. It was no record-setter in terms of the number of flights, but it was perfect for the national television audience who once again were treated to a 1-hour special from the *Science Channel* after the launch.

Gary Tortora traveled to LDRS-30 to launch several rockets, including the flight of his *Albatross*, a 110-pound vehicle loaded with a research motor O4300 built by Al Goncalves. The *Albatross* roared into the sky to an altitude of more than 12,000 feet before settling back under parachute for a perfect dual deployment-style recovery. Other early flyers included Kevin Cornelius with his carbon fiber airframe *Mongoose* on an M745 to more than 16,000 feet and Kevin Trojanowski's Level 3 certification flight—this was his fifth attempt—on a 35-pound rocket that looked like a stack of five-gallon pails stacked one on top of the other.

There were several flights dedicated to the memory of motor-making legend Frank Kosdon, including one by young Charlie Ogino of North Carolina. When Kosdon was experimenting with his first reloadable prototypes in the late 1980s, Ogino had not even been born yet. On day 4 at LDRS-30, Ogino launched his *Little John* rocket, a Kosdon L3000. The rocket ripped off the pad and disappeared high in the sky.

Television personality Kari Byron was at LDRS again, interviewing flyers while also presiding over events like the fastest-to-10,000-feet competition, where N-powered rockets drag raced off the pad, and the second annual odd roc competition. This year's odd rocket winner was a 180-pound, 6-foot-wide replica of a Tiki hut with four M1882 motors secured to its roof. The flying hut was built by several flyers who called themselves the Cajun Coalition. The vehicle raced into the air for several hundred feet and, surprisingly, recovered well under parachute.

There were several flights by international Tripoli members at LDRS-30, including Frank DeBrouwer and Peter Mueller of the Netherlands. The pair launched two October *Sky*-inspired *Ms. Riley* rockets. Their 55-pound missiles were drag-raced together, each powered by M1505 motors. Both had good flights and recoveries.

Also, in attendance was a large contingent of flyers from Australia, which was experiencing plenty of high-power growth in the 2000s. The Australians launched their *Outback Thunda* twice during LDRS-30. The giant blue rocket had the lyrics of *Waltzing Matilda* printed on one side. The rocket took off Sunday on an N2600 and three J-impulse Sparky motors for a perfect flight and great recovery. The rocket was then flown again on Tuesday for another good flight. Fellow Australian David Wilkins—Tripoli's longtime Internet guru—was elected to the Tripoli Board of Directors on Saturday night. When Wilkins took his seat, he became the first person outside the United States to hold a board position with the association.

There were several heavy projects at LDRS in 2011, including Charlie Ogino's 260-pound full-scale *WAC Corporal*. The big *WAC* was reminiscent of the missile flown by Tripoli board member Dennis Lamothe in the early 1990s. Ogino's flight was just as memorable. With the help of his father, Guy, and several other family members, Ogino launched the rocket with a Cesaroni O25,000 motor. The 6-inch-diameter motor was the most powerful commercial motor available in high-power rocketry. On ignition, it lifted the *WAC Corporal* effortlessly and carried the rocket to an altitude of 12,000 feet. It was a majestic ride up, with a bright green flame and plenty of smoke in its wake.

Unfortunately, there was no separation at apogee. The rocket turned over and descended, undeployed. It was coming in fast and ballistic. Then at approximately 1,500 feet, the back-up ejection charges for the main parachute fired. The charges blew the nose cone off, the main parachute appeared for a fraction of a second, and then the chute was stripped away like a piece of tissue paper out of its box. It floated off into the distance while the rest of the rocket, sans the nose cone, plunged into the ground and was obliterated.

The biggest rocket of LDRS-30 was the *Udder Madness Project*, also known as the *Cow Rocket* because of its black-and-white dairy cow paint scheme. The rocket was built primarily by MDRA's Neil McGilvray and was a direct descendant of a similar rocket that had sustained a launch pad CATO during the Discovery Channel's filming of LDRS-22 in Amarillo in 2003.

The new rocket, just under 13 inches in diameter, tipped the scales at 460 pounds with a P11,000 motor. McGilvray had help from many flyers in completing the project, as well as a dozen people who helped him load the rocket on the launcher on Monday, September 5. The rocket had a good takeoff and climbed to 9,000 feet before deploying its recovery gear as planned, landing several hundred yards away in a sorghum field.

The *SORTA* rocket was another large project by several Tripoli members from around the country, including David Reese—a high-power member who was not born yet when Chris Pearson hosted the first LDRS in Medina in 1982. The 320-pound *SORTA* lifted nicely on a homemade O8000, which carried the rocket nearly a mile high before it settled back to a perfect recovery. An even larger rocket was the result of another team effort, dubbed the *Safety Rocket* because in one member's view, its bright fluorescent colors of pink, green, and yellow would allow it to stand out in a crowd. This missile was 22 feet tall and weighed nearly 400 pounds. It was powered by three research motors created by Pat Gordzelik: a central P5600 and three outboard L803 motors in strap-on boosters. The rocket shook the ground on takeoff. But shortly into its flight, a bright fire erupted out of the lower side of the airframe. McGilvray of *Rockets* magazine described what happened next:

> A micro second later the booster of the rocket was in a million pieces, engulfed in the expanding flame of the released propellant. The payload section and the strap on boosters were blown clear of the flaming carnage. Pieces of the brightly-colored rocket began to rain back down into the field. Parachutes were deployed but didn't function. Long arcs of smoke trailed the burning segments of the rocket and pieces of propellant that streamed down on the range.
>
> The team was in shock. This was not the outcome they anticipated, despite the fun-loving jocularity prior to the launch. They were serious [flyers] with a serious project. When the motor was examined after the range was deemed safe, it had been split open like a can of tuna. Pat Gordzelik suspected it might have been a thread for the closure that had been cut too deep that led to the demise of the rocket.
>
> One thing is for sure, sometimes bad things happen to good rockets.

43

"Let's Punch a Hole in the Sky"

From the first LDRS at Medina in 1982, where participants launched model rockets to an altitude of perhaps a couple of thousand feet high, to this modern age where individual flyers strive for 100,000 feet and more, this hobby has always been about reaching for the heavens. The AP composite motor may be the beating heart of high-power rocketry, but altitude has always been its soul.

Sometimes this quest for altitude takes the form of comedy—like the lifting of bowling balls, wooden spools, and Tiki huts into the sky. But the high priests of rocketry have always affixed their gaze on the ultimate high: outer space. They don't admit that, of course, unless you're Ky Michaelson, yet that is surely where this hobby has been heading for almost forty years, today with its computer simulation programs, carbon fiber and aluminum airframes, minimum diameters, and thousands upon thousands of pounds of gravity-defying thrust.

Altitude contests were a prominent feature of high-power rocketry during the early years of the hobby. One of the better-known competitions was organized by William "Bill" Wood of California. For several years, Wood published the *Rocket Newsletter*, a periodical to keep track of altitude records in high power. The newsletter sponsored a Perpetual Altitude Trophy to be awarded the first flyers to top the flight of the Fort Rocket Team of San Jose, who allegedly reached 22,070 feet with a flight at the Smoke Creek Desert in 1984.

During the late 1980s, Wood and a few other dedicated flyers at the Lucerne range in Southern California kept a close watch on the progress of high-altitude flights. In this era before onboard altimeters, the technology for determining altitude was still by optical tracking. The highest tracked flight in 1986 was a 2-inch-diameter rocket built by Gary Rosenfield and Wood. It streaked to 15,312 feet on a K motor. By 1988, the highest track went to Tim Brown, who achieved the dizzying height of 3 miles, also on a K motor. Several more flyers exceeded the 3-mile mark in 1989, including Bill Barber (3.15 miles), Bruce Kelly (3.37 miles), and Steve Buck (3.99 miles).

The Fort Rocket Team was finally eclipsed in 1989 by Korey Kline and Chuck Rogers. Using a 2.26-inch-diameter vehicle just over 5 feet long, their missile was optically tracked using two Centuri trackers to the incredible altitude of 24,763 feet (4.69 miles) on an Ace Aeronautics L500 motor at Lucerne. Kline and Rogers won the Perpetual Altitude Trophy. But their rocket was never recovered. After becoming a tiny blip at apogee, it simply disappeared. It was never seen again. (As late as early 1994, the Black Rock record stood at 27,576 feet in a two-stage rocket called *Way High* by Pius Morozumi.)

The *Rocket Newsletter* did not survive long into the 1990s. Yet the quest for higher altitudes continued. By the end of that decade, it was not uncommon for experienced flyers at Black Rock to reach 20,000–25,000 feet if they knew what they were doing.

In 1999, Richard King set a new Tripoli N-motor record using an AeroTech N2000 in a scratch-built four-inch-diameter fiberglass rocket called *Gambling Money Ain't Got No Home*. King's 10-foot-tall rocket weighed 28 pounds and carried dual altimeters by Adept as well as a Walston tracking transmitter. It was launched on September 11, 1999, at Black Rock. The weather that day was the best he had ever seen there, recalled King. His rocket reached the then-astounding altitude of 32,049 feet. And he recovered the entire vehicle intact.

King's N altitude record would remain unbroken for seven years.

Rocketry's final climb to the upper atmosphere began in the early 2000s at Tripoli's annual experimental launch—the BALLS launch.

"This is where would-be Werner von Brauns get to show off the most powerful, fastest missiles they can make," wrote *Forbes* magazine after sending a team of writers to Black Rock to cover the event in the fall of 2000. "[I]t is the purpose of this little gathering to shoot off some of the most powerful [rockets] outside the Pentagon and certain paranoid

Third-World dictatorships," observed *Forbes*, which led its feature story with a familiar remark, long made by flyers on the Northern Nevada desert floor:

Let's Punch a Hole in the Sky

By 2000, the BALLS launch was nearly a decade old, and it had settled into a familiar routine. Flyers from around the world met in Northern Nevada every year in the early fall to take advantage of good weather and the highest altitudes waiver in amateur rocketry: 100,000 feet and higher at the Black Rock Desert. They would bring minimum-diameter rockets with airframes made of everything from fiberglass to Kevlar to carbon fiber to T-6061 aluminum. They launched their homemade missiles on commercial and experimental motors, M through Q power.

Most of the time, these flights failed miserably. Leading-edge high-power rockets built by experienced Level 3 flyers were destroyed in spectacular fashion every year at BALLS. Some were destroyed by white-hot fires at the pad, some shredded like tissue paper during flight, some were lost in high-pitched screams back to Earth. This last method of destruction was the fate of countless high-dollar rockets, including the *OuR Project* in 1996, which reached an estimated altitude of 94,000 feet and, except for a hole in the ground, was swallowed up without a trace by the desert on its return.

And so it continued for the next several years. For those pushing the altitude envelope in high power, the reward was typically failure. High-power rocketry at BALLS in the 2000s was in many ways as Norman Mailer described in his written history of the early days of professional rocketry, like the burning of witches:

> One sorceress did not burn at all, another died with horrible shrieks, a third left nothing but a circle of ash and it rained for eight days. At the Raketenflugplatz outside Berlin in the early Thirties, rocket engines exploded on their stands, refused to fire, or when tested in flight lifted in one direction, then all but refired at right angles and took off along the ground. Indeed, the early history of rocket design could be read as the simple desire to get the rocket to function long enough to give an opportunity to discover where the failure occurred. Most early debacles were so benighted that rocket

engineers could have been forgiven for daubing the blood of a virgin goat on the orifice of the firing chamber.

"So dread inhabited the technology of rockets," observed Mailer, recounting the many failures of *Thor* and *Jupiter, Juno* and *Vanguard*, and even the *Atlas*—before the knowledge attained from so many disasters led to success—and man's successful trip to the moon aboard *Apollo 11* in 1969.

There were at least three engineering hurdles to ultrahigh-altitude flights in high-power rocketry. The first was building reliable motors capable of propelling rockets to great heights. By the late 1990s, commercial motors in the M- and N-impulse range allowed flyers to reach altitudes approaching 30,000 feet and perhaps a little more. To go even higher, some flyers were making their own motors. "If we can reach 30,000 feet on a commercial motor," went the adage, "just imagine how high we will go on a P motor!"

Once the stigma of the basement bomber passed in the mid-1990s, a small group of Tripoli members dove into motor-making at home. Yet there was an immutable learning curve for these makers of homemade N, O, P, and Q motors. And more often than not, just around the bend of that learning curve, lay destroyed rockets either in ashes on the pad or scattered all over the rocketry range.

"It took us one-half year to build this *A-1* rocket," said von Braun after the first static test of one of his large liquid-fueled rockets in 1934.

"And exactly one-half second to blow it up."

The seat-of-your-pants rocketry education for high-power flyers was the same. There was a lot of trial and error before the average flyer could properly formulate, mix, and cast APCP into a viable, working motor. It was one thing to take a Thunderflame course and build a J motor in class under the supervision of Jim Mitchell or Sonny Thompson. It was another matter altogether to create from scratch your first N motor, unsupervised, in your garage at home. Like baking a cake, people needed time to get good at it. And to get really good at making high-power motors, people needed a lot of time.

In the process, scores of sophisticated rockets were annihilated at Black Rock and on rocketry ranges all over America.

One of the more spectacular yet typical research motor failures was in 2002 with a project built by a Tripoli team from Las Vegas known as NASSA. NASSA's *Group Project Q* was a beautifully finished 225-pound, 6-inch-diameter vehicle powered by a Q-10,000. The aluminum rocket held thousands of dollars' worth of altimeters, computers, and video equipment. It was erected on an equally impressive custom-built launch trailer. Flight simulations put *Group Project Q's* estimated altitude at 75,000 feet. It would have been a new high-power standard had it reached that altitude. Yet like so many other ambitious altitude attempts of its day, the result was inglorious. On ignition, a newly designed started grain plugged the core of the nozzle, and soon, the entire vehicle was on fire. Witnesses said that fuel grains began ejecting out of the forward end of the rocket like some kind of giant Roman candle. One of the ignited grains rolled under the launch pad itself. In the end, the rocket and even the launch pad were destroyed, leaving pools of molten aluminum in their wake on the desert floor.

One of the technical issues plaguing homemade motors was over-pressurization of the motor case at or near the moment of ignition. Too much propellant surface area is exposed to too much heat, and kaboom, the motor CATOs like a huge firecracker.

At Black Rock, the CATO usually followed one of three paths. The first two paths were closely related. Too much pressure from the burning propellant inside the motor caused the metallic case to expand ever so slightly at either end, where the closures were attached. If the aft end of the case expanded even a millimeter or two, the closure securing the nozzle end would loosen, and then the closure would literally fall out of the rocket, taking the nozzle and sometimes everything else within the motor with it. Of course, if the nozzle dropped out of the case, the pressures inside the motor would be immediately reduced. But the burning propellant would spill out of the rocket and create a fire that would engulf the entire vehicle right there on the pad. Sometimes the nozzle did not fall out until the rocket was hundreds of feet high. In this scenario, burning propellant slugs would drop from the rocket like white-hot flares dropped from a fighter jet, burning in the air all the way back to Earth and then smoldering on the ground until the fuel was completely exhausted. If the deployment system was not damaged and the parachutes came out, the rest of the rocket might be spared.

In the second scenario, over-pressurization of the case would loosen the forward end of the case, leaving the nozzle end intact and functioning. As the rocket climbed high into the sky under thrust, the forward metal closure would suddenly be ejected upward inside the vehicle, like a projectile. The closure would travel straight up through the interior airframe into the heart of the rocket, ripping through bulkheads, electronics bays, and recovery gear before blasting the nose cone off the rocket. Witnesses described one such P-power incident at BALLS in 2003: "The rocket leapt from the pad trailing about 20 feet of fire. But shortly into the flight, the motor CATO'd, blowing the forward bulkhead up through the rocket like a cannonball. The stripped chute and bits of the nose cone landed in and around the flight line, while the reminder of the rocket plowed into the playa about a mile to the southwest."

The third path of destruction was probably the worst. Over-pressurization would happen so fast that the closures didn't have time to loosen at either end. The metal body of the case would simply come apart. This would manifest in one of two ways, either as a blowtorch-like eruption venting flaming fuel out of the side of the motor case and right through the rocket's airframe or the most spectacular motor failure of all, where the case would come apart so quickly it seemed like an uncontrolled explosion. In a millisecond, twenty-foot-tall rockets disintegrated on the pad, or perhaps they climbed a few hundred feet in the air first before deconstructing, leaving nothing but a scrap metal rain shower.

The precise causes of over-pressurization varied. Sometimes it was poor propellant mixing techniques or inadequate curing or packing procedures during casting of the individual motor grains. The quality of the chemicals used to make the fuel or their chemical composition could factor in as well. In one detailed analysis of motor failures, a leading cause of over-pressurization was flyers choosing the incorrect nozzle size for their motors. The bottom line was that the more a flyer made his or her fuel, the better they got at it.

In the early 2000s, only a handful of experienced flyers outside the traditional commercial manufacturers were proficient at making their own propellant, M through Q power. However, within a few years, journeymen motor builders seemed to be everywhere. CATOs of research motors still occurred. But the high-impulse homemade motor was slowly being tamed. By 2008, if a flyer with high-altitude ambitions could not make their own reliable motor in the N to Q range, he or she could easily find someone who could do it for them.

As more homemade motors became available in the upper impulse ranges, there were more high-altitude attempts at Black Rock. These attempts increased when commercial motor manufacturers such as AeroTech and Cesaroni began to offer their own long-burning M-and N-impulse motors that specifically catered to the high-altitude market.

And so the next engineering hurdle then presented itself. Many of the airframes designed to carry high-impulse motors turned out to be too fragile to handle the power. Rockets were coming apart in the air everywhere. By the mid-2000s, N-, O-, P-, and Q-powered rockets would blast off the pad just fine. Then seconds into flight, their airframes would collapse under thrust, or even more commonly, the fins would be ripped off at speeds approaching Mach 2. The result was a spectacular flight to a few thousand feet. Then the smoke trail—a rocket's signature in the sky—would turn to cartwheels or somersaults or other impossible gyrations in the sky. The rocket was gone. It was raining parts again.

Inspection of the debris fields following these flights revealed that even carbon fiber and aluminum airframes could be zippered like paper by a nylon shock cord under the right circumstances. The techniques for putting together minimum-diameter airframes that worked for L- or even M-powered rockets were simply inadequate for larger motors. When rockets reached speeds in excess of Mach 2, fin alignment became absolutely critical. If a fin was not perfectly aligned, it was going to come off. When a rocket loses a fin and becomes unstable at Mach 2 and then deviates from a straight trajectory, the airframe is ripped to shreds. The means of attaching fins had to be reevaluated. Simple epoxy was out of the question. Flyers bought exotic composite glues and epoxies or covered their fin cans in carbon fiber or Kevlar wrap. Some flyers went to bolt-on fins with aluminum airframes. This was no guarantee of success either. If bolt-on fins were not seamless, or if air penetrated along the root edge, then it was coming off too.

And it would be raining parts again.

But after so many more disasters, people learned. Fin attachment techniques improved. Carbon fiber fin cans and other preassembled fin can assemblies became more popular. Some flyers hired professionals to weld their fins to the airframe. Fins began staying on even at speeds in excess of Mach 2. Airframes were strengthened.

Soon, the only thing coming off rockets at Mach 3 was the paint.

There was still one final engineering problem that dogged high-altitude flyers longer than anything else. Even when their high-thrust motors performed well and their carefully built airframes reached high altitudes intact, too many rockets were turning over at apogee, undeployed, and coming in hot. And with rockets reduced to tiny bits of rubble on the desert floor, or buried deep underground, postflight evaluations revealed few clues as to the cause of these disasters. What was happening? Maybe this was as high as high-power rocketry was destined to go?

"Worse by far than failure," wrote Mailer in *Of a Fire on the Moon*, "was failure for undetermined reasons."

Tom Rouse discovered rocketry in the mid-1990s at a Bay Area chapter of the NAR called LUNAR. Rouse lived in San Jose, and his proximity to the Northern Nevada desert soon brought him to his first Black Rock launch. AERO-PAC leader Pius Morozumi was one of Black Rock's foremost flyers at the time.

"When I met Pius, I was telling him how much our family enjoyed rocketry and how it was the cheapest hobby I could think of for so much fun," recalled Rouse.

"You haven't been in rocketry very long, have you?" replied Morizumi.

In the early 2000s, Rouse noticed that many ultrahigh-altitude rockets impacted the desert floor undeployed. No one knew why this was happening, although there seemed to be a direct relationship between destruction and altitude. The higher a rocket traveled, the greater the chances it would come back to the ground a screaming dart. Flyers lucky enough to find the hole in the desert created by their incoming rocket would linger for a while at the scene, not sure what to do. Their high-dollar vehicles were deep underground, and usually, no amount of digging was going to reach it. Sometimes there was considerable swearing done, but that would not recover the rocket either.

This pain was felt by novices and experienced flyers alike. Fortunately for high-power rocketry, Rouse decided to take a closer look at the problem.

"One of the years I went to BALLS, I remember everyone was having a hard time getting over 30,000 feet successfully. Everyone that tried, their rocket would crash," said Rouse during an interview in 2006. "Someone would guess, 'Well, you didn't set the computer up right,' or 'You didn't set your ejection charge up right.' Everyone was trying to figure out what the heck was causing all these rockets to fail. These were talented people using state-of-the-art recovery systems. One year I was there with Ken Biba when we launched his rocket to 24,000 feet, and it crashed. After we dug it out of the ground, we looked at the ejection charges and saw that the charge did go off, but not all the black powder burned. We asked ourselves, 'How could some of the black powder burn but not all of it?'"

Rouse surmised that something might be happening up there—way up there at high altitude—that was preventing deployment systems from functioning properly. He and some friends designed a vacuum chamber to simulate air pressures at higher altitudes. Then they conducted a series of experiments where they placed electric matches, typically connected to altimeters in a rocket to trigger black powder charges, into the vacuum chamber with the charges.

"The results were astonishing," said Rouse. "Each experiment over 20,000 feet showed a drop-off in the burn of pyrogens, including black powder, smokeless powder, and even electric matches."

Rouse discovered that at altitudes of 20,000–25,000 feet, black powder did not burn as effectively as it did at or near sea level. Above 30,000 feet, the problem grew even worse, he found. A much larger, oversize black powder ejection charge was needed to force an airframe apart as compared with the charges sufficient at lower altitudes. And even then, the bigger ejection charge was no guarantee of success. At simulated altitudes above 60,000 feet, Rouse found black powder ejection charges would not burn at all; even electric matches sometimes failed to ignite. If this happened in an actual rocket, the results would be disastrous. The rocket would simply turn over and race back to the ground undeployed.

Was this the hidden monster destroying high-altitude rockets at Black Rock?

Rouse discussed his findings with a professor friend who worked at nearby Stanford University. Rouse's friend referred Tom to NASA to discuss his test results. He learned the most likely cause of the drop-off in the burning of black powder at higher altitudes was simply thinning air; the higher the altitude, the less air that was present. At ultrahigh altitudes, there was not enough air to transfer the heat necessary to sustain burning of any kind.

"The way they explained it is if you were in space and there was a 4,000-degree flame and you were a foot away, you wouldn't feel anything other than the radiation from it because there are no air molecules to transfer the heat," said Rouse. "So if there are no air molecules to transfer heat from one molecule to the next, there is not enough heat to keep the combustion temperature going. Then the thing goes out."

Rouse had made a profound discovery for high-power rocketry: The culprit in many of the high-altitude deployment failures was not inexperience. It was not poor altimeter programming. It was not incorrectly sized black powder ejection charges. It was not a phantom lurking in the sky at high altitude.

It was just a lack of air.

Rouse decided to do something about it. He created his CD3 system, an ejection charge package for high-altitude flights based on small onboard cylinders of compressed carbon dioxide. Instead of burning black powder to over-pressurize a rocket's airframe, carbon dioxide gas was released at apogee to over-pressurize and separate the airframe. No air was needed. There was nothing to burn. Rouse's CD3 cartridges allowed flyers to deploy recovery gear at altitudes up to 100,000 feet and higher—theoretically, all the way to space. The final engineering obstacle was overcome for the average flyer. And by the mid-2000s, everything was in place for a real run to the edge of the upper atmosphere of Earth.

Between 2000 and 2005, the annual BALLS launch was both a heavy vehicle and a high-altitude rocketry showcase. The launch was now unquestionably the most exciting high-power event in the world. Even the *New York Times* took notice:

> Members of a gonzo subculture, the hobbyists have been known to launch Weber grilles, Port-A-Potties, bowling balls and pink flamingos. But once a year, on this bleak, 400-square-mile dry lake bed, they meet for the Indy 500 of rocketry, with waivers from the Federal Aviation Administration.

LARGE AND DANGEROUS ROCKET SHIPS

In 2000, a 35-foot-tall, 500-pound vehicle dubbed *Max-Q* blasted off on Q motor to 15,000 feet. Similar big projects followed every year. In 2004, there were two rockets at BALLS that weighed more than 700 pounds each. The first, built by Wedge Oldham, was a 25-foot-tall scale replica of a *Black Brant* rocket. Empty without the motors, the *Black Brant* weighed more than 400 pounds. It took more than twenty-five men to carry the airframe on their shoulders to the launch pad. The rocket was powered with three P-impulse motors designed by Darren Wright. On ignition, all three motors worked perfectly, lifting the 740-pound rocket to an apogee of more than 18,000 feet. The rocket recovered well under several parachutes.

The second mega rocket that same weekend was a team effort led by Kimberly Harms of Washington: a full-scale replica of a U.S. Army *Honest John Missile*, powered by four O-impulse motors built by John Lyngdal. "Like most large team projects, we split the cost, which was great since no one person could afford to build such a project," said Harms, who was captain of the ship during the build of the 27-foot-tall rocket. She added,

> There is so much more involved in a large project than the simple stuff like propulsion. Integration becomes really important. In a larger project like the *Honest John*, you have just plain logistical problems of how to move the parts around. It involves a whole different way of thinking than people are used to. Most projects that people do, you pick it up and put it in the back of your car, you take it to the launch, and you launch it. Well, you can't do that when a project gets to a certain size. Things become a challenge, like how to get it from horizontal to vertical. These are serious issues. We spent an entire day figuring out how we could lift the *Honest John* to vertical. Almost everything about the project turned out to be a challenge. How do we find the right color of paint? (The team leader discovered Laura Ashley designer paint makes a perfect olive drab.) How do we construct a nose cone that doesn't weight too much? How do we construct large fins that don't weight too much? It was almost comical at one point. It was like the story of the person who built a boat in their garage and then couldn't get it through the door. At one point in time during the *Honest John* build, someone said, "Gee, is this thing going to fit through the door?" We got out the tape measure, and indeed, it would not fit through one of the

doors in the workroom. The other door was just barely large enough. One of the other difficult tasks was simply moving the large parts from the ground level to the upper garage level. It's hard when you have a big, ungainly thing to move around. The thing was really heavy. Sitting on the pad, it was over 700 pounds. Dry weight was in the neighborhood of 500 pounds. The rocket stood about 27.5 feet tall. So it was really big to move about.

The *Honest John* had a spectacular liftoff. As it climbed into the air, eight single-use model rocketry G motors mounted horizontally in the upper airframe were also ignited, causing the massive rocket to rotate slowly as it climbed higher. Suddenly, one of the main O motors suffered an in-flight CATO. In the blink of an eye and only seconds into flight, the rocket simply disintegrated, scattering thousands of parts, big and small, in a debris field that stretched for hundreds of yards. Some burning chunks of propellant continued to climb another few hundred feet high, trailing white smoke in their wake. The booster section tumbled to the ground on fire. The upper end of the rocket fell without a parachute, landing in a crumpled pile on the barren ground with unopened recovery harness parts resting nearby.

"The motor failure was so catastrophic that it really ripped things up," said Harms afterward. "But what can one do?"

Rocketry photographer and writer McGilvray sums it up another way: "It's a rocket. What could possibly go wrong?"

The era of the big rocket continued at BALLS for a few more years. Large-diameter rockets weighing many hundreds of pounds were launched each year through 2008. They were crowd favorites with their imposing size, and the visceral high-thrust power plants needed just to get such rockets into the air. Yet BALLS remained the ultimate high-altitude destination in the world of rocketry. And the excitement of the minimum-diameter flight to the upper atmosphere began to eclipse the crowd-pleasing and visually appealing heavy lifters.

During the first five years of the new decade, rocketry's most ambitious high-altitude junkies took their shot at Black Rock, including people such as Frank Kosdon, Les Derkovitz, Tony Alcocer, Ken Biba, and John Coker, to name but a few. In 2000, a team from the

United Kingdom Rocketry Association of Great Britain led by Ben Jarvis propelled a boosted dart on a Kosdon O10,000 to 34,579. Both the booster and dart were successfully recovered, setting a new record for European flyers. There was more carnage and also some success in 2003, including a flight by Robin Meredith to nearly 25,000 feet and a team effort headed by Woody Wood, Jason Gimble, and Missile Works's owner, Jim Amos, to more than 30,000 feet.

In 2004, Kosdon, David Triano, and Mike Hobbs launched a high-altitude rocket on a Q12,000 motor made by Kosdon. The team was looking for 80,000 feet. The missile had a great liftoff and disappeared high in the sky. It may have had a separation event at apogee, or it may have come in undeployed and ballistic. No one really knows. After about 20,000 feet, even the largest rockets disappear to the naked eye. Only the wispy trace of a smoke trail remains. The rocket was never seen again. That same weekend, Jeff Jacobs launched an all-aluminum hybrid-powered rocket with an R10,000 motor. Jacobs was looking for 60,000 feet. Yet like the Kosdon-Triano-Hobbs rocket, Jacobs's rocket also disappeared. Only his rocket vanished right there on the pad. At ignition, the rocket sustained an immediate CATO of its hybrid power plant, reducing the 20-foot-tall rocket to bits of shiny aluminum confetti in a millisecond.

"There is a strong correlation between wisdom and scar tissue," observed veteran flyer and future Tripoli president Stu Barrett of the annual carnage at BALLS.

Then, almost imperceptibly, things began to change. Some of the change was simple math. The number of high-altitude attempts at Black Rock increased each year. At one moment at BALLS in 2005, there were seven rockets being readied for flight at the same time, in pads one next to the other, with a dazzling array of motors: a single P motor in one rocket, a three M cluster in the next one, three more rockets on separate pads each carrying a P motor, an O-powered rocket on another pad, and finally, a Q motor in the last rocket. All seven vehicles were being prepared at the same time in a shallow arc of launch pads only a few hundred yards long. Most of these flights were not successful. But one of the seven—launched by a Missouri Tripoli member who had already seen more than his fair share of failures—reached almost 60,000 feet on a P motor and was recovered.

In September 2006, several more teams arrived at Black Rock, each anxious and ready to set a new high-power rocketry altitude record.

Jim Wilkerson of Washington brought a four-inch-diameter rocket powered by a long-burning Cesaroni N1100 motor. "A high-altitude rocket should be as low drag as possible," explained Wilkerson. "The fins need to be aligned well to reduce rolling and off-vertical trajectory. Nose cones and fins should be appropriately shaped for high-speed flights. Fins should be sufficiently robust to eliminate flutter, and the rocket's finish must be as smooth as possible. Anything you can do to make the rocket slip through the air easier is a good thing. Wilkerson's N-power rocket reached more than 45,000 feet and was recovered intact. "The flight was supersonic for over 20 seconds, and it hit a maximum velocity of Mach 1.8," he said Wilkerson.

Richard Hagensick of Wisconsin launched a minimum-diameter rocket carrying a P2700 motor. The rocket weighed 124 pounds on the pad, ready to go. "High-altitude attempts require you to solve the greatest number of engineering challenges," believes Hagensick. "You have the aerodynamic design challenge, the propulsion challenge, the high-altitude deployment challenge, and the recovery challenge. The altitude achieved provides direct feedback on how good a job you did in the design, and it provides you with a goal to try and reach on your next project." Hagensick's missile reached 47,000 feet on P-power, followed by an outstanding recovery.

Next up were Alex McLaughlin and his Team Numb, with their minimum-diameter all-aluminum vehicle weighing more than 400 pounds. "I have a core team of folks that I rely heavily on to get stuff into the air," said McLaughlin shortly before his untimely passing in 2007. "Collaboration between friends makes your projects more fun to build and more fun to fly."

For their 2006 flight, Team Numb chose an eight-inch-diameter experimental Q18,800 that weighed almost 160 pounds. The rocket carried state-of-the-art high-power altimeters and a Rouse CD3 ejection system. The crew was looking for 65,000 feet. They dubbed their creation *Questionable Mental Health*. At the moment of ignition, the 20-foot-tall rocket roared off the pad, leaving a shallow crater below the launch tower. The missile was trailing a long brightly colored deep blue flame and was several thousand feet high when it simply disintegrated. "We had actually static tested this motor prior to flight with no problems at all," said McLaughlin afterward. "At this point, we're thinking that there was a faulty liner to blame for the burn through. Next time we build an eight-inch motor, we are going to spend the extra money and get a convolute-wound composite liner."

Other flyers looking for ultrahigh altitudes in 2006 included veterans Robert DeHate, Dave Leninger, Rob Grygar, and Tim Covey. In the end, however, it was an engineer from Kansas City who stole the show.

A year earlier at BALLS in 2005, Gene Nowaczyk and friend Mark Brown launched Nowaczyk's *Scion* to 59,718 feet—the highest altitude achieved by any rocket in high power that year. The *Scion* was a six-inch-diameter all-aluminum rocket that weighed 135 pounds. It carried a P10,480 motor Nowaczyk built in his Missouri garage.

"I became interested in high-altitude flights trying to improve on my personal goals," explained Nowaczyk. "It was not trying to be better than others but trying to be better than I was yesterday. In my early days of high power, a 2,000-foot flight was great, a 5,000-foot launch was amazing, and a 20,000-foot flight was like a white elephant. Just talking about reaching 40,000 feet was crazy."

Yet in 2005, Nowaczyk reached 40,000 feet and then some. *Scion's* recovery from nearly 60,000 feet was perfect. But it was not easy. The rocket inadvertently deployed its 18-foot-diameter main parachute at apogee. Upper level winds were up to 80 miles per hour, said Nowaczyk. "It was one of those days when at launch time, it was dead calm and then just as fast as the rocket took off the winds came in. In retrospect, a 6-foot-diameter parachute would have been just fine. But for whatever reason, we stuffed a parachute into the rocket that was big enough to bring a tank back to the ground. With the upper level winds and an empty weight of only 45 pounds, the 18-foot parachute carried the rocket 35 miles!"

The arduous recovery aside, the *Scion* flight in 2005 was one of the most successful flights in the history of high-power rocketry up to that time.

Now, a year later in 2006, Nowaczyk returned to Nevada, hoping to set a new standard for high power. His *Piper* was an all-aluminum rocket that "looked like something that NASA would produce after spending a gazillion dollars," said one reporter at the event. *Piper* was 8 inches in diameter and 14 feet tall. It weighed 318 pounds on the pad. Nearly 80 percent of that weight—245 pounds—was a Q20,000 AP composite motor, also built by Nowaczyk. Three fins were used on the rocket, cut from 3/16-inch aluminum and secured to the airframe with ten to thirty-two screws tapped directly into the motor case. The rocket was packed with altimeters,

GPS tracking units, and video downlink equipment to provide live video from the rocket back to the range head throughout the entire flight.

At noon on Saturday, September 20, 2006, and under perfect blue skies at the Black Rock Desert, Nowaczyk's *Piper* roared into the air, trailing a 40-foot-long blue flame in its wake. "The motor rumble across the desert playa all the way to the flight line was astounding," said high-power veteran John Lyngdal. "It was well worth 20 hours of driving just to be here for that," said another observer of the takeoff of the 345-pound missile. "It left the pad like an Estes rocket on a D motor."

The rocket disappeared from sight in seconds. But under a darkened tent set up along the flight line, a live video stream out of the side of the rocket gave all those present a bird's-eye view from the missile as it climbed higher and higher. *Piper* exceeded the speed of sound almost immediately, and then it passed Mach 2, and then it topped three times the speed of sound, reaching a top speed of Mach 3.45.

It took 81 seconds for *Piper* to reach apogee at precisely 93,342 feet.

"It was beautiful," said Geoff Huber, one of the lucky flyers watching the ascent of *Piper* on the range head television screen. "At one point, you could see the curvature of the earth and outer space."

With the data to prove it, Gene Nowaczyk of Missouri had just broken every altitude record high-power rocketry had ever dreamed of in 25 years of flight. Perhaps what made *Piper* even more remarkable was its recovery. It was perfect. Both sections of the single-stage rocket returned to Earth as planned, under parachute, and nearly unscathed. The booster and upper airframe were recovered less than 4 miles from the launch pad and only a few hundred yards apart from each other. The rocket halves had some nicks and scratches from being dragged across the desert floor by parachutes after landing, and the paint was blistered by heat generated at Mach 3. Otherwise, it looked fine.

"He recovered the entire vehicle intact, and it was ready to fly again," said veteran flyer Bob Fortune. "All he needed was another 150 pounds of propellant."

Nowaczyk had done what no individual flyer in high-power rocketry had ever even come close to. He built a rocket that reached nearly 100,000 feet, followed by a successful recovery. This was the new standard for high-power rocketry. And suddenly, altitudes of 100,000 feet or more seemed within reach.

It would be five more years before anyone would eclipse Gene Nowaczyk's *Piper* flight to nearly 94,000 feet. Before Nowaczyk's record in 2006, few flyers cleared even half that altitude and recovered their rocket intact. Now everyone knew it could be done, and there was a sudden rush of flyers determined to try.

In 2007, Richard Hagensick launched his 8-inch-diameter *Q-Reaction* powered by a Q6102 motor—with a burn time of nearly 20 seconds—to more than 85,000 feet. Although the booster was never found, the upper airframe of the 287-pound rocket returned intact.

In 2008, Curt Newport and Jeff Taylor launched Newport's *Proteus 6* on a Loki Research P motor. The 6-inch-diameter all-aluminum rocket reached 88,240 feet. The rocket's deployment gear included a CD3 carbon dioxide ejection system. *Proteus 6* was recovered intact after flight.

The number of flyers looking for ultrahigh altitudes increased in 2009 and 2010. But there were few successes. In 2009, Geoff Huber took aim at 100,000 feet with his O-powered minimum-diameter booster that sat below a metal dart designed to separate at motor burnout and race for the upper atmosphere. "The speed machine was performing well until it appeared to slam into an invisible brick wall, jackknifed, and come apart into a million pieces," wrote one reporter afterward.

Kent Newman, Denny Smith, and Brad Wright launched a minimum-diameter P-powered rocket that looked fine as it cleared the launch tower but not for long. At about 2,500 feet, the forward closure blew, sending hot propellant spilling into the sky. The rocket cartwheeled, falling end over end, with fire belching out of both ends of the motor as the payload section hurtled back to Earth.

Robert DeHate was hoping for 140,000 feet with a two-stage rocket holding a P10,000 in the booster and an N4000 in the sustainer. Both sections of the 180-pound rocket came in ballistic, burying themselves deep in the desert after in-flight problems. However, one flyer—Adrian Carbine—found success with his 25-foot-tall three-stage rocket *Triple Threat*, which performed flawlessly on two Ns and one M motor, reaching apogee just under 60,000 feet with a good recovery. Curt Newport and Jeff Taylor returned to Black Rock in 2009 to launch *Proteus 6.5* to 74,884 feet, with another good recovery.

There were similar attempts in 2010. Jerry McKinlay successfully recovered his 4-inch-diameter O-powered *Prometheus* from 55,000 feet, Curt Newport cleared more than 72,000 feet with another flight of *Proteus 6.5*, and hybrid-motor expert Jeff Jacobs returned yet again with an R10,000-powered machine that looked as though it could reach 100,000 feet or even higher. The aluminum rocket weighed 350 pounds, carrying 200 pounds of nitrous oxide. It roared off the pad and climbed straight and true, but then it disappeared. Onboard GPS sent back signals from 45,094 feet. Then the track was lost. The rocket simply vanished. There was no incoming whine indicated a high-speed ballistic failure, no sighting of parachutes in the distance, no tracking from onboard locating devices, and no mushrooming dirt cloud on the horizon indicating a downrange impact. It was just gone.

Gene Nowaczyk's 2006 launch remained the high-power rocketry record.

On July 8, 2011, Space Shuttle *Atlantis* blasted off from the Kennedy Space Center in Florida. This would be the last flight of the space shuttle. The four-person crew lifted off in the presence of a crowd of more than a million spectators who gathered to see the end of another era in the American space program. Since 1981, the shuttle program had been planet Earth's primary program of human space flight. On this final flight of *Atlantis*, also known as STS-135, the crew spent more nearly 13 days in low-Earth orbit before returning to a perfect landing on the morning of July 21, 2011. For nearly the entire history of high-power rocketry, going back to just before the first LDRS, the space shuttle was the prominent feature of America's space program. Now it was over.

In the summer of 1982 at Chris Pearson's LDRS-1 in Medina, there were no carbon fiber airframes, no altimeters, no aluminum rockets, and no reloadable motors. The modern-day Tripoli Rocketry Association did not yet exist. Large and dangerous rocket ships, other than those flown at places like Smoke Creek, were a figment of rocketry's imagination, a catchy phrase uttered by an enthusiastic flyer in the Southern California desert and then grafted to an in-your-face launch event in the suburbs of Cleveland, Ohio. There was nothing large or dangerous about that first LDRS. The launch was about an attitude. And in that attitude lay the future of amateur rocketry in America and the world.

Thirty years and a million flights later—with thousands of destroyed rockets along the trail—high-power rocketry's destiny was about to arrive. And in September 2011, that destiny revealed itself, not far from the Smoke Creek Desert where this story began.

On Friday, September 30, 2011, BALLS-21 went forward at Black Rock with near-perfect launch conditions for a run at a high-altitude record: blue skies and nary a trace of wind. And for the first time at any high-power event anywhere, more than half-a-dozen flyers each had the right motors, the right rockets, the right experience, and the right weather conditions to be legitimate contenders to clear 100,000 feet. Any one of them, or perhaps all of them, could make this event the most memorable weekend in high-power rocketry history.

Amazingly, almost every one of them had a different means by which they were going to reach for the edges of space.

Geoff Huber's preferred choice to clear 100,000 feet was with the boosted dart concept tested by the U.S. military in the 1950s. In this type of rocket, a large booster pushes a heavy yet tiny sustainer called a dart that does not carry any propellant. The dart rests atop the booster without any firm attachment to the rest of the vehicle. When the fuel in the booster is exhausted, gravity, along with the booster's large fin section, causes the lower half of the vehicle to immediately slow down. The dart separates from the booster naturally, continuing its climb like a bullet shot out of a gun without the added drag of the booster, now discarded.

Huber's rocket ship was a fine piece of engineering built at his home near Chico, California. The booster was 4 inches in diameter and almost eight feet long. It held an O6000 motor. The tiny dart on top of the booster was a mere 1.5 inches in diameter and weighed less than 10 pounds. The dart was crammed full with the latest in high-power electronics technology and locating devices. The entire rocket weighed only 58 pounds. This put the thrust-to-weight ratio of the vehicle at more than 23:1. It would scream off the pad. Preflight computer simulations put Huber's dart at a potential altitude just over 100,000 feet. "I have always wanted to go that high as it has always seemed to be the magic number to reach," he said. "I have seen lots of people try and decided a few years ago that I wanted to take a stab at it. I have been working on this project for about four years now."

At ignition, Huber's rocket roared off the pad, the tiny dart poised to jump off the booster at motor burnout. However, it never reached the magic 100,000-foot mark. Tens of thousands of feet high, and 5.3 seconds into the motor burn, the O6000 motor failed.

"The case suffered a hole in the side just above the fin can, and the sudden drop in motor pressure caused the remaining [propellant] to extinguish," said Huber. The rocket went from steadily increasing speeds up through Mach 2.4 to an unstable, tumbling mass of metal, carbon fiber, and fiberglass. The debris found on the desert floor was unrecognizable as any rocket. Huber would have to wait for another day.

Curt Newport of Maryland was another legitimate contender for more than 100,000 feet. With Loki Research motor-making guru Jeff Taylor, Newport's *Proteus* rockets had already cleared 70,000 feet at least twice in the previous five years. His highest altitude was just shy of 90,000 feet in 2009. Now with his *Proteus 7*, Newport was confident he would reach 20 miles high, if not higher.

"The vehicle we will be flying is generally identical to what has been flown before, except for a more powerful motor and a redesigned fin structure," said Newport. The 6-inch-diameter nearly all-aluminum rocket weighed 90 pounds unloaded. The nose cone was made of laminated birch plywood covered with several layers of composite cloth. With a Taylor-designed Loki Q motor, *Proteus 7*'s weight was almost 200 pounds. Onboard electronics included Ozark altimeters and telemetry system, Rouse CD3-controlled ejection charges, and locating devices from Adept and Walston. It took six men to carefully lift the shiny three-finned rocket out of the bed of a transport truck and to line up its rail buttons with the launch pad. *Proteus 7* lifted off Friday trailing a 20-foot-long flame, pushed by more than 1,700 pounds of initial thrust from the Q7800 motor. The rocket climbed straight and true. It looked like a sure bet to reach a six-figure altitude.

Yet disaster struck again. This time at Mach 2, a fin tore loose high in the sky, the vehicle careened back and forth above the range, tearing off more fins, and *Proteus 7* reduced itself to long strips of shiny shredded aluminum at twice the speed of sound.

New Englander Robert DeHate preferred a two-stage design to clear 100,000 feet, and he had been trying for several years to do it. His personal best so far was 85,000 feet. The upper stage of that rocket was destroyed on its return. This year, DeHate was looking for 150,000 feet on a rocket carrying a P10,000 motor in a six-inch-diameter booster and an N4000 in the bright orange four-inch-diameter sustainer. DeHate's rocket weighed 45 pounds dry, loaded with two motors it reached 185 pounds. Takeoff was impressive. The rocket rose quickly in a mighty burst of flame and smoke. High in the sky, the yellow sustainer fired perfectly, separating cleanly from the booster. But the second stage fired too early. It did not get the full benefit of the push from the big booster. The rocket had a terrific flight, and recovery was successful, but apogee was at 80,000 feet. For DeHate, the 100,000-foot mark would also have to wait.

Adrian Carbine took a shot at the record too with the return of his impressive three-stage rocket *Triple Threat*. The rocket recovered successfully after a perfect flight to more than 75,000 feet on an N4100 to another N4100 to a final M1540.

Richard Hagensick launched a rocket at BALLS in 2007 that reached nearly 90,000 feet. Computer simulations put his 2011 rocket project at 110,000 feet if all went as planned. The all-aluminum vehicle was a single-stage three-finned design.

"The rocket took about six months to build," said Hagensick. "It has Raven 2 and ARTS2 altimeters. There is a GPS flight and telemetry unit in the nose cone, and it uses a floating piston design with ARRD for recovery." The rocket also relied on CO_2-activated ejection charges for deployment at high altitude. Hagensick's solid fuel motor was not made up of multiple, individual fuel grains as most motors were at the time. Instead, it was cast as a one-piece grain, 65 inches long, which was slowly poured into a casting tube over a four-day period. The bottom one-third of the grain had a six-inch-diameter finocyl design, meaning the central core had a five-legged starlike shape, with the upper two-thirds of the motor grain having a more conventional 1.25-inch circular- or bates-type grain. This design would theoretically allow the motor to slowly build up thrust as it burned. The longer the motor burned, the more powerful it would become in the air.

Hagensick had previously tested his new grain design on an N-class motor that had worked well. His motor today was a Q6100. "The motor will start with 800 pounds of thrust, but by 9 seconds into the burn, the motor will be delivering 2,700 pounds of thrust," he said. "So if anything goes wrong, it will probably be 9 seconds into the flight." Richard estimated his six-inch-diameter rocket might achieve Mach 3.

Unfortunately, it never got the chance.

At the very moment of ignition, the rocket nearly disintegrated at the pad. The Q motor sustained a thundering CATO, reducing the lower half of the rocket to little more than twisted junk metal. The rest of the airframe was thrown straight up the launch rail and into the air by the force of the disaster: The half-bodied rocket rose in a slow-motion dance, revolving lazily around itself, to an apogee of perhaps a hundred feet or so before it started to sink back down. The deployment charges then fired on cue, separating the nose cone from the airframe and releasing a small ballistic parachute. The rocket then dropped straight back to the ground, nearly landing on the damaged pad. Around it rested the still-on-fire, smoking pieces of the lower half of the vehicle.

Jeff Jacobs also attempted 100,000 feet this year, with another R10,000 hybrid-powered all-aluminum rocket. The rocket appeared to have a near-perfect flight. It blasted off from Jacobs's state-of-the-art launcher to extreme altitude. But there was no successful tracking of the vehicle. It was never seen again.

Three students from the Massachusetts Institute of Technology—Andrew Wimmer, Christian Velledor, and Julian Lemus—also took a shot at ultrahigh altitude with a three-stage rocket powered solely by commercial motors: an N5800 to an M2020 to another M2020. The four-inch-to-three-inch-diameter rocket weighed 93 pounds and stood more than 15 feet tall. It had a good takeoff but soon veered off a vertical ascent. With the subsequent successful ignition and separations of each of the upper stages, the rocket's path tilted closer and closer to being parallel with the ground. Wimmer estimated the rocket traveled 87,003 feet, which seemed likely. However, much of that distance was in a flight trajectory parallel to the desert floor.

Several well-built and highly engineered rockets had attempted to reach 100,000 feet and failed. There were two rockets left.

"I started in high-power rocketry as a result of the Discovery Channel rocketry shows in 2003," explained Jim Jarvis of Austin, Texas, as he was preparing for his high-altitude shot at Black Rock in 2011. "I had a rocket within a few days after the show, and I attended my first launch a few weeks after that."

Since that time, Jarvis had become a well-known high-altitude flyer at Black Rock. In 2006, he briefly held the Tripoli N-impulse altitude record at 34,917 feet, breaking the seven-year-old record set by Richard King in 1999. The record set by Jarvis lasted barely three months. It was then broken by Ken Biba of California, who reached 41,457 feet on an N motor later that year.

After his N-record flight, Jarvis began building multistage high-altitude projects. Between 2007 and 2011, he launched at least seven two-stage projects, each powered by a combination of N and M motors. During the course of these four years, Jarvis destroyed a lot of rockets, typically at high altitude when his airframes buckled, or his fins were torn off, or for reasons undetermined. Along the way, he became one of high power's foremost experts in the construction of homemade carbon fiber rockets that weighed next to nothing. With each failure, Jarvis took valuable negative information forward, improving the next rocket for a better performance. "The key enhancements included thicker airframes that were rolled under tension, better fin attachment procedures with vacuum bagging, and upgraded, thicker coupler tubes," he said.

At BALLS-21 on September 30, 2011, Jarvis hauled his latest two-stage creation to an away cell at Black Rock. The black carbon fiber rocket was created from nine layers of rolled carbon fiber. The booster and sustainer were just slightly larger than minimum diameter. Jarvis provided a tiny air gap between the motors and the inner walls of the rocket, thereby protecting the airframe from the intense heat generated by burning propellant during flight. When the rocket was vertical on the pad, it stood more than 15 feet tall, and yet the entire assembly weighed only 33 pounds unloaded. It would carry an array of modern altimeters and GPS devices and two powerful Cesaroni commercial N motors: an N5800 C-Star motor in the booster and a long-burning N1100 in the sustainer.

"One of the most challenging aspects of flying at Black Rock is just to get your infrastructure to the playa," explained Jarvis, who, like the other flyers looking for 100,000 feet this day, spent countless hours at home estimating every conceivable need right down the last spare igniter. "I think a lot of people would be surprised at how much planning is required to do this," he continued. "I had lots of help in the process, but the overall effort is largely me, my wife, and my wallet."

By 10:00 a.m., Jarvis's rocket, called *FourCarbonYen*, was vertical on the pad. The launch pad was a red four-legged tower held firmly to the desert floor by four guide wires that fanned out in different directions from a cross beam on top of the tower, each wire attached to a metal rod anchored deep in the desert floor a short distance away. Jarvis climbed a ladder that rested against a brace installed near the top of the pad, and from there, he leaned toward the rocket and armed the electronics in both the booster and the sustainer. Then he scurried back down the ladder. He and his crew backed away. Computer flight simulations for *FourCarbonYen* predicted an apogee over 100,000 feet.

As Jarvis was making final preparations for the launch of *FourCarbonYen*, Derek Deville of Florida was busy working at another away cell only a few hundred yards away.

A former member of the Tripoli Board of Directors and member of the *GoFast* team project in 2004, Deville traveled to Black Rock in 2011 to put his own rocket above 100,000 feet. He created an 8-inch-diameter all-aluminum missile that he dubbed *Qu8k* (pronounced "quake"). The single-stage vehicle was 15 feet tall and weighed more than 300 pounds. Most of that weight was propellant: a Q18,000 motor made entirely by Deville. The motor was designed to generate more than 5,000 pounds of peak thrust during flight. It was a finocyl design—a six-sided star grain poured and cast by Deville in Florida and then transported to Northern Nevada.

Deville and his crew spent the prior day setting up a rectangular launch tower made entirely of unistrut rail, the same ordinary unistrut found in the electrical conduit section of any home improvement store in America. The tower was engineered by Deville in Florida, disassembled for transport, shipped to Black Rock, and reassembled on the desert floor. It weighed nearly 300 pounds. To get the tall rocket inside the tower, the launcher was placed flat on the ground. It looked like the exoskeleton of a 16-foot-long rectangular box. With the tower in horizontal position on the ground, the rocket was slid inside, section by section, from the open top end of the launcher. Then the entire 600-pound assembly—launcher and rocket together—was lifted into vertical using a small winch attached to the back end of Deville's rental car.

Like the Jarvis rocket, *Qu8k* represented the pinnacle of modern high-power rocketry. It was loaded with state-of-the art altimeters, tracking equipment, GPS, ballistic parachutes, and video gear. Included in this array were two Tommy Billings–designed Adept timers.

Once in the vertical position, the launch tower was secured to the ground by four guide wires. Deville climbed the launcher and reached through his Unistrut tower to arm his electronics array. It was 10:25 a.m. Deville and his team were almost ready.

But now all attention shifted back to Jim Jarvis. His *FourCarbonYen*, loaded with two N motors, was ready to fly.

All other activity on the range came to a halt.

At 10:30 a.m., the two-stage carbon fiber rocket came to life. It raced away from the tower fast and clean, powered by the flaming N5800 motor in the booster. Within seconds, the rocket exceeded the speed of sound. The booster performed perfectly. There were no motor problems at ignition, there was no CATO in flight, and the fins held firm as the rocket broke through Mach 2. Then the booster shut down, followed by the perfect separation of the two stages and the ignition of the N1100 motor in the sustainer. As the booster dropped slowly under its own parachute, the rocket's upper stage streaked into the stratosphere quickly even the smoke trail vanished beyond sight.

Moments later, *FourCarbonYen* reached apogee at 100,418 feet.

Jarvis had done it. He had broken the six-figure barrier, becoming the first member in Tripoli history to do so on his own and without corporate sponsorship.

The onboard GPS confirmed the rocket performed flawlessly to an altitude of 97,000 feet. Then an error in one of the altimeters caused the ejection charges to fire prematurely while the rocket was still traveling in excess of Mach 1. The vehicle continued higher to clear 100,000 feet, but the tumult of the ejection charge mishap caused the nose cone, which contained the onboard electronics and trackers, to separate from the rest of the sustainer. The nose cone with all its onboard electronics was recovered downrange later that afternoon. The sustainer, unfortunately, was never seen again. Jarvis had done what no other Tripoli member had done before him, yet he had not recovered his *entire rocket* from near space.

The door to that accomplishment was still open. And the next high-power flyer in line was determined to step through it.

At 11:18 a.m., Derek Deville's *Qu8k* blasted off. At ignition, the 300-pound launch tower was completely obscured by smoke, dust, and debris as the all-metal missile roared into the air. In only a few moments, the 15-foot-tall rocket was a tiny speck in the sky. Then it disappeared at high altitude. Deville's Q18,000 was generating more than 4,000 pounds of thrust, propelling the rocket to a peak velocity of more than 3,200 feet per second—faster than a rifle bullet.

In less than 90 seconds, the rocket cleared 100,000 feet.

And it kept going higher.

At an apogee of 121,000 feet, Deville's rocket turned over gently, allowing onboard video cameras to capture a stunning view of the curvature of the deep blue earth against the pitch blackness of outer space.

Recovery of the rocket was as perfect as its ascent. The nose cone separated from its airframe at apogee, the two halves of the vehicle connected together by a long Kevlar tether. The entire assembly then dropped from near space under a ballistic parachute manufactured by the Rocketman Company. *Qu8k* landed barely 3 miles from its launch pad, right in the middle of the desert floor under the bright sunshine. The recovery team drove right up to the rocket. They found it entirely unscathed except for the partial melting of a plastic camera shroud and some singed decals along the edge of the rocket's airframe. The entire ride from the desert floor at Black Rock to near space and back again took 8.5 minutes.

For high-power rocketry, the future had arrived.

30

Above: Wedge Oldham's 740-pound *Black Brant* represented the pinnacle of the heavy rocket period at Black Rock in the early 2000s. The thirty-foot-tall rocket held 3 P-impulse motors designed by Tripoli member Darren Wright. Here, the rocket is carefully loaded onto the launch rail prior to flight. The trio of six-inch-diameter P motors are yet to be installed in the aft end of the rocket.

31

Left: Oldham's *Black Brant* at launch, with a combined impulse equal to a Q-21,685 motor, the massive rocket reached more than 18,000 feet and was successfully recovered.

32

33

Above: Kimberly Harms led a team of nineteen flyers in the design and construction of a 700-pound full scale *Honest John* rocket that carried four O-impulse motors designed by John Lyngdal. "Like most large team projects we split the cost, which was great since no one person could afford to build such a project," said Harms just before flight.

Opposite page: With their rocket vertical and ready to go on the pad, members of the *Honest John* team take a moment for pictures prior to flight at BALLS in 2004. The four O-impulse motors were almost the equivalent of a Q motor.

Left: The *Honest John* in flight a second after takeoff. Notice the darker smoke trailing down the side of the airframe. This was the result of single-use G motors mounted horizontally on the upper airframe. The thrust from these small motors caused the rocket to rotate slightly on ascent. A split second after this photo was taken, one of the primary O motors suffered a failure, leading to the destruction of the vehicle.

35

Extereme Rocketry publisher, writer, and graphic designer Brent McNeely at the Black Rock Desert in the mid-2000s. McNeely's magazine covered the hobby all over America, and captured high-power rocketry's rise to the upper atmosphere from 2000-2009.

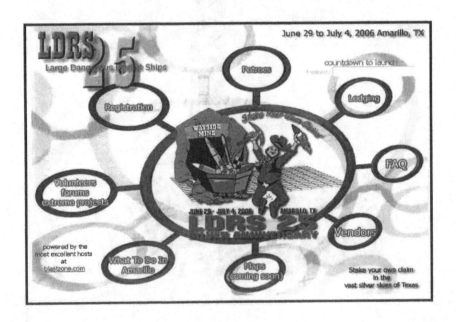

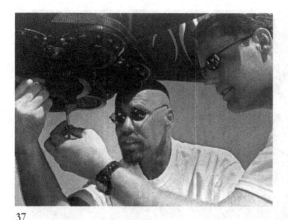

37 Erik (left) and Dirk Gates were a high-power rocketry phenomenon. They arrived on the national scene at LDRS-20 in Lucerne in 2001. They remained at the forefront of the hobby for nearly a decade. Their clustered motor combinations drew large crowds wherever they appeared for a launch. Tragically, Eric Gates lost his life in a work-related accident in 2009. At the time, he was also a member of the Tripoli Board of Directors. Here, the two brothers prepare their 620-pound *Porthos II* with a central P-impulse motor surrounded by multiple N motors. *Porthos II* reached an apogee of 25,000 feet and was successfully recovered.

38 A meeting of the rocketry minds. (from left to right): Tripoli's first TAP and AeroTech representative Karl Baumann (in white cap); high-power motor-maker and MIT-trained rocket scientist Frank Kosdon; AeroTech founder and longtime high-power rocketry leader Gary Rosenfield; and the developer of high-altitude carbon dioxide recovery systems known as CD3, Tom Rouse. All four men share stories of the day following a high-power rocketry launch in California in 2006.

Jack Garibaldi's full-scale *Nike Smoke* was powered by five Animal Motor Works N motors, including a central N2800 Sparky motor that rained down a brilliant shower of magnesium sparks on takeoff. The beautiful twenty-foot-tall rocket had a magnificent takeoff at Black Rock in 2007 and roared to an estimate apogee of 10,000 feet. However, there was no deployment. The rocket was destroyed on impact.

High-power rocketry often displays its lighter side, too: Here, Pat Easter's six-foot-tall *The Tin Man* takes off on an M motor at LDRS-23 outside of Geneseo, New York, in 2004. Easter's "rocket" had an excellent flight and recovery.

Top: Members of the Coyote Team prepare their scratch-built *Orbiter* for flight at Southern Thunder on a sod farm in central Tennessee.

Bottom: Dan Michael's ¾-scale *Patriot Missile* takes to the sky above the Higgs Farm at a Red Glare launch in Maryland. The 345-pound rocket had a perfect flight and recovery.

43

Ken Good, a member of the Tripoli Federation in the mid-1960s, was elected in 2004 as the sixth President of the Tripoli Rocketry Association, a post he would hold until 2010. Between 2004 and 2009, Good was Tripoli's point man in the federal lawsuit filed by Tripoli and the National Association of Rocketry against the Bureau of Alcohol Tobacco Firearms & Explosives. Rocketry prevailed in that case in April of 2009. Here, Good prepares one of his own rockets for flight at Black Rock in 2011.

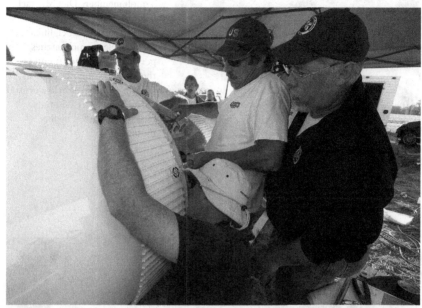

44 Steve Eves (center, white shirt and dark cap) and crew prepare to join the upper and lower sections of his 1,600-pound *Saturn V* prior to the being raised to vertical on the pad. The 1/10-scale rocket was launched to commemorate the 40th Anniversary of the flight of *Apollo 11* in 1969.

The 43-foot-tall *Saturn V* is raised by a commercial crane on Friday afternoon, April 24, 2009. The flight would take place the next day. As of 2018, this rocket remains the largest rocket in high-power rocket history.

45

Tripoli members Dan Michael (right) and his son Andrew methodically insert eighteen separate electronic matches, two per motor, into the aft end of the *Saturn V* minutes prior to launch. The rocket held a central six-inch diameter P-impulse motor surrounded by 8 N motors.

46

47

With national and local media and more than 5,000 spectators on hand for the event, Steve Eves prepares for launch of his four-story-tall *Saturn V* from the Higgs Farm in Price, Maryland. Here, Eves takes time to answer a question posed by a young enthusiast prior to launch.

48

At just after 1:00 p.m., on April 25, 2009, Steve Eves' *Saturn V* blasts off into high-power rocketry history. The 1600-pound vehicle had a near-perfect flight and recovery. The nine high-power motors on board the *Saturn V* were manufactured by Jeff Taylor of Loki Systems. The combined cost of the motors was more than $13,000. The cost of each individual motor was sponsored by rocketry flyers from around the nation. After its successful flight, the rocket was transported to Northern Alabama and placed on temporary display at the U.S. Space and Rocket Center in Huntsville.

The high-altitude rush is on: Gene Nowaczyk (far left) and crew prepare his eight-inch diameter *Piper* for its historic flight on September 20, 2006 at Black Rock. The 345-pound rocket build by Nowaczyk at his home in Missouri held a Q20,000 motor that pushed the all-aluminum vehicle to a peak altitude of 93,342 feet. *Piper* was recovered intact barely four miles from its launch pad, making Nowaczyk the first solo high-power flyer to achieve nearly 100,000 feet with a perfect recovery. After this flight, the floodgates were opened for other flyers to reach for an altitude in the six figures. However, it would be another five years before Nowaczyk's record was broken.

Gene Nowaczyk (top, ladder) and Mark Brown make final preparations for the launch of *Piper* at Black Rock in 2006.

51

Left: Gene Nowaczyk stands beside his rocket *Piper* shortly before flight in September of 2006.

Below: Alex McLaughlin of Washington and his Team Numb were among the small but growing fraternity of ultra-high-altitude flyers who in the early 2000s were convinved that high power could reach the upper limits of the Earth's atmosphere. Here, they load their rocket, *Questionable Mental Health*, on the rail at Black Rock in September of 2006. The 400-pound, eight-inch diameter vehicle held a research Q16,000 motor weighing nearly 160 pounds. The projected altitude was 65,000 feet. The rocket had a textbook takeoff but disintegrated when the motor failed several thousand feet in the air.

52

Left: Improper assembly of a rocket motor leads the fiery destruction of this high-power rocket on the pad at Red Glare in Maryland in 2007.

Below: The remains of a beautiful *Patriot Missile* after a failed deployment at the annual Hellfire launch held on the Bonneville Salt Flats in Utah in 2009. As of 2019, the Hellfire event was the longest-running regional launch in all of high power, dating back to 1994.

55

"Research is what I am doing when I don't know what I am doing," said Werner von Braun as an explanation for all of the destroyed rockets that littered his path to the *Saturn V*. Above: The remains of a Tripoli member's research rocket wait for cleanup after the rocket was destroyed in spectacular fashion at Black Rock.

Right: Another rocket bites the dust moments after ignition. As Robert Goddard would say, every rocketry mishap provided scientists with additional "valuable negative information."

56

Robert Utley (left) and Neil McGilvray (below). The two flyers were co-founders of one of the most successful prefectures in Tripoli, the Maryland Delaware Rocketry Association (MDRA). In 2006, they created *Rockets Magazine*, which would go on to cover more events in more places than any other publication in high-power rocketry history. Among their many other rocketry-related accomplishments, Utley and McGilvray not only secured a site for the launch of the largest high-power rocket ever flown (The Steve Eves *Saturn V* in 2009), they also arranged for financing the powerplant for the 1600-pound rocket. *Rockets Magazine* covered the annual LDRS and BALLS events for 13 years. The magazine was published from 2006 to 2018.

A scratch-built, 1/9-scale version of the McDonnell Douglas Delta II launches at Plaster Blaster in the fall of 2009. Powered by an M1939 and four additional motors in the booster, the 150-pound rocket held an L-motor in the sustainer that separated cleanly in flight for a nearly flawless two-stage launch and recovery. At LDRS-30 in Lucerne in 2010, the rocket would have an accidental ignition of the sustainer on the ground, leading to serious injury of bystanders. The accident in 2010 would lead to safety range operation changes by Tripoli to help make the hobby safer for flyers and observes alike.

A flyer climbs high inside a custom-built launch tower to arm electronics and make final adjustments minutes before the flight of his sophisticated high-power rocket.

Left: New England flyer Robert DeHate was among those high-altitude experts who preffered a two-stage design in the quest to clear 100,000 feet in altitude. Here, crew members get DeHate's favorite design ready for flight. This rocket carried a research P10,000 motor in the booster and an N motor in the sustainer and reached an altitude of 80,000 feet.

Below: DeHate's favorite design takes to the sky again.

61

62

63

Texan Jim Jarvis makes final adjustments to one of his launch pad's guide wires as he prepares for the historic flight of his *FourCarbYen* on September 30, 2011 at Black Rock. Jarvis created the all-carbon-fiber airframe for his rocket from scratch at his home in Austin. The two-stage vehicle held an N5800 motor in the booster and an N1100 in the narrow sustainer. The four-inch diameter rocket had a spectacular flight to 100,418 feet, making Jarvis the first Tripoli member to clear a six-figure altitude with a rocket designed and created by himself. The nose cone of the two-stage rocket with all of its electronics was recovered intact and undamaged. However, portions of the vehicle, including the sustainer, were never seen again.

Right: Derek Deville's *Qu8k* is lifted into vertical position after being installed in its homemade launch tower. The "lift" was provided by a small automotive store winch Deville had attached to the front end of his rental car. The launch tower was constructed from ordinary uninstrut available at any home improvement store.

Below: Deville secures the nose cone of *Qu8k* prior to flight at the Black Rock Desert on September 30, 2011. The single-stage all metal vehicle was eight inches in diameter and carried a Q motor built by Deville. On the pad, the rocket weighed more than 300 pounds.

66

67

Derek Deville's mighty *Qu8k* blasts off from the desert floor in Northern Nevada at 11:18 a.m. on September 30, 2011. The rocket's Q18,000 motor, designed and constructed by Deville, generated more than 4,000 pounds of peak thrust and propelled the vehicle to speeds in excess of Mach 3. At an apogee of more than 121,000 feet, the rocket's on-board video cameras captured hundreds of square miles of desert below and the curvature of the Earth set against the black void of outer space just above. The rocket landed barely three miles from the launch paid in perfect condition. With this flight, Deville became the first solo high-power flyer to clear 100,000 feet and recover his entire rocket intact. Right: Deville makes some igniter adjustments just before flight.

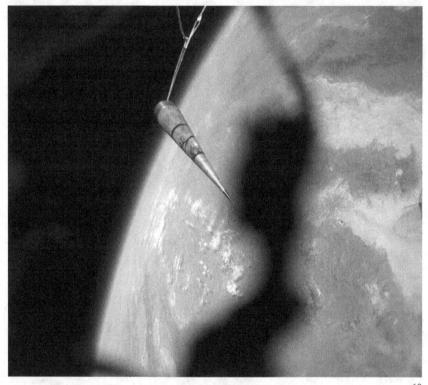

68

The view from near-space: The separated nose cone and tether of Derek Deville's *Qu8k* is captured by the rocket's on-board video camera at an apogee of 121,000 feet. The 15-foot-tall homemade rocket reached a peak velocity of more than 3,200 feet per second on its way to high-power rocketry history. Clearly visible in this photo is a downward-looking view of the rocket's airframe (far right). The shadowy substance in the center of the photo is the melted remains of the on-board video camera's plastic shroud. The Black Rock Desert can be seen directly below. Deville's rocket descended without incident and landed barely three miles from its launch pad, totally unscathed except for the paint that blistered away during the rocket's heated ascent.

Epilogue

It had been two years since Derek Deville and Jim Jarvis successfully cleared 100,000 feet at Black Rock in 2011. Now on September 20, 2013, Jarvis had another rocket high in the air. Would it clear 100,000 feet? Would he get it back intact? Would he even find it?

LDRS-32 had been held at Black Rock a few months prior, in July 2013. It was the first time Tripoli's national launch had been to the Northern Nevada desert since 1995. Attendance was low in no small part due to a scorching heat wave that blasted the playa all summer long. But Jarvis was there with two multistage rockets, ready for another high-altitude attempt.

"I had a couple of flights planned, and I wanted a longer launch window to fly them, which is why I chose LDRS-32 instead of BALLS," said Jarvis. His first flight was a two-stage vehicle that carried an N2500 in the 4-inch-diameter booster and an M745 in the 3-inch-diameter sustainer. That rocket called *TooCarbYen* had a great flight to 74,000 feet and a successful recovery. His second rocket, staging an N to another N, was not so perfect. The sustainer failed to ignite. There would be no 100,000-foot flight for Jarvis at LDRS-32.

"I got the rocket back undamaged, but the idea of waiting fourteen months for BALLS in 2014 to try the flight again was not making me happy." So he went back to Texas, got everything ready again, and returned to the desert for BALLS 22, barely sixty days later. His *FourCarbYen* rocket, a rebuilt version of his record-setting design in 2011, staged an N5800 in the booster to an N1100 in the sustainer. The 91-pound rocket was just shy of 16 feet long. Both the booster and the sustainer were barely 4 inches in diameter. The airframe was all carbon fiber that Jarvis created from scratch at home. This flight could not have been better. *FourCarbYen* reached apogee at more than 118,000 feet, and the booster and sustainer were recovered intact to fly another day.

"One nice thing about this flight is that I had a decent onboard video," said Jarvis. "This was my first video of a complete two-stage flight, and it is really interesting to see how things unfold throughout the flight and at the higher altitudes. A film excerpt from near apogee by the onboard video camera was akin to an astronaut's view from outer space. The curvature of the earth and the blackness of space above and below, a near-satellite view of not only Lake Tahoe, Reno, and Pyramid Lake but also the full length of the Smoke Creek Desert, where the remains of rockets long past rest beneath the desert soil.

Between 2012 and 2019, several more people joined the ranks of those who have cleared 100,000 feet at the Black Rock Desert.

In September 2012, a team of flyers led by veteran high-altitude expert Ken Biba and including Erik Ebert, Becky Green, Jim Green, David Raimondi, Steve Wigfield, and Tom Rouse, launched a two-stage vehicle to an altitude of 104,659 feet. The rocket was powered by an AeroTech N1000 staging to an AeroTech M-685 and was recovered fully intact.

In 2014, Tom Rouse and Neil Anderson were among those aiming for near space with separate rockets. Rouse fell just short, with both a perfect ride up and recovery from 90,100 feet. His two-stage rocket held three N motors in the booster and another N in the sustainer. Meanwhile, Anderson's rocket, the *Money Pit*, roared past 100,000 feet to an apogee of 118,000 feet.

At Black Rock in 2015, Curt von Delius fell just short of six-figures with the flight of his two-stage 4-inch-diameter *Phoenix*, which reached just over 96,000 feet on a Cesaroni M3400 in the booster and an N1100 long burner in the sustainer. The rocket had a good recovery, touching down barely 3 miles from the pad. And James Donald joined the 100,000-foot club with his two-stage *MinFlous*. The 5-inch-diameter booster held an O3400 motor, and the three-inch sustainer carried an M motor. The rocket reached an apogee of 110,000 feet 88 seconds after takeoff. Both halves were recovered unscathed. The accomplishment of Donald's flight was only slightly overshadowed by a return to even more extreme altitudes by Jim Jarvis, who successfully reached more than 130,000 feet in his new three-stage vehicle, *ThreeCarbYen*. The first and third stages of Jarvis's incredible flight were recovered. The second stage is still out there, waiting to be found, in or perhaps beneath the Black Rock Desert.

Then in 2016, Kristofer "Kip" Daugirdas of Utah broke things wide open again, setting a new high-altitude mark with his rocket *Workbench 2.0*. The thirty-two-year-old Daugirdas had been involved in rocketry since the eighth grade and was a flyer at LDRS-24 in Lethbridge, Canada, in 2005. "I started getting bored with Estes, but I found a hobby shop on the north side of Chicago in 1999, and they had copies of Bruce Kelly's *High Power Rocketry*," said Daugirdas. "I was sold."

Workbench 2.0 took approximately three months to design and build. "I based it off of the Aeropac 100K ARLISS designed rocket," explained Daugirdas. "But I tried my best to improve on the design." The rocket held a 4-inch-diameter Cesaroni N2500 in the booster and a three-inch-diameter AeroTech M685 in the sustainer. "The flight was mostly successful, but the onboard flight computers had an issue with their programming, causing the timer-based apogee charges not to fire," said Daugirdas. The rocket reached a confirmed apogee of 144,589 feet but then plummeted back undeployed. "The upper stage came in ballistic until at 10,000 feet the main charge fired and saved everything," he said. Well, almost everything. The main parachute was destroyed during the high-speed deployment, but the sustainer's Kevlar harness held firm, leading the rocket's two halves to slow down and finish their descent in a flat spin. The airframe was severely damaged on impact, but the avionics and payload bay, including two GoPro video cameras, were unscathed.

In 2017, there were no successful flights over 100,000 feet at Black Rock.

But to illustrate how rapidly old altitude records were still falling, and by a large margin, a new N record was set by Tripoli member Nic Lottering of Australia in 2013. Lottering became the first person to clear 50,000 feet on a single N motor, reaching 51,228 feet on an N5800. Lottering's impressive record lasted barely four years. In September 2017, Greg Morgan reached 59,448 feet at Black Rock on an N1560. That same weekend, Matt Orsack shattered the record again, reaching an unbelievable 63,818 feet on a single N5800. In mid-2018, Morgan reportedly cleared 70,000 feet on a single O motor.

Then on July 4, 2018, Kip Daugirdas returned to Black Rock again with a reworked version of his rocket *Workbench 2.0*. His two-stage rocket reached 154,068 feet. The entire rocket was recovered intact. (A snapshot obtained from an onboard video camera in the Daugirdas rocket is on the back cover of this book.) Less than ninety days later, on September 21, 2018, Jim Jarvis went even higher. His three-stage rocket reached

175,000 feet. Jarvis recovered the first and third stages intact, less than 3 miles from the launch pad.

The second stage came in ballistic. It rests under the desert floor at Black Rock, with the remains of the 1996 *OuR Project*, and the scores of other vehicles entombed forever in the history of high-power rocketry.

"I love the excitement of these types of flights," explains California high-power flyer and computer programmer James Daugherty, who briefly held the high-altitude N record at more than 46,000 feet in 2016. "There have been a lot of cool companies that have started in garages here in Silicon Valley, and I like the idea of building a rocket in my garage that can go higher than an airliner. It's a real rush!"

Today even the K altitude record, held by Jarvis, stands at the dizzying altitude of 29,266 feet.

Did the pioneers of high-power ever dream that ordinary flyers would ever breach the 100,000-foot mark? Is it realistic to believe that individual flyers or perhaps a small group of hobbyists without corporate support can reach 200,000–300,000 feet, the very edge of outer space?

How high can high-power rocketry really go?

"I'm not comfortable calling what we do a hobby," Kimberly Harms told *Air & Space Magazine* in 2004. "A hobby is fun. Well, we don't come out here in the heat to fly rockets just for fun. This is a mission."

Former Tripoli vice president Pat Gordzelik couldn't agree more. "My mission in high-power rocketry is to offer folks an opportunity to achieve not only something educational but to also encourage them to push their personal limits to achieve a goal. Throughout history, success in any endeavor meant striving to do what had not been done before. And yes, that may lead to failures—something the risk-adverse crowd shrink away from. But I, for one, am sick and tired of Americans being dumbed down with expectant protection from risk via the Nanny State." Gordzelik, and those like him in high power, prefers a world where individuals are encouraged to takes chances, an attitude best summarized by American president Theodore Roosevelt more than a century ago— one of Gordzelik's favorite quotes on life:

> It is not the critic who counts; nor the man who points out how the strong man stumbles, or where the doer of deeds could have done them better. The credit belongs to the man

who is actually in the arena, whose face is marred by dust and sweat and blood; who strives valiantly; who errs, who comes up short again and again, because there is no effort without error and shortcoming; but who actually strives to do the deeds, who knows great enthusiasms, the great devotions; who spends himself in a worthy cause; who at best knows in the end the triumph of high achievement and who at worst, if he fails, at least fails while daring greatly, so that his place shall never be with those cold and timid souls who know neither victory nor defeat.

"I knew that flyers would reach very high altitudes," said rocketry pioneer Gary Rosenfield in 2019, looking back to the early days of the hobby more than forty years ago. "But not as high as they are achieving now. I also did not know in the beginning that the regulatory environment would expand successfully to permit the legal launching of 40,960 N-sec rockets."

"I remember we were always very focused on the next altitude record to be gained," recalls fellow high-power pioneer Chuck Rogers, describing the early altitude attempts of the 1980s. "But it took forever to beat even the Fort Rocket Team altitude record of 22,070 feet in 1984. The next record, set by Korey Kline and myself, then stood for another five years at 24,771 feet!

"For a high-power rocket with rocket motors smaller than an R or an S motor to reach 200,000 feet, it's going to have to be a two-stage rocket," believes Rogers. "But to get a two-stage high-power rocket to 200,000 feet, we're going to have to improve the mass ratios of our stages. We need even lighter motor cases, rocket body tubes, nose cones, and fins. This will probably be the next area of advancement in high-power rocketry. We're already seeing the beginning of this with carbon fiber airframes."

Since the 1950s, the test pilots working at NASA and with the United States Air Force have traditionally received their "astronaut wings" at just above 50 miles or 264,000 feet in altitude. High-power rocketry is now closing in on this mark. But what about the international edge of outer space, defined by some as the Kármán line, an altitude of 328,000 feet? Can that lofty goal be reached by talented hobbyists in high-power rocketry?

"For reaching space, the rocket will have to be either a large S-impulse single-stage rocket which would be much like the 2004

CSXT *GoFast* rocket or a high mass ratio two-stage rocket of lower total impulse," insists Rogers. "And yes, I think some advanced individuals or university teams can do it."

"I'm waiting for the next team to make it to space. I would say it's possible and, with enough time, probable."

Acknowledgments

This book would not have been possible without contributions in both time and resources from many people involved in high-power rocketry. The project started with a gift from flyer George Truett of South Carolina who, in 2006, gave me a box of old rocketry magazines, manufacturer's brochures, and other written rocketry memorabilia he had been collecting since the mid-1980s. Reading through those materials and getting a firsthand look at the early days of the hobby set me on a path that ended with this book. I lost track of George over the years. I hope a copy of this history finds its way into his hands.

Scores of rocketry pioneers were generous with their time and allowed me to interview them either in their homes or on rocketry ranges all over America and also online. Their names are all set forth in the sources section of this book. Thanks especially to Gary Rosenfield, Bruce Kelly, Tom Blazanin, Chris Pearson, Ken Good, Chuck Rogers, J. Patrick Miller, Pat Gordzelik, John Baumfalk, Jim Mitchell, Curt Hughes, Frank Uroda, Tom Binford, Francis Graham, Terry McCreary, Steve Buck, Jerry Larson, and Tommy Billings for their insights on not only the early days of this hobby but also its continued evolution to the present.

Bruce Kelly's *High Power Rocketry* magazine and his availability for interviews and questions and his generous loan of *Tripolitan* back issues stretching back to the mid-1960s were all important parts of the history included in this work. Without back issues of Kelly's magazine and Blazanin's *Tripolitan*, there would be no written record of the first two decades of this hobby. Kelly also provided back issues of William "Bill" Wood's 1980s *Rocket Newsletter* and many photographs of the early days of high power, some of which are included in this book. Jerry Irvine's *California Rocketry* should be included in this list too.

Another invaluable resource was *Extreme Rocketry* magazine, especially the scores of rocketry interviews conducted by publisher Brent McNeely between 2000 and 2009. McNeely created a historical interview record unparalleled in this hobby, publishing in-depth interviews from almost every pioneer in high power and many in model rocketry too. As set forth in the bibliography, I relied on McNeely's history of interviews for background information throughout this book. Rocketry video pioneer Earl Cagle's excellent video and DVD presentations of launches all over the country from the early 1990s to the mid-2000s, carried under the banner of *Point 39 Productions*, were another invaluable resource for this book. The *Model Rocketeer*, *American Spacemodeling*, and *Sport Rocketry* were also important resources for this book. The editors and publishers of all these publications are unsung heroes in the history of this hobby.

I wish I had finished this work sooner as I did not get a chance to thank again LOC founder and high-power pioneer Ron Schultz, whom I interviewed at his home in Ohio with his wife, Debbie, in 2011. "You could always count on the 'Doc from Loc' Ron Schultz, to make specialized parts or help you solve technical problems," wrote flyer Art Markowitz in 2002, describing the early days of high power. "He spent many of his nights on the phone providing technical support for us." Ron and Debbie Schultz were extremely generous with their time, and I had hoped to speak to Ron again before this book was finished. He passed away in 2017.

Special thanks also to another rocketry pioneer who passed away far too young in 2011, Darrell Mobley, creator of *Rocketry Online* and later *Rocketry Planet*. Darrell's extraordinary hard work in the creation of *Rocketry Planet* and his creative vision for a world-class Internet site devoted to high power was greatly needed and has been sorely missed. Information from his website, where it can still be found, was also useful to this book.

Rockets Magazine cofounders Robert "Bob" Utley and Neil McGilvray produced high-power's contemporary record from 2006 to 2018. Their coverage of rocketry events around the country is without match in the history of this hobby. Simply put, they attended and reported on more events, in more locations, in more detail, than anyone else in the history of high-power rocketry. Their collection of stories and statistics and events ranging from Steve Eves's *Saturn V*, to the annual BALLS event, to the last fourteen years of LDRS helped make this book possible. The end of *Rockets Magazine* in late 2018 is a real loss to the hobby and an opportunity for the next generation of writers and photographers. As of this time, there is no contemporary record being written about the continued evolution of high-power rocketry.

I wish to also thank William "Bill" Lewis of San Jose, California, one of the founders of AERO-PAC. Lewis shared not only his personal recollections of the early days of AERO-PAC but also an incredible written record of original high-power documents he maintained for nearly thirty years. These documents include official programs and details of the first Black Rock launches in the late 1980s, the first LDRS at Black Rock, and the first BALLS events held there. His collection also included, among many other interesting items, early AERO-PAC newsletters, launch brochures, Tripoli insurance documents, and the first high-power rocketry waivers ever secured at the Black Rock Desert. Bill passed away at his home in San Jose in September 2017. He was seventy-four.

Many other people provided VHS rocketry event tapes, DVDs, photographs, or other memorabilia or historical information that helped piece together the events in this book. These people include, but are not limited to, David Rose, Neal Baker, Art Upton, Bob Yanecek, Art Markowitz, Dick Embry, Steve Roberson, Ken McGoffin, Brad Ream, Kent Newman, Scott Eakins, Gary Rosenfield, Chris Pearson, Lee Brock, Stuart Barrett, Bill Spadafora, Tom Binford, Robert Utley, David Reese, Nadine Kinney, Mark Clark, Jim Mitchell, the Kloudbusters, ROC, AERO-PAC, and Chuck Rogers.

I also want thank my editors for their long and tireless efforts and input, including Lee Brock, Valerie Smith Canepa, and Karen Ray. Ken Good and Chuck Rogers spent an enormous amount of time reviewing the final drafts, making invaluable suggestions and corrections.

This history is far from perfect or complete and reflects only the opinions of the author based on the written and oral records available. Much of the story ends in 2011, and outside the more recent high-altitude stories in the Epilogue, I was not able to add some of the latest innovations and changes in the hobby, including the recent return of the single-use motor and the rise of the Regional Launch. LDRS has become smaller in recent years, and large regional launches such as Red Glare, Midwest Power, and ROCstock have grown larger than ever before. More coverage of these regional events would have made for a more complete picture. I also did not include all the interesting flights or events that have occurred in this hobby over the past forty years. I had to eliminate stories that were worthy of mention but were left on the editing desk because there was simply not enough room in this book. For example, more should be written about people who have helped run the Tripoli and NAR motor certification programs over the years, like Paul Holmes or Sue McCreary, as well as the many other organization members who chair or oversee club sections. I did try to cover the highlights. But there are many unsung heroes. Any errors are mine.

Finally, this hobby would not be possible without the commercial manufacturers and local rocketry vendors who invest their time and energy into the often-thankless task of keeping flyers supplied with motors and airframes and the related parts necessary to put high-power rockets into the air. And the hobby would never have survived without the local volunteers who prepare for each and every high-power rocketry launch held nearly every weekend of the year somewhere in remote locations all over North America, Europe, and Australia.

Appendix One
LDRS Statistics[1]

Event	Location	Date	Flyers	Flights
LDRS-1	Medina, OH	July 24–25, 1982	49[2]	N/A
LDRS-2	Medina, OH	July 30–31, 1983	N/A	N/A
LDRS-3	Medina, OH	August 11–12, 1984	N/A	N/A
LDRS-4	Medina, OH	July 13–14, 1985	N/A	N/A
LDRS-5	Medina, OH	August 1–3, 1986	N/A	N/A
LDRS-6	Hartsel, CO	August 7–9, 1987	66	232
LDRS-7	Hartsel, CO	August 5–7, 1988	91	231
LDRS-8	Hartsel, CO	August 4–6, 1989	N/A	N/A
LDRS-9	Hartsel, CO	August 16–19, 1990	N/A	N/A
LDRS-10	Black Rock	August 16–18, 1991	111	N/A
LDRS-11	Black Rock	August 14–16, 1992	146	383[3]
LDRS-12	Argonia, KS	August 13–15, 1993	260	500
LDRS-13	Argonia, KS	August 11–14, 1994	200	600

---

[1] Except as noted, these statistics were derived from contemporary magazine articles/launch reports.

[2] According to most reports, there were forty-nine people total at LDRS-1. The actual number of flyers is unknown.

[3] These numbers are from the letter of William Lewis to the BLM following LDRS-11 (William Lewis Collection).

LDRS-14	Black Rock	August 10–13, 1995	350	1,000
LDRS-15	Orangeburg, SC	July 4–7, 1996	N/A	995
LDRS-16	Hartsel, CO	August 7–10, 1997	276	730
LDRS-17	Bonneville, UT	August 6–9, 1998	300	900[4]
LDRS-18	Argonia, KS	July 29–August 2, 1999	381	1,133
LDRS-19	Orangeburg, SC	July 1–4, 2000	399	1,379
LDRS-20	Lucerne, CA	July 19–22, 2001	N/A	N/A
LDRS-21	Wayside, TX	July 11–16, 2002	301	1,259
LDRS-22	Argonia, KS	July 17–23, 2003	318	1,197
LDRS-23	Geneseo, NY	July 1–6, 2004	N/A	1,200
LDRS-24	Lethbridge, Ca	July 14–19, 2005	250	750–1,000
LDRS-25	Wayside TX	June 29–July 4, 2006	248	893
LDRS-26	Jean, NV	July 12–17, 2007	N/A	827
LDRS-27	Argonia, KS	August 28–September 2, 2008	317	1,059
LDRS-28	Potter, NY	July 2–6, 2009	N/A	1,193
LDRS-29	Lucerne, CA	June 11–16, 2010	N/A	1,349
LDRS-30	Argonia, KS	September 1–6, 2011	374	936
LDRS-31	Potter, NY	July 12–16, 2012	N/A	1,076
LDRS-32	Black Rock	July 17–20, 2013	N/A	N/A
LDRS-33	Kenosha, WI	July 17–20, 2014	N/A	742
LDRS-34	Potter, NY	June 25–28, 2015	N/A	798
LDRS-35	Lucerne, CA	June 8–12, 2016	N/A	728
LDRS-36	Price, MD	April 6–9, 2017	N/A	719
LDRS-37	Helm, CA	May 16–20, 2018	268	875
LDRS-38	Argonia, KS	August 29–September 2, 2019	483	1101

---

[4] These numbers are from the actual flight cards of the event (Neal Baker Collection).

Appendix Two
Tripoli Rocketry Officers/ Board Members (1985–2018)

Tripoli Rocket Society (January 1985)
President: Allen James "AJ" Reed
VP: Curtis Hughes
Secretary: Francis Graham
Treasurer: Vacant

Tripoli Rocket Society (February 1985)
President: Allen James "AJ" Reed
Secretary: Francis Graham
Treasurer: Bill Kust
Librarian: Tom Blazanin

Tripoli Rocket Society (April 1985)
(Preincorporation and FirstNationalBoard)
President: J.P. O'Connor
VP: Mark Webber
Secretary: Tom Blazanin
Treasurer: Francis G. Graham

Tripoli Rocketry Association (1986)

(Preincorporation and First Elected Board)
President: Tom Blazanin
VP: Bill Barber
Secretary: Francis G. Graham
Treasurer: Ed Tindell
Board: Chuck Mund, Gary Fillible, Gary Strickland, Philip Motte, and Curt Hughes

Tripoli Rocketry Association, Incorporated (1986–1987)
(First Corporate Board)
President: Tom Blazanin
VP: Bill Barber
Secretary: Francis G. Graham
Treasurer: Ed Tindell
Board: Chuck Mund, Gary Fillible, Gary Strickland, Chuck Rogers, and Curt Hughes

1987–1988
President: Tom Blazanin
VP: Bill Barber
Secretary: A.J. Reed
Treasurer: Ed Tindell
Board: Chuck Mund, Gary Fillible, Gary Strickland, John O'Brien, and Curt Hughes

1988–1989
President: Ed Tindell
VP: Bill Barber
Secretary: A.J. Reed
Treasurer: Tom Blazanin
Board: Chuck Rogers, Bill Wood, Glenn Strickland, John O'Brien, and Curt Hughes

1989–1990
President: Chuck Rogers
VP: Bill Barber
Secretary: Marc Lavigne
Treasurer: Bill Wood

Board: Gary Rosenfield, A.J. Reed, Glenn Strickland, John O'Brien, and Curt Hughes

1990–1991
President: Chuck Rogers
VP: Gary Price
Secretary: Marc Lavigne
Treasurer: Ken Vosecek
Board: Bruce Kelly, Richard Zarecki, Dennis Lamothe, Bill Wood, and Gary Rosenfield

1991–1992
President: Chuck Rogers
VP: Richard Zarecki
Secretary: Marc Lavigne
Treasurer: Ken Vosecek
Board: Bruce Kelly, Dennis Lamothe, Bill Wood, Gerald Kolb, and Gary Rosenfield

1992–1993
President: Chuck Rogers
VP: Dennis Lamothe
Secretary: Richard Zarecki
Treasurer: Dennis Lamothe
Board: Bruce Kelly, Dennis Wacker, Bill Wood, Gerald Kolb, and Gary Rosenfield

1993–1994
President: Chuck Rogers
VP: Dennis Lamothe
Secretary: Bill Maness
Treasurer: Dennis Lamothe
Board: Bruce Kelly, Dennis Wacker, Bill Wood, Gerald Kolb, and Gary Rosenfield

1994–1995
President: Bruce Kelly

VP:	Dennis Lamothe
Secretary:	William Maness
Treasurer:	Dennis Lamothe
Board:	Chuck Rogers, Dennis Wacker, Scott Bartel, Henry Holzgrefe, Dick Embry, and Gary Rosenfield

1995–1996

President:	Bruce Kelly
VP:	Dennis Lamothe
Secretary:	Bruce Lee
Treasurer:	Dennis Lamothe
Board:	Chuck Rogers, C.R. "Sonny" Thompson, Scott Bartel, Korey Kline, Dick Embry, and William Maness

1996–1997

President:	Bruce Kelly
VP:	Dennis Lamothe
Secretary:	Damian Russo
Treasurer:	Bruce Lee
Board:	Chuck Rogers, Scott Bartel, Korey Kline, Dick Embry, and Robin Meredith

1997–1998

President:	Bruce Kelly
VP:	Dennis Lamothe
Secretary:	Damian Russo
Treasurer:	Bruce Lee
Board:	Chuck Rogers, Scott Bartel, Korey Kline, Dick Embry, and Robin Meredith

1998–1999

President:	Bruce Kelly
VP:	Damian Russo
Secretary:	Bill Davidson
Treasurer:	Bruce Lee

Board: Chuck Rogers, Scott Bartel, C.R. "Sonny" Thompson, Dick Embry, and Robin Meredith

1999–2000
President: Bruce Kelly
VP: Scott Bartel
Secretary: Ken Good
Treasurer: Bruce Lee
Board: Chuck Rogers, Bill Davidson, C.R. "Sonny" Thompson, Dick Embry, and Robin Meredith

2000–2001
President: Bruce Kelly
VP: Dick Embry
Secretary: Ken Good
Treasurer: Bruce Lee
Board: Chuck Rogers, Bill Davidson, C.R. "Sonny" Thompson, Ken Biba, and Robin Meredith

2001–2002
President: Bruce Kelly
VP: Dick Embry
Secretary: Ken Good
Treasurer: Bruce Lee
Board: Chuck E. Rogers, Bill Davidson, Derek Deville, Ken Biba, and Robin Meredith

2002–2003
President: Dick Embry
VP: Ken Good
Secretary: Robin Meredith
Treasurer: Bruce Lee
Board: Chuck E. Rogers, Bill Davidson, Derek Deville, Ken Biba, and Terry McCreary

2003–2004

President:	Dick Embry
VP:	Ken Good
Secretary:	Bob Schoner
Treasurer:	Bruce Lee
Board:	Chuck E. Rogers, Bill Davidson, Derek Deville, Terry McCreary, and Pat Gordzelik

2004–2005

President:	Ken Good
VP:	Pat Gordzelik
Secretary:	Bob Schoner
Treasurer:	Bruce Lee
Board:	Chuck E. Rogers, Bill Davidson, Derek Deville, Terry McCreary, and Dick Embry

2005–2006

President:	Ken Good
VP:	Pat Gordzelik
Secretary:	Bob Schoner
Treasurer:	Bruce Lee
Board:	Chuck E. Rogers, Bill Davidson, Derek Deville, Terry McCreary, and Darren Wright

2006–2007

President:	Ken Good
VP:	Pat Gordzelik
Secretary:	Bob Schoner
Treasurer:	Bruce Lee
Board:	Chuck E. Rogers, Bill Davidson, Derek Deville, Terry McCreary, and Darren Wright

2007–2008

President:	Ken Good
VP:	Pat Gordzelik
Secretary:	Bob Schoner
Treasurer:	Bruce Lee
Board:	Dick Embry, Terry McCreary, Darren Wright, Robin Meredith, and Erik Gates

2008–2009
President: Ken Good
VP: Pat Gordzelik
Secretary: Bob Schoner
Treasurer: Bruce Lee
Board: Dick Embry, Terry McCreary, Robin Meredith, Erik Gates, and Stu Barrett

2009–2010
President: Ken Good
VP: Terry McCreary
Secretary: Stu Barrett
Treasurer: Bruce Lee
Board: Dick Embry, Robin Meredith, Erik Gates, Bob Schoner, and Debra Koloms

2010–2011
President: Terry McCreary
VP: Pat Gordzelik
Secretary: Stu Barrett
Treasurer: Bruce Lee
Board: Dick Embry, Robin Meredith, Kenneth Good, Bob Schoner, and Debra Koloms

2011–2012
President: Terry McCreary
VP: Pat Gordzelik
Secretary: Stu Barrett
Treasurer: Bruce Lee
Board: Dick Embry, Robin Meredith, Bob Schoner, Debra Koloms, and David Wilkins

2012–2013

President:	Stu Barrett
VP:	Debra Koloms
Secretary:	David Wilkins
Treasurer:	Bruce Lee
Board:	Dick Embry, Robin Meredith, Terry McCreary, Pat Gordzelik, and Bob Brown

2013–2014

President:	Stu Barrett
VP:	Bob Brown
Secretary:	David Wilkins
Treasurer:	Bruce Lee
Board:	Dick Emery, Debra Koloms, Tom Rouse, Gerald Meux Jr., and Burl Finkelstein

2014–2015

President:	Debra Koloms
VP:	David Wilkins
Secretary:	Gerald Meux Jr.
Treasurer:	Bruce Lee
Board:	Burl Finkelstein, Dick Emery, Stu Barrett, Tom Rouse, and Tom Blazanin

2015–2016

President:	Debra Koloms
VP:	David Wilkins
Secretary:	Gerald Meux Jr.
Treasurer:	David Rose
Board:	Stu Barrett, Burl Finkelstein, Tom Blazanin, Dick Emery, and Steve Shannon

2016–2017

President:	Steve Shannon
VP:	Debra Koloms
Secretary:	Dick Emery
Treasurer:	David Rose
Board:	Gerald Meux Jr., Burl Finkelstein, Tom Blazanin, Dick Embry, and Gary Rosenfield

Endnotes

Book I

Chapter 1 (1981)

1. Lake Lahontan covered more than 8,000 square miles in Western Nevada, California, and Oregon. Wheeler, *Nevada's Black Rock Desert*, 25.
2. Carlson, *Nevada Place Names: A Geographical Dictionary*.
3–6. Evanoff, *Nevada History: Smoke Creek Desert*.
7. Ammonium perchlorate, also called AP, is a manufactured substance. It can be manufactured via electrolysis using ordinary table salt (sodium chloride).
8–9. Irvine, "Exclusive Report: Smoke Creek Amateur Launch." *California Rocketry Quarterly*.
10. Pearson interview with MBC.
11. There were some commercial manufacturers such as Composite Dynamics, selling G, H, and perhaps I motors. But these sales were few and far between.
12. NFPA 1122; see also Federal Aviation Act, Federal Regulations, Part 101.1(a)(3).
13. Irvine redesigned the newsletters several times during its brief history and generally improved the format as he went along. *California Rocketry Quarterly*, January 1982.
14–16. Rosenfield, "Feature Article: Power Freaks Arise." *California Rocketry Quarterly*.
17–20. Rosenfield interview with MBC.
21. Rosenfield interview with McNeely.
22. Rosenfield interview with MBC.
23–24. Johnson, "Indiana's Composite Motor Man." *Launch Magazine*.
25–27. Ream, "Enerjet-8: The Original AP Composite Model Rocket Motor." *Sport Rocketry*.
28–29. Rosenfield interview with McNeely; also, Rosenfield interview with MBC.
30. Perdue and Mayfield, "The Enerjet Story."
31–33. Rosenfield interview with MBC.
34. Nelson, *Rocket Men*, 96–97.
35–37. Mayfield, "Power Ascent." *Launch Magazine*.
37–41. Rosenfield interview with MBC.
42. Rosenfield interview with McNeely.
43–46. Rosenfield interview with MBC; also, Mayfield, "Power Ascent." *Launch Magazine*.

47–48. Rosenfield interview with MBC; see Parts I, II, and III, *California Rocketry Quarterly*, 1981.
50. Rosenfield, "Feature Article: Trends in High-Power Model Rocket Motor Design for the 80's—Part 3 Performance Optimization." *California Rocketry Quarterly*.
51. http://en.wikipedia.org/wik/STS-2, accessed April 24, 2012; http://science.ksc.nasa.gov/mirrors/images/pao/STS2/10060503.htm, accessed April 24, 2012.
52. Irvine, "Exclusive Report: the Second Flight of the Space Shuttle *Columbia*—Landing." *California Rocketry Quarterly*.
53. In contrast, between 1975 and 1980, the former Soviet Union sent manned vehicles into space nearly twenty times on a variety of missions. America's retreat from manned exploration in the mid-to-late 1970s was due, in no small part, to the periodic lack of political will, national vision, and pride that plague the country from time to time.
54. http://www.reagan.utexas.edu/archives/speeches/1981/40981f.htm, accessed on July 16, 2014.
55. The range was originally an air force bombing range. It is named for Army Air Corps Officer Arno H. Luehman, who surveyed the range in the 1930s. http://en.wikipedia.org/wiki/Edwards_Air_Force_Base, accessed on April 24, 2012.
56–57. *California Rocketry Quarterly*, January 1982.
58. Blazanin interview with MBC.

Chapter 2 (The Basement Bomber)

1. http://www.rrs.org/main.1.0/index.php?option=com_content&task=view&id=20&Itemid=33; accessed on April 25, 2012.
2. Pyrotechnics is the science of using materials capable of undergoing self-contained and self-sustained exothermic chemical reactions for the production of heat, light, gas, smoke, and/or sound. Pyrotechnics include not only the manufacture of fireworks but also items such as safety matches, oxygen candles, explosive bolts and fasteners, components of the automotive airbag and gas pressure blasting in mining, quarrying, and demolition. Source: *Wikipedia*, accessed on March 16, 2012.
3. See, for example, *Scientific American*, "The Amateur Scientist: About the Activities and the Trials of Amateur Rocket Experiments." June 1957.
4. See, for example, *Popular Mechanics*, "These Rocketeers Play Safe." September, 1958; *American Modeler*, "Rocket College for Model Makers." August, 1959; *Scientific American*, "The Amateur Scientist: About the Activities and the Trials of Amateur Rocket Experiments." June 1957.
5. *Scientific American*, "The Amateur Scientist: About the Activities and the Trials of Amateur Rocket Experiments." June 1957.
6. *Life Magazine*, December 16, 1957.
7. Brown, "These Rocketeers Play Safe." *Popular Mechanics*.
8. Amateur Rocketry: A Delineation of the Problem, a Policy Statement and a Program for Action by the American Rocket Society (1959), 51.
9. *Life Magazine*, December 16, 1957. "From Coast to Coast, U.S. Youth Gets Its Rockets Up in the Air." page 32; *Popular Mechanics*, September 1958. "These Rocketeers Play Safe." page 75; "Rocket Expert Warns Against Grocery Fuel," *Rocky Mountain News*, April 22, 1958, page 16, cited in *Youth Rocket Safety*, March, 1967.

10–13. "Amateur Rocketry: A Delineation of the Problem, a Policy Statement and a Program for Action by the American Rocket Society. American Rocket Society, 1959, 28.
14. Estes, Youth Rocket Safety. 1967.
15. Stine, *The New Model Rocketry Manual*. 15–16.
16. Craddock, "The Last Interview." *Launch Magazine*.
17. Brown, "These Rocketeers Play Safe." *Popular Mechanics*.
18–19. "Amateur Rocketry: A Delineation of the Problem, a Policy Statement and a Program for Action by the American Rocket Society. American Rocket Society, 1959, 66.
20. Neufeld, *Von Braun*. 76.
21. Letter by G. Harry Stine to J. Patrick Miller, May 24, 1989; see also Stine column "Rocket Trails" in *American Modeler* 1963–1964.
22–24. See, generally, www.nfpa.org.
25–26. See, for example, Stine, "Forty Years of Model Rocketry: A Safety Report"(1997).
27–28. See Forward to the Code for Unmanned Rockets, 1982 (NFPA).
29. For example, the NAR was instrumental in exempting all model rockets from the requirement of an FAA waiver prior to launch. However, to qualify for that exemption, the rocket and motor had to satisfy the requirements of the Code for Model Rocketry. Federal Aviation Act Rules limited model rockets to 1 pound. Below this weight limit, no special permission or waiver from the FAA was required to launch rockets into controlled airspace. This exemption was helped along its way in 1961 by G. Harry Stine. See, generally, Stine, "Working with the FAA. *American Spacemodeling*, January 1987, 16.
30. NARAM-23, *Model Rocketeer*, November 1981, 6–14.
31–32. Miller, *American Spacemodeling*. March 1986; see also Miller, *American Spacemodeling*. January 1989.
33. The membership numbers would drop again to 1,870 in 1982, which NAR leaders attributed to the 1982 recession. See Miller, *American Spacemodeling*, "President's Corner," March 1986.
34–35. Miller, *Model Rocketeer*, May 1981.
36. *California Rocketry Quarterly*, "Letters to and from the Editor." July 1981, 15.
37–38. *California Rocketry Quarterly*, April 1982, "NAR Politics," 6.
39. The Newton second is the standard unit of impulse in rocketry.
40–42. *Model Rocketeer*, "Board Defines Policy on High-Powered Engines." May 1982.

Chapter 3 ("Large and Dangerous Rocket Ships")

1. "Comet Halley Recovered." http://xeus.esa.int/science-e/www/object/index.cfm?fobjectid=13795.
2. Through a quirk of fate, the author briefly drove the superstar's vehicle. Regrettably, he did not meet Mr. Mays.
3. The compact disc player was conceived in 1969 but did not enter the commercial market in the United States until 1984. See, generally, http://www.ehow.com/facts_4866914_history-cd-players.html.

4. The term "dual deployment" is generally credited to Tommy Billings of Adept Rocketry. Billings began using the phrase in connection with the sale of his altimeters in the early to mid-1990s.
5. The first high-powered rocket launch to be held at Black Rock took place in the summer of 1989. The pagan festival known as Burning Man, which started at Baker Beach in San Francisco in 1986 with eighty participants, moved to Black Rock in 1990. By 2019, more than sixty thousand people attended the annual event, usually held in early September. For a timeline of Burning Man, see, generally, www.burningman.com.
6. Ron and Debbie Schultz interview with MBC.
7. Billings interview with MBC.
8. *Model Rocketeer*, "Con Calendar." April 1982. Chris Johnston was another SNOAR member.
9. The phrase "large and dangerous rocket ships" had likely been in existence for several years before 1981 and is generally attributed to flyer Roger Johnson, who is said to have uttered it as a catch phrase at Western United States rocket launches beginning in the mid-1970s. Early usage of the phrase can be found in a videotape of Johnson at a rocket launch hosted by the RRS at their Mojave Test Area (MTA) near Garlock, California, in 1977. Johnson, facing a camera and holding a microphone and wearing a white T-shirt and a ball cap, said, "Here we are at MTA. We are going to blast off large dangerous rocket ships, so look out. The cameras could get obliterated in the shower of broken and damaged and battered rocket ships." See www.rocketmotorparts.com/newsarchive.html. A similar phrase, "large and dangerous rocket activities," was used by Jerry Irvine to describe rockets at Smoke Creek in *California Rocketry Quarterly* in early 1982 (January issue at page 27). Ultimately, however, it was Chris Pearson who used the phrase to describe what is now known as LDRS.
10. An announcement for LDRS-1 was also printed in the April 1982 issue of *California Rocketry Quarterly* at page 16. The ad stated that the launch was "by invitation of power freaks only" and stated that Mark Webber or Chris Pearson could be contacted for more information.
11. *SNOAR News Quarterly*, Volume 8, Number 2 (1982) page 23; the advertisement read as follows: "SNOAR'S first annual high-power sport launch. Four unofficial events (no national contest points); greatest number of engines flown successfully, greatest total impulse, best overall flight and best prang. Equally large prizes will be awarded to the winner of each category."
12. Pearson interview with MBC.
13. Wolfe, *The Right Stuff*. 17.
14. http://en.wikipedia.org/wiki/Apollo_11; accessed on April 25, 2012.
15–18. Pearson interview with MBC.
19. Rogers interview with MBC.
20–21. Pearson interview with MBC.
22. *Model Rocketeer*, "Con Calendar." May 1982, page 18 (Coincidentally, this was the same issue of the *Model Rocketeer* that announced the NAR's official decision to ban G motors). Pearson's ad also ran in the *SNOAR News*.
23. Pearson interview with MBC.

24. *Model Rocketeer*, "Con Calendar." June 1982, page 19; the same warning ran again in the July issue of *Model Rocketeer*.
25. http://www.imb.com/title/ttoo84827; http://en.wikipedia.org/wiki/Blade Runner, accessed on April 24, 2012.
26. http://en.wikipedia.org/Star_Trek_(film_series), accessed on April 24, 2012.
27. "[T]he march of freedom and democracy which will leave Marxism-Leninism on the ash heap of history as it has left other tyrannies . . ." See, generally, http://www.reaganshearitage.org/html/reagan_panel_spalding.shtml, accessed on April 30, 2012. The speech was given on June 8, 1982. See also http://teachingamericanhidstory.org for a text of the entire speech. Accessed on April 24, 2012.
28. http://en.wikipedia.org/wiki/list_of_hot_100, accessed on April 24, 2012.
29–30. *Model Rocketeer*, The President's Corner, July 1982.
31–32. Pearson, "What the Hell Happened in California?" *SNOAR News Quarterly*. In fact, members were purportedly suspended following the NARWIN launch at Lucerne for flying, among other things, rockets that carried gasoline on board.
33. The Wagner farm is located at the intersection of Spieth Road and Township Road 23.
34–43. Pearson interview with MBC. See also Irvine, "Filing an FAA Clearance." *Tripolitan*, October 1986. ("LDRS only had a 2,000 clearance.")
44. *Model Rocketeer*, June 1982, page 19; *Model Rocketeer*, July 1982, 19.
45. Pearson interview with MBC.
46. *Medina County Gazette*, July 22, 1982, Calendar section, page 7 (now called the NASA John H. Glenn Research Center at Lewis Field).
47. "Briefs." *Medina County Gazette*, July 23, 1982, 5.
48. Pearson interview with MBC.
49. "More Sun this weekend." *Medina County Gazette*, July 24, 1982, 2.
50. Steele, "Opinion." *High-Power Rocketry*, July 1996.
51. Gary Rosenfield would attend his first LDRS the following year. He has attended every LDRS since LDRS-2; Chris Pearson said years later that these first high-power motors at LDRS-1 may have been by Vulcan or Irv Wait. See LDRS-19 video interview of Pearson by Earl Cagle Jr., July 2000 (Point 39 Productions).
52. Hughes, "Product Review—ACE Mongrel." *California Rocketry Quarterly*.
53. Hughes, Curt. "The Early Tripoli Experience: A Personal Reflection." http://www.rimworld.com/pgh/history/root2/triproot2.html, accessed on August 24, 2010; also, Hughes telephone interview with MBC. See also *Tripolitan*, Summer 1981, and discussion regarding 1981 Kent State Space Modeling Convention sponsored by *SNOAR*.
54. Good interview by MBC.
55. The rocket purportedly reached about 3,000 feet—much higher than the waiver; see *Tripolitan*, Autumn 1982, page 3.
56. Good interview by MBC. According to Good, Hughes and Good were not part of the famous LDRS-1 group photograph as they were off searching for the *Ace Mongrel* at the time the picture was taken.
57–59. *California Rocketry Quarterly*, October 1982. The award was also listed as "Best Overall Flight" of LDRS or "Most Impressive Flight" or simply the "Best Flight Award." Interestingly, the rocket was thought at the time to have exceeded 5,000 feet

in altitude. Id., at 16; the award was also called the Most Spectacular Launch in the Autumn 1982, *Tripolitan.*
60–61. *Pearson,* "How It All Began—LDRS-1." *High Power Rocketry,* 1996; see also Pearson interview by Earl Cagle, wherein he states there were forty-three people total; in this article, there are forty persons mentioned in attendance. Other articles put the number at forty-seven.
62. *Irvine,*"Large and Dangerous Rocket Ships LDRS-1 Medina, Ohio, July 1982." *California Rocketry Quarterly.*
63. Good, "The Road to the Modern Tripoli—A Personal Recollection." http://railtech.com/How Tos/GM/ken/triproot1.html, accessed on March 16, 2010.
64. *Irvine,* "Large and Dangerous Rocket Ships LDRS-1 Medina, Ohio, July 1982." *California Rocketry Quarterly.*
65. *California Rocketry Quarterly,* Letters to and from the Editor, April 1983.
66. Pearson interview by Earl Cagle Jr., at LDRS-19 in 2000 at Orangeburg, South Carolina (Point 39 Productions).
67. *Medina County Gazette,* Entertainment, July 26, 1982, 8.

Chapter 4 (The Aftermath)

1. *High-Power Rocketry,* December 2002, 8.
2. Hughes online interview with MBC.
3. The club was renamed in the late 1970s as the Tripoli Science Association.
4–5. *California Rocketry Quarterly,* April 1982, "Letters to and from the Editor," page 5.
6. The author was unable to confirm that any NAR member was actually suspended or expelled for LDRS-related activities. In an interview with Miller by the author, the NAR president said that few if any of these threats were actually carried out.
7. Hughes interview by McNeely, *Extreme Rocketry.*
8–9. Pearson interview with MBC.
10–11. See, for example, *California Rocketry Quarterly,* October 1982, "Letters," 22.
12. *Model Rocketeer,* September, 1982, 3.
13. Irvine's U.S. rockets were allowed back into the advertising pages of the *Model Rocketeer* no later than September 1982, so the two must have reached some sort of accommodation. See the *Model Rocketeer,* September 1982, 8. Meanwhile, the NAR's request for higher-powered articles went unheeded for a time as members apparently did not send in enough material for any issue to be produced for some time. See, generally, the *Model Rocketeer,* January 1983, 19.
14–15. *California Rocketry Quarterly,* January 1983, 3.
16. *California Rocketry Quarterly,* January 1983. "NAR Politics," 10.
17. *Miller,* "The President's Corner." *Model Rocketeer.*
18. *California Rocketry Quarterly,* April 1983, 3.
19–20. Pearson interview with MBC.
21. The system used motor ejection to deploy a drogue and then remote control to deploy the main parachute. Rocket-based altimeters that would perform both functions were still many years away.
22–33. Koenn, "LDRS-2 Experimental Rocket Launch." *California Rocketry.* Also, Koenn online interview by MBC.
34. *High Power Rocketry,* December 2002, 8.

35. *California Rocketry Magazine*, October 1983, LDRS Meet Summary, 10.
36. "Dual deployment" was a term likely coined by Adept founder Tommy Billings in the early 1990s. Billings would produce the first truly mass-marketed, inexpensive dual-deployment commercial altimeters for High-Power Rocketry.
37. *California Rocketry Magazine*, October 1983, LDRS Meet Summary, 10.
38. Schultz interview with MBC.
39. *California Rocketry Magazine*, October 1983, LDRS Meet Summary, 10.
40–41. Hughes interview with McNeely.

Chapter 5 (Enter the NAR)

1–2. Miller interview with MBC.
3. Pearson, Chris. "The LDRS Story," www.northernohiotra.com, accessed on July 11, 2018.
4. Pearson, "What the Hell Happened in California?" *SNOAR News*; see also *California Rocketry Quarterly*, January 1982, Calendar of Events, 20, and for general coverage of this event, *California Rocketry Quarterly*, April 1982, 16–17.
5–6. Rogers interview with MBC.
7. Jay Apt would fly in shuttle mission numbers STS 37, 47, 57, and 79. His first flight on the shuttle was in 1991 aboard *Atlantis*.
8–15. Miller interview with MBC; also, Pearson, "What the Hell Happened in California?" *SNOAR News*.
16. Nelson, *Rocket Men*, 124–125.
17–25. Miller interview with MBC.
26. *American Spacemodeling*, January 1979, 2.
27–28. Stine, "A Tribute to Pat Miller." *Model Rocket News Magazine*.
29–31. Miller interview with MBC. Harry Stine and Trip Barber were also instrumental in securing the exemption, and they were both recognized for their efforts when the NAR awarded them the H. Galloway Service Award, the NAR's highest honor, in 1979. See, generally, *American Spacemodeling*, June 1989, 12.
32–39. *Model Rocketeer*, November 1983, 3; Trip Barber would become president of the NAR in 2009.
40. Miller interview with MBC.
41–48. Pearson, "LDRS 3 11/12 August 1984." *SNOAR News*; reprinted in *High Power Rocketry*, December 2002, 10–13.
49–50. Ron and Debbie Schultz interview with MBC.
51. "In the Mail." *High Power Research Magazine*, Fall, 1984.
52. Miller interview with MBC.
53–57. *High Power Research/Sport Rocketry Journal*, Winter 1984/Spring 1985, 30.
58–61. Geier, "NAR Board Meets in Washington." *SNOAR News*.
62. By 1985, Miller was openly looking for a way to allow NAR members to enjoy high power without impacting their NAR membership. See, generally, *Tripolitan*, December 1985, 4.
63–64. "Final Report of the Blue Ribbon Commission on *High Power Rocketry*." February 16, 1985, 1.
65. *SNOAR News*, Volume 11, Number 2, 9–11.

66. "NAR Board Meets in Washington, D.C.", *SNOAR News*, Volume 11, Number 2 (1985), 10.
67–69. Final Report of the Blue Ribbon Commission on *High Power Rocketry*." February 16, 1985, Appendix D-1.
70–71. Rogers, B. "Val-Sun Members Conduct High-Power Launch." *American Spacemodeling*.
72. *American Spacemodeling*, January 1987, 2; see also Zsidisin. "The 1985 Pearl River Modroc Seminar." *American Spacemodeling*.
73. See, generally, *American Spacemodeling*, January 1987, 2.
74. *American Spacemodeling*, January 1987, Model Rocket Safety Code, 5.
75–77. *American Spacemodeling*, January 1987; also, "NAR Board Meets in Washington, D.C." *SNOAR News*, Volume 11, Number 2 (1985), 10; also, *American Spacemodeling*, May 1985, 4.
78. See *High Power Research Sport Rocketry Journal* (Winter 1984/Spring 1985) 30; also, *American Spacemodeling*, May 1985, 4.

Chapter 6 (The Tripoli Rocket Society)

1–15. Blazanin interview with MBC.
16. Blazanin wrote a letter to *California Rocketry* in 1984, correcting the magazine on some of its math equations. See *California Rocketry Magazine*, April 1984, 3.
17. This club should not be confused with the *American Rocketry Association*, another short-lived venture based in Wisconsin in the early 1980s. That club failed too. See, *SNOAR News Quarterly*, Volume 8, number 3, 1982, 17.
18. Blazanin interview with MBC.
19. Regarding the selection of the name *Tripoli* for the club: "To help finance experiments and projects one of the members donated some gold coins he had received from his father. These coins came from Tripoli, Lebanon during World War II. Since the members came from three towns [East Pittsburgh, Braddock, and Irwin, Pa.], and 'Tripoli' meant three cities, the name was accepted . . ." See *High Power Rocketry*, "Of Words, Stories, and Memories Forgotten," by Bruce Kelly, Francis Graham, and Tom Blazanin, March 2001 (33 years, 168 issue index edition); see also *Extreme Rocketry* interview of Ken Good, May–June 2007, wherein Good suggests that the Tripoli name could also have come from the word "Tripolis," which means three cities. Good said these three cities could be Pittsburgh, Braddock, and Irwin, Pennsylvania—the three founding cities of Tripoli. But he defers to Francis Graham on this matter.
20. The Tripoli Federation was renamed to the Tripoli Science Association on December 30, 1979. One of the reasons for the name change was to make the Tripoli name "less mysterious" while preserving the Tripoli moniker for the future. See *Tripolitan*, Spring 1980, 1.
21. See *Tripolitan* issues from that time period; see also footnote 17.
22. The second issue is dated September 1, 1968, and was called *The R. A.T.C. Tripolitan*. RATC stood for the "Radio and Aerospace Team of Conneautville." See *Tripolitan*, December 25, 1970.
23. Graham interview with MBC.
24. The full Latin phrase is "sic transit Gloria mundi."

25. Hughes interview with McNeely.
26–27. Good interview with MBC.
28. Hughes interview with McNeely; see also *Tripolitan*, April 1969.
29–30. Good interview with MBC.
31. This is based on Tripoli Federation membership, which peaked in the late 1960s at 100–150 members; see *Tripolitan*, October 1970, 2 (graph); Francis Graham estimated the peak membership at 127 in 1968; *Tripolitan*, August 1971, 2.
32. For example, the Summer 1982 issue was the first *Tripolitan* in more than a year; *Tripolitan*, Summer 1982, 2.
33. Hughes interview with McNeely.
33–34. Hughes, "The Coming of the 'New' Model Rocket Engine." *Tripolitan*.
35. The Tripoli Rocket Society name was formally approved February 7, 1985; see *Tripolitan*, Winter 1985; March 1985.
36–38. Blazanin interview with MBC.
39–40. The Code for Model Rocketry was the original name as designated in 1968. The code was renamed in 1976 to the Code for Unmanned Rockets. Later, the code was returned to its original name, the Code for Model Rocketry. G. Harry Stine—one of the authors of NFPA 1122—said he wrote this exemption into the Code "to allow a means for the eventual expansion of rocketry into more advanced forms."
41. Rosenfield, "Rocketry Elitism." *Extreme Rocketry*.
42. See *High Power Research Sport Rocketry Journal* (Winter 1984/Spring 1985), 29; also, Blazanin interview with MBC; *Tripolitan*, May 1985, 3–4.
43. Blazanin interview with MBC; Weber and O'Connor had their own commercial rocket company as did several of High-Power Rocketry's early pioneers.
44. *SNOAR News*, March–July 1989, "Really Serious Letters to the Editor," letter by Don Carter to Matt Steele, pages 7–8. "Tripoli, before that night in Mark's living room, was a group of people in the Pittsburgh area who were interested in astronomy," said Carter later; this was an oversimplification. Tripoli was both astronomy- and rocketry-related.
45. The Tripoli bylaws had only recently been fully amended in February 1985; *Tripolitan*, March 1985, 1–5.
46. *Tripolitan*, May 1985, 3–4; the first ten members of the modern Tripoli Rocketry Association were assigned membership numbers as follows: J.P. O'Connor (1), Mark Webber (2), Tom Blazanin (3), Francis Graham (4), A.J. Reed (5), Curtis Hughes (6), John O'Brien (7), Scott Dixon (8), Korey Kline (9), and Ramona Dixon (10). Ron Schultz and Gary Rosenfield would become members 19 and 22, respectively. See, generally, *Tripolitan* December 1988, also, February 1987. In later years, early Tripoli membership numbers were sometimes reassigned to newer members when the original members left Tripoli.
47. Graham interview with MBC. Graham wrote in the early 1980s that he believed it was possible for small groups to reach out into space, although in a very small way. "It would provide the inspiration and impetus for more substantial and scientifically important ventures in the years to come," he said. *Tripolitan*, Winter 1981, 2.
48. Blazanin interview with MBC.
49–51. *Tripolitan*, "Issues for 1988." December 1987, 6.
52–53. *Tripolitan*, May 1985, 3–4.

54–55. *High Power Research Sport Rocketry Journal* (Winter 1984/Spring 1985), 4; the new version of 1122 would be NFPA 1127, which was written in the 1990s.
56–61. *High Power Research/Sport Rocketry Journal* (Winter 1984/Spring 1985), 30.

Chapter 7 (The Rocket Builder)

1. Generally speaking, the center of gravity should be one body tube diameter forward of the center of pressure. This helps ensure, but does not guarantee, a stable rocket in flight.
2–6. Rosenfield interview with MBC.
7. *Tripolitan*, December 1985, inside cover.
8–11. Rosenfield interview with MBC.
12–14. "LDRS 4 13/14 August 1985." *SNOAR News*, Volume 11, Issue 8, 1985.
15–26. Ron and Debbie Schultz interview with MBC.
27. Pearson, "LDRS-3." *SNOAR News*.
28–30. Ron and Debbie Schultz interview with MBC.
31. See also "L.O.C./AeroTech *Magnum*, *High Power Research/Sport Rocketry Journal*, Volume 3, 1985, 20–21.
32. Nelson, "Large and Dangerous Rocket Ships 4." *High Power Research Sport Rocketry Journal*.
33–34. Ron and Debbie Schultz interview with MBC.
35. Nelson, "Large and Dangerous Rocket Ships 4." *High Power Research Sport Rocketry Journal*.
36–37. Ron and Debbie Schultz interview with MBC.
38. "*Big Bertha*—A Feasibility Study." *Tripolitan*, October 1985.
39. Pearson and Schultz, "LDRS 4." *SNOAR News*; also, Nelson, "Large and Dangerous Rocket Ships 4." *High Power Research Sport Rocketry Journal*. Also, "*Big Bertha*—A Feasibility Study." *Tripolitan*.
40. Nelson, "Large and Dangerous Rocket Ships 4." *High Power Research Sport Rocketry Journal*.
41. The *Big Bertha* was eventually destroyed at Lucerne. See *Tripolitan*, August 1986, 24–25.
42. *Tripolitan*, October 1985.
43. Nelson, "Large and Dangerous Rocket Ships 4." *High Power Research Sport Rocketry Journal*.
44. Pearson and Schultz, "LDRS 4." *SNOAR News*.
45. Nelson, "Large and Dangerous Rocket Ships 4." *High Power Research Sport Rocketry Journal*. See also Meyer, "Observations—LDRS Report." *Tripolitan*.
46–47. Nelson, "Large and Dangerous Rocket Ships 4." *High Power Research Sport Rocketry Journal*.
48–55. Meyer, "Observations—LDRS Report." *Tripolitan*. Meyer was replaced as chairman of the commission by Jim Barrowman in early 1986; see *American Spacemodeling*, April 1986, 11.
56–57. *High Power Research Sport Rocketry Journal* (1985), Number 12, Letters, 23 and 28.

Chapter 8 (The End of the Beginning)

1–3. *Tripolitan*, December 1985, 12.
4. Pearson interview with MBC.
5–7. See *SNOAR News*, January/February 1986, 19.
8. http://commerce.Alaska.gov/CBP/Main/CorporationDetail.aspx?id=38589D, accessed on June 21, 2012. The bylaws were accepted by the board and the membership at LDRS in August 1986. The last set of Tripoli bylaws were in the unincorporated version found in the *Tripolitan* in early 1985.
9. Tripoli Bylaws, included in mailed-out membership packets—circa mid-1990s.
10–13. *Tripolitan*, October 1986, 4.
14. *Tripolitan* October, 1986. Blazanin would continue as president, Bill Barber was vice-president, Ed Tindell was treasurer and Graham continued on as secretary.
15–16. Graham, "Tripoli Board of Directors Meeting." *Tripolitan*.
17–18. Schultz, "The *Esoteric*." *Tripolitan*; also, Schultz interview with MBC.
19. Steele, "Highlights of LDRS-5." *SNOAR News*.
20. *Tripolitan*, October 1986, "LDRS-V," author unknown (but likely compiled by Tom Blazanin), 17–21. For another general story on LDRS-5, see *High Power Research, Sport Rocketry Journal*, Number 14, "LDRS-5—Goodbye to Medina," 1986.
21. Schultz, Ron, and Debbie. "*Esoteric*: Maxi-Sport Rocket." *High Power Research*. Also, Schultz interview with MBC.
22. *Tripolitan*, October 1986, "LDRS-V," author unknown (but likely compiled by Tom Blazanin), 17–21; *High Power Research, Sport Rocketry Journal*, Number 14, "LDRS-5—Goodbye to Medina," 1986.
23–24. Schultz, Ron, and Debbie. "*Esoteric*: Maxi-Sport Rocket." *High Power Research*. Also, Schultz interview with MBC.
25. Tripoli had held general assembly meetings going back to the 1960s when it was the Tripoli Confederation. These meetings were not associated with any launches.
26–28. Graham, "General Assembly. *Tripolitan*.
29–32. Johnson, "NAR Blue Ribbon Commission at LDRS-5." *SNOAR News*.
33–34. Pearson interview with MBC; see also "Official Announcement from LDRS Director Chris Pearson." *SNOAR News*, October/November 1986, 8.

Chapter 9 (The Move West)

1. Miller, "President's Corner." *American Spacemodeling*.
2–3. Some former critics of the NAR now did an about-face. The NAR was suddenly "vital" and possessed of "foresight and courage" and was applauded for advising the FAA, the NFPA, and the Department of Transportation as to how hobby rocketry should be managed. The still-influential *SNOAR News*, which in 1982 labeled Miller as a "prosecutor out for blood" because he investigated NAR members for using gasoline in their rockets, changed its tune. The NAR president was now "instrumental in developing and implementing the strategies that allowed changes to take place in the hobby." *SNOAR News*, January/February 1987, 3.
4. Russell. "High Power Model Rocket Motors." *American Spacemodeling*.
6–9. "Advanced Rocketry Safety Code." *Tripolitan*, February 1987, 28–29.

10. See, for example, *Tripolitan*, February 1987, 9.
11. "Tripoli Update." *Tripolitan*, February 1987, 6.
12–13."Launch Announcement." *Tripolitan*, February 1987, 4.
14. *Tripolitan*, February 1987, 7; see also interview with Tom Blazanin by the author.
15. See, for example, *Tripolitan*, August 1987, 12.
16–17. "Launch Announcement." *Tripolitan*, May 1987, 4.
18. Rogers, "Lucerne Test Range Tracking Results." *Tripolitan*, May 1987.
19–24. *Tripolitan*, June 1987, 8; see also *Tripolitan*, August 1987, 11.
25–29. Blazanin interview with MBC.
30. "Opinions." *Tripolitan*, October 1987, 11–12.
31. "LDRS Recap." *Tripolitan*, October 1987, 44.
32. See, generally, *Tripolitan*, August 1987 (inside front cover). At the time, the H180 was $27; a K250 was $225.
33. "Large and Dangerous Rocket Ships VI." *Tripolitan*, October 1987, 30–31.
34–35. "LDRS Memories." *Tripolitan*, October 1987; see also, "Large and Dangerous Rocket Ships VI." *Tripolitan*, October 1987, 21.
36–39. O'Brien, "*Warp Factor II*." *Tripolitan*, October 1988, 77–81; see inside front cover LOC Precision advertisement.
40–44. *Tripolitan*, October 1987, 2, 11, 29, 42–45.
45. *Tripolitan*, December 1987, 11.
46. LDRS Recap." *Tripolitan*, October 1987, 44.
47–48. "2nd General Assembly." [mistitled in the original text] *Tripolitan*, October 1987, 5.
49–51. *Tripolitan*, December 1987, 6.
52. "Opinions." *Tripolitan*, October 1987, 12.
53. Redd online interview with MBC.
54. "Tripoli Membership Update." *Tripolitan*, October 1987, 48.
55. "2nd General Assembly," *Tripolitan*, October 1987, 5.
56–59. "Opinions." *Tripolitan*, October 1987, 13.
60. *Tripolitan*, October 1987, 13 (untitled article).
61–62. "Opinions." *Tripolitan*, October 1987, 13.
63. By 1987, the NAR was no longer actively pursuing members who were participating in high-power activities.
64. Miller,"President's Corner." *American Spacemodeling*, October 1987.
65. Tavares, "No Mixing Rule Deserves Your Support." *American Spacemodeling*.
66. Barrowman was president of the NAR 1970–1973. The commission was originally chaired by Dan Meyers. Meyers stepped down in January 1986. "NAR Notes." *Tripolitan*, February 1986, 6.
67–69. Miller, "President's Corner." *American Spacemodeling*, October 1987; Tavares, "No Mixing Rule Deserves Your Support." *American Spacemodeling*; "NAR Notes." Tripolitan, February 1986.
70–71. Miller, "President's Corner." *American Spacemodeling*, October 1987, 2.
72. "Motor Policy." *Tripolitan*, February 1988, 14; Tripoli's motor policy did not use the term "basement bombers." But it prohibited homemade motors. For the time being, only commercial motors manufactured by recognized motor makers were permitted to be flown at Tripoli events.
73. "Z4." (Multiple authors), *Tripolitan*, December 1987, 28–44.

74. Sicker. "Editor's Thermal: Tripoli Motor Certification." *Star-Date 10.91* [ASTRE Newsletter].
75–77. Schultz interview with MBC; see also "Z4 Aftermath." *Tripolitan*, December 1987, 42–45.
78. There was at least one published claim that the field was lost because of the number of motor CATOs at the event. See Sicker. "Editor's Thermal: Tripoli Motor Certification." *Star-Date 10.91* [ASTRE Newsletter].
79. Safford. "Fort A.P. Hill." *Star-Date 12.88* [ASTRE Newsletter].
80. Blazanin interview with MBC.
81. Blazanin, "NFPA Meeting." *Tripolitan*, December 1987, 8.
82. "NFPA Update." *Tripolitan*, April 1988, 10.
83. "Tripoli Membership Update." *Tripolitan*, December 1987, 49.
84. See, for example, "President's Corner," *American Spacemodeling*, March 1986 (with graph), 4–5.

Chapter 10 (A New President)

1–2. *Tripolitan*, February 1990, 6.
3–4. *Tripolitan*, February 1988, 5; the address change was for day-to-day operations; Tripoli remained an Alaska corporation.
5. *Tripolitan*, December 1987, 4.
6–7. *Tripolitan*, June 1987, 5–6.
8–9. *Tripolitan*, June 1987, 5; also, *Tripolitan*, February 1988, 23.
10–11. *Tripolitan*, December 1987, 5.
12–15. Piper. "Ammonium Perchlorate: NH4CLO4: Myth, Mystery, Misconception, and the Henderson Syndrome." *High Power Rocketry*, July 1998, 10–19; related story, 42–51; also http://en.wikipedia.org/wiki/PEPCON_disaster, accessed on July 12, 2013; see also, YouTube videos of fire and explosions.
16. Kelly. "Editorial: The Devil Is in the Details." *High Power Rocketry*.
17. http://en.wikipedia.org/wiki/PEPCON_disaster, accessed on July 12, 2013. At the time of the PEPCON fire, a local camera crew was filming an unrelated event from an elevated position not far from the plant. The film captured the moment of two of the largest explosions that followed. That film is available on YouTube.
18. Rogers, "Lucerne Test Range Tracking Results." *Tripolitan*.
19. Rogers, "USXRL-88 Tracking Results." *Tripolitan*.
20. McGoldrick, "Opinion." *Tripolitan*.
21. Hall, "Opinions." *Tripolitan*.
22. "Recommended Motors." *Tripolitan*, June 1988, 35.
23–24. "Insurance Obtained." *Tripolitan*, June 1988, 5.
25. "The Tripoli Safety Code." *Tripolitan*, June 1988, 9–10.
26–29. Stine, "Orville H. Carlisle NAR #1." *American Spacemodeling*.
30. "Insurance Obtained." *Tripolitan*, June 1988, 5–7.
31. *Tripolitan*, June 1988, 8 (list of contests at the time).
32. *Tripolitan*, October 1988 (inside cover page).
33–35. Schultz, *"Mother Lode." Tripolitan.*
36. Rogers, "LDRS-7 Tracking Results." *Tripolitan*. Strangely, some observers said that the rocket actually disappeared from view into the ceiling, estimated to be as low as

3,000 feet in some areas around the field. However, a different vantage point provided Rogers with a view to the top of the flight.

37–39. *Tripolitan*, October 1988.
40. "*Warp Factor II.*" *Tripolitan*, October 1988, 77–81.
41. "*Eye in the Sky.*" *Tripolitan*, October 1988, 73–76.
42–43. "LDRS-7/The Nationals." *Tripolitan*, October 1988; Binstock, "Showdown at LDRS." *Tripolitan*.
44–45. Rogers, "Why Did the Rockets Fly So High at LDRS?" *Tripolitan*.
46–49. Tindell, "LDRS-7 Statistics Report." *Tripolitan*.
50–52. "Board of Directors Meeting" and "3rd National General Assembly." *Tripolitan*, October 1988, 10–12.
53–56. Safford, "Fort A.P. Hill-1." *Star-Date 12.88*; see also Manganaro, "Fort A.P. Hill-1." *Tripolitan*.
57. "Octoberfest 88." *Tripolitan*, December 1988, 46–57.
58. Rogers, "Octoberfest Tracking Results." *Tripolitan*; the rocket was launched by Tim Brown. It was a 2-inch-diameter LOC kit powered by a J-class motor. The margin of error for the reported altitude, according to Rogers, was 1.3 percent.
59. "Octoberfest 88." *Tripolitan*, December 1988, 53–54.
60. Kelly interview with MBC.

Chapter 11 (The Rocket Scientist)

1–3. *SNOAR News*, January/February 1989.
4. *SNOAR News* (Editorial with Ed Tindell reply letter) January/February 1989, 20–21.
5. "Election Redo." *Tripolitan*, December 1988 (not published until the early summer of 1989), 4.
6–7. Rogers, "Presidential Campaign." *Tripolitan*.
8. *SNOAR News*, March–July 1989, 6.
9. "Fourth Annual Board of Directors Meeting." *Tripolitan*, December 1989, 8–9.
10. Rogers had the proxy votes of board members Bill Woods, A.J. Reed, and Bill Barber. He also had a verbal proxy made from a telephone booth in the hallway from John O'Brien. See *Tripolitan*, December 1989, 8.
11. Meyerrieks, "LDRS-8."(VHS to DVD) 1990.
12–13. The February 1991 issue of the *Tripolitan* ran a photo essay on the event.
14–24. Meyerrieks, "LDRS-8." (VHS to DVD) 1990; see also *Tripolitan*, February 1991.
25. Redd interview with MBC.
26. *Tripolitan*, December 1989, 8.
27. "Outgoing President's Message." *Tripolitan*, November 1989, 4.
28. Tindell would stay active with the local NAR chapter in Houston and was for a time the editor of the monthly NAR newsletter called *Rendezvous*. See *American Spacemodeling*, April 1988, 11.
29–30. *Tripolitan*, December 1989, 9–10.
31–50. Rogers interview with MBC.
51. As of 2018, this was the longest run for any board member. Board member Bruce Lee is next.
52. *Tripolitan*, November 1989, 5.

53–54. Rogers interview with MBC.
55–57. *Tripolitan*, December 1989, 9–10.
58–59. *Tripolitan*, February 1990, 5.
60–63. *Tripolitan*, December 1989, 5–14.
64. *Tripolitan*, December 1988 (released June 1989), 10.
65–66. *Tripolitan*, December 1989, 5.
67–69. Letter from G. Harry Stine to J. Patrick Miller, May 24, 1989.
70. Miller, "President's Corner." *American Spacemodeling*, December 1989, 3 (list of members). Committee members included several high-profile model rocketry flyers and also high-power enthusiasts, including Gary Rosenfield and Chuck Mund. Mund was on the commission as an official representative of Tripoli. It was reported that G. Harry Stine first petitioned the board to incorporate programs for advanced high-power rocket flyers in May 1989; see also Miller, "President's Corner." *American Spacemodeling*, October 1989, 14; see also *American Spacemodeling*, April 1990, 5.

Chapter 12 (Black Rock)

1–2. Buck interview with MBC.
3. Signpost at entrance to the desert off Jungo Road; see also John Fremont's history in *Wikipedia*.
4. Emerson. *The Applegate Trail of 1846*.
5. In the early 1980s, archeologists unearthed the remains of a Columbian mammoth near Black Rock. The animal stood over 10 feet high and roamed the ground here, which at the time was a shallow bay with extensive marshland, up to twenty-two thousand years ago.
6. Wolfe. *The Right Stuff*.
7. Kelly, "BALLS 101: Separating Myth from Fact about BALLS." *High Power Rocketry*.
8. *Tripolitan*, December 1988, 45.
9. See, for example, Buck's letters to authorities dated May 4, 1989 (Bill Lewis Collection).
10. The author was unable to confirm that the forms were sent to the FAA for Black Rock I.
11. The flyers packet stated the launch was located at 40 degrees, 49 minutes long, 119 degrees, 10 minutes lat. The launch would run from seven o'clock in the morning to sunset. There would be no night launches (Bill Lewis collection).
12. "Black Rock One." *Tripolitan*, November 1989, 27–39 (likely written by Tom Blazanin).
13. The flyers packet for Black Rock I contained the detailed rules for the contest, which included penalties and awards. The rules were established by Tripoli flyer Scott Pearce. It was $5 to enter the contest (Bill Lewis collection).
14–18. "Black Rock One." *Tripolitan*, November 1989, 27–39 (likely written by Tom Blazanin).
19. See, for example, *Tripolitan*, February 1990, 8.
20–21. *Tripolitan*, December 1990, 16–29, and tracking results, 30–33.
22. For more information on the history of AERO-PAC and the first waivers at Black Rock, see chapter 14.
23–24. Binford online interview with MBC.

25–28. *Tripolitan*, December 1990, 16–33.
29–31. Binford online interview with MBC.
32. The man in the shorts was Frank Kosdon.
33–36. "The P Motor." *Tripolitan*, December 1990, 26–30.
37. Sicker, "LDRSIX." *Star Date 2.91*.
38–44. *Tripolitan*, "LDRS IX." February 1991, 13.
45–47. Sicker, "LDRS IX." *Star Date 2.91*.
48. The Black Rock Two Launch report would not come out for months.
49–50. Rogers interview with MBC.
51–60. "LDRS-10 Site Selection." *The Tripoli Report*, October 1990.
61. *Tripolitan*, December 1990, 13.
62. The next time an eastern LDRS would be held would be in Orangeburg, South Carolina, in 1996.
63–64. *The Tripoli Report*, March 1991, 6–8.
65. Miller, "President's Corner." *American Spacemodeling*, April 1990.

Chapter 13 (NAR WARS)

1. See ASTRE *Star-Date 1.90*: "The NAR and the Tripoli Rocketry Association are communicating in a way that they haven't done in years . . ."
2. "NARAM-31 Overview." *American Spacemodeling*, November 1989, 5.
3–4. Miller, "President's Corner." *American Spacemodeling*, December 1989.
5–6. *American Spacemodeling*, November/December 1990, Miller, 34–35; see also "News Online," *Star-Date 11.90* [ASTRE Newsletter], 3.
7. See, generally, *American Spacemodeling*, March/April 1991, 22–23.
8–11. *Tripolitan*, November 1989, 5.
12. The author was unable to locate any documents that attest to the claim that the NAR was trying to outlaw Tripoli.
13–15. *Tripolitan*, December 1990, Editorial (coauthored by Rogers and Bruce Kelly), 5–7.
16. "An Open Letter." *The Tripoli Report*, January 1991, 1–3.
17–18. Miller interview with MBC.
19. Rogers interview with MBC.
20. Miller interview with MBC.
21. *Tripolitan*, December 1990, 3.
22. Moose Lavigne either appeared by telephone or gave a proxy as his flight from Florida was grounded because of weather conditions. See *The Tripoli Report*, March 1991, 7.
23–24. Miller interview with MBC.
25–26. "From the Prez." *The Tripoli Report*, March 1991, 2.
27. Miller interview with MBC.
28. See, for example, *American Spacemodeling*, March/April 1991, 24; *The Tripoli Report*, March 1991, 1.
29. *The Tripoli Report*, March 1991, 1–3.

Chapter 14 (LDRS X and BALLS)

1. "LDRS-10." Point 39 Productions; see also Blazanin, "Black Rock LDRS Advice." *Tripolitan*: "We also witnessed three Air Force B-52s literally diving over the mountains to about 500 feet above the lake bed surface, open the bomb bay doors and proceed to perform full throttle bombing runs across the lake bed. Yes, the road maps and posted signs all over the lake bed stating the Black Rock Desert as a military bombing range hold some truth!"
2. Spinelli, "The Race to 1,000 MPH." *Popular Science*.
2. The waiver was obtained by AERO-PAC's Bill Lewis and was granted on August 8, 1991. The basic waiver was 20,000 feet AGL on August 16 and 30,000 feet AGL on August 17–18; windows on these days run from 40,000 to 60,000 feet AGL (Bill Lewis collection). Lewis also obtained permission from the BLM for the waiver at LDRS-10.
3. "LDRS-10." Point 39 Productions.
4. *High Power Rocketry Magazine*, October/November 1991.
5. "LDRS-10." Point 39 Productions.
6. *High Power Rocketry Magazine*, October/November 1991.
7. AERO-PAC's Bill Lewis compiled statistics for LDRS-10. There were 115 flyers. Some flyers launched up to forty rockets that weekend; presumably, these were model rocket flights as no detailed list of motors used or rocket type was kept (Bill Lewis collection).
8. *Tripolitan*, August 1990, 25.
9. Lewis interview with MBC.
10. See Certificate of Waiver or Authorization dated April 21, 1990 (Bill Lewis collection). The precise location set forth in the waiver was 40 52' 55"/longitude 119 8' 10."
11. See Certificate of Waiver or Authorization dated June 26, 1990 (Bill Lewis collection). This waiver was signed by FAA manager Sabra W. Kaulia.
12. "1991 Launch/Event Calendar." *Aeronaut*, September 1990, 5. ("A special launch for K motors [or equivalent] and above, requiring 1300 N-s minimum total impulse per flight. Send $30 to BALLS, P.O. Box 5127, Reno, NV 89513.")
13. Lewis, "Fireballs 001: AERO-PAC's First Experimental Unlimited Rocket Launch." *Aeronaut*.
14. AERO-PAC officials such as Bill Lewis obtained the necessary FAA waivers and BLM permits for LDRS-10.
15. *Tripolitan*, August 1990, 25. Although the magazine had an August publication date on the cover, it is likely it did not appear until later that year.
16. Lewis interview with MBC.
17. Bill Lewis was afraid to use "BALLS" on the application," recalled AERO-PAC's Dave Bucher, another active member in the early 1990s. "So he decided, with the club officers' approval, to rename the proposed experimental launch, 'FIREBALLS 001.'"
18. Lewis interview with MBC. The name for the launch ("BALLS") may also have been coined by Tom Blazanin, who purportedly said that one had to have balls to launch a big rocket.
19. Launch registration forms were initially sent to *West Coast Rocketry's* Tim Brown in Sacramento. Brown was designated the launch director. See *The Tripoli Report*, May

1991, 21–23; AERO-Pac members Ron Devine, Dean Lewis, Bob Baker, Jay Orr, and Kelly Badger also assisted as RSO/LCO support. *Aeronaut*, April 1992; see also Lewis, "Fireballs 001: AERO-PAC's First Experimental Unlimited Rocket Launch."
20. The application or registration forms were also made available in *The Tripoli Report* in May 1991.
21. Fireballs 001 flyers information packet, 2 (Bill Lewis collection).
22. Bill Lewis letter to Thomas Moody, FAA-AWP, dated July 5, 1991 (Bill Lewis collection). The proposed location for Fireballs 001 was set forth in the FAA application as 40 degrees 55' 30" latitude, 119 degrees 7' 30" longitude.
23. Lewis interview with MBC. In a letter to Lewis dated March 13, 1991, Dean Oberg gave a rough description of the proposed rocket. It was to have a weight under 75 pounds and thrust of 500 pounds. The burn time would be 20 seconds, and the predicted apogee was 11–16 miles. "There will hopefully be a parachute recovery system developed in time for the flight," wrote Oberg (Bill Lewis collection). See also Lewis, "Fireballs 001: AERO-PAC's First Experimental Unlimited Rocket Launch." *Aeronaut*.
24. Letter from FAA Manager Sabra W. Kaulia, including waiver from and conditions to Bill Lewis, dated August 15, 1991 (Bill Lewis collection).
25. *The Tripoli Report*, July 1991, 2.
26. "Minutes for the Board of Directors Meeting." *The Tripoli Report*, October 1991, 2.
27. Sicker, "Editor's Thermal: Tripoli Motor Certification." *Star-Date 10.91*.
28–30. *The Tripoli Report*, October 1991, 2.
31. Lewis interview with MBC.
32. Flyers information packet for Fireballs 001 (Bill Lewis collection).
33–35. Lewis, "Fireballs 001: AERO-PAC's First Experimental Unlimited Rocket Launch." *Aeronaut*.
36. Clark, Meredith, and the AHPRA would take over BALLS in 1998 and run it for the next seventeen years through BALLS 24 in September 2015.
37. *Lewis*,"Fireballs 001: AERO-PAC's First Experimental Unlimited Rocket Launch." *Aeronaut*.

BOOK II

Chapter 15 (The Reloadable Revolution)

1–2. Rosenfield interview with MBC; see also Rosenfield interview with McNeely.
3–4. http://www.jetex.org/history/heydey-wm.html, accessed on November 9, 2012.
5. Rosenfield interview with MBC; see also Rosenfield interview with McNeely.
6. *American Spacemodeling*, November 1991, 18.
7. Rosenfield interview with MBC.
8. Shenosky, "Futureshock: ISP's New Reloadable Motors." *American Spacemodeling*.
9–14. Rosenfield interview with MBC; see also Rosenfield interview with McNeely.
15–16. *American Spacemodeling*, November/December 1991, 18–21.
17. *American Spacemodeling*, November/December 1990, 34–35.

18–19. Shenosky, "Futureshock: ISP's New Reloadable Motors." *American Spacemodeling*.
20–27. Kosdon interview with McNeely.
28–29. A "case-bonded" motor is one in which the fuel grain(s) is directly bonded to the case.
30–32. Kosdon interview with McNeely.
34. Rosenfield interview with MBC; Rosenfield interview with McNeely.
35. *Tripolitan*, April 1991.
36–40. "Minutes from the Tripoli Board of Directors Meeting." *The Tripoli Report*, October 1990, 3–4; see also *The Tripoli Report*, March 1991, 7.
41. *The Tripoli Report*, March 1991, 7–8.
42. *The Tripoli Report*, May 1991, 7.
43. See, for example, *American Spacemodeling*, November/December 1991, 5.

Chapter 16 (Showtime)

1. http://www.westegg.com/inflation/infl.cgi, accessed on December 31, 2013; the actual conversion is $1,000 in 1988 to $1,910.98 in 2012.
2–3. Mary Roberts's letter to G. Harry Stine, April 4, 1991 (Gary Rosenfield collection).
4–5. NAR/Tripoli Aquarius Commission Report, June 1991.
6. Additional footage for the video was purportedly shot at Vulcan Systems.
7–9. "Reloadable Model Rocket Motors: An Accident Waiting to Happen." (Author not stated). December 1991, 2; the author suggested that even cigarettes or cigarette ash could ignite a motor.
10. Letter from G. Harry Stine, Chairman, NAR Model Rocket Regulations Committee and NAR Representative to the NFPA Committee on Pyrotechnics, to U.S. Consumer Product Safety Commission, November 8, 1991.
11–13. Tavares, "Reload Madness Video Rocks NFPA." *The Sentinel*.
14–15. "NFPA Committee Meets in Phoenix." *American Spacemodeling*, July/August 1992, 29–31.
16. See, for example, NAR Letter by G. Harry Stine to Barry Tunick, General Manager of Estes, March 5, 1992 (Gary Rosenfield collection).
17. Letter to Harry Stine from Barry Tunick, March 13, 1992 (Gary Rosenfield collection).
18. Vulcan Systems Inc. fax transmission memo dated February 25, 1992, from Scott Dixon to Harry Stine, 1–2 (Gary Rosenfield collection).
19–20. Stine, "Model Rocket Motor Burn Tests: A Report to the National Association of Rocketry," April 4, 1992, 1.
21–22. Tavares, "Reload Madness Video Rocks NFPA." *The Sentinel*.

Chapter 17 (Black Rock, Argonia, and *Down Right Ignorant*)

1–2. Sienkiewicz, "LDRS 11: Groundwork for the Future." *High Power Rocketry*, November/December 1992; at the time, Tripoli had nearly two thousand members.

3–4. Cooper, "Desert Reactions." *High Power Rocketry*, November/December 1992, 23–33; the night launch was canceled because of the sandstorm event.
5–6. Anthony, "LDRS XI & Fireballs II." *American Spacemodeling*, January/February 1993.
7. Cooper "Desert Reactions." *High Power Rocketry*, November/December 1992, 24.
8. Anthony, "LDRS XI & Fireballs II." *American Spacemodeling*, January/February 1993.
9. Cooper "Desert Reactions." *High Power Rocketry*, November/December 1992, 32.
10–11. Anthony, "LDRS XI & Fireballs II." *American Spacemodeling*, January/February 1993, 4–5 and 46.
12. There were 146 registered flyers at LDRS-11.
13–14. Sienkiewicz, "LDRS 11: Groundwork for the Future." *High Power Rocketry*, November/December 1992, 13; see also Anthony, "LDRS XI & Fireballs II." *American Spacemodeling*, January/February 1993.
15–16. Tindell, "T-17:00:00/A Personal Account of the Launch of the *SK3J*." *High Power Rocketry*, August/September 1991, 40–51.
17. Dexheimer, "Let's Do Launch." www.phoenixnewtimes.com, accessed on February 18, 2014.
18–25. Sackett, "*Down Right Ignorant.*" *High Power Rocketry*, January/February 1993, 10–17; the rocket was 34 feet tall.
19–25. Id; see also Anthony, "LDRS XI & Fireballs II." *American Spacemodeling*, January/February 1993.
26–28. Tripoli Minutes, *Tripoli Report*, January 1993; Baumfalk online interview with MBC.
29–33. Baumfalk and Swayze, "Kountdown in Kansas." *High Power Rocketry*, July/August 1992.
34. Lamothe, "My Visit to Argonia." *High Power Rocketry*, July/August 1992, 34.
35. Baumfalk online interview with MBC.
36. Dexheimer, "Let's Do Launch." www.phoenixnewtimes.com, accessed on February 18, 2014.
37. Baumfalk online interview with MBC.
38–43. "LDRS-12." Point 39 Productions; see also Meredith and Sharpe, "The Making of the Mega-Skeeter." *High Power Rocketry*, November/December 1993, 18.
41–43. "LDRS-12." Point 39 Productions.
44–46. Ezell and Davis, "The Rocket Tree." *High Power Rocketry*, November/December 1993, 43–46; see also "LDRS-12." Point 39 Productions.
47–48. Dexheimer, "Let's Do Launch." www.phoenixnewtimes.com, accessed on February 18, 2014.
49. Rogers, "The Aerodynamic Drag of Rockets." *High Power Rocketry*, August 1995, 11–34; see also http://www.ahpra.org/o10k.htm, "The O-10,000 Page," accessed on January 31, 2013.
50. "LDRS-12." Point 39 Productions.
51. Lee online interview with MBC.
52. Rogers, "The Aerodynamic Drag of Rockets." *High Power Rocketry*, August 1995.
53–54. "LDRS-12." Point 39 Productions.
55–62. Dexheimer, "Let's Do Launch." www.phoenixnewtimes.com, accessed on February 18, 2014.

63–65. Lamothe, *"Aerobee II."* *High Power Rocketry*, February 1995; see also "LDRS-13," Point 39 Productions.
66. Clary, David A. *Rocket Man*. 186–187.

Chapter 18 (The Visionary)

1. Kelly, "Of Words, Stories, and Memories Forgotten." *High Power Rocketry*, March 2001 (thirty-three-year index).
2. Without the *Tripolitan*, there would be little, if any, contemporary record of the events in high power in the 1980s.
3. Kelly, "Of Words, Stories, and Memories Forgotten." *High Power Rocketry*, March 2001 (thirty-three-year index).
4. *Tripolitan*, October 1988, 10–11.
5. *The Tripoli Report*, June 1990, 1.
6–7. *Tripolitan*, August 1990, 5.
8–9. Kelly interview with MBC.
10. Kelly Interview with McNeely.
11–16. Kelly interview with MBC.
17. *High Power Rocketry*, October 1995, 7.
18. Kelly interview with McNeely.
19. *High Power Rocketry*, December 1996, 55.
20. *The Tripoli Report*, October 1990.
21–25. Kelly interview with MBC; Kelly interview with McNeely.
26. *The Tripoli Report*, May 1991.
27. *High Power Rocketry*, December 1996, 55.
28. Kelly, "Of Words, Stories, and Memories Forgotten." *High Power Rocketry*, March 2001 (thirty-three-year index).
29. *The Tripoli Report*, July 1992.
30. Leninger online interview with MBC.
31. Ejma online interview with MBC.
32. Inman interview with McNeely.
33. Miller online interview with MBC.
34. Scholl online interview with MBC.
35. Eiszner online interview with MBC.
36. Swenson online interview with MBC.

Chapter 19 (A Code for High Power Rocketry)

1. Brinley, *"Rocket Manual for Amateurs."*
2–3. Tripoli Articles of Incorporation, Section III (d).
4. *Tripolitan*, February 1987, 28–29; Ed Tindell said later that Gary Fillible was the actual author of the code. See *Tripolitan*, December 1988, 15–16.
5–7. *Tripolitan*, June 1988, 9–10.
8. "NFPA Meeting," *Tripolitan*, December 1987, 8.
9–11. "NFPA Update," *Tripolitan*, April 1988, 10.
12. "NFPA Meeting," *Tripolitan*, December 1987, 8.

13. "NFPA Update," *Tripolitan*, October 1988, 6.
14–19. *The Tripoli Report*, March 1991, 1–4.
20. See, generally, *The Tripoli Report*, January 1991, 3–9.
21. The Tripoli Board voted to adopt the draft in October 1991. See *The Tripoli Report*, January 1993, 12.
22. NFPA 1127 (1994 Edition—Draft Version 1.3).
23. *The Tripoli Report*, February 1992, 31.
24. In 1985, Dan Meyer had suggested this very thing—that Stine help high power draft a safety code of its own or amend NFPA 1122 to include high power. See *Tripolitan*, October 1985, 12–14.
25. NAR High Power Rocket Safety Code—Interim, *American Spacemodeling*, January/February 1992, 15–19.
26. See NFPA 1127 (1995 Edition).
27–28. NFPA 1127 (1994 Edition—Draft Version 1.3).
29. *The Tripoli Report*, February 1992, 26–29; Wood and at least one other board member suggested Tripoli reject this version of NFPA 1127 because of deficiencies in the draft.
30–32. *The Tripoli Report*, February 1992, 31–32.
33–34. *The Tripoli Report*, January 1993, 12.
35. The final version did not change the dimensions table. The table was, in practice, ignored. The table was incorporated into the final 1995 version and then subsequent versions in 2002. But it was dropped in 2008.
36. *The Tripoli Report*, April 1994, 10–11. The NFPA draft adopted was dated November 12, 1993. The dissenting board member was Dennis Wacker. It is not clear if all the board members were present. But apparently, there was a quorum.
37. NFPA 1127 (1995 Edition).
38. Kelly, "Knowing When to Say No: The NFPA Code Writing Process." *High Power Rocketry Tripoli Report*, Special Combined Issue, April 2004, 37–47; see also *High Power Rocketry*, December 2003.
38–39. *American Spacemodeling*, May/June 1991, 23–24; see also *American Spacemodeling*, January/February 1992, 15–19.
40–42. *The Tripoli Report*, May 1995, 4.
43. *The Tripoli Report*, January 1991, 12.
44. *The Tripoli Report*, January 1993, 19.
45. http://www.imdb.com/title/tt0049473/trivia?ref_=_.
46–53. Bartel interview with McNeely; also, candidate resume in *The Tripoli Report*, June 1993, 3.
54. See, for example, *The Tripoli Report*, December 1993.
55. Bartel was initially appointed to fill the empty seat that opened up when the board removed one of its members. He was later elected to the board. See *The Tripoli Report*, December 1993, 23.
56–60. *The Tripoli Report*, December 1993.
61–63. *The Tripoli Report*, April 1994.
64–67. *The Tripoli Report*, May 1995, 11–21.
68. *The Tripoli Report*, June 1998, 25–26.

69. Kelly, "Knowing When to Say No: The NFPA Code Writing Process." *High Power Rocketry Tripoli Report*, Special Combined Issue, April 2004; see also *High Power Rocketry*, December 2003.

Chapter 20 (Theodolite's Demise)

1. Stine, "The Old Rocketeer," October 1969, 37–38.
2. See, for example, Steele and Fleischer, "Electronic Staging of Rockets." *Tripolitan*, January 1991, 25–29; see also Benson, "Direct Entry Ejection Timer." *Tripolitan*, August 1990, 24–25 (modifying a Radio Shack electronic timer for rocket use); see also Buxton, "Simple Timer," *Tripolitan*, December 1988, 28; see also "Altitude Record Update," *Tripolitan*, December 1987, 11 (discussing homemade barometric altimeter on rocket that reached more than 13,000 feet at LDRS-6); and *Tripolitan*, April 1991, 35 (advertisement for digital barometric altimeter by Rosemary Rocket Research); and *Tripolitan*, October 1988, 81 (advertisement for Space Dynamics "Digital Flight Sequencer," which was a timer for staging or recovery to augment or replace current in-motor pyrotechnic delays).
3. Fleischer, "Rocket Altimeter." *Radio Electronics*.
4. North Coast Rocketry Catalog of Products for 1989, 32; by the mid-1990s, Fleischer would improve his original altimeter and make it smaller and less expensive. It would also record altitude up to 9,000 feet. Transolve would one day be a well-known manufacturer of rocketry electronics.
5. See, for example, *American Spacemodeling*, September/October 1990, 39.
6–7. Flight Control Systems, advertising brochure for Flight Pack One (FP1) Data Logger/Controller (1990).
8–17. Billings interview with MBC.
18. Gianokis, "The Launch of the Freedom Seven." *The Tripolitan . . . America's High Power Rocketry Magazine*, August/September 1991, 14–17; also, Billings interview with MBC.
19–21. Gianokis, "The Flight of SA-205." *The Tripolitan . . . America's High Power Rocketry Magazine*, March/April 1992, 18–25; also, Billings interview with MBC.
22. http://www.ehow.com/about_5047422_invented-altimeter.html, accessed on January 9, 2014.
23–26. Billings interview with MBC.
27. See, for example, *American Spacemodeling*, NARAM Issue 1992, 33.
28. Adept Catalog 1992 Price sheet, 3.
29. Adept Rocketry Catalog #1, 1992, 2.
30. See also Buxton, "Simple Timer." *Tripolitan* (timer to be used in conjunction with motor ejection to control parachute deployment from a separate payload section).
31–35. Billings interview with MBC.
36. Theodolite devices are still marketed for model rocketry, although their use in model rocketry has also declined over the years.

Chapter 21 ("My Own Private *Hindenburg*")

1–2. This was the first time that an LDRS/BALLS event was scheduled for five days in a row.
3. *The Tripoli Report*, May 1995, 52.
4. There were 146 flyers and perhaps a few hundred flights at LDRS in 1992 (Black Rock) and 260 flyers and 500 flights in 1993 at LDRS-12 (Kansas). There were 600 flights at LDRS-13 in Kansas in 1994.
5. Registration fees for LDRS-14 were $17, plus $8 for a BLM fee. It was another $10 for launch at BALLS; *The Tripoli Report*, May 1995, 59–60.
6–9. Blatzheim and Sienkiewicz, "Large and Dangerous Rocket Ships XIV, Part 1." *High Power Rocketry*, January 1996, 29–31; also by Martha Sienkiewicz, 31–39.
10–14. Davidson, "Flight of the WAC." *High Power Rocketry*, January 1996, 40–41, 57–58.
14–15. Arakaki and Cello, "Large and Dangerous Rocket Ships XIV, Part 2." *High Power Rocketry*, February 1996, 25–40.
16. Reports at the time are conflicting as to whether Paul Robinson flew this rocket at LDRS or on Monday at BALLS. However, it seems likely this rocket was launched at BALLS.
17. Arakaki and Cello, "Large and Dangerous Rocket Ships XIV, Part 2." *High Power Rocketry*, February 1996, 25–40.
18. Binford interview with MBC.
19. Blatzheim, "Fireballs 005." *High Power Rocketry*, April 1996, 38–43, 66.
20–22. Clark, "Project: Thunderbolt." *High Power Rocketry* Magazine, April 1996, 44–49.
23. Blatzheim, "Fireballs 005." *High Power Rocketry*, April 1996, 38–43, 66.
24. *The Tripoli Report*, November 1995, 19.
25. By 1995, *Down Right Ignorant* was resting in the rafters of the Sackett Machine Shop in Orlando, Florida.
26–27. Cagle, "Monster Rockets: *Project 463* and *Stratospheric Dreams*." Point 39 Productions(Video 1996).
28. Blatzheim, "Fireballs 005." *High Power Rocketry*, April 1996; see Sackett,"*Project 463*." *High Power Rocketry*, February 1996, 49–53.
29–30. "Cagle, "Monster Rockets: *Project 463* and *Stratospheric Dreams*." Point 39 Productions(Video 1996).
31. Sackett, "Project 463." *High Power Rocketry*, February 1996.
32–37. Cagle, "Monster Rockets: *Project 463* and *Stratospheric Dreams*." Point 39 Productions(Video 1996).

Chapter 22 (1996: Orangeburg and the O*u*R Project)

1. Eiszner, "LDRS 15 Flight Facts." *The Tripoli Report*, October 1996, 27; Bruce Kelly would later report that the Orangeburg statistics were flawed because apparently, a number of flyers failed to fill out flight cards before their launches. Others have echoed this sentiment. Thus, the actual tally may have been well over 1,000 flights, making this LDRS the largest to date. However, the official number remains 995.

This was only 5 less than at Black Rock in 1995. See, generally, editorial by Bruce Kelly, *High Power Rocketry*, April 1997, 7.

2. It is possible that special events at Lucerne had similar or even larger spectator numbers. But no counts survive from that period either.

3–5. http://en.wikipedia.org/wiki/orangeburg,_South_Carolina, accessed on July 9, 2013.

6. http://www.supersod.com/about, accessed on July 9, 2013.

7–10. Eiszner online interview with MBC; Roberson, "Coastal Rocketry Association and Launch Complex 39." *High Power Rocketry*, March/April 1993, 64–67; also *High Power Rocketry*, July/August 1993, 4–5 (first listing of Tripoli South Carolina, with Larry Smith as Prefect. Tim Eiszner is listed as prefect of Tripoli Atlanta).

11. Daulphin, "ICBM IV Tripoli West Palm Invades South Carolina." *High Power Rocketry*, October 1995, 29–31.

12. The Orangeburg bid was the only one received for LDRS-15. It was approved at LDRS-14 in Gerlach in the late summer of 1995. See *The Tripoli Report*, November 1995, 10–13.

13–14. These Tripoli prefectures included (in addition to Tripoli South Carolina) Central Virginia, Connecticut, Atlanta, and Coastal Georgia; see, generally, *High Power Rocketry*, May 1996, 53.

15. Wilkins, "LDRS XV—Part 1: An Event to Remember." *High Power Rocketry* magazine, April 1997, 30–33.

16–27. Conn, "LDRS XV—Part 2: LDRS Fifteen An East Coast Event." *High Power Rocketry* magazine, April 1997, 66–69; Eiszner, "LDRS 15 Flight Facts." *The Tripoli Report*, October 1996, 27.

28. *The Tripoli Report*, October 1996, 3.

29. There are several definitions for the beginning of outer space. Most fall in the range of approximately 252,000 feet.

31–33. *High Power Rocketry*, July 1997, 29–33.

34. *High Power Rocketry*, July 1997, 36–48; some characterized the motor in the *OuR Project* as an R-27,000; see *The Tripoli Report*, October 1996, 28.

35. *High Power Rocketry*, July 1997, 29–33.

36. *High Power Rocketry*, July 1997, 2.

37–42. *The Tripoli Report*, October 1996, 28; see also Chuck Rogers analysis in *High Power Rocketry*, July 1997, 29–48; see also *The Tripoli Report*, October 1996, 28 (by Paul Robinson and Ken Mizoi).

Chapter 23 (The Rise of the Basement Bomber, Part 1)

1. Carter, "So You Want to Build Your Own Motors?" *Tripolitan*, February 1986, 15–16.

2. Carter and Webber were the president and vice president, respectively, of the modern Tripoli Rocketry Association formed in early 1985, prior to incorporation. They eventually made motors that were sold under the name ACS Reaction Labs.

3–10. *Tripoli Report*, May 1991, 7.

11–28. Kline interview with McNeely.

29–30. Cotriss, "Hybrids Fly at Fireballs 004." *High Power Rocketry*, February 1995.

31–32. *High Power Rocketry*, December 1994.

33. Miller, "Danville Eleven." *High Power Rocketry*, March 1995.
34. *High Power Rocketry*, May 1995.
35. *Tripoli Report*, May 1995.

Chapter 24 (The Rise of the Basement Bomber, Part 2)

1. *Tripoli Report*, November 1995; approximately 10 percent of the membership replied to the survey.
2–6. *Tripoli Report*, May 1995.
7–8. *Tripoli Report*, May 1995.
9–11. *Tripoli Report*, March 1996; the prefectures were in South Carolina, Nevada, Kansas, and Tennessee.
12. *Tripoli Report*, October 1996.
13–15. *Tripoli Report*, June 1997.
16. *High Power Rocketry*, March 1985.
17. *High Power Rocketry*, August 1985.
18. See, for example, *High Power Rocketry*, October 1995, 9, 41, 46; see also *High Power Rocketry*, July 1996.
19. *High Power Rocketry*, January 1997.
20–34. Mitchell interview with MBC; the company was Firefox; Mitchell credits Thompson for coming up with the name *Thunderflame*.
35. See, for example, *High Power Rocketry*, March 1998.
36–37. Forward written by Sonny Thompson to *Thunderflame* course (1997 edition).
38. *Rockets Magazine*, April 2012.

Chapter 25 (Return to Colorado)

1–2. Cochrane, "LDRS '97 XVI: Thunder in the Rockies." *High Power Rocketry*, May 1998.
3. Jeff Taylor would become the founder of *Loki* in the 2000s.
4–7. Cochrane, "LDRS '97 XVI: Thunder in the Rockies." *High Power Rocketry*, May 1998; see also Lee,"Nebraska Heat: Anatomy of a Group Project." *High Power Rocketry*, May 1998.
8. Daley, "The Making of *Return to Sender*." *High Power Rocketry*, May 1998.
9–12. Cochrane, "LDRS '97 XVI: Thunder in the Rockies." *High Power Rocketry*, May 1998; there were 15 Ls and 3 Ns at LDRS-16. There were 71 K-powered flights and 115 J flights. H- and I-powered rockets accounted for nearly 300 launches.
13. Miller, "A Tribute to G. Harry Stine." *Sport Rocketry*, January/February 1998.
14. Stine, "The Roots of Model Rocketry." *Model Rocketeer*, September 1977.
15. Stine, "World's Safest Business." *Mechanix Illustrated*, February 1957.
14. Stine, "Shoot Your Own Rockets." *Mechanix Illustrated*, October 1957.
15. Stine, "The Do-It-Yourself Rocketeers." *American Modeler*, April 1958.
15. Stine, "The Roots of Model Rocketry." *Model Rocketeer*, September 1977.
16. Stine, "Orville Carlisle's Model Rockets." *Model Rocketeer*, May 1977.

17–18. Stine, "The Beginning of the Model Rocket Industry." *Model Rocketry*, November 1977; Craddock, "The Last Interview," [of G. Harry Stine]. *Launch Magazine*, May/June 2007.
19. Stine, *Earth Satellites and the Race for Space Superiority*.
20. Stine, "The Birth of the MMA/NAR." *Model Rocketry*, December 1977 (article contains the names of the first twenty NAR members, including Werner Von Braun and author Robert Heinlein).
21. Stine, "Special Report from NARAM-1." *Model Rocketeer*, August 1977.
22. Craddock, "The Last Interview," [of G. Harry Stine]. *Launch Magazine*, May/June 2007.
23. Beach, "Owing Harry." *Sport Rocketry*, January/February 1998.
24–25. Craddock, "The Last Interview," [of G. Harry Stine]. *Launch Magazine*, May/June 2007.
26. Miller, "A Tribute to G. Harry Stine." *Sport Rocketry*, January/February 1998.
27. Kelly, "G. Harry Stine." *Tripoli Report*, June 1998, 22.
28. See, generally, *Tripolitan*, June 1986, 7.
29. Kelly, "G. Harry Stine." *Tripoli Report*, June 1998, 22.
30–31. https://en.wikipedia.org/wiki/Bonneville_Salt_Flats, accessed on December 26, 2016.
32. https://en.wikipedia.org/wiki/Blue_Flame_(automobile), accessed on December 27, 2016.
33. As of 2018, Hellfire was among the longest-running annual events in all of high-power rocketry. It has been hosted since 1995. The only other event that comes close in terms of longevity is ROCstock at Lucerne, which began a year later.
34. LDRS-17: Thunder on the Flats," by Earl Cagle (VHS Video) (Point 39 Productions 1999).
35. Review of actual flight cards from LDRS-17 (Neal Baker collection).
36. McGough online interview with MBC.
37–50. "LDRS-17: Thunder on the Flats." Point 39 Productions (1999); also, this does not include the Canadian LDRS in 2005; Kinney, "LDRS 17: Focus on the Flats." *High Power Rocketry*, March 1999, 27–42.
51. The vote was on March 11, 1998, *Tripoli Report*, June 1998, 22; previously, experimental and commercial launches were kept separate, and Tripoli Research Rule 19 also prohibited experimental flyers from going above their commercial certification levels. The board did not articulate the rationale for their vote to make an exception for the rules at LDRS-17.
52–55. "LDRS-17: Thunder on the Flats." Point 39 Productions (1999).

Chapter 26 (The Showman)

1. Michaelson, "The Joe Boxer Launches." *High Power Rocketry*, August 1997, 14–17.
2–4. Michaelson interview with McNeely.
5. Michaelson, *Rocketman: My Rocket-Propelled Life and High-Octane Creations*.
6–7. http://en.wikipedia.org/wiki/User:Ky_Michaelson, accessed on June 24, 2013.
8–9. Michaelson interview with McNeely.
10. See, for example, *High Power Rocketry*, June 1995, 31.

11–12. See, for example, *High Power Rocketry*, December 1995, 8–9; the *Rocketman* line also included hats and shirts and other rocketry-inspired apparel.
13–16. Michaelson, J., "My First Black Rock Experience." *High Power Rocketry*, April 1996, 42–43.
17–22. Michaelson, "The Joe Boxer Launches." *High Power Rocketry*, August 1997, 14–17.
23–25. Michaelson, "The Rockets of October Sky." *High Power Rocketry*, February 1999, 33–34.
26–27. *Tripoli Report*, June 1998, 9–10.
28. The web page for LDRS-19 was www.ldrs99.0rg; LDRS-17 at Bonneville was likely the first LDRS to have its own Internet web pages. These pages were originally accessible via the UROC web site at www.uroc.org.
29. *Tripoli Report*, December 1999, 14.
30–32. "LDRS XVIII." Point 39 Productions (2000); Gerrard, *"My Mind's Eye." High Power Rocketry*, December 1999, 28–36; Nowaczyk and Kinney, "LDRS—Argonia Kansas." Kinney, *High Power Rocketry*, November 1999, 31–43.
33. The reported number was 447. *Tripoli Report*, December 1999, 14 (statistics compiled by Chuck Mies).
34. Nowaczyk would launch a rocket to nearly 100,000 feet at Black Rock in 2006; see chapter 43.
35–36. "LDRS XVIII." Point 39 Productions (2000); Gerrard, *"My Mind's Eye." High Power Rocketry*, December 1999; Nowaczyk and Kinney, "LDRS—Argonia Kansas." Kinney, *High Power Rocketry*, November 1999.
37–38. *Tripoli Report*, December 1999, 14.
39–40. "LDRS XVIII." Point 39 Productions (2000); Gerrard, *"My Mind's Eye." High Power Rocketry*, December 1999; Nowaczyk and Kinney, "LDRS—Argonia Kansas." Kinney, *High Power Rocketry*, November 1999.
40. Birchett, "Tripoli Oklahoma's *V-2* Project." *High Power Rocketry*, February 1999.
41–43. "LDRS XVIII." Point 39 Productions (2000); Gerrard, *"My Mind's Eye." High Power Rocketry*, December 1999; Nowaczyk and Kinney, "LDRS—Argonia Kansas." Kinney, *High Power Rocketry*, November 1999.
44–45. Lee, "LDRS Project: *Mercury Redstone*—Overview and Construction (Part 1)." *High Power Rocketry*, December 1999, 38–46.
46–47. Ritz, "LDRS Project: *Mercury Redstone*—Glassmeister's Delight (Part 2)." *High Power Rocketry*, December 1999.
48. Lee, LDRS Project: *Mercury Redstone*—Overview and Construction (Part 1)" by Bruce Lee. *High Power Rocketry*, December 1999, 38–46.
49. Davis, "LDRS Project: *Mercury Redstone*—Detailing and Finishing (Part 3)." *High Power Rocketry*, December 1999, 50–51. Lee, "LDRS Project: *MercuryRedstone*—Overview and Construction (Part 1)." *High Power Rocketry*, December 1999.
52–55. Burney, "LDRS Project: *Mercury Redstone*—Final Countdown (Part 4)." *High Power Rocketry* Magazine, December 1999, 54–60.
56. "LDRS XVIII." Point 39 Productions (2000).
57–59. Burney, "LDRS Project: *Mercury Redstone*—Final Countdown (Part 4)." *High Power Rocketry* Magazine, December 1999.

Chapter 27 (The Biggest LDRS Ever)

1. The count according to *High Power Rocketry* was 1,379. See, generally, Lee, "LDRS Chronicles." *High Power Rocketry*, January 2001.
2. Schoner, "View from Down Range." *High Power Rocketry*, January 2001, 35.
3. "LDRS 19." Point 39 Productions(2000); see also "LDRS Chronicles." *High Power Rocketry*, January 2001.
4. O'Sullivan, "LDRS-19." *Maryland Tripoli Report*, July/August 2000, 2–3.
5–15. "LDRS 19." Point 39 Productions(2000); see also "LDRS Chronicles." *High Power Rocketry*, January 2001; Kinney,"LDRS-19." *Extreme Rocketry*, September/October 2000, 30–35.
16. At LDRS-29 in Lucerne in 2010, there were 1,349 flights, making it the second largest flight total. Some have suggested that the LDRS in Lucerne in 2001 also had similar numbers. But there are no surviving statistics, and the fact that the event shut down early almost every day because of high wind makes it unlikely one of the top launches in terms of total rockets flown. It may have seen the largest single-day spectator crowd, though, on Saturday.
17. O'Sullivan, "LDRS-19." *Maryland Tripoli Report*, July/August 2000.
18–20. McGilvray, "MDRA at Small Balls." July/August 2000, 6; see also *High Power Rocketry*, February 2001.
21–22. Kinney, "Inferno at Whitakers: Small Balls." *High Power Rocketry*, February 2001.
23. McGilvray, "MDRA at Small Balls." July/August 2000.

Chapter 28 (Return to Lucerne)

1. See, generally, Kelly, "Words from the President." *Tripoli Report*, December 2000.
2. *Tripoli Report*, December 2000; https://groups.google.com/forum/#!topic/rec.models.rockets/ZQfOf1DB1GQ, accessed on December 27, 2016 (LDRS-20 press release).
3–4. See *High Power Rocketry*, August 2002; see also Alcocer, "LDRS XX." *Extreme Rocketry*, September–October 2001; see also http://www.ahpra.org/ldrs20.htm, accessed on December 27, 2016 (AHPRA photo essay of LDRS-20); see also http://david.tdkpropulsion.com/launch.php?year=2001&event=LDRS+20, accessed on December 27, 2016.
5. Good, "LDRS-20." http://members.tripod.com/Tripoli_Rocketry_PGH/ldrs20.html, accessed on December 20, 2016.
6. See *High Power Rocketry*, August 2002; see also Alcocer, "LDRS XX." *Extreme Rocketry*, September–October 2001.
7–8. "The Gates Brothers Hit LDRS XX," by Dirk and Erik Gates as told to Nadine Kinney, *High Power Rocketry*, August 2002; see also Alcocer, "LDRS XX." *Extreme Rocketry*, September–October 2001; also, "LDRS 20." Point 39 Productions (2001).
9. McConaghy, "LDRS 20." http://www.lunar.org/docs/LUNARclips/v8/v8n4/LDRS20.shtml, accessed on December 27, 2016.

10–13. "The Gates Brothers Hit LDRS XX," by Dirk and Erik Gates as told to Nadine Kinney, *High Power Rocketry*, August 2002; also, Alcocer, "LDRS XX." *Extreme Rocketry*, September–October 2001.
14–18. The first stage had six K1050s and twenty-four G38 motors. The second stage had a central J motor and eight H124s. The third and final stage carried four H124 motors; see Alcocer, "LDRS XX." *Extreme Rocketry*, September–October 2001; "The Gates Brothers Hit LDRS XX," by Dirk and Erik Gates as told to Nadine Kinney, *High Power Rocketry*, August 2002.
19. The seven-motor rocket that careened all over the sky was launched by the author.
20. "The Gates Brothers Hit LDRS XX," by Dirk and Erik Gates as told to Nadine Kinney, *High Power Rocketry*, August 2002, 24. How this number of spectators was counted or determined is unknown.
21. *Tripoli Report*, April 2002, 19.
22. "The Gates Brothers Hit LDRS XX," by Dirk and Erik Gates as told to Nadine Kinney, *High Power Rocketry*, August 2002, see also Alcocer, "LDRS XX," by Tony Alcocer, *Extreme Rocketry*, September–October 2001; "LDRS-20." Point 39 Productions (2001).

Chapter 29 (Fuse Is Lit)

1. http://www.federallawenforcement.org/atf/what-is-atf/, accessed on December 24, 2016.
2. See, for example, *The Tripoli Report*, June 1994, 2–3 ("Joint Communiqué of the High Power Rocket Manufacturers and Dealers Association and the Tripoli Rocketry Association to the High-Power Community").
3. See, for example, 1995 Edition of NFPA 1127, 2; see also 1987 Edition of NFPA 1122, 2 and 1982 Edition of NFPA 1122, 2.
4–6. *Tripoli Report*, June 1994, 2–3.
7. 18 U.S.C. chapter 40.
8. Section 555.11.
9–11. *Tripoli Report*, June 1994, 2–3; see also Plaintiff's Motion for Summary Judgment filed on August 30, 2002, and supporting declarations and exhibits.
12. *Tripoli Report*, June 1994, 3.
13. This is sometimes also referred in federal and other rules as a "User of Low Explosives Permit."
14–17. *Tripoli Report*, June 1994, 5–6.
18. *Tripoli Report*, November 1994, 6.
19. *Tripoli Report*, May 1995, 29.
20–22. *Tripoli Report*, January 1996, 7–10.
23. *Tripoli Report*, May 1995, 5.
24. See NFPA 1129 and TIA (1995).
25. *Tripoli Report*, October 1996, 24–25.
26–27. *Tripoli Report*, January 1997, 1–4.
28–30. Kosdon interview with McNeely.
31. *Tripoli Report*, June 1998, 16.
32–33. *Tripoli Report*, June 1999, 4.
34. *Tripoli Report*, April 2000, 11; see also *Tripoli Report*, December 1999, 7.

35. *Tripoli Report*, July 1999, 7–8.
36–37. *Tripoli Report*, December 1999, 6–7.

BOOK III

Chapter 30 (Texas)

1–10. Gordzelik interview with MBC.
11. https://en.wikipedia.org/wiki/Rick_Husband, accessed on March 3, 2017.
12–24. Gordzelik interview with MBC.
25–26. See, generally, Gary Rosenfield's open letter to the rocketry community, printed in *Extreme Rocketry*, January–February 2002; see also verified federal court complaint in *AeroTech, Inc. v. Clark County et al.*, CV-S-02-0384-PMP-PML, filed in the United States District Court of Nevada, March 20, 2002, paragraphs 18–34.
27. http://www.8Newsnow.com/story/509511/fire-at-model-rocket-company-burns-for-hours, accessed on April 24, 2014.
28. See, generally, "AeroTech's lawsuit against fire department dismissed." *Las Vegas Sun*, December 9, 2003, accessed on April 24, 2014, at http://www.lasvegassun.com/news.2003.dec/09/AeroTechs-lawsuit-against-fire-department-dismisse./.
29. See, for example, *Extreme Rocketry*, January–February 2002.
30. http://www.lasvegassun.com/news/2003/feb/10/AeroTech-facing-no-criminal-charges/, accessed on April 24, 2014; see also http://www.lasvegassun.com/news/2003/dec/09/AeroTech-lawsuit-against-fire-department-dismissed/, accessed on April 25, 2014; the dismissal of AeroTech's complaint was not on the merits; rather, AeroTech lost its attorneys in a fee dispute that ultimately led to the dismissal of the case.
31. See Declaration of Gary Rosenfield in AeroTech Incorporated's Opposition [regarding counsel's motion to withdraw] filed on July 7, 2003, in *AeroTech, Inc. v. Clark County et al.*, CV-S-02-0384-PMP-PML.
32–40. Gordzelik, "LDRS 21 Texas Style." *High Power Rocketry*, December 2002, 18–59; McNeely, "LDRS 21: Amarillo, TX, July 11–16." *Extreme Rocketry*, September–October 2002, 16–30.
41. Gurstelle was the author of, among other things, *Backyard Ballistics*, published in 2001; Quentin Wilson was one of the characters in Homer Hickam's book *Rocket Boys* (movie: *October Sky*).
42–49. Gordzelik, "LDRS 21 Texas Style." *High Power Rocketry*, December 2002, 18–59; McNeely, "LDRS 21: Amarillo, TX, July 11–16." *Extreme Rocketry*, September–October 2002, 16–30.
50–51. See, generally, http://www.rietmann.net/gbrocketry.com/motor_usage.htm, accessed on March 13, 2017; see also GBR website, accessed on November 2013 (original web page now replaced).
52. Gordzelik, "Mini BALLS 2002." *High Power Rocketry*, January/February 2003, 30–32.
53. GBR website, accessed on November 2013.

54–55. GBR website, accessed on November 2013; see also Gordzelik, "Mini BALLS 2002." *High Power Rocketry*, January/February 2003, 31–32; Gordzelik, "LDRS 21 Texas Style." *High Power Rocketry*, December 2002, 59.

Chapter 31 (Passing the Baton)

1. At the time, there were more than 1,350 Tripoli numbers assigned to date; 835 were active.
2–3. *Tripoli Report*, April 1994, 49.
4. *Tripoli Report*, January 1994, 15.
5. *Tripoli Report*, April 2002, 5; the total number of ballots mailed was 3,636; ballots received were 1,125; 31 percent turnout per Guy Soucy; the popular vote was eliminated on motion by Kelly in 2002 at LDRS; see *Tripoli Report*, December 2002, 16.
6. See *High Power Rocketry*, July 2000, 5.
7. See *High Power Rocketry*, August 2002.
8–9. See *High Power Rocketry*, July 2002.
10. *Tripoli Report*, December 2002, 6; McCreary received 460 votes to Kelly's 439 votes.
11–15. Embry interview with McNeely; see also *The Tripoli Report*, June 1993, 4–5.
16. Confirmation card from Dick Embry (Dick Embry collection).
17. Embry interview with McNeely; see also *The Tripoli Report*, May 2005, 3.
18. Embry online interview with MBC.
19. See, for example, *The Tripoli Report*, May 1996, 3; *The Tripoli Report*, June 1997, 13–14; *The Tripoli Report*, April 2000; *The Tripoli Report*, May 2005, 3.
20–25. Embry online interview with MBC; also, Embry interview with McNeely.
26. *The Tripoli Report*, December 2002, 4–6.

Chapter 32 (The Rocket Challenge)

1–2. http://en.wikipedia.org/wiki/Scrapheap_Challenge; accessed on November 8, 2013.
3. Michaelson was a judge, and Lee was a "team expert."
4. Lee interview with McNeely.
5. See, generally, Lee, "Junkyard Wars." *High Power Rocketry*, November 2003, 38–44.
6. http://en.wikipedia.org/wiki/BattleBots; accessed on November 8, 2013.
7. Good, "LDRS-20." www.memberstripod, accessed on August 13, 2013.
7. *Tripoli Report*, April 2002, 22.
8–13. *Tripoli Report*, April 2002, 22; also, Good interview with MBC.
14. McNeely, "LDRS 22." *Extreme Rocketry*, September–October 2003, 16–27.
15–18. McGilvray, "*Udder Madness.*" *High Power Rocketry*, June 2004.
19–28. Uroda interview with McNeely.
29–36. Uroda online interview with MBC.
37–45. *The Rocket Challenge*, the *Discovery Channel*, November 2003, accessed on YouTube in 2013.
46. The flyer was Richard Hagensick.
47. Gordzelik, "LDRS-22 [Part One]." *High Power Rocketry*, May 2004, 12–30.

48. Stroud, *"Aurora, Goddess of the Sky."* *Extreme Rocketry*, December 2003, 18–23.
48–51. *The Rocket Challenge*, the *Discovery Channel*, November 2003, accessed on YouTube in 2013.
52. *The Tripoli Report*, January 2004, 3.

Chapter 33 (The Space Shot)

1. The precise altitude of the 1996 Michaelson rocket is unclear. One article written states it was optically tracked to 46,500 feet. See *High Power Rocketry*, July 1997, 64–65; another article states it reached 50,000 feet but does not mention how the altitude was determined. See *High Power Rocketry*, August 1997, 16–17; in yet another reference, the rocket was said to have reached 60,000 feet—with no details provided. See *Extreme Rocketry*, April 2001, 33. The quote is from the 1997 *High Power Rocketry* article on BALLS.
2–5. Michaelson, *Rocketman: My Rocket-Propelled Life and High-Octane Creations*, 149.
6. Michaelson interview with McNeely; for the story on how Eric Knight became a part of the CXST team, see Knight, *The NEW Race to Space*.
7–10. Larson, *"Space Shot 2000."* *High Power Rocketry*, July 2002, 11; also, Michaelson, *Rocketman: My Rocket-Propelled Life and High-Octane Creations*, 160–161.
11. Michaelson interview with McNeely.
12. Knight, *The NEW Race to Space*, 23–34.
13. For a complete list of the members at that time, see Knight, *The NEW Race to Space*, 75.
14. http://www.the-rocketman.com/CSXT/news/n6_01_01_major_sponsor.htm, accessed on March 29, 2014.
15. The name of the rocket came from one of its major sponsors, Primera Technology, Inc.; see Knight at 46.
16–18. Knight, *The NEW Race to Space*, 47, 76; Larson, *"Space Shot 2000."* *High Power Rocketry*, July 2002; Knight said the motor was an S 20,000; Larson reported it as an S 20,000; various reports put the rocket's diameter and the *Space Shot 2000* rocket's diameter at between 8.6 and 9 inches; in his article in *High Power Rocketry*, Larsen put the weight of the rocket at 510 pounds, Knight at 551.
19–23. Michaelson, *Rocketman: My Rocket-Propelled Life and High-Octane Creations*, 181–189; this rocket was originally set to fly in the fall of 2001. That launch was scrubbed after 9/11. It was next set to fly in the early summer of 2002. That launch too had to be scrubbed—this time because of weather and also the passing of Ky Michaelson's mother in June.
23–25. Michaelson at page 189–190; Knight, *The NEW Race to Space*. 122–124, citing an e-mail by Steven McMacken.
26. Knight, *The NEW Race to Space*, 114. Knight estimated that June launch attempt cost the team $130,000 and that another $30,000-40,000 would be needed to make the September flight. Other team members believe these numbers were much lower—under six figures for the entire project.
27–29. Knight, *The NEW Race to Space*, 124, 136–137,149-150; see also Clary, *Rocket Man*,115.
30–31. Deville, "Putting the 'S' in CXST." *Rockets Magazine*, April 2012.
32. Knight, *The NEW Race to Space*, 144.

33–40. Deville, "Putting the 'S' in CXST." *Rockets Magazine*, April 2012. Also, Michaelson, *Rocketman: My Rocket-Propelled Life and High-Octane Creations*, 200.
41. Rogers interview with MBC. See also Rogers article on erosive burning in *High Power Rocketry*, January 2005.
42. Rogers interview with MBC.
43. Deville, "Putting the 'S' in CXST." *Rockets Magazine*, April 2012, 16.
44–49. Michaelson, *Rocketman: My Rocket-Propelled Life and High-Octane Creations*; Deville, "Putting the 'S' in CXST." *Rockets Magazine*, April 2012.
50–51. Knight, *The NEW Race to Space*, 157; Deville, "Putting the 'S' in CXST." *Rockets Magazine*, April 2012.
52. Michaelson, *Rocketman: My Rocket-Propelled Life and High-Octane Creations*, 208.
53–55. Deville, "Putting the 'S' in CXST." *Rockets Magazine*, April 2012. The propellant weighed 435 pounds out of the 607; the motor case made up the rest of the weight.
56–58. Michaelson, *Rocketman: My Rocket-Propelled Life and High-Octane Creations*, 217.
59. Knight, *The NEW Race to Space*, 195.
60–61. As quoted in Michaelson, 218.
62–64. Knight, *The NEW Race to Space*, 200–214. The *GoFast* payload bay was found at 40.945260/-119449810 (Knight at 199).
65. In a subsequent press released by the team, it was stated that the booster's parachutes ripped off the vehicle at 50,000 feet, causing the 211-pound booster to impact the ground nose first at more than 500 miles per hour. See Knight, *The NEW Race to Space*, 214.
67. The location was 40.911037/-119.515079 at 6,404 feet in altitude 4.14 miles from the payload bay. See Knight, *The NEW Race to Space*, 215.
68. See statement released by the team in March 2005 and quoted in Knight at 218–219 and Michaelson at 225–225.
69–70. Rogers interview with MBC.
71. http://www.rocketryforum.com/showthread.php?125609-2004-CSXT-GoFast-flight-Data, accessed on March 7, 2017.

Chapter 34 (Rocketry v. the U.S. Government, *Part One*, the Lawsuit)

1–4. *Tripoli Report*, April 2000, 4; see also the Complaint, 4–10.
5. See, generally, *Tripoli Report*, December 2000, 7.
6. The limitations argument was likely anticipated by defense counsel when they filed the action and was probably the reason why the TRA/NAR complaint tried to tie their claim to ATF's republication of the explosives list in 1999 rather than the original date of 1971.
7. Memorandum of Decision, filed on June 24, 2002, by Judge Walton on Defendant's Motion to Dismiss.
8–9. See, generally, *Tripoli Report*, December 2000, 7; also, Bundick interview by McNeely.
10. Leonnig, "Libby Jurist's Career Built on Toughness," *The Washington Post*, June 5, 2007.
11–15. Transcript of Proceedings before the Hon. Reggie B. Walton, *Tripoli Rocketry Assoc. et al. v. Bureau of Alcohol, Tobacco and Firearms*, C.A. No. 00-272, April 30, 2002.

16–20. See Memorandum Opinion, *Tripoli Rocketry Assoc. et al. v. Bureau of Alcohol, Tobacco and Firearms*, C.A. No. 00-272, June 24, 2002, 5–22.
21–22. Plaintiff's Memorandum of Points and Authorities in Support of Motion for Summary Judgment, August 30, 2002, 9–15.

Chapter 35 (Rocketry v. the U.S. Government, *Part Two*, SB-724 and Other Setbacks)

1–3. Senator Enzi speech on April 1, 2003, on Senate floor.
4. Good interview with MBC.
5. See, for example, *High Power Rocketry*, November 1999, 47.
6. Senator Enzi speech on April 1, 2003, on Senate floor.
7–8. Good interview with MBC.
9–10. Letter from William E. Moschella, Assistant Attorney General, U.S. Department of Justice, Office of Legislative Affairs, to Chairman of the Senate Judiciary Committee, sent on June 10, 2003.
11–17. Monoson, "Hobby Rocket Plan Under Fire." *Star Tribune*, July 30, 2003; Lichtblau, "Threats and Responses: New Regulations; Rocket Bill Stirs Debate on Potential for Terror." *New York Times*, July 30, 2003.
18–19. Good interview with MBC.
20–21. See, generally, *High Power Rocketry*, April 2004, 12–14.
22. *The Tripoli Report*, May 2004, 4.
23. For more information on *Project 463*, see chapter 21.
24–25. Wright interview with McNeely. The booster section weighed 444 pounds; the payload bay weighed 425 pounds; the motors weighed 232.6 pounds; the nose cone weighed 155 pounds and miscellaneous components added another 110.2 pounds, bringing the total weight of *The Liberty Project* to 1,366.8 pounds. *High Power Rocketry*, September 2004, 34.
26–32. McGilvray, "Give Me Liberty—or Give Me a Crane and Some Scaffolding: The Liberty Project." *High Power Rocketry*, September 2004, 30–44.

Chapter 36 (Geneseo)

1. There were two proposals submitted this year, one from Geneseo and the other for Las Vegas, Nevada. LDRS would eventually go to Las Vegas in 2007. See *Tripoli Report*, January 2004, 8.
2. Canepa, "Teamwork at Launches." *Extreme Rocketry*, August 2004, 46.
3–8. Canepa, "LDRS 23 2004." *Extreme Rocketry*, August 2004, 20–33; see also "LDRS-23." DVD by Chuck Rudy.
9–10. Clark, "AHPRA Bowling Ball Loft at LDRS XIX." *High Power Rocketry*, January 2005, 30–35.
11–17. Canepa, "LDRS 23 2004." *Extreme Rocketry*, August 2004, 20–33.
18–19. *The Tripoli Report*, September 2004, 3–4.
20. Canepa, "LDRS 23 2004." *Extreme Rocketry*, August 2004, 20–33.
21. "LDRS-23." DVD by Chuck Rudy.
22–26. Canepa, "The River Ate My Rocket." *Extreme Rocketry*, March 2006, 22–23.

27–41. Good interview by MBC; Good interview by McNeely.
42. Good, "Words from the President." *The Tripoli Report*, September 2004, 3.
43–48. Good interview by MBC; Good interview by McNeely.

Chapter 37 (The International Launch)

1–6. Canadian Association of Rocketry website, history section, accessed on October 29, 2013; see also Illerbrun "Sullivan Lake: High Power Rocket One." *High Power Rocketry*, April 1994, 67–71; for a follow story on Sullivan Lake III, see story by Mr. Illerbrun in *High Power Rocketry*, July 1996, 43–47.
7. For prefecture listing in 1998, see *High Power Rocketry*, August 1998, 6.
8–9. Baines online interview with MBC.
10. Lethbridge rests at just over 3,000 feet and allegedly has more sunny days than any other city in Canada; its dry climate provides mild weather compared with other parts of the country.
11. Baines online interview with MBC.
12–13. CAR website; NOTAM is short for "notice to airmen."
14. See, for example, Baines, "ROC Lake Six: The Future of LDRS." *High Power Rocketry*, December 2003, 10–15.
15. Baines online interview with MBC.
16. Baines and Stephens, "Editorial: Important LDRS 25 Info." *High Power Rocketry*, January 2005, 14, 38–39.
17. Other team members who assisted in running LDRS-24 were Barry Mackadenski, Brad Derzaph, Simon Stirling, Bruce Aleman, Kyle Baines, and Brad Baines.
18. Baines and Stephens, "Editorial: Important LDRS 25 Info." *High Power Rocketry*, January 2005.
19–21. Clapp, "LDRS 24." *Extreme Rocketry*, November 2005, 20–23.
22. http://www.vernk.net, accessed on October 30, 2013.
23–24. Clapp, "LDRS 24." *Extreme Rocketry*, November 2005.
25. *Tripoli Report*, September 2005, 9.
26–27. https://en.wikipedia.org/wiki/Robert_Thirsk, accessed on March 11, 2017.
28. Dunseith online interview with MBC.
29. Clapp, "LDRS 24." *Extreme Rocketry*, November 2005.
30. http://www.vernk.net, accessed on October 29, 2013.
31. No flight cards survived the event. The estimated flight was probably in the range of 750 based on the number of registered flyers.
32 http://www.vernk.net, accessed on October 30, 2013.
33. *Tripoli Report*, September 2005, 3.
34–35. *Tripoli Report*, December 2002, 12.
36–45. McNeely interview with Trojanowski.
46. See, for example, *High Power Rocketry*, February 1999.
47–51. McNeely interview with Trojanowski.
52. *Extreme Rocketry*, September/October 2002, 38.
53. *High Power Rocketry*, October/November 2002, 4.
54–55. *High Power Rocketry*, May 2003, 10.
56. *Tripoli Report*, May 2004, 12.
57. Good interview with MBC.

58–60. *Tripoli Report*, September 2005.
61. http://www.magazinedeathpool.com/photos/museum_of_dead_magaz/watermagazine.html, accessed on April 3, 2014, for a list of magazines and periodicals that have gone under in the last fifteen years.
62. McGilvray, "The Final Countdown—*Rockets Magazine* Last Issue," Fall 2018.

Chapter 38 (Rocketry vs. the U.S. Government, *Part III*, the Appeal)

1. However, according to Attorney Lyons, ATF believed that the judge's ruling on PADs applied only to rocket motors with less than 62.5 grams of propellant—model rocketry only.
2. Letter from Joseph Egan to Jane Lyons, May 6, 2004.
3. Message to Tripoli prefects from Pres. Ken Good, October 2004.
4. Joint Statement, December 22, 2004.
5. Appellate court docket, accessed by PACER on October 17, 2013.
6–10. See, generally, Tripoli Rocketry Association, Inc. and National Association of Rocketry vs. Bureau of Alcohol, Tobacco, Firearms and Explosives, No. 04-5433 (D.C. Cir.) Appellants' Opening Brief, filed on August 5, 2005.
11. Appellant's Reply Brief, September 23, 2005, 2–3.
12. https://en.wikipedia.org/wiki/Harry_T._Edwards, accessed on March 2, 2017.
13. Slavin, "A Judge of Character: Although He's Blind, David Tatel Skis, Runs, and Climbs Mountains. By Summer's End, He May Be a Top Jurist, Too." *Los Angeles Times*, July 28, 1994, accessed on October 22, 2013.
14. https://en.wikipedia.org/wiki/Merrick_Garland, accessed on March 3, 2017.
15. Joint Statement, January 18, 2006.
16–20. *Tripoli Rocketry Association, Inc. and National Association of Rocketry vs. Bureau of Alcohol, Tobacco, Firearms and Explosives*, No. 04-5433 (D.C. Cir. Feb. 10, 2006).

Chapter 39 (The March of Time)

1–2. *Rockets Magazine*, September 2006.
3. There was one significant point of bad news over the weekend. The rocketry range used by POTROCS for LDRS and their customary launches was 7 square miles of land owned by rancher Wylie A. Byrd, who generously provided POTROCS with not only his land but also electricity and access to well water—all at no charge. Sadly, Mr. Byrd passed away unexpectedly on July 1, 2006.
4. The *Event Horizon* team included Art Hoag, Troy Hummel, and Joe Cowan.
5. *Extreme Rocketry*, October 2006, 28.
6–9. *Rockets Magazine*, September 2006.
10. Rothman and Trojanowski, "LDRS 25: Tripoli's 2006 National Event." *Extreme Rocketry*, October 2006, 19–29, 28; see also *Rockets Magazine* DVD of LRS-25 by Liberty Launch Systems.
11–14. *Tripoli Report*, December 2005, 3.
15. See, generally, Gordzelik, "Perspective from Tripoli Rocketry Association." *Rockets Magazine*, March/April 2006, 7.
16. *Rockets Magazine*, March/April 2006, 4.

17. *Rockets* Magazine, July/August 2006, 4.
18. http://en.wikipedia.org/wiki/Jean,_Nevada.
19–23. McNeely and LaPanse, "LDRS 26 Launch Report." *Extreme Rocketry*, September–October 2007, 21; Catalano, "Prelude." *Rockets Magazine*, October 2007, 4.
24. http://www.weather.com/weather/wxclimatology/monthly/graph/89019.
25–28. *Rockets Magazine*, July/August 2006.
29. *Rockets Magazine*, October 2007, 32.
30. Lickteig, "Thanks for the Memories." *Rockets Magazine*, October 2008, 58–59.
31. "LDRS-27." *Rockets Magazine* DVD.
32–33. *Extreme Rocketry Magazine*, December 2008, 21; see also *Rockets Magazine*, October 2008, 41–42.
34. *Rockets Magazine*, October 2008, 32–33; see "LDRS-27." *Rocket Magazine* DVD.

Chapter 40 (Rocketry vs. the U.S. Government, *Part IV*, Friday the Thirteenth)

1–2. Plaintiff's Memorandum of Points and Authorities in Support of Motion for Summary Judgment on Count One, January 31, 2007, 5.
3. See *Tripoli Rocketry Assn., Inc. v. Bureau of Alcohol, Tobacco, Firearms & Explosives*, 437 F.3d 75, 97 (D.C. Cir. 2006).
4. Plaintiff's Reply to Defendant's Opposition to Motion for Summary Judgment, April 10, 2007, 1–2.
5–6. Affidavit of Dr. Terry McCreary, filed on January 1, 2007; signed on January 29, 2007.
7. Plaintiff's Reply to Defendant's Opposition to Motion for Summary Judgment, April 10, 2007, 12.
8–11. Affidavit of Gary Rosenfield, filed on April 10, 2007.
12. Libby's prison term was commuted by Pres. George W. Bush on July 2, 2007.
13–14. Good online interview with MBC.
15. Mobley, "Where in the World is Reginald Walton?" *Rocketry Planet*, September–October 2008, 46; reprinted from www.rocketryplanet.com.
16. Good online interview with MBC.
17. Tyson online interview with MBC.
18–19. Order of Judge Walton, filed on March 16, 2009, 2–3.

Chapter 41 (*Saturn V*)

1–22. Canepa, "One Man's Quest to Honor America's *Saturn V*." *Rockets Magazine*, February 2009, 40–47.
23. See, generally, *Rockets Magazine*, February 2009, page 41 (advertisement for *Saturn V* Launch.)
24. Canepa, "Steve Eves's Field of Dreams: The Launch of the *Saturn V*." *Rockets Magazine*, June 2009, 22–37.
25. The motor cases were made by Jeff Taylor and also MDRA member Tim Wilsey.
26. The motor sponsors of the *Saturn V* were Jeff and Anne Scott (P motor), Dave Weber, Erection Associates, Tim Wilsey, Howard and Ben Ullman, Peter and

Benjamin Abresch, Shepherd Stein, and Mark Canepa (N motors); *Rockets Magazine*, June 2009.
27. McGilvray, "By the Numbers: What it took to Launch Steve Eves' 1:10 Scale *Saturn V.*" *Rockets Magazine*, June 2009, 52–55.
28–40. Canepa, "Steve Eves's Field of Dreams: The Launch of the *Saturn V.*" *Rockets Magazine*, June 2009, 22–37.
41–43. Koloms, "LRS-28: From the Inside Looking Out." *Rockets Magazine*, April 2009, 26.
44. http://en.wikipedia.org/wiki/Torrey_Farms, accessed in 2013.
45–47. Online interview with the author of Mike Dutch, November 2013.
48. McGilvray, "LDRS-28." *Rockets Magazine*, August 2009, 6–7.
49. Terri Lamothe was also the wife of Tripoli board member Dennis LaMothe.
50–53. *Tripoli Report*, March 2006.
54. Koloms received 201 votes in 2006 (*Tripoli Report*, September 2006, 19); she received 234 votes in 2007 (*Tripoli Report*, March 2008, 6).
55–58. *Tripoli Report*, October 2009.
59–60. *Rockets Magazine*, December 2009, 58.
61–65. *Rockets Magazine* February 2010, 32.

Chapter 42 (LDRS Turns Thirty)

1. There were 1,379 flights at LDRS-19 in four days of flying at Orangeburg.
2. *Rockets Magazine*, August 2010; *Project Odyssey* was powered by an O 6900 and had a good flight and recovery from 16,777 feet; the *Talon* was destroyed during its ascent on P power; Lehr's *Dark Star* had a good flight and recovery as shown on the Science Channel.
3–5. "Large Dangerous Rocket Ships." DVD (2010) Discovery Communications LLC.
6–7. Author's personal observations of the event.
8. The Science Channel was present at LDRS-29, LDRS-30, and LDRS-31. The contract between Tripoli and the television show was not renewed for 2013, purportedly at the request of Tripoli.
9. As a result of the accident at the away cells, the final segment of the drag race was postponed until the following day.
10–11. See *Rockets Magazine*, June 2010, 49–51.
12. Ken Good editorial in *Tripoli Gerlach* newsletter.
13–17. See the complaint.
18. The dismissal with prejudice was filed on August 28, 2012.
19. The author has not had any confirmation or discussion with anyone in authority regarding the settlement amount. This is speculation based on the publicly disclosed injuries and the nature of the case.
20–22. See Good Editorial; see also NFPA 1127 Section 4.13.7, which provides, in pertinent part, that "[f]iring circuits shall not be armed with the rocket in other than a launching position."
23. "Code for High Power Rocketry—Tripoli Rocketry Association." July 31, 2012.
24. Tripoli Gerlach *Research Rocket News*, July 2013.
25–30. Interview with the author 2014.

31. Terry W. McCreary resume, attached to the affidavit of Terry W. McCreary filed in federal court case (Document 100), January 29, 2007.
32–42. McCreary interview with MBC; McCreary interview with McNeely; McCreary, "Spears 2." *High Power Rocketry*, November 1995, 35–39.
43. McCreary, "Does the Lawsuit Suit?" *Extreme Rocketry*, April 2002, 50.
44. McCreary interview with MBC.
45. Affidavit of Terry W. McCreary filed in federal court case (Document 100), dated January 29, 2007, 2–3.
46–49. McCreary interview with MBC; see also Darrell Mobley obituary on *Rocketry Planet*.
51. See discussion of the *OuR Project* in chapter 22.
52–54. Frank Kosdon obituary by Darrell Mobley on *Rocketry Planet*.
55–65. *Rockets Magazine*, February 2012.
66. Whitmore, Alan. "WELD 2002 Whitakers Experimental Launch Days." *Extreme Rocketry*, July 2002.

Chapter 43 (Let's Punch a Hole in the Sky)

1–2. *Rocket Newsletter*, December 1990, 2–5.
3. Fisher, Daniel. "Let's Punch a Hole in the Sky." *Forbes*. November 13, 2000.
4. This altitude was an estimate only. There were no onboard altimeters that were recovered and no live downlinks during flight. The altitude was based on a flight simulation.
5–6. Mailer, *Of a Fire on the Moon*.
7–8. Neufeld, *Von Braun*.
9. Zeppin, "BALLS 2002: And Sometimes This Happens." *High Power Rocketry*, March 2003.
10. Taylor, "BALLS 2003." *Extreme Rocketry*, December 2003, 28–34.
10. See, for example, McBurnett and Whitmore, "Rocket Autopsies Part One: Analysis of Motor Failures." *Rockets Magazine*, December 2010.
11. Mailer, *Of A Fire on the Moon*.
12–16. Rouse interview with McNeely; also, Canepa, "Chasing the N Record: Pursuing stratospheric Dreams." *Rocketry Planet*, 2010.
17. Brown, "A Cult of Backyard Rocketeers Keeps the Solid Fuel Burning." *New York Times*, October 14, 2006.
18. For a photo of *Max-Q*, see cover of *Extreme Rocketry*, March 2002.
19. Canepa, "Balls 2004." *Extreme Rocketry*, December 2004.
20. *High Power Rocketry*, September 2002.
21–22. *High Power Rocketry*, December 2004.
23. Harms interview with McNeely.
23–25. The rocket was owned by Gene Nowaczyk; see *Extreme Rocketry*, December 2005.
26–31. http://www.aeroconsystems.com/Gene_Nowaczyk_Balls2006/balls2006.html, accessed on June 15, 2014; also Extreme Rocketry, December 2005.
32. *Rockets Magazine*, December 2008, 32.
33–34. *Rockets Magazine*, February 2010, 51.

35–37. *Rockets Magazine*, February 2010, 47.
38–44. *Rockets Magazine*, April 2012.
45–57. Canepa, "Black Rock 2011: 100,000 Feet or Bust." *Rockets Magazine*, April 2012, 32–45.

Sources

Adept Product Catalog 1992.
Aeronaut. "1991 Launch/Event Calendar." September 1990.
AeroTech Consumer AerospaceCatalog 1988& 1992 & 1998–1999.
AeroTech Consumer Aerospace website. www.AeroTech.rocketry.com, accessed on 9-26-13.
AeroTech Consumer Aerospace News. "Remembering Ray Goodson 1952-2005." August 31, 2005.
Alcocer, Tony. "LDRS XX." *Extreme Rocketry,* Sept–Oct 2001.
Amateur Rocketry: A Delineation of the Problem, a Policy Statement and a Program for Action by the American Rocket Society." *American Rocket Society* 1959. (K.C. McGoffin Collection).
American Modeler. "Rocket College for Model Makers." August 1959.
American Spacemodeling. Journal of the National Association of Rocketry. 1984–1992.
———. An Interview with NAR Creator G. Harry Stine. January/February 1993.
———. NAR High Power Rocketry Safety Code (Interim). January/February 1992.
———. NFPA Action. April 1986.
———. NFPA Committee Meets in Phoenix. July/August 1992.
Anthony, Wayne. "LDRS XI & Fireballs II." *American Spacemodeling,* January/February 1993.
Arakaki, Jimmy. "Large and Dangerous Rocket Ships XIV, Part 2." *High Power Rocketry,* February 1996.
Baines, Max. "ROC Lake Six: The Future of LDRS." *High Power Rocketry,* December 2003.
———. "Roc Lake 5." *Earthrise* [Canadian Association of Rocketry] Volume 3, Number 1, 2002.
Baker, Neal. LDRS 17: Flight Cards and Registration Materials. (Neal Baker Collection).
Bangs, Gary L. "Letter to Tim Lehr at Wildman Hobbies re ATF Policies." April 13, 2007.
Barber, Bill. "Welcome to Lucerne: Eastern Rocketeer's First Impressions of Lucerne Test Range." *High Power Research,* Fall 1984.
Bartel, Scott. "Resume for Scott Bartel." *The Tripoli Report,* April 1994.
Baumfalk, John. "LDRS Update." *The Tripoli Report,* January 1993.
———. "A Weather Update on K.L.O.U.D. Busters." *The Tripoli Report,* January 1993.
———, and Sawyze, Allen. "Kountdown in Kansas." *High Power Rocketry,* July/August 1992.
Beach, Thomas. "Owing Harry." *Sport Rocketry,* January/February 1998.
Benson, Carl. "Direct Entry Ejection Timer." *Tripolitan,* August 1990.

Bierlein, Lawrence W. "Letter to Alan Roberts [Department of Transportation] [on behalf of AeroTech/ISP], December 29, 1992.
Binstock, Mort. "Showdown at LDRS." (LDRS-7) *Tripolitan,* October 1988.
Birchett, Dale. "Tripoli Oklahoma's *V-2* Project." *High Power Rocketry,* February 1999.
Blatzheim, Jason. "Large and Dangerous Rocket Ships XIV, Part 1." *High Power Rocketry,* January 1996.
_____. "Fireballs 005." *High Power Rocketry,* April 1996.
Blazanin, Tom. "NFPA Meeting." *Tripolitan,* December 1987.
_____. "Octoberfest 88." *Tripolitan,* December 1988.
_____. "NFPA Update." *Tripolitan,* April 1988.
_____. "Black Rock One." *Tripolitan,* November 1989.
_____. "Tripoli Name Change?" *The Tripoli Report,* October 1990.
_____. "The Road to Black Rock." *Tripolitan,* June 1991.
_____. "Black Rock LDRS Advice." *Tripolitan,* June 1991.
Brinley, Captain Betrand R. *Rocket Manual for Amateurs.* Ballantine Books. 1960.
Brown, Joseph E. "These Rocketeers Play It Safe." *Popular Mechanics,* September 1958.
Brown, Patricia Leigh. "A Cult of Backyard Rocketeers Keeps the Solid Fuel Burning." *New York Times.* October 14, 2006.
Buck, Steve. "Letter to the Gerlach Empire Volunteer Ambulance& Fire Company." May 4, 1989. (William "Bill" Lewis Collection).
Burney, Richard C. "LDRS Project: *Mercury Redstone* Final Countdown. (Part 4)." *High Power Rocketry,* December 1999.
Buxton, Glen. "Simple Timer." *Tripolitan,* December 1988.
Cagle, Earl. "BALLS 001." *Point 39 Productions* (VHS) 1991.
_____. LDRS X. *Point 39 Productions* (VHS) 1992.
_____. LDRS XI. *Point 39 Productions*(VHS) 1993.
_____. LDRS XII. *Point 39 Productions*(VHS). 1994.
_____. LDRS XIII. *Point 39 Productions*(VHS) 1995
_____. LDRS XIV. *Point 39 Productions*(VHS) 1996.
_____. BALLS 005. *Point 39 Productions* (VHS) 1996.
_____. Monster Rockets: *Project 463* and *Stratospheric Dreams. Point 39 Productions* (VHS)1996.
_____. LDRS XV. *Point 39 Productions* (VHS) 1997.
_____. LDRS XVII. *Point 39 Productions* (VHS) 1999.
_____. LDRS XVIII. *Point 39 Productions* (VHS) 2000.
_____. LDRS 19. *Point 39 Productions* (VHS) 2000.
_____. LDRS XX: A High Power Odyssey. *Point 39 Productions* (VHS) 2001.
California Consumer Aeronautics Catalog 1991–92.
California Rocketry. Jerry Irvine, Editor. 1981–1985.
_____. LDRS Meet Summary. October 1983.
_____. Letters to and from the Editor. July 1981.
_____. Letters to and from the Editor. April 1982.
_____. Letters to and from the Editor. October 1982.
_____. NAR Politics. April 1982.
_____. NAR Politics. January 1983.
_____. Letters to and from the Editor.April 1983.
_____. Letters to and from the Editor. April 1984.

Camp, Dusty. "Von Delius's *Phoenix 4:* Two-stage Sweeps 96,000 Feet." *Rockets Magazine.* Volume 9 (2015).

Canepa, Mark B. "Teamwork at Launches." *Extreme Rocketry,* August 2004.

———. "LDRS 23 2004." *Extreme Rocketry,* August 2004.

———. "Balls 2004." *Extreme Rocketry,* December 2004.

———. "BALLS 2005." *Extreme Rocketry,* December 2005.

———. "The River Ate My Rocket." *Extreme Rocketry,* March 2006.

———. "Black Rock Experimental: Balls 2006." *Extreme Rocketry,* December 2006.

———. "Aiming for the Stars: A Q&A with some of High-Power Rocketry's Most Ambitious Flyers." *Extreme Rocketry,* August 2007.

———, and McGilvray, Neil. "LDRS 27." *Rockets Magazine,* October 2008.

———. "One Man's Quest to Honor America's *Saturn V.*" *Rockets Magazine,* February 2009.

———. "Steve Eves's Field of Dreams: The Launch of the *Saturn V.*" *Rockets Magazine,* June 2009.

———. "Chasing the N Record: Pursuing Stratospheric Dreams." *Rocketry Planet,* May 1, 2010.

———. "Plaster Blaster 2009." *Rockets Magazine,* June 2010.

———. "Eves's Mighty *Saturn V* Celebrates One Year in Huntsville." *Rocketry Planet,* July 1, 2010.

———. "High-Power Rocketry's Top Ten Biggest Launches." *Rocketry Planet,* November 18, 2010.

———. "Black Rock 2011: 100,000 Feet or Bust." *Rockets Magazine,* April 2012.

———. Interview with Max Baines (online via e-mail) (October 2013).

———. Interview with Tom Blazanin. Gerlach, Nevada (September 26, 2010).

———. Interview with Tommy Billings. South Carolina (June 10, 2011).

———. Interview with Tom Binford (online via e-mail) (Summer 2013)

———. Interview with Steve Buck. Reno, Nevada (May 2018).

———. Interview with Paul Davis (online via e-mail) (February 2013).

———. Interview with Rock Dunseith (online via e-mail) (Fall 2013)

———. Interview with Tim Eiszner (online via e-mail) (June 2013).

———. Interview with Randall Ejma (online via e-mail).

———. Interview with Dick Embry (online via e-mail) (November 2013).

———. Interview with Ken Good. Monroeville, Pennsylvania (April 6, 2014).

———. Interview with Patrick Gordzelik. Cross Plains, Texas (January 23, 2014).

———. Interview with Francis Graham. Gerlach, Nevada (September 30, 2011).

———. Interview with Curtis Hughes (online via e-mail) (2011).

———. Interview with Ben Jarvis (online via e-mail) (2014).

———. Interview with Jim Jarvis (online via e-mail) (2013).

———. Interview with Bruce Kelly. Orem, Utah (November 15, 2013).

———. Interview with Bob Koenn (online via e-mail) (2010).

———. Interview with Jerry Larson (online via email) (2018).

———. Interview with Bruce Lee (online via e-mail) (2013).

———. Interview with Dave Leninger (online via e-mail).

———. Interview with William "Bill" Lewis, San Jose, California (March 28, 2015).

———. Interview with Art Markowitz (online via e-mails) (January 2013).

———. Interview with Terry McCreary. Murray, Kentucky (April 5, 2014).

_____. Interview with David Miller (online via e-mail) (September 2013).
_____. Interview with J. Patrick Miller. Berkeley, California (August 16, 2011).
_____. Interview with Jim Mitchell. Murray, Kentucky (April 4, 2014).
_____. Interview with Chris Pearson, Gerlach, Nevada (September 2010).
_____. Interview with Randall Redd (online via e-mails) (December 2011).
_____. Interview with Charles A. "Chuck" Rogers. Lancaster, California (June 2011); second interview (March 2014).
_____. Interview with Gary Rosenfield. Cedar City, Utah (March 7, 2011).
_____. Interview with Curtis Scholl (online via e-mails.)
_____. Interview with Ron and Debbie Schultz. Brunswick, Ohio (June 7, 2011).
_____. Interview with Tom Swenson (online via e-mails).
_____. Interview with Mike Tyson (online via e-mails) (October 2013).
_____. Interview with Art Upton (online via e-mails)(June 2013).
_____. Interview with Frank Uroda (telephone interview) (November 2013).
_____. Interview with Robert Utley. Gerlach, Nevada (September 2013).
Carlson, Helen S. *Nevada Place Names: A Geographical Dictionary*. University of Nevada Press, 1974.
Carter, Don. "The Basics of Composite Motor Construction." *High Power Research*, Fall 1984.
_____. "Composite Motor Construction Part 2." *High Power Research/Sport Rocketry Journal*, Winter 1984 Spring 1985.
_____. "So You Want to Build Your Own Motors?" *Tripolitan*, February 1986.
_____. "Letter to Matt Steele [Re Early Tripoli History]" *SNOAR News*, March–July 1989.
Catalano, Melinda. "Prelude." *Rockets Magazine*, October 2007.
Cato, John H. Jr. *Problems with Motor Certification of the Tripoli Rocketry Association*. October 1988.
Cello, Steve. "Sunday, August 13, 1995." [LDRS XIV] *High Power Rocketry*, February 1996.
Clapp, Rick. "LDRS 24." *Extreme Rocketry*, November 2005.
Clark, Mark. "Project: Thunderbolt." *High Power Rocketry*, April 1996.
_____. "AHPRA Bowling Ball Loft at LDRS XIX." *High Power Rocketry*, January 2001.
Clary, David A. *"Rocket Man—Robert H. Goddard and the Birth of the Space Age."* New York: Hyperion, 2003.
Colburn, William H. *Civilian Rocketry: The History of Selected Amateur Rocketry Groups 1942-2007 in Text, Photographs, Drawings, Charts and Tables*. (Unpublished manuscript) 2007.
Cochrane, Tony. LDRS 97 XVI: Thunder in the Rockies." *High Power Rocketry*, May 1998.
Cole, Vivian. "Secret City United Missile Society: Report from LDRS in Orangeburg, South Carolina." *High Power Rocketry*, January 2001.
Code for *High Power Rocketry*. Tripoli Rocketry Association. July 31, 2012.
Conn, Jim. "LDRS XV Part 2: LDRS Fifteen; an East Coast Event." *High Power Rocketry*, April 1997.
Cooper, Alan. "Desert Reactions." *High Power Rocketry*, November/December 1992.
Cotriss, David. "Hybrids Fly at Fireballs 004." *High Power Rocketry*, February 1995.
Craddock, Robert A. 2007."The Last Interview." *Launch Magazine*, May/June 2007.
Daley, Greg. "The Making of *Return to Sender*." *High Power Rocketry*, May 1998.

Daulphin, Bill. "ICBM IV Tripoli West Palm Invades South Carolina." *High Power Rocketry,* October 1995.
Davidson, Bill. "Flight of the WAC." *High Power Rocketry,* January 1996.
Davis, Arley. "LDRS Project: *Mercury Redstone* Detailing and Finishing. Part 3." *High Power Rocketry,* December 1999.
Deville, Derrek. "Putting the 'S' in CSXT." *Rockets Magazine,* April 2012.
Dexheimer, Erick. "Let's Do Launch." *www.phoenixtimes.com,* accessed February 18, 2014.
Dixon, Scott. "An Introduction to Vulcan Systems, Inc. *High Power Research,* Fall 1984.
_____. Fax Letter to G. Harry Stine [re Burn Tests and Reloadables] February 25, 1992.
Eiszner, Tim. "LDRS 15 Flight Facts." *The Tripoli Report,* October 1996.
Embry, Dick. "Words from the President." *Tripoli Report,* December 2002.
_____. "Judgment Day." [Editorial re ATF lawsuit]*High Power Rocketry,* April 2004.
Emerson, William. *"The Applegate Trail of 1846."* Ember Enterprises. 1996.
Estes, Vernon. *Youth Rocket Safety: A Report to the Model Rocket Manufacturers Association.* 1967.
Evanoff, John. 2007. *Nevada History: Smoke Creek Desert.* http://visitreno.com/evanoff/feb07.php. 20 Dec. 2011, accessed January 30, 2012.
Extreme Rocketry. Brent McNeely, Publisher, 2000–2009.
_____. LDRS 21. *Rocketeer Media.* (DVD) 2002.
_____. LDRS 22. *Rocketeer Media* (DVD) 2003.
Ezell, Barbara R., and Neil Davis. "The Rocket Tree." *High Power Rocketry,* November/December 1993.
Federal Explosives Law and Regulations (Orange Book). Department of the Treasury, Bureau of Alcohol, Tobacco and Firearms. 2000.
Final Report of the Blue Ribbon Commission on High Power Rocketry. Trip Barber, Chair, February 16, 1985.
Fisher, Daniel. "Let's Punch a Hole in the Sky." *Forbes.* November 13, 2000.
Fleischer, John. "Rocket Altimeter." *Radio Electronics,* October 1990.
Flight Control Systems Flight Pack 1(FP1) Advertisement, 1990.
Freeman, Don. "Capt. Kirk Sails on Looking for Star Trek III." *Medina County Gazette,* July 26, 1992.
Furnas, Dr. C.C. "Why Did the U.S. Lose the Race? Critics Speak Up." [quoting Harry Stine on *Sputnik*]. *Life.* October 21, 1957.
Gates Brothers Rocketry (website), accessed November 7, 2013.
Gatto, Paul Del. "All About JETEX." *Rocket Science Institute*: New York, 1957.
Geier, Bob. "NAR Board Meets in Washington." *SNOAR News,* Volume 11, No. 2, 1985.
Gerrard, Doug. *"Eye in the Sky."* *Tripolitan,* October 1988.
_____. "My Mind's Eye." *High Power Rocketry,* December 1999.
Gianokis, David P. "The Launch of the Freedom Seven." *The Tripolitan . . . America's High Power Rocketry Magazine,* August/September 1991.
_____."The Flight of SA-205." *The Tripolitan . . . America's High Power Rocketry Magazine*(March/April 1992).
Good, Ken. "The Road to the Modern Tripoli—A Personal Reflection." http://railtech.com/How Tos/GM/ken/triproot1.html, accessed on March 16, 2010.

_____. "LDRS-20." www.members.tripod, accessed August 13, 2013.
_____. "Editorial - Tattooing a Fish and Other Adventures." *Tripoli Report*, April 2002.
_____. "Editorial - Elections, Changing the Guard, and Tripoli's Organizational Direction." *Tripoli Report*, December 2002.
_____."Editorial-Elections, Voting, and Regulatory Relief." *Tripoli Report*, May 2004.
_____. "Words from the President." *Tripoli Report*, September 2004.
_____. "Words from the President." *Tripoli Report*, February 2005.
_____. "Words from the President - Election Issue." *Tripoli Report*, May 2005.
_____. "Words from the President." *Tripoli Report*, September 2005.
_____, and Bundick, Mark. "Joint Statement on BATF Litigation." November 30, 2005. *Tripoli Report*, December 2005.
_____. "Words from the President." *Tripoli Report*, December 2005.
Gordzelik, Pat. "LDDRS 21 Texas Style." *High Power Rocketry*, December 2002.
_____. "Mini BALLS 2002." *High Power Rocketry*, Jan–Feb 2003.
_____. "LDRS-22 (Part One)." *High Power Rocketry*, May 2004.
_____. "Perspective from Tripoli Rocketry Association." *Rockets Magazine*, March/April 2006.
_____. "LDRS-25 Report from the Launch Director." *Rockets Magazine*, September/October 2006.
_____. "LDRS-25 Banquet Report." *Rockets Magazine*, September/October 2006.
Graham, Francis. "Tripoli Board of Directors Meeting," *Tripolitan*, October 1986.
_____. "General Assembly." *Tripolitan*, October 1986.
_____. Letter to the Editor. [Re Ed Tindell] *SNOAR News*, March–July 1989.
Hall, Stephen C. "Opinions." *Tripolitan*, October 1988.
High Power Research, Sport Rocketry Journal. Editor-Mike Nelson. 1983–1986.
_____. LDRS-5—Goodbye to Medina. Volume 14 (1986).
_____. Letters. Volume 3 (1985).
_____. L.O.C./AeroTech*Magnum*. Volume 3 (1985).
_____. 1985 High Power Motor Guide. Winter 1984/Spring 1985.
_____. 1987 High Power Motor Guide. 1986, Number 13.
_____. Notice. [Announcement of Formation of Tripoli Rocketry Association] Winter 1984/Spring 1985.
High Power Rocketry. Bruce Kelly, Publisher. 1991–2005.
_____. Prefecture Listings. 1991–2005.
_____. Make Your Own Motors! [Advertisement CP Technologies] *High Power Rocketry*, March 1995.
_____. Thunderflame Basic Propellant School [Advertisement] High Power Rocketry, March 1998.
_____. How to Make Amateur Rockets. CP Technologies. *High Power Rocketry*, November 1999.
High Sierra Rocketry Catalog. 1990.
Hughes, Curtis. "The Coming of the 'New' Model Rocket Engine." *Tripolitan*, Winter 1981.
_____. "Product Review—ACE Mongrel." *California Rocketry Quarterly*, October 1982.
_____. "The early Tripoli Experience: A Personal Reflection." http://www.rimworld.com/pgh/history/root2/triproot2.html; accessed on August 24, 2010.

Huntley, J.D. "The History of Solid Propellant Rocketry: What We Do and Do Not Know." American Institute of Aeronautics and Astronautics (1999).

Illerbrun, Garth. "Sullivan Lakes: High Power Rocket One." *High Power Rocketry*, April 1994.

_____. "Sullivan Lake: High Power rocket Three." *High Power Rocketry*, July 1996.

Irvine, Jerry. "Exclusive Report: Smoke Creek Amateur Launch." *California Rocketry Quarterly*, July 1981.

_____. "Exclusive Report: The Second Flight of the Space Shuttle—Landing." *California Rocketry Quarterly*, January 1982.

_____. "Letter to Pat Miller." *California Rocketry Quarterly*, April 1982.

_____. "Meet Summary." [LDRS-2] *California Rocketry Quarterly*, October 1982.

_____. "Letter to C.D. Tavares." *California Rocketry*, October 1982.

_____. "Filing an FAA Clearance." *Tripolitan*, October 1986.

John C. Fremont. www.wickepedia.com, accessed on 9-23-13.

Johnson, Chris. "Blue Ribbon Commission at LDRS-5." *SNOAR News*

Johnson, Mark. "Indiana's Composite Motor Man." *Launch Magazine*, March/April 2007.

Keith, Paul. "AeroTech Officials Criticize Handling of Blaze." *Las Vegas Sun*, Feb 8, 2002.

Kelly, Bruce. *High Power Rocketry*. Publisher. (1991–2005).

_____. "Editorial." *The Tripoli Report* (October 1990).

_____. "The Tripolitan: Guidelines, Deadlines, Policy, and Procedure." *Tripolitan*, December 1990.

_____. "Editorial." [Regarding Tripolitan and Magazine Direction]*The Tripoli Report*, May 1991.

_____. "Letter from the President." *The Tripoli Report*, May 1995.

_____. "Something's missing."[Editorial] *High Power Rocketry*, October 1996.

_____. "Rumors." [Editorial] *High Power Rocketry*, April 1997.

_____. "From the Prez." [Editorial] *The Tripoli Report*, June 1997.

_____. "A Project worth Documenting." [Editorial; *OuR Project*] *High Power Rocketry*, July 1997.

_____. "G. Harry Stine." *High Power Rocketry*, June 1998; also, *Tripoli Report*, June 1998.

_____. "Editorial: The Devil is in the Details," *High Power Rocketry*, July 1998 (Crash and Burn Issue).

_____, and Bundick, Mark B., and Platt, Mike. "Addressing Your Regulatory Concerns." *Tripoli Report*, July 1999.

_____. "Time for a Change." [Editorial] *High Power Rocketry*, July 2000.

_____. "Words from the President." *Tripoli Report*, December 2000.

_____. "Of Words, Stories, and Memories Forgotten." *High Power Rocketry*, March 2001 (33-year index).

_____. "Experimenting Again." *High Power Rocketry*, Oct–Nov 2002.

_____. "BALLS 101: Separating Myth from Facts about BALLS." *High Power Rocketry*, March 2003.

_____. "Thick or Thin." *High Power Rocketry*, May 2003.

_____. "Knowing When to Say No!" [NFPA Code issues]*High Power Rocketry*, April 2004.

Kinney, Nadine. "Focus on the Flats." *High Power Rocketry*, March 1999.

_____. "LDRS-19." *Extreme Rocketry*, September/October 2000.
_____. "Inferno at Whitakers: Small Balls." *High Power Rocketry*, February 2001.
_____. "The Gates Brothers Hit LDRS XX." *High Power Rocketry*, August 2002.
Kirk, Douglas. "An Interview with NAR Creator G. Harry Stine." *American Spacemodeling*, January/February 1993.
Knight, Eric. *The NEW Race to Space*. Eric Knight. 2010.
Koenn, Bob. "LDRS-2 Experimental Rocket Launch." *Space Coast Rocketry*, September 1983; reprinted in *California Rocketry Magazine* (October 1983).
Koloms, Deborah. "LDRS-28: From the Inside Looking Out." *Rockets Magazine*, April 2009.
Lamothe, Dennis. "My Visit to Argonia." *High Power Rocketry*, July/August 1992.
_____."*Aerobee II*." *High Power Rocketry*, February 1995.
Large and Dangerous Rocket Ships. (DVD) *Discovery Channel* (Discovery Communications LLC 2010).
Larson, Jerry. "*Space Shot 2000*." *High Power Rocketry*, July 2002.
Launch Magazine. MM Publishing, Inc. (Mark Mayfield-Editor) 2006–2008.
Launching Safely in the 21st Century: Final Report of the Special Committee on Range Operations and Procedure to the National Association of Rocketry. October 29, 2005.
Lee, Bruce. "Nebraska Heat: Anatomy of a Group Project." *High Power Rocketry*, May 1998.
_____. "BALLS 006." *High Power Rocketry*, July 1997.
_____. "LDRS Project: *Mercury Redstone* Overview and Construction (Part 1)." *High Power Rocketry*, December 1999.
_____. "My LDRS Experience." *High Power Rocketry*, January 2001.
_____. "Junkyard Wars." *High Power Rocketry*, November 2003.
Leonnig, Carol D. "Libby Jurist's Career Built on Toughness." [Profile of Judge Reginald Walton] *The Washington Post*, June 5, 2007.
Lerner, Preston. "It's All About Fire, Smoke, and Noise." *Air & Space Magazine*, January 2004.
Letter by G. Harry Stine to J. Patrick Miller, May 24, 1989.
Letter from Department of Justice to The Chairman of the Senate Judiciary Committee. [Re SB 724] June 10, 2003.
Lewis, William "Bill.""Awesome Black Rock II: The World's Highest Power Launch." *Tripolitan*, December 1990.
_____. Letter to the FAA [re securing a waiver at Black Rock] April 27, 1990 (William "Bill" Lewis Collection).
_____. Letter to Bureau of Land Management [re LDRS X] March 22, 1991 (William "Bill" Lewis Collection).
_____. Letter to FAA-AWP [re Waiver Request for LDRS X] July 5, 1991 (William "Bill" Lewis Collection).
_____. Letter to Washoe County Sheriff's Department [re LDRS X] July 5, 1991 (William "Bill" Lewis Collection).
_____. Letter to Charles Rogers, President Tripoli [re BLM and future rockets launches at Black Rock] April 3, 1992 (William "Bill" Lewis Collection).
_____."Fireballs 001: AERO-PAC'S First Experimental Unlimited Rocket Launch." *Aeronaut* (April 1992).

_____. Letter to Gus Overstrom, Oakland Center flight Control [re waivers at Black Rock in 1992] May 5, 1992 (William "Bill Lewis Collection).

_____. Certificate of Waiver or Authorization. April 12, 1990 (William "Bill" Lewis Collection).

_____. Certificate of Waiver or Authorization. June 26, 1990 (William "Bill" Lewis Collection).

_____. Certificate of Waiver or Authorization. [re LDRS X]August 8, 1991 (William "Bill" Lewis Collection).

_____. FIREBALLS 001 Experimental-Unlimited High Power Rocket Launch [Aero-Pac Information Packet] August 19, 1991 (William "Bill" Lewis Collection).

Lichtblau, Eric. "Threats and Responses; Rocket Bill Stirs Debate on Potential for Terror." *The New York Times* (July 30, 2003).

Lickteig, Lance. "Thanks for the Memories." [LDRS 27] *Rockets Magazine*, October 2008.

Life. "A Safe Toy Missile." April 7, 1958.

_____. "From Coast to Coast, U.S. Youth Gets Its Rockets Up in the Air." December 16, 1957.

Mailer, Norman. *Of A Fire On the Moon*. Boston: Little, Brown and Company. 1969.

Malsch, Martin G. Letter to Acting Director of ATFE re Court Judgment. June 9, 2009.

Manganaro, William. "Fort A.P. Hill-1." *Tripolitan*, December 1988.

Markowitz, Art. "LDRS-22 Old Timers Reunion." *Tripoli Report*, December 2002.

Mayfield, Mark. "Power Ascent." [Interview with Gary Rosenfield] *Launch Magazine*, November/December 2006.

McBurnett, Michael. "Mach Madness at LDRS 27." *Rockets Magazine*, October 2008.

McCreary, Terry. *Experimental Composite Propellant*. McCreary. 2000.

_____. "Spears 2." *High Power Rocketry*, November 1995.

_____. "Does the Lawsuit Suit?" *Extreme Rocketry*, April 2002.

McGilvray, Neil"MDRA at Small Balls." *High Power Rocketry*, February 2001.

_____. "Udder Madness." *High Power Rocketry*, June 2004.

_____. "Give Me Liberty or Give Me a Crane and Some Scaffolding." [*The Liberty Project*] *High Power Rocketry* (September 2004).

_____, and Utley, Bob. "Editorial: Who are You?" *Rockets Magazine*, March/April 2006.

_____, and Utley, Robert. "Editorial." *Rockets Magazine*, July/August 2006.

_____. "LDRS 25." *Rockets Magazine*, September/October 2006.

_____."By the Numbers: What it took to Launch Steve Eves' *Saturn V*." *Rockets Magazine*, June 2009.

_____. "LDRS-28." *Rockets Magazine*, August 2009.

_____. "Paul Robinson: His Legacy Will Live On." *Rockets Magazine*, December 2009.

_____. "In Memoriam Erik Gates." *Rockets Magazine*, February 2010.

_____. "Bad Day at Black Rock: Balls 18." *Rockets Magazine*, February 2010.

_____. "LDRS 29." *Rockets Magazine*, August 2010.

_____. "BALLS 19." *Rockets Magazine*, December 2010.

_____. "LDRS XXX." *Rockets Magazine*, February 2012.

_____. "BALLS 24." *Rockets Magazine*, 2016.

_____. "BALLS 25." *Rockets Magazine*, 2017.

McGoffin, Kenneth C. "Thoughts on the ARS Report." [Unpublished Article re the American Rocket Society 1959 Policy Statement on Amateur Rocketry] 2004 (Kenneth C. McGoffin Collection).
_____. "Notes on the ARS Report." Undated.
McGoldrick, Kevin. "Opinion." *Tripolitan*, June 1988.
McNeely, Brent. An Interview with Gary Rosenfield. *Extreme Rocketry*, March–April 2000.
_____. Interview with Scott Bartel. *Extreme Rocketry*, Sept–Oct 2001.
_____. Interview with Mark Bundick. *Extreme Rocketry*, November 2002.
_____. Interview with Earl Cagle. *Extreme Rocketry*, March 2006.
_____. Interview with Mark Clark. *Extreme Rocketry*, May–June 2001.
_____. Interview with John Coker. *Extreme Rocketry*, April 2001.
_____. Interview with Derek Deville. *Extreme Rocketry*, March 2005.
_____. Interview with Dick Embry. *Extreme Rocketry*, May–June 2006.
_____. Interview with Ken Good. *Extreme Rocketry*, May–June 2007.
_____. Interview with Pat Gordzelik. *Extreme Rocketry*, August 2004.
_____. Interview with Francis Graham. *Extreme Rocketry*, May 2004.
_____. Interview with Kimberly Harms. *Extreme Rocketry*, Jan–Feb 2006.
_____. Interview with Paul Holmes. *Extreme Rocketry*, Sept–Oct 2008.
_____. Interview with Curt Hughes. *Extreme Rocketry*, Sept.–Oct. 2007.
_____. Interview with Bill Inman. *Extreme Rocketry*, December 2002.
_____. Interview with Bruce Kelly. *Extreme Rocketry*, Sept–Oct 2002.
_____. Interview with Korey Kline. *Extreme Rocketry*, July 2003.
_____. Interview with Frank Kosdon. *Extreme Rocketry*, July–August 2000.
_____. Interview with Bruce Lee. *Extreme Rocketry*, Sep–Oct 2003.
_____. Interview with Barry Lynch. *Extreme Rocketry*, May–June 2005.
_____. Interview with Terry McCreary. *Extreme Rocketry*, Jan–Feb 2002.
_____. Interview with Sue McMurrray. *Extreme Rocketry*, March 2001.
_____. Interview with Ky Michaelson. *Extreme Rocketry*, April 2001.
_____. Interview with Randall Redd. *Extreme Rocketry*, November–December 2000.
_____. Interview with Gary Rosenfield. *Extreme Rocketry*, March–April 2000.
_____. Interview with Tom Rouse. *Extreme Rocketry*, November 2006.
_____. Interview with Kevin Trojanowski. *Extreme Rocketry*, August 2005.
_____. Interview with Frank Uroda. *Extreme Rocketry*, Jan–Feb 2001.
_____. Interview with Darren Wright. *Extreme Rocketry*, Sep–Oct 2004.
_____. "LDRS 21: Amarillo, TX, July 11-16." *Extreme Rocketry*, September–October 2002.
_____. "LDRS 22." *Extreme Rocketry*, Sept–Oct 2003.
_____, and LaPanse, Ray. "LDRS 26 Launch Report." *Extreme Rocketry*, September/October 2006.
McNeil, J.D. Letter to Francis Graham [Re Ed Tindell] *SNOAR News*, March–July 1989.
_____. "Editorial/Tripoli: Will It Stay, Or Will It Go?" [Problems with Tripoli Organization] *SNOAR News*, August 1990.
Meredith, Robin, and Sharp, Brandy. "The Making of the Mega-Skeeter." *High Power Rocketry*, November/December 1993.
Meyer, Dan. "Observations—LDRS Report." *Tripolitan*, October 1985.

Meyerrieks, Will. LDRS-8. (VHS). 1989. (David Rose Collection).
_____. LDRS-9. (VHS). Date Unknown. (David Rose Collection).
Michaelson, Ky. *Rocketman: My Rocket-Propelled Life and High-Octane Creations.* Motorbooks: St. Paul, Minnesota. 2007.
_____. "The Joe Boxer Launches." *High Power Rocketry,* August 1997.
_____. "The Rockets of October Sky." *High Power Rocketry,* February 1999.
_____. "Resume." *The Tripoli Report,* June 1998.
_____. "*Space Shot 2000*: CSXT." *High Power Rocketry,* July 2002.
Meinhardt, Scott. "Launch Insurance changes for 2000." *Tripoli Report,* December 1999.
Mies, Chuck. "Flight and Participation Data LDRS XVII." *The Tripoli Report,* December 1999.
Miller, David J. "Danville Eleven." *High Power Rocketry,* March 1995.
Miller, J. Patrick. "A Tribute to G. Harry Stine." *Sport Rocketry,* Jan/Feb 1998.
_____. The President's Corner. *Model Rocketeer,* May 1981.
_____. Letter to the Editor [to Jerry Irvine]. *California Rocketry Quarterly,* July 1981.
_____. The President's Corner. *Model Rocketeer,* July 1982.
_____. Letter to Jerry Irvine. December 1981. Reprinted in *California Rocketry,* April 1982.
_____. The President's Corner. *Model Rocketeer,* April 1983.
_____. The President's Corner. *American Spacemodeling,* May 1985.
_____. Letter to Tom Blazanin. May 7, 1985. Reprinted in *High Power Research/Sport Rocketry Journal,* Winter 1984/Spring 1985.
_____. The President's Corner. *American Spacemodeling,* March 1986.
_____. The President's Corner. *American Spacemodeling,* January 1987.
_____. The President's Corner. *American Spacemodeling,* October 1989.
_____. The President's Corner. *American Spacemodeling,* January 1989.
_____. The President's Corner" *American Spacemodeling,* December 1989.
_____. The President's Corner. *American Spacemodeling,* April 1990.
_____. The President's Corner. *American Spacemodeling,* March/April 1991.
_____. Letter to John W. Sicker. [Re NAR's first high power event] June 12, 1991. Star-Date 7.91 (Astre Newsletter)
_____. Non-Professional Consumer Rocketry-Historic Report. *American Spacemodeling,* November/December 1990.
Model Rocketeer. Official Journal of the National Association of Rocketry.
_____. Board Defines Policy on High-Powered Engines." May 1982.
_____. NARAM-23. November 1981.
_____. Con Calendar. April 1982.
_____. Con Calendar. June 1982.
_____. Con Calendar. July 1982.
_____. High Power Regulations. [Trip Barber announcement] *Model Rocketeer,* November 1983.
Monoson, Ted. "Hobby Rocket Plan under Fire." *Star Tribune,* July 30, 2003.
Morozumi, Pius. "News about LDRS XIV." *The Tripoli Report,* May 1995.
Murray, Bob. "*Tripolitan High Power Rocketry* Eleven Year Index." *High Power Rocketry,* December 1996.
NAR/Tripoli Aquarius Commission Report. Las Vegas Tests June 6, 1991. (Rosenfield collection).

Nelson, Craig T. *Rocket Men: The Epic Story of the First Men on the Moon.* New York: Penguin Books. 2009.
Nelson, Mike. Letter to Matt Steele. *High Power Research,* Fall 1984.
_____. Large and Dangerous Rocket Ships 4. *High Power Research/Sport Rocketry Journal,* Volume 3, 1985.
Nestor, Art. "A Visit with Irv Wait." *Pittsburgh Space Command* [NAR Section Newsletter] (Nov–Dec 2009).
Neufeld, Michael J. *Von Braun.* New York: Vintage. 2007.
Newman, Kent. "10 Questions." [Hobby rocketry and the FAA's Class 3 Rules].
Newport, Curt. "Proteus Experimental Rocket at BALLS 19: Part 2." *Rockets Magazine,* December 2010.
National Fire Protection Association (NFPA). Code for Unmanned Rockets. *Forward.* 1982.
_____. NFPA § 1122 (1982 edition).
_____. NFPA § 1122 (1987 edition).
_____. NFPA § 1122(1992) (Proposed).
_____. NFPA § 1122 (2008 edition).
_____. NFPA § 1125 (2012 edition).
_____. NFPA § 1127 (1994 draft version 1.3).
_____. NFPA § 1127 (1995 edition).
North Coast Rocketry catalog (1989).
Nowaczyk, Gene, and Kinney, Nadine. "LDRS Argonia Kansas." *High Power Rocketry,* November 1999.
O'Brien, John. *"Warp Factor II."* *Tripolitan,* October 1988.
O'Sullivan, Jerry. "LDRS-19." *Maryland Tripoli Report,* July/August 2003.
"Paavo John Rahkonen Dies at Age 79." *Rocketry Planet,* January 18, 2010.
Palmer, Kris. "The Rocketman." [Michaelson Biography] http://rocketman.com, accessed on June 25, 2013.
Pearson, Chris. "What the Hell Happened in California." *SNOAR News,* Volume 8, Number 2, 1982.
_____, and Shultz, Deborah. "LDRS-4." *SNOAR News,* Volume 11, 1985.
_____. "Official Announcement from LDRS Director Chris Pearson." *SNOAR News,* October/November 1986.
_____. "The LDRS Story." *Tripolitan.* 1987; reprinted in *High Power Rocketry,* December 2002.
_____. "Rocketry in the 80's: A Retrospective View." *SNOAR News,* Late 1989–Early 1990.
_____. "LDRS 3." *SNOAR News,* Volume 10, No. 6 1984, reprinted in *High Power Rocketry,* December 2002.
_____. "How it all Began—LDRS-1." *High Power Rocketry,* July 1996.
Pearson, Ed. "NARAM-31 Overview." *American Spacemodeling,* November 1989.
Perdue, Mario and Mayfield, Mark. 2010. *The Enerjet Story.* http://www.modelrocketryhistory.com/history/centuri/4-the-enerjet-story?showall=1
Piper, Chuck. "Recycled Rockets." (Abstract).
_____. "RRI Smoke Creek Desert Launches and the Roots of LDRS (Abstract)

Piper III, Charles J. "Ammonium Perchlorate: NH4CLO4: Myth, Mystery, Misconception, and the 'Henderson Syndrome,'" *High Power Rocketry*, July 1998 (*Crash and Burn* Issue).
Platt, Mike. "April 95 ATF Update." *Tripoli Report*, January 1996.
_____. "AFT LEUP Filing Instructions." *Tripoli Report*, October 1996.
Prior, M.D., Phil. "*OuR Project.*" *High Power Rocketry*, July 1997.
Pursely, John. "Flight Test: AeroTech RMS Reloadable Motors." *American Spacemodeling*, November/December 1991.
Raden, Lewis P. Correspondence to Mr. Joseph Egan. October 13, 2006. (Defining APCP as an Explosive under 18 U.S.C., Chapter 40) [With accompanying internal ATF memorandum on APCP Classification].
Reaction Research Society (History) www.rrs.org. Accessed June 21, 2011.
Reloadable Model Rocket Motors: An Accident Waiting to Happen. (December 1991).
Ream, Bradley N. "Model Rocketry Remembers P. John Rahkonen." (Unpublished work 2014).
_____. "Enerjet-8: The Original AP Composite Model Rocket Motor." *Sport Rocketry*, July/August 2015.
Ritz, John. "LDRS Project: *Mercury Redstone* Glassmeister's Delight (Part 2)." *High Power Rocketry*, December 1999.
Roberson, Steve. "Coastal Rocketry Association and Launch Complex 39." *High Power Rocketry*, March/April 1993.
Roberts, Mary. Letter to G. Harry Stine [re separate codes for model and high-power] April 4, 1991 (Gary Rosenfield Collection).
_____. Letter to G. Harry Stine [re Reloadable Motors] July 3, 1991. (Gary Rosenfield Collection).
_____. Letter to G. Harry Stine [re Support for reloadable metal motors] October 4, 1991 (Gary Rosenfield Collection).
Robinson, Paul, and Mizoi, Ken. "*OuR Project.*" *The Tripoli Report*, October 1996.
Rochester, Teresa. "Man Killed in Accident T.O. Businessman." [Erik Gates] *Ventura County Star*, December 21, 2009.
The Rocket Challenge television program. *Discovery Channel* (November 2003), accessed via YouTube, 2013.
Rocket Dyne Systems (Catalog-1993).
Rocky Mountain News, "Rocket Expert Warns Against Grocery Fuel." April 22, 1958, cited in *Youth Rocket Safety*, March 1967.
Rocket Newsletter. Edited by William Wood (1987–1990) (Bruce Kelly Collection).
Rockets Magazine. Robert Utley and Neil McGilvray, Publishers. 2006-2018. (Liberty Launch Systems).
_____. LDRS-25. DVD. 2006.
_____. LDRS-26. DVD. 2006.
_____. LDRS-27. DVD. 2007.
_____. LDRS 28. DVD. 2008.
_____. Steve Eves 1:10 Scale *Saturn V*. DVD. 2009.
_____. LDRS-29. DVD. 2010.
_____. LDRS-XXX. DVD. 2012.
_____. LDRS-31. DVD. 2013.
Rocket R&D Catalog. 1992–1993.

Rocket Research Catalog. 1991.
Rogers, B. "Val-Sun Members Conduct High Power Launch." *American Spacemodeling*, January 1987.
Rogers, Charles E "Chuck". "USXRL-88 Tracking Results." *Tripolitan*, June 1988.
_____. Octoberfest Tracking Results. *Tripolitan*, December 1988.
_____. Lucerne Test Range Tracking Results. *Tripolitan*, April 1988.
_____. LDRS-7 Tracking Results. *Tripolitan*, October 1988.
_____. Why Did the Rockets Fly So High at LDRS (or did they…?). *Tripolitan*, October 1988.
_____. Presidential Campaign. *Tripolitan*, December 1988.
_____. The Prez. *Tripolitan*, November 1989.
_____. Black Rock II Tracking Results. *Tripolitan*, December 1990.
_____. 1991 LDRS-10 Site Selection. *The Tripoli Report*, October 1990.
_____. An Open Letter. [From Charles Rogers to J. Patrick Miller]*The Tripoli Report*, January 1991.
_____. From the Prez. *The Tripoli Report*, March 1991.
_____. From the Prez. *The Tripoli Report*, July 1991.
_____. Letter to Tripoli Membership [re Tripoli Legal Fund for high power] *Tripoli Report*, January 1993.
_____. The Aerodynamic Drag of Rockets. *High Power Rocketry*, August 1995.
_____, and Kelly, Bruce. Editorial. *Tripolitan*, December 1990.
_____. Postflight Analysis of the *OuR Project* R Rocket Flight. *High Power Rocketry* July 1997.
_____. Erosive Burning Design Criteria for High Power and Experimental/Amateur Solid Rocket Motors, *High Power Rocketry*, January 2005.
_____. 2004 CSXT *GoFast* Flight Data. www.rocketryforum.com, accessed on October 9, 2015 (Data from Chuck Rogers).
Rogers, John D. [Office of Compliance & Enforcement; CPSC] Letter to Gary Rosenfield [re Inquiry on Safety of Reloadable Motors]. March 25, 1992.
Rose, David. LDRS-4. VHS. 1985. (David Rose Collection).
_____. LDRS-6. VHS. 1987. (David Rose Collection).
_____. Winterfest 88. VHS. 1988. (David Rose Collection).
Rosenfield, Gary. "Feature Article: Power Freaks Arise." *California Rocketry Quarterly*, April 1981.
_____."Feature Article: Trends in High-Power Model Rocket Motor Design for the 80's—part 3." *California Rocketry Quarterly*, October 1981.
_____. Letter to Barry Tunick [President-Estes Industries] [re Reloadable Motors] April 29, 1991 (Gary Rosenfield Collection).
_____. Letter to Martha H. Curtis [NFPA] [re TIA and Reloadable Motors] April 30, 1991 (Gary Rosenfield Collection).
_____. Petition for Rulemaking Amendment to the Consumer Product Safety Commission. November 15, 1991.
_____. Letter to David Schmeltzer of CPSC [re Petition for Rulemaking on Reloadable Motors]. November 18, 1991.
_____. Letter to John Rogers of CPSC re Reloadable Motors. April 2, 1992.
_____. Letter to Ray Lamagdalane of Office of Hazardous Materials Transportation [re Anonymous Reports and Reloadable Motors]. April 16, 1992.

_____. Letter to David Schmeltzer of CPSC re Petition for Rulemaking on Reloadable Motors. April 21, 1992.
_____. "Rocketry Elitism." [Aft Closure-Opinion] *Extreme Rocketry*, Jan–Feb 2001.
_____. "Open Letter to the Rocketry Community." *Extreme Rocketry*, Jan–Feb 2002.
_____. Remembering Ray Goodson. [Obituary] *AeroTech News* (August 24, 2005) (Gary Rosenfield Collection).
Rothman, Greg, and Trojanowski, Kevin. "LDRS 25: Tripoli's 2006 National Event." *Extreme Rocketry*, October 2006.
Russell, Chas. "High Power Rocket Motors." *American Spacemodeling*, January 1987.
Rudy, Chuck. "LDRS 23." *Voo Doo Digital Productions*. (DVD) 2004.
Sackett, Charles III. Sackett, "*Down Right Ignorant.*" *High Power Rocketry*, January/February 1993.
_____. Project 463. *High Power Rocketry*, February 1996.
Safford, Will. "Fort A.P. Hill." *Star-Date12.88* [ASTRE Newsletter] 1989.
Schoner, Bob. "View from Down Range." *High Power Rocketry*, January 2001.
Schultz, Ron. "The *Esoteric*." *Tripolitan*, October 1986.
_____. "*Esoteric*: Maxi-Sport Rocket." *High Power Research, Sport Rocketry Journal* (Number 13, 1986/1987) (co-author Deborah Schultz).
_____. "Recommended Motors." *Tripolitan*, June 1988.
_____. *"Mother Lode." Tripolitan*, October 1988.
Shenosky, Larry. "Futureshock: ISP's New Reloadable Motors." *American Spacemodeling*, September/October 1990.
Sicker, John. "Editor's Thermal: Tripoli Motor Certification." *Star-Date 10.91* [Astre Newsletter] 1991.
_____."LDRS IX." *Star-Date 2.91* [Astre Newsletter] February 1991.
Sienkiewicz, Martha. "LDRS 11: Groundwork for the Future." *High Power Rocketry*, November/December 1992.
_____. "Friday, August 11, 1995." [LDRS XIV]*High Power Rocketry*, January 1996.
Slavin, Barbara. "A Judge of Character: Although He's Blind, David Tatel Skis, Runs, and Climbs Mountains. By Summer's End, He May Be a Top Jurist, Too." *Los Angeles Times*, July 28, 1994, accessed on October 22, 2013.
SNOAR News. Official Newsletter of the Suburban Northern Ohio Association of Rocketry (NAR Section 337).
_____. Bull Sheet [Change in NAR Pledge] 1982.
_____. LDRS-1. [Advertisement] (1982).
_____. LDRS-5 Large and Dangerous Rocket Ships. [Advertisement] January–February 1986.
_____. LDRS 4.August 1985 (reprinted in *High Power Rocketry*, December 2002).
_____. NAR Board Meets in Washington, D.C. Volume 11, Number 2, 1985.
_____. Bull Sheet [News] [Demise of *California Rocketry*] May–July 1986.
_____. Bull Sheet [News] [IRA Terrorists & Tripoli Incident] August 1990.
_____. Bull Sheet [News] [Tripoli Political Fireworks at LDRS-8] March–July 1989.
_____. Letters to the Editor. January–February 1989.
_____. Letters to the Editor. March–July 1989.
_____. Really Serious Letters to the Editor. March–-July 1989.
Soucy, Guy. "1995 Tripoli board Elections." *The Tripoli Report*, May 1995.
_____."Board of Director Candidate Resumes." *Tripoli Report*, July 1999.

_____. "1999 Board of Directors Election Results." *Tripoli Report,* December 1999.
_____. "2001 TRA Board of Directors Election Results." *Tripoli Report,* April 2002.
_____. "TRA 2002 Board of Directors Election Results." *Tripoli Report,* December 2002.
_____. "2004 Board of Directors Candidate Resumes." *Tripoli Report,* May 2004.
_____. "2006 BOD Elections Candidate Resumes." *Tripoli Report,* March 2006.
_____. "TRA Board of Directors Election Results." *Tripoli Report,* September 2006.
_____. "2009 BOD Election Results." *Tripoli Report,* October 2009.
Spinelli, Mike. "The Race to 1,000 MPH." [Breaking the Land Speed Record] *Popular Science* October 2009.
Sport Rocketry. Official Journal of the National Association of Rocketry. Thomas Beach-Editor. 1992–2018.
Steele, Matt. "Highlights of LDRS-5," *SNOAR News,* October/November 1986.
_____, and Fleischer, John. "Electronic Staging of Rockets." *Tripolitan,* January 1991.
_____. Opinion. *High Power Rocketry,* July 1996.
Stephens, Ian, and Baines, Kyle A. "Editorial: Important LDRS 25 Info." *High Power Rocketry* (January 2005).
Stine, G. Harry. *Earth Satellites and the Race for Space Superiority.* New York: Ace Books. 1957.
_____. 1977. *The New Model Rocket Manual* (New York: Arco Press).
_____. "World's Safest Business." *Mechanix Illustrated,* February 1957.
_____. "Shoot Your Own Rockets." *Mechanix Illustrated,* October 1957.
_____. "Rocket Trails: Large Power Plants Tested." *American Modeler,* May/June 1963.
_____. "Rocket Trails." *American Modeler,* September/October 1963.
_____. "Rocket Trails." *American Modeler,* November/December 1963.
_____. "Rocket Trails." *American Modeler,* January/February 1964.
_____. "The Old Rocketeer." *Model Rocketry,* October 1969.
_____. "Orville Carlisle's Model Rockets." *Model Rocketeer,* May 1977.
_____. "Special Report from NARAM-1." *Model Rocketeer,* August 1977.
_____. "The Roots of Model Rocketry." *Model Rocketeer* (September 1977); see also *American Spacemodeling,* August 1993.
_____. "The Beginning of the Model Rocket Industry." *Model Rocketeer,* November 1977.
_____. "NAR Roots, Part II: The Beginning of the Model Rocket Industry." *Model Rocketry,* November 1977.
_____."NAR Roots, Part III: The Birth of the MMA/NAR." *Model Rocketry,* December 1977.
_____. "NAR Roots, Part IV: Reminiscences of Green Mountain." *Model Rocketeer,* January 1978.
_____. NAR Roots, Part V: The First Model rocket Kits." *Model Rocketeer,* February 1978.
_____. "Observations on High Power." [Written for the Barber Commission] *SNOAR News,* January–February 1986.
_____. "Working with the FAA." *American Spacemodeling,* January 1987.
_____."A Tribute to Pat Miller." *Model Rocket News Magazine,* Fall 1987.
_____."Orville H. Carlisle NAR #1." (Obituary by G. Harry Stine, NAR #2) *American Spacemodeling,* September 1988.

_____. Letter to J. Patrick Miller [re NAR's Failure to Embrace New Technology] May 24, 1989 (Gary Rosenfield Collection).
_____. Letter to John D. Rogers [CPSC] [re Reloadable Motors meet all NFPA and NAR Standards] November 8, 1991. (Gary Rosenfield Collection).
_____. Letter to Scott Dixon [re Proposed Reloadable Burn Tests] February 19, 1992.
_____. Letter to Scott Dixon [re Reloadable Burn Tests] March 8, 1992.
_____. Rocket Motor Burn Test Plan [Draft Copy] March 8, 1992.
_____. Letter to Barry Tunick [Estes] [re NAR Burn Tests Plans] March 16, 1992. (Gary Rosenfield Collection).
_____."Model Rocket Motor Burn Tests: A Report to the National Association of Rocketry." April 4, 1992.
_____."The Roots of Model Rocketry." *American Spacemodeling*, August 1993.
_____."Forty Years of Model Rocketry: A Safety Report." (1997).
Tavares, C.D. Letter to Jerry Irvine. Letters. *California Rocketry*, October 1982.
_____. Letter to Ed Tindell [Re LDRS-8 Location and Other Issues] *SNOAR News*, January/February 1989.
_____."No Mixing Rule Deserves Your Support." *American Spacemodeling*, February 1989.
_____. "Reload Madness Video Rocks NFPA." *The Sentinel* [Central Mass. Spacemodeling Society] (1992).
"The Amateur Scientist: About the Activities and the Trials of Amateur Rocket Experiments." *Scientific American*, June 1957.
"The Golden Days of Model Rocketry: An Interview with Vern and Gleda Estes." *Sport Rocketry* [Date unknown].
The Tripoli Report. 1990–2018.
_____. Minutes from Tripoli Board of Directors' Meetings: 1990–2006.
_____. Joint Statements/Communiqués with NAR re ATF and litigation. 1994–2010.
_____. Tripoli's 1989 Top Ten Altitude Holders. October 1990.
_____. Tripoli Rocketry Association Certification Process (Draft Proposal). January 1991.
_____. Tripoli Rocketry Association Certified Motor List. February 1992.
_____. LDRS XIV and FireBalls 005 Registration Form. May 1995.
_____. Tripoli Motor Testing Committee Detailed Approved Motor Listing. April 1994.
_____. Tripoli Rocketry Association Projected 1994 Budget. April 1994.
_____. Tripoli Middle Tennessee (Barrens Test Range) Experimental Motors Use Guidelines. March 1996.
_____. Certified Member List. May 1996.
_____. ATF Issue, January 1997.
_____. LDRS XVI General Membership Meeting. June 1998.
_____. Tripoli Altitude Records (Official). June 1998.
_____. Certified Member List. June 1998.
_____. 1998 TRA Board of Elections Resumes. June 1998.
_____. Minutes of the Board of Directors Vote Approval for a Contract with *First TV Productions* for a HP Rocketry Competition television Program. April 2002.
_____. Request for Proposal (for new rocketry magazine) December 2005.
Thompson, Sonny. *Forward to Thunderflame Course* (1997 Ed).

Tindell, Ed. "Tripoli Houston: The True History of the Forming of Tripoli Houston." *Tripolitan,* June 1987.
_____. "Have I Got Jobs for You?" *Tripolitan,* December 1988.
_____. "LDRS-7 Statistics Report." *Tripolitan,* October 1988.
_____. Letter to Chris Tavares [Re LDRS-8 and Other Issues] *SNOAR News,* January/February 1989.
_____. "Outgoing President's Message." *Tripolitan* (November 1989).
_____. "T-17:00:00/A Personal Account of the Launch of the *SK3J*." *High Power Rocketry,* August/September 1991.
Tripoli Bylaws (date uncertain/circa 1990s and circulated in membership packets).
Tripoli Rocketry Assoc., Inc., and National Association of Rocketry, Plaintiffs, v. Unites States Bureau of Alcohol, Tobacco and Firearms, Defendant. Civil Action No. 00-273 (RBW).
_____. Complaint filed February 11, 2000.
_____. Defendant's Motion to Dismiss Complaint, filed March 5, 2000.
_____. Joint Motion for Temporary Stay of Action, filed August 9, 2000.
_____. Case Reassignment to Judge Reggie B. Walton (entered November 6, 2001).
_____. General Order and Guidelines for Civil Cases, filed February 14, 2002.
_____. Amended Complaint, filed February 27, 2002.
_____. Order Granting Defendant's Unopposed Motion for Enlargement of Time to File Opposition to Plaintiffs' Motion for Preliminary Injunction, filed March 5, 2002.
_____. Transcript of Proceedings before the Honorable Reggie B. Walton, dated April 30, 2002. [Re Plaintiffs' Motion for Preliminary Injunction].
_____. Memorandum Opinion, filed June 24, 2002. [Judge Walton's Ruling on Defendant's Motion to Dismiss].
_____. Scheduling Order, filed July 22, 2002.
_____. Plaintiffs' Memorandum of Points and Authorities in Support of Motion for Summary Judgment, dated August 30, 2002.
_____. Plaintiffs' Response in Opposition to Defendant's Renewed Motion for Summary Judgment, dated September 30, 2002.
_____. Memorandum Opinion, dated March 19, 2004. [Order by Judge Walton Granting ATF Summary Judgment of the Issue of Whether APCP is an Explosive] Cited also as 337 F. Supp. 2d 1 (D.D.C. 2004).
_____. Plaintiffs' Motion for Leave to Supplement Complaint, dated October 28, 2004.
_____. Notice of Appeal, dated December 22, 2004.
_____. Appellants' Opening Brief, filed August 5, 2005 (United States Court of Appeals D.C. Circuit).
_____. Appellants' Reply Brief, filed September 23, 2005 (United States Court of Appeals D.C. Circuit).
_____. Order, filed November 18, 2005 (United States Court of Appeals D.C. Circuit). [Setting Appellate Oral Argument for January 10, 2006].
_____. Order, filed December 23, 2005 (United States Court of Appeals D.C. Circuit). [Setting Appellate Oral Argument for January 10, 2006 at 10 minutes argument per side].

_____. Decision/Opinion of the Appellate Court, decided February 10, 2006 (United States Court of Appeals D.C. Circuit). [Reversing the Order Granting Summary Judgment to ATF by Judge Walton] Cited also as: 437 F.3d 75 (D.C. Cir. 2006).

_____. Order, dated April 20, 2006. [Judge Walton received Case back on Remand from the Appellate Court and Instructions thereon].

_____. Defendant's notice of Agency Decision, dated October 13, 2006 [ATF Designation that APCP is an Explosive].

_____. Scheduling Order, filed October 17, 2006.

_____. Plaintiffs' Third Amended Complaint for Judicial Relief of Agency Action, Declaratory Judgment, and Injunctive Relief, filed December 18, 2006.

_____. Plaintiffs' Motion for Summary Judgment on Count One of Their Third Amended Complaint, filed January 31, 2007. [And Accompanying Memorandum of Points and Authorities, and other supporting papers].

_____. Affidavit of Mark Bundick (filed in Support of Plaintiffs' Motion for Summary Judgment) filed January 31, 2007.

_____. Affidavit of Dr. Terry McCreary (filed in Support of Plaintiffs' Motion for Summary Judgment) filed January 31, 2007.

_____. Defendant's Memorandum of Points and Authorities in Support of Renewed Motion for Partial Summary Judgment on Counts One, Four and Five, filed January 31, 2007. [And supporting papers].

_____. Plaintiffs' Opposition to Defendant's Renewed Motion for Summary Judgment, filed March 16, 2007.

_____. Defendant's Opposition to Plaintiffs' Motion for Summary Judgment on Count One, filed March 16, 2007. [And supporting papers].

_____. Affidavit of David S. Shatzer, filed March 16, 2007.

_____. Plaintiffs' Reply to Defendant's Opposition to Motion for Summary Judgment, filed April 10, 2007.

_____. Affidavit of Gary Rosenfield, filed April 10, 2007.

_____. Defendant's Reply in Support of Renewed for Partial Summary Judgment, filed April 10, 2007.

_____. Notice of Withdrawal for Joseph R. Egan, Counsel for Plaintiffs [Notice to Court and Parties of the Death of Mr. Egan], filed October 1, 2008.

_____. Order [Granting Plaintiffs' Motion for Summary Judgment] filed March 16, 2009.

Tripolitan. Francis Graham and others, Publishers. 1968–1984. (Bruce Kelly Collection).

Tripolitan. Journal of the Tripoli Rocketry Association. Tom Blazanin, Publisher. 1985–1991.

_____. *Big Bertha*—a Feasibility Study. October 1985.

_____. By-Laws Set [of the Tripoli Rocketry Society] March 1985.

_____. Tripoli Notes. *Tripolitan*, December 1985.

_____. Rocketeers Reorganize Tripoli. Winter 1985.

_____. February Minutes [of the Tripoli Rocketry Society] March 1985.

_____. Tripoli Member Listing. February 1987.

_____. Tripoli Membership Update. October 1987.

_____. Large and Dangerous Rocket Ships VI. October 1987.

_____. LDRS Memories. October 1987.

_____. LDRS Recap. October 1987.

———. NFPA Meeting. December 1987.
———. Z4. [Multiple authors] December 1987.
———. NFPA Update. October 1988.
———. LDRS-7 Competition. June 1988.
———. New Officers and Address. February 1988.
———. Motor Policy. February 1988.
———. LDRS-7: The Nationals. October 1988.
———. Black Rock I. [Advertisement] December 1988.
———. The Tripoli Safety Code. June 1988.
———. Fourth Annual Board of Directors Meeting. December 1989.
———. Black Rock II. [Advertisement] February 1990.
———. LDRS IX. February, 1990.
Trojanowski, Kevin. "Brent McNeely Interview." *Extreme Rocketry*, July 2004.
Tunick, Barry. Memorandum to Dane Boles & Bill Stine [re Reloadables and TIA Comments and Tapes] March 9, 1992. (Gary Rosenfield Collection).
———. Letter to G. Harry Stine [re Upcoming Testing of Motors] March 13, 1992 (Gary Rosenfield Collection).
———. Letter to Harry [Stine] and Pat [Miller] [re Vulcan, Tape, and Reloadables] March 31, 1992. (Gary Rosenfield Collection).
Van Milligan, Tim. *Model Rocket Propulsion*. Apogee Components. 1996.
Wald, Matthew L. "Joseph Egan, Lawyer Who Fought Nuclear Waste Site, Is Dead at 53. *The New York Times*. May 12, 2008.
Weber, Mark. "Solid Propellant Motor Technology." *High Power Research*, Fall 1984.
Wheeler, Sessions S. *Nevada's Black Rock Desert*. Idaho: Caxton Press. 2003.
Whitmore, Alan. "WELD 2002 Whitakers Experimental Launch Days." *Extreme Rocketry*, July 2002.
———, and McBurnett, Michael. "Rocket Autopsies Part One: Analysis of Motor Failures." *Rockets Magazine*, December 2010.
Wilkins, Mick. "LDRS XV Part 1: An Event to Remember." *High PowerRocketry*, April 1997.
Wolf, Tom. 1979. *The Right Stuff*. New York: Picador.
Wood, Bill. "Comments and Suggested Changes to NFPA 1127, 1993, version 1.0." *The Tripoli Report* (February 1992).
Wright, Brad. "Four Projects, Five Friends and a Shared Goal Converge on the Playa." *Rockets Magazine* (December 2010).
"You Can Fly High Power Now." [With NAR High Power rocket Safety Code-Interim] *American Spacemodeling* (May/June 1991).
Zeppin, Ron. "And Sometimes This Happens." *High Power Rocketry* (March 2003).
Zsidisin, Greg. "The 1985 Pearl River Modroc Seminar." *American Spacemodeling* (August 1985).
18 U.S.C. Chapter 40.
"The 0-10,000 Page." www.ahpra.org/010k.htm, accessed on January 31, 2013.

Illustration Credits

Book I: Large and Dangerous Rocket Ships

1: By Gary Rosenfield
2, 15: By the Author
3, 4, 6, and 10: By Deborah Schultz
5, 12: By Tom Blazanin
7: By John Valasek
8, 11: By Steve Buck
9: Courtesy Dick Embry
13: By Nadine Kinney
14: Courtesy William "Bill" Lewis

Book II: The Reloadable Revolution

16: By Deborah Schultz
17: By Mark Clark
18. By the Author
19: By Ky Michaelson
20: By Mikal Mitchell
21, 26 and 28: Courtesy of the Kloudbusters
22: Courtesy AERO-PAC
23: Courtesy Ky Michaelson
24. Courtesy UROC
25 and 27: Courtesy of ROC
29: Courtesy Neil McGilvray

Book III: Here Is Your Temple

All photos by the Author except:

36. Courtesy Pat Gordzelik
37. By Nadine Kinney
68. By Derek Deville

Index

A

A-1, 303, 628, 634
A-1 altimeter, 303
Abbott, Mike, 557
Above and Beyond, 160
accelerometer, 103–4, 185, 311, 482–83
accident, 19, 25, 27, 47, 57, 63–64, 131, 133, 138, 169, 245–46, 290, 359, 432–33, 477, 488, 565, 588, 609, 611–14, 667, 711, 731, 747
Ace Mongrel, 38, 40, 697, 740
Ace Rockets, 80, 338
Achilles, 519
A-Crane Service, 593
Adept, 93, 303, 306–7, 309–10, 312, 315, 379, 383, 626, 644, 648, 696, 699, 715, 735
Advanced Rocketry Safety Code, 117–18, 120, 141, 286, 703
Advanced Rocketry Society, 81–82
Aerobee, 263–64, 266–67, 315, 359, 713, 742
Aerojet, 94–95, 230
Aeronaut I, 178, 206–7
AERO-PAC, 177–78, 205–10, 212, 314, 392, 681
Air & Space, 456, 676, 742
Airfest, 2
Air Force Flight Test Center, 159, 164, 220, 483
Akron, Ohio, 586
Alabama, 407, 515–17, 560, 569–70, 593
Albatross, 622

Albuquerque, 61, 443–44
Alcocer, Tony, 636
Alien with an Attitude, 456
Allegro, 80
Allen, Ken, 316
Allentown, 24
Alpha, 6, 368
Alpha III, 161, 535
ALTS1A, 309
Altus 40, 529
Amarillo, 401, 429–34, 439, 441–42, 526, 559, 563, 623
Amateur Rocket Motor Construction, 433
American Heritage, 542
American Lunar Society, 85
American Modeler, 18, 157, 362
American Spacemodeling, 192, 234, 241, 680
ammonium perchlorate composite propellant (APCP), ix, xi, 4, 7–8, 30, 94, 329, 334, 402, 404, 409, 411–14, 474, 485–88, 492–98, 501–4, 523–24, 530, 544, 546–47, 549–56, 564, 566, 571–76, 578–84, 618–19, 621, 628, 747, 752–53
Amos, Jim, 383, 637
Anderson, Neil, 674
Anthony, Wayne, 252, 367
Apollo, x, 5, 15, 24, 31, 61–62, 108, 165, 237, 266, 370, 456, 585, 589, 595
Apollo 11, 593, 601, 628
Apollo 12, 593
Apollo 17, 15, 594
Applegate brothers, 172

Applegate Trail, 172
Applied Physics Lab, 126
Apt, Jay, 56, 699
Aquarius Commission, 244, 246
Aramis, 395–96
Aramis II, 442
ARC-Polaris, 60
Argonia, 29, 255–56, 258, 262–63, 265–66, 325, 378, 382, 387, 407, 429, 456–57, 463, 465, 499, 511, 513, 568–70, 578, 609, 620, 622
ARGOS, 395
Arianne, 395
Arianne 4, 395, 397
Arias, 387
Aries, 380–81
Aries Space Plane, 608
ARLISS, 675
Armstrong, Neil, 31, 161, 165, 500
Army Hawk, 401
Arrowhead Regional Medical Center, 611
Articles of Incorporation, 285
ARTS, 458, 506, 645
ASP, 391
Associated Press, 258
Astrobee-G, 204
Astron Scout, 60, 99, 296
Athena, 123, 468
Athos II, 443–45
Atlantis, 642
Aurora, 144, 203–4, 465
Australia, 623, 675, 682
axial acceleration, 482–83

B

Backyard Ballistics, 441
Bacon, Chet, 471
Badger, Kelly, 156
Baines, Kyle, 528
Baines, Max, 526
Baker, Bob, 212, 239, 252, 710
Bales, Tom, 339
balls, 205, 207–13, 255, 314, 316–17, 319–21, 328, 331, 340, 375, 390, 437, 467, 512, 538, 626–27, 630, 632, 634–37, 639, 645, 673, 680–81
BALLS-4, 340
BALLS-5, 317–18, 321, 329, 374–75, 620
BALLS-21, 643, 647
BALLS 22, 673
Barber, Arthur "Trip," 64, 71–72, 291, 537, 553, 579, 583, 699, 745
Barber, Bill, 67, 110–11, 626
bar code, 281
barium, 603
Barnes & Noble, 284
barometric, 126, 185, 308–9, 311
Barrens Test Range, 346
Barrett, Stu, 370, 637
Barrowman, James, 65
Barstow, 15, 276, 565
Bartel, Scott, 295, 345, 528, 530
basement bomber, 19–20, 22, 25, 27, 132, 216, 230, 306, 332–34, 344, 348, 355, 359, 446, 588, 628
Bassette, James, 594
Battlebots, 454
Baumann, Karl, 301, 318
Baumfalk, John, 255–56, 262–63, 569, 679
Beach, Thomas, 561
Bean, Alan, 593
Beast, The, 466
Bell X-1, 78
Bennett, Stephen, 379
Benson, Carl, 526
Bermite, 12–13, 29, 94, 337
Biba, Ken, 370, 633, 636, 647, 674
Big Bertha, 102–3
Big Daddy, 440, 456, 465
Bigger Dawg, 507
Big Kahuna, 374, 376–77
Big Screaming Banana, 356
Big Wahoo, 390
Billboard Magazine, 34
Billings, Tommy, 29, 303–4, 313, 648, 679, 696, 699

Biloxi, Mississippi, 78
Binford, Tom, 179, 201, 251, 317, 325, 328, 355, 387, 679, 681
Binion, Paul, 145
Binstock, Mort, 145
Black Brant, 155, 635
Black Brant II, 386
Black Eye, 559
Black Jack, 239, 617
Black Rock Desert, ix, 29, 167, 171–74, 176, 181, 186–90, 200–201, 206–8, 210, 215, 254, 262, 325, 330–31, 364, 374, 376, 390, 392, 441, 469–70, 473, 477–78, 481, 512, 562, 587, 603, 627, 640, 674, 681, 754
Black Rock One, 174–78, 205–6
Black Rock Two, 177–82, 185–87, 201, 206–7, 233, 280
blacksky, 310, 318, 379
Blade Runner, 33
Blanca, Walter, 318
Bland, James, 593
Blatzheim, Jason, 315
Blazanin, Becki, 79–80, 122
Blazanin, Tom, 78, 82, 85, 104, 110, 114, 117, 120, 132, 134, 136, 147–48, 152, 165, 176, 181, 186, 269, 272–73, 286–87, 338, 448, 679
Blue Flame, 365
Blue Jay, 67
blue-ribbon commission, 64–65, 68–69, 71, 75–76, 86, 88, 91, 104–5, 113–14, 130
Blue Thunder, 239, 244, 318, 370, 567, 617
Boeing *B-52 Stratofortress*, 201
Boesenberg, Mike, 103
Bolduc, Dave, 213
Bolene, John, 379
Boles, Dane, 291
Bomarc, 567
Bomb Pop, 387
Bonham, Paul and Andy, 386
Bonneville, Benjamin, 364

Bonneville Salt Flats, 215, 354, 364
Book of the Dead, 535
Booster Bruiser, 557
Booster Vision, 557–58
Bower, Art, 82, 521
Bowled Over, 512
Bowling Ball Loft, 464, 511
Boyette, Rick, 325, 388
Boy's Life, 296
Braye, Randy, 456
Brechbul, Christian, 595
Brekke, Robert, 206
Brewloft, 176
Brigham Young University, 275, 535
Brinley, Bertrand R., 83
Britain, Mark, 559
Brock, Lee, 570, 681
Brown, Bob, 462, 568
Brown, James, 411
Brown, Mark, 639
Brown, Tim, 207, 626, 709
Bruiser, 557
Bruno's Country Club and Casino, 177, 201
Bryant, Ray, 356
Buck, Steve, 171, 173, 205, 207, 276, 393, 459, 626, 679
Buffalo Rocketry Society, 509, 598
Buhler, David, 529
Bullpup, 443
Bundick, Mark, 291, 412, 489, 491, 524, 547–48, 577, 583
Bureau of Land Management (BLM), 174, 742
Burney, Richard, 385
Burning Man, 29, 201, 223, 696
burn rate, 7, 475–76, 549–50, 553–54, 572–76
Burris, Darrell, 567
Butterworth, Manning, 61
Byrd, Wylie A., 435, 729
Byron, Georgia, 324
Byron, Kari, 607, 609, 622
Byrum, Gary, 567

C

Cagle, Earl, Jr., 262, 388
Cailletet, Paul, 308
Cajun Coalition, 622
Caldwell, Doug, 181
California, 8–9, 12, 15, 19, 29, 34–35, 47, 56, 132, 147–48, 163, 171–72, 177, 205–6, 233, 235, 272, 276, 297, 337, 373, 392, 440, 563, 565, 681
California fire marshal, 297
California Rocketry, 3, 14–16, 25–26, 37, 39–41, 43, 45–46, 48–50, 53, 80, 85, 110, 118, 163, 165, 174, 338, 679
California Rocketry Quarterly, 3, 14, 43, 49, 80
Calisto, 460
Camp Pendleton, 273
Canada, 128, 148, 159, 314, 349, 525–26, 528–29, 549, 675
Canadian Association of Rocketry, 525
Canadian Explosives Act, 525
Cape Canaveral, 59–60, 108, 161, 612
Cape Kennedy, 33, 37, 449, 487
Capital Rockets, 360
Car and Driver, 281
Carbine, Adrian, 641, 645
carbon fiber, 125, 180, 365, 465, 515, 622, 625, 627, 631, 642, 647
Carlisle, Orville H., 142
Carter, Don, 87, 332
Carter, Scott, 357
Castle Air Force Base, 201
Catalano, Melinda, 565
Cato, 67, 149, 157, 183, 244, 267, 317, 322, 330, 367, 370, 376, 384, 390, 465, 467, 479, 507, 558, 600, 608, 616–17, 623, 629–30, 636–37, 646, 649
Cato, John, 202, 251, 262, 319, 325, 406
Cat's Eye, 512
Cats in the Cradle, 514
CD3, 634, 638, 641, 644, 655

Cedar City, Utah, 139, 438
cellulose, 342
center of gravity, 92, 101, 128, 140, 260, 295, 300–301, 702
center of pressure, 92, 101, 128, 140, 260, 295, 300–301, 702
Centuri, 8–9, 39, 63, 80, 99, 114, 336, 350, 626
Centuri Engineering, 8, 114
certification, 211
Cesaroni, Anthony, 526
Chaffee, Roger, 108
Challenger, 108, 147, 432
Chandler, John, 559
Chevrolet, 274, 565
Christ, Lyle, 256
Cincinnati, 86–88, 91
Citation, 567
Civilian Space eXploration Team (CSXT), 467–70, 472–73, 475–76, 481, 484, 678, 748
Clapp, Rick, 529
Clark, Mark, 213, 265, 316–17, 365, 514, 559, 681
Clark County, 437–39
Clemons, Lynn, 174
Clemson University, 304
Cleveland, 2, 30–31, 34, 36–37, 116, 126, 400, 591, 642
Cloudbuster, 179–80, 251, 317
Cloudbuster 54, 202
Cloudbuster 110, 179
Cochrane, Tony, 358
Code for High Power Rocketry, 2, 285, 289–90, 293, 301, 346, 363, 403, 408, 446
Code for Model Rocketry, 22–23, 38, 64, 73, 86–87, 105, 117, 134, 163, 230, 243–44, 248, 286–87, 289, 695, 701
Coker, John, 394, 455, 636
Collier's Encyclopedia, 410
Collier's Wonder Book, 372
Colorado, 18, 29, 119–22, 125, 132, 142, 147–48, 150, 153, 167, 179, 182,

186–87, 204–5, 233, 246, 249, 277, 304, 309, 356, 358, 362, 411
Colorado Springs, 61, 116, 119–22, 154, 158, 182, 185, 245
Columbia, 15, 33, 326, 432, 511
Commercial Space Transportation Office, 450, 468
Commission on Advanced High-Power Rocketry, 170, 190–91
Committee on Pyrotechnics, 21, 63, 73, 105, 134, 136, 142, 194, 198, 243, 245, 247–49, 270, 286–87, 291–92, 346, 348, 363, 403, 412
Commodore VIC 20, 269
Community 8, 440
Composite Dynamics, 11–12, 38, 81, 337
Concept 75, 557
confirmation, 111–12, 114, 120, 136, 141, 146, 156, 179, 193, 240–41, 251, 276–77, 294–95, 299, 446, 450
Conn, Jim, 324–25
Connecticut, 20–21, 481, 557
Connecticut Tripoli, 327
Consumer Product Safety Commission (CPSC), 23, 198, 243–45, 249, 287, 291, 363, 492, 748–49, 751
Contests and Records Committee, 318
Convair *B-58 Hustler*, 79, 362
Convair *F-102 Delta Dart*, 296
Cordova, Bill, 367
core sample, 140, 176, 341
Corey, Lee, 364
Cornwell, Jim, 262, 328
Corpuz, Avelino, 437
Cotriss, David, 314
Cotriss, Jim, 213
Covey, Tim, 639
Cow Rocket, 623
Cox, Charlie and Joe, 607
CP Technologies, 349, 501
Crisalli, David, 350
Cross, Lannie, 355
Crossbow CXL, 482

cruise missile, 139, 357, 476, 530, 588
Cruiser, 99
Cuban Missile Crisis, 41
Cullen, Jackie, 595
Culpepper, Virginia, 374, 587
Cumberland College, 615
Cummings, Chris, 367

D

Daedalus III, 515, 517
DairyAire, 621
Daley, Greg, 357
Dallas, 196–98, 234, 288
Damerau, Drake, 513
Danville Dare, 41
Danville III, 290
Dark Star, 607, 731
dart, 140, 632, 637, 641, 643–44
Data Logger, 303
Daugherty, James, 676
Daugirdas, Kip, 675
Dauntless, 315
Davidson, Bill, 328, 353, 358, 369–70
Davidson, Damian, 315
Davis, Arley, 383
Davis, John, 11, 13, 38, 55, 337
DC-3 from Hell, 262
DeBrouwer, Frank, 622
Deep Space Transporter, 184
deflagrate, 497, 504, 549–50, 553–56, 571–74
DeHart, Dan, 559
DeHate, Robert, 514, 641, 668
Delta, 560
Delta III, 560
Delta II Project, 609
Delzell, Carl, 401
Dennet, Mike, 526
Dennett, Bill, 113
Department of Defense, 138, 292
Deputy, Greg, 461, 463
Derkovitz, Les, 636
Der Red Max, 79
detonate, 497, 504, 549, 571, 573
Detroit, 30, 115, 459

Deville, Derek, 355, 379, 464, 473, 522, 540, 648–49, 673
Dickinson, Rick, 568
Dinan, Robert J., 572
Discovery, 147, 432, 496
Discovery Channel, 378, 380–81, 383–85, 453–54, 458–59, 461–63, 499, 511, 604, 607, 623
Discovery Magazine, 258
Disney Magazine, 542
Dixon, Ramona, 107, 119–20, 701
Dixon, Scott, 52, 66, 110, 143, 181, 229, 245, 247, 253, 701
Dizzy Dog Rocket Ranch, 251
Donald, James, 674
Donatelli, Jeff, 87
Donner Party, 364
DOT Section 7887, 63
Douglas, Georgia, 324–25
Dow Jones Industrial Average, 28
Down Right Ignorant, 253–55, 263, 284, 318–20
Draco, 389
Dragon Breath motor, 268
Drake, Terry, 567
DR Hero, 375
dual deployment, 2, 52, 93, 102, 144, 251, 310–11, 401, 622, 696, 699
Dunakin, Ray, 265
Dunbar, John, 329, 620
Dunlap, Jim, 53, 123
Dunseith, Rick, 510
DuPont, Blair, 510
Dutch, Mike, 599
Dynacom *Tarantula*, 317
dynamic propulsion systems, 476

E

Early, Ron, 316
Earth Satellites and the Race for Space Superiority, 361
East Freedom, Pennsylvania, 614
Ebert, Erik, 674
Edison, Thomas, 302
Edmunds, Sterling, 560

Edwards, Harry Thomas, 552
Edwards Air Force Base, 13, 15, 78, 147, 159, 161, 165, 176
Egan, Dick and Lucy, 577
Egan, Joseph, 412
18 U.S.C. § 841(d), 549
Eisenhower, Dwight D., 296
Eiszner, Tim, 284, 320, 324, 617
Ejma, Randall, 283
Elders, Geoff, 511, 513
El Dorado Lake Bed, 342
election redo, 152
E-legal, 8
Eleventh Frame, 512
Ellis, Hal, 441
Embry, Dick, 298, 437, 449, 499, 518, 522, 547, 583, 606, 681
Encyclopedia of Explosives, 554–55
Endeavor, 441, 487
Energon, 254, 318
Enerjet, 8–10, 99, 296, 336
Enerjet 2560, 103
Environmental Aeroscience Corporation, 340, 473
Enzi, John, 500
Enzi, Mike, 451
Erb, Tom, 595
erosive burning, 475–76, 479
Esoteric, 112–13, 124, 144, 156, 253, 276
Estes, 3, 27, 39–40, 62–63, 79–80, 84, 99, 102, 243–44, 247–48, 296, 306, 336, 379, 387–88, 397
Estes, Vern, 17, 65, 112, 142, 196, 278, 291, 537
Estes Industries, 65, 243, 288
Euclid, Ohio, 29
Event Horizon, 559
Eves, Donald, 585
Eves, Steve, 327, 587–89, 596–97, 601–2, 680
exoskeleton, 259, 560, 648
Experimental Composite Propellant, 618
Experimental Development of an Isocyanate Solid Propellant, 620
Experimental Ionosphere, 213

Explorer, 60
Explorer 400, 67
explosives, 292, 343, 402–9, 411–14, 452, 486, 496–97, 499, 503–4, 523–26, 528, 550, 553–56, 572, 580, 582–83, 619, 621
Extreme Machines, 383
Extreme Rocketry, 236, 389, 439, 462, 534, 537–39, 561–63, 679
Eye in the Sky, 144
EZI-65, 103, 201, 276–77

F

FAA, 17, 23, 35–36, 178, 263–64, 366, 433, 435, 437, 451, 468–69, 489, 492
fastest-to-1000-feet, 608–9
Fat Boy, 379
Federal Aviation Administration, 17, 634
Federal Register, 408, 494
Federal Reserve Bank, 521
Federal Rules of Civil Procedure, 485
Fernley, 200
Fillible, Gary, 110, 117, 241, 286
Fireballs, 208–10, 212, 252–53, 318
Firefly, 374
Firestarters, 184, 203
First TV, 454–56, 464–65
Fleischer, John, 303
Flight Control Systems, 303
Flight Systems Incorporated, 162
Flink, David, 52
Florida, 37, 50, 67, 126, 148, 155, 268, 319, 324, 336, 340, 440, 456, 476, 478, 545, 572, 642
Flury, Daniel, 441
Flying Pyramid of Death, 328
Flying Space Cat of Death, 397
Flyrockets.com, 461–62
Forbes, 626–27
Forrestor, Doug, 67
Fort AP Hill-1, 148
Fort Carson, 119–20
Fort Rocket Team, 625–26, 677

Fortune, Bob, 640
FourCarbonYen, 647–49
Fourth Amendment, 405, 486
Frank-en-bolts, 478
Freedom Launch, 2, 324
Freedom Phiter, 388–89
Fremont, John C., 172
Fresno, 276, 283
Fullerton Junior College, 10
Full Metal Jacket, 264–65, 620
Fuscient, 473
Future Directions Committee, 293, 328, 334–35, 344–45

G

G-10, 46, 215, 460
Galando, Joe, 521
Galaxy Hobby, 284
Gallinger, Wayne, 529
Gambling Money Ain't Got No Home, 626
Gardner, Dale, 109
Garibaldi, Jack, 567, 656
Garland, Merrick B., 552
Gates, Eric, 395, 655
Gates Brothers, 443, 567
Gates Brothers Rocketry, 395, 442, 604
Geier, Bob, 69, 74
Gemar, Charles Donald "Sam," 457, 462
Gemini, 5, 31, 60–61, 82, 566
Gemini DC, 566
general assembly, 114, 120, 126, 128, 147, 158, 167, 185, 265, 411, 703–4
General Electric, 466
Geneseo, 509–10, 513–17, 598–99
Gentry, Jim, 184
Georgia, 140, 179–80, 317, 324–25, 355
Gerlach, 32, 171–72, 174–75, 177, 189, 200–201, 205–6, 255, 300, 307, 330, 478
Germany, 20, 266
Gerrard, Doug, 144, 378, 607
Ghiz, Scott, 317
Gianokis, David, 306

Gila Monster, 465, 558
Gilliand, Kathy, 391
Gimble, Jason, 637
Glenn, John, 31, 236
Global Positioning System (GPS), xi, 296, 311, 469, 482, 557, 640, 642, 645, 647–49
Gloor, Mathias, 441
Gloria Mundi, 83–84, 334, 519
Goddard, Robert, 9, 261, 472, 484, 665
Goessel, Kansas, 256
GoFast, 474, 476, 479–83, 485, 505, 514, 648, 678
Go Fast Sports & Beverage Company, 473
Goldstein, Ken, 357
Good, Ken, 38–39, 83, 334, 394, 452, 454, 501, 503, 518, 532, 537, 544, 547, 553, 561, 578–83, 598, 602, 606, 609, 612, 619, 679, 681
Goodin, Kairee, 610
Goodson, Ray, 13
Goodsprings Junction, 563
Gordzelik, Edward Joseph, 429
Gordzelik, Lauretta, 433–34
Gordzelik, Michael, 429, 595, 601, 743, 754
Gordzelik, Patrick, 429–30, 432, 439, 462, 465, 524, 547, 557, 568, 582, 605, 624, 676, 679, 744, 756
Gormley, John, 471
Gort, Herbert, 441
Gosner, Tom, 368
Graduator, 297
Graf, Christoph, 441
Graham, Francis, 46, 82, 87–88, 110, 152, 281, 521, 679
Graham, Nicholas, 376
Grand Island, Nebraska, 586
Grand Slam, 67
Granite Range, 201
Green, Becky and Jim, 674
Green Gorilla, 356, 386, 456, 603
Green Mountain, 362, 750
Green River, Wyoming, 588

Green Tree Motel, 277
Grippo, Andrew, 569
Grissom, Gus, 108, 382
Group Project Q, 629
Grubelich, Mark, 126
Grygar, Rob, 639
Gugisberg, Kurt, 397
Gurstelle, William "Bill," 441, 559
Gustavsen, John, 506
G-Wiz, 310

H

Haas, Bob, 569
Hagensick, Richard, 513, 570, 638, 641, 645
Hall, Darren and Eric, 457
Hall, Rick, 600
Halley's Comet, 28
Halm, Ray, Jr., 380, 387
Handbook of Model Rocketry, The, 363
Hanson, Doc, 568
Harms, Kimberly, 397, 440, 635, 676
Hart, Jim, 317
Hartsel, 29, 120–23, 127–29, 144–47, 154, 167, 182, 184–87, 202, 233, 240, 325, 356, 411
Hartsel, Samuel, 121
Hathaway, Phillip, 510
Hawaiian Punch, 181
Hawk, 97
Hawk, John, 572
Hawk-Dinan Report, 572–74, 619
Hawk Missile, 401, 567
Hayburner, 41
Hayes, Mark, 567, 608
Hayti, Missouri, 350
Heinze, John, 317
Heller, Mac and Steve, 557
Hellfire, 179–80, 203, 365, 719
Henderson, Nevada, 138
Herrick, Ken, 570
Hickman, Homer, Jr., 377
Hicks, Carl, 570
Higgs Farm, 507–8, 592–94, 596
"High Flight," 108

High Power Research, 68
High Power Rocketry, 203, 240, 253, 268, 273, 282–84, 297, 307, 316, 334, 339, 341–42, 348–49, 351, 358, 373, 377–78, 446–49, 522, 528, 532–34, 536, 538, 540–43, 561, 616–17, 679
High Sierra Rocketry, 171, 174–75, 177, 351
High Thrust Video, 374–75
High Venture, 181
Highway 101, 532
Hill Air Force Base, 366
Hillbilly One, 567
Hillbilly Rocketry, 465, 558, 567
Hilton Hotel, 158
Hindenburg, 314, 320
Hi-Tech H45, 149
Hoag, Vern, 570, 601
Hobbs, Mike, 316, 637
Hobby Horse, 283
HobbyTown USA, 283
Hoburg, Woody, 514, 531, 558
Hoey, Bob, 164
Holmboe, John, 102, 125, 132, 181
Holmes, Paul, 514, 682
Home, 542
Home Depot, ix
Honest John, 266, 441, 635–36
Houston, 26, 111, 126, 128, 136–37, 150
Howard Johnson Motel, 120
How to Make Amateur Rockets, 501, 740
Hubble spacecraft, 296
Huber, Geoff, 640–41, 643
Hughes, Curt, 38, 45, 49, 54, 81–82, 110, 153, 520, 679
Hughes Aircraft, 11, 296
Hungarter, Scott, 594
Huntsville, Alabama, 407, 593
Hurlburt, Charles, 185
Husband, Rick, 432
Hyatt Regency, 457
Hybrid, 342
Hybrid Mama, 374
Hyneman, Jamie, 604

Hypertek, 336, 340–43, 388, 401, 440, 515

I

IBM, 269, 303–4
Idaho, 148, 531
Illerbrun, Garth, 526
Industrial Solid Propulsion (ISP), 230–32, 235, 238–40, 244, 279–80, 736, 749
Inman, Bill, 283
Internet, 22, 89, 167, 194, 269, 462, 534, 536, 539, 542, 612, 680
Interstate 80, 200, 366, 564, 588
Intruder, 510
Io, 460
Iris, 386
I-ROC, 327
Irvine, Jerry, 3, 14, 25, 37, 40, 43, 48, 80, 113, 163, 181, 338, 393, 679, 696

J

Jacobs, Jeff, 637, 641, 646
James, Earnest, 159
James, George, 363
Jarvis, Ben, 637
Jarvis, Jim, ix, 649, 673–75
Jayhawk, 397–98
Jean Dry Lake Bed, 563
Jefferies, Blaine, 560
JETEX, 231
Joe Boxer, 376–77, 467
Johnson & Johnson, 98
Johnson, Chris, 30, 69
Johnson, Don, 559
Johnson, Lyndon, 236
Johnson, Roger, 32, 337, 696
Johnson Space Center, 296
Jones, Matt, 283
JPL, 7
Junkyard Wars, 453
Juno, 628
Jupiter, 628

Justus, Robert, 370

K

Kansas, 122, 255–56, 258, 262–64, 314–15, 325, 380, 429, 440, 457–58, 463, 465, 568–69
Kansas Cosmosphere and Space Center, 462
Kármán line, 467, 481, 677
Kaulia, Sabra W., 709–10
Kelly, Bruce, 121, 126, 138–39, 174, 180, 186, 194, 255, 272–73, 275, 282, 284, 287, 291, 293, 301, 325, 334, 338, 341, 352, 373, 434, 446, 449, 452, 491, 521–22, 532, 536–38, 542–43, 583, 606, 616–17, 619, 626, 679
Kelly, Geraldine, 273–74
Kelly, John, 273
Kelly, Lisa, 277
Kelly, Phyllis, 275
KentCon, 39
Kerr-McGee, 138
Ketchum, Mark, 512
Kevlar, 52, 215, 365, 627, 675
Kilby, Bruce, 388
King, Richard, 283, 626, 647
King Viper, 126, 173
King Viper III, 173
Kinney, Nadine, 368, 391, 681
Kline, Korey, 39, 52, 80–81, 103, 110, 163, 308, 336, 339, 343, 345, 393, 401, 473, 478, 626, 677
Kloudburst I, 256
Kloudburst II, 256–57
Kloudbusters, 255–58, 261–62, 264, 266, 429, 434, 457, 568, 681
Knight, Eric, 469
Knowles, Vern, 357, 531, 567
Koenn, Bob, 50
Kolb, Gerald, 184, 459
Kolis, Tom, 161, 164
Kollsman, Paul, 308
Koloms, Deborah "Deb," 598, 601
Kondi, Christopher, 597

Kopplin, John, 41
Kosdon, Frank, 212, 235, 239, 252, 264, 312, 317, 322, 329, 333, 373, 375, 387, 410–11, 454, 537, 621–22, 636, 708
Kosdon East, 356–57, 386–87, 410, 603
Krell, John, 339
Kust, Bill, 85
Kyte, John, 502

L

LaCroix, Ed, 380
Lake Erie, 34
Lake Lahontan, 172, 226, 693
lake stake, 140, 608
Lake Township, 586
Lamothe, Dennis, 210, 253, 263, 266, 268, 318, 623
Lamothe, Terry, 253
L&K Motel, 37, 67, 96, 101, 110, 112, 114, 121
land shark, 139, 608
Lappert, Dennis, 510
Larson, Jerry, 468, 470, 480, 482, 484, 679
Las Cruces, New Mexico, 142, 360
LaserLoc, 252
LaserLoc 3.1, 252
Las Vegas, 8, 12, 63, 109, 128, 138, 150, 167, 229, 233, 244, 299–300, 342, 394, 437, 535–36, 563
launch control officer (LCO), 119, 123, 129, 154–57, 183, 203, 212, 251, 259–60, 316, 320, 366, 368, 377, 397–400, 429, 441
Launch Magazine, 543, 566
Lautenberg, Frank R., 502
Lavigne, Moose, 204, 708
lawn dart, 140
Lawson, Greg, 564
lawsuit, 437–38, 448, 487–90, 492–93, 495–96, 499, 504, 523–24, 547–48, 553, 556, 576–80, 583–84, 588, 602, 606, 612, 619

LDRS-1, 29, 33, 35–47, 51, 53–54, 63, 81, 115, 176, 184, 204, 323, 520, 642
LDRS-2, 49–50, 53–54, 81, 99, 338, 520, 721
LDRS-3, 53–54, 65–70, 86, 100, 338
LDRS-4, 89, 91, 96, 101–2, 104, 107, 109, 137
LDRS-5, 107, 109, 112, 114–16, 124, 276
LDRS-6, 107, 116, 119–21, 123–29, 132, 134, 144, 150
LDRS-7, 128, 140–42, 144–46, 155, 168
LDRS-8, 147, 150, 152–54, 157–58, 167–68, 179
LDRS-9, 167, 182–84, 186–87, 205, 233–34, 240, 277
LDRS-10, 157, 167, 186–90, 193, 201–4, 207–8, 210–11
LDRS-11, 250–51, 255–57, 282, 295
LDRS-12, 255, 257–58, 261–64, 314, 620
LDRS-13, 265–66, 407
LDRS-14, 314, 316, 319, 345, 374
LDRS-15, 323, 328, 345–46, 376
LDRS-16, 357–58, 411
LDRS-17, 354–55, 365–66, 369
LDRS-18, 378–80, 382, 429
LDRS-19, 386–89, 512, 607
LDRS-20, 392–97, 399–400, 442, 448, 454–55, 563, 604
LDRS-21, 436–37, 441, 443, 445, 456, 495, 526, 539
LDRS-22, 429, 456–57, 461, 464, 466, 499, 539, 623
LDRS-23, 509–10, 513–15, 517–18, 539, 598
LDRS-24, 527–29, 531–32, 541, 549, 675
LDRS-25, 557–59, 563
LDRS-26, 563–66, 569, 576
LDRS-27, 568–70, 578
LDRS-28, 598–602, 605
LDRS-29, 183, 607, 609–12, 614
LDRS-30, 622–23
LDRS-32, 673
LDRS-HPR Committee, 69–71, 74–75, 90–91, 105, 114
LDRS WEST, 109
Leander, Jussi, 135
Lee, Bruce, 265, 284, 328, 382, 387, 440, 453, 455, 468, 601
Lehr, Tim, 441, 466, 528, 559, 564, 607
Lemus, Julian, 646
Leninger, David, 283
Lethbridge, Canada, 526, 549, 675
Level 1, 294, 299, 301, 316, 327, 380
Level 2, 295, 299–301, 326–27, 353, 356
Level 3, 295, 299–301, 327–28, 355, 357, 367–68, 370, 379, 386, 395, 397, 440–41, 463, 510, 515–17, 536, 570, 613, 622
Level 3 Certification Committee, 515
Level 4, 295
Level 5, 295
Lewis, William "Bill," 178, 205, 207, 212, 252, 681, 736–37, 742–43, 755
Lewis Research Center, 37
Liberty Project, 505, 562, 727, 743
Life Magazine, 18, 296, 361, 431
Liggett, Larry, 204
Little Blue Pill, 557
Lloyd, Mark, 391, 507, 560
Lockheed *SR-71*, 79
Loehr, Rick, 212, 473
Logan, Gary, 355
Loki Research, 511, 591–92, 596, 641, 644
Long Burn Express, 370
Long Gone II, 356
Long Island Advance Rocketry Society, 509
Long March, 513
Lord, Dan, 510
Los Angeles Times, 552
Lots of Crafts (LOC), 29, 53, 68, 96, 99–103, 112–13, 121, 124, 126,

137, 140, 142–44, 146, 149, 156, 173–74, 184, 201, 244, 252–53, 262, 276, 297, 319, 327, 357, 366–67, 369, 389, 394, 459, 557, 586, 602, 617, 680, 706
Lottering, Nic, 675
Lousma, Jack, 511
low explosives users permit (LEUP), 405–6, 408–9, 414, 486, 492–93, 497, 499, 505, 523–24, 528, 545–46, 747
Lucerne, 12–13, 25, 35, 56, 71, 93, 120, 129, 139, 148–49, 163, 167, 171, 173, 239, 337, 339, 392–93, 395, 397, 399, 607–8
Luehman Ridge, 15
LUNAR, 632
Lusty Corn Maiden, 439
Lyngdal, John, 635
Lyons, Jane, 492, 580

M

Mach 2, ix, 181, 322, 530, 631, 640, 644
Mach 3, 570, 631, 640, 645
Mach 5, 479–80, 483
Mach Buster, 213
Mach Madness, 570
Mad Max, 369
Magnalite, 316
magnesium, 437–38, 603
magnetometer, 482
Magnum, 100–102, 369, 381, 459
Mailer, Norman, 627
Malsch, Martin, 578
Man in Space, 82, 295
March, Dale, 525
Markielewski, Bruce, 262
Mark II, 360
Markowitz, Art, 155, 257, 344, 681
Martin Marietta, 296
Maryland Delaware Rocketry Association (MDRA), 505, 507–10, 562, 591–93, 598, 623, 743

Massachusetts Institute of Technology (MIT), 185, 236, 411, 491, 620, 646
Maxi-Alpha, 388
Maxi-Mosquito, 155
Max-Q, 617, 635
Mays, Willie, 28
McBurnett, Mike, 560, 570
McConaghy, Chuck, 396
McConaughy, Patrick, 316
McCreary, Harry and Charlotta, 614
McCreary, Terry, 352, 390, 449, 573, 679
McCune, Larry J., 403
McDonnell Douglas, 609
McGilvray, Neil, 355, 390, 457, 505, 514, 542, 559, 562, 567–68, 591, 603, 605, 623, 680
McKinlay, Jerry, 641
McLaughlin, Alex, 316, 638, 663
McMacken, Steven, 471
McNally, Kerry, 463
McNeely, Brent, 439, 441, 462, 534, 536, 542, 561–62, 679
Mechanix Illustrated, 18, 142, 359
Medicine Wheel, 41
Medina, 34–38, 41–42, 45, 49, 53, 66, 71, 81, 86, 89, 93, 107, 109–10, 112–13, 115–16, 121, 125, 127, 165, 182, 215, 258, 323, 325, 336, 338, 518, 624–25, 642
Mega-Motor-Eater, 328
Megg, 559
Meinhardt, Scott, 356
Memphis, Tennessee, 273
Memphis Belle, 509
Mercer, Kelly, 391, 507
Mercury, 5, 31, 60, 82, 383, 614
Mercury device, 215
Mercury Redstone, 125, 132, 306, 319, 382, 440, 567
Meredith, Robin, 213, 317, 365, 559, 637
Merrill, Claude, 13
METRA, 598

Meyer, Dan, 104, 230, 232, 234
Michael, Andrew, 595
Michael, Dan, 510, 595, 601, 657
Michaelson, Jodi, 374–76
Michaelson, Ky, 372–73, 383, 414, 453, 467, 476, 484, 514, 588, 625
Midwest Power, 2, 681
Mile Square Park, 161
Milky Way Galaxy, 32
Miller, David J., 284
Miller, Ed, 514
Miller, Jamie and Rosemary, 59
Miller, J. Patrick, 24, 55, 679
MinFlous, 674
Minnesota, 4, 19, 372–73, 383, 470, 476–77
Minuteman, 6, 94, 380
Missile Works, 93, 310, 383, 506, 515, 637
Mitchell, James, 350
Mitchell, Kevin, 386, 390
Mobley, Darrell, 536, 579, 621, 680
Model Missile Association, 362
Model Missiles Inc., 361
Model Rocketeer, 24–27, 33–34, 37, 48, 362, 680
Mojave Desert, 2, 13, 17, 78, 350, 564
Monai, Ken, 184
Money Pit, 674
Mongoose, 622
Monocopter, 103
Morgan, George, 210
Morgan, Greg, 675
Mormon Church, 274–75
Morozumi, Pius, 300, 307, 626, 632
Morrow, Bill, 212
Mosquito, 262, 328, 514
Mother Lode, 143, 145
Motor Listing Committee, 117
Motor Testing Committee, 751
M Project, 459–60
MRC2, 383
Ms. Riley, 622
Mueller, Peter, 622

Mund, Chuck, 69, 81, 102, 110, 125, 132, 145, 181, 191, 252, 393
Mund, Eric, 181
Murdock, Keith, 181
Murray State University, 615
My Mind's Eye, 378–79
Mythbusters, 604, 607

N

N1, 394–95, 397, 455, 638, 674
Nafzinger, Rick, 257, 568, 622
Nafzinger farm, 257, 264
Nancy One, 180
NAR 33
NARAM, 24, 37, 42, 49–50, 119, 150–51, 167, 198, 234, 352
NARAM-31, 191–92
NARAM-38, 352
NAR board of trustees, 26, 56, 65, 70, 74, 117, 190, 195–96, 235
NAR burn tests, 247–48
NAR Safety Code, 26, 31, 34, 45, 55, 130, 359
NARWIN 3, 56–58
NAR winter nationals, 56
NASA, 4–5, 50, 61–62, 162, 165, 261, 296, 336, 378, 382–83, 432, 443, 457, 468, 514, 519, 560, 587, 595, 612, 633, 677
NASA Dryden Flight Research Center, 161
National Association of Rocketry (NAR), 3, 16–17, 20–27, 29–31, 33–35, 37–39, 41–51, 55–59, 61–66, 68–76, 79–81, 84–86, 88–91, 93–94, 104–6, 112–15, 117, 119, 129–37, 141–42, 146–48, 150–51, 161–63, 166–70, 182–83, 190–99, 230–31, 234–35, 243–49, 285–86, 288–91, 294–95, 300–302, 337–38, 345–46, 362–64, 406–7, 411–12, 461–62, 485–93, 495–500, 504–5, 546–49, 553–54, 561–62,

577–79, 581–84, 598–99, 745–46, 749–51
National Fire Protection Association (NFPA), 21–22, 26, 64, 105, 117, 234, 243, 247–48, 286–87, 289–91, 294, 363, 407–8, 492
National Geographic, 280, 456
National Hot Rod Association, 373
national launch, 116, 182, 207, 215, 250, 255, 323, 325–26, 358, 366, 386, 393, 429, 509, 540, 557, 563, 565, 568–69, 598, 609, 673
National Sport Launch, 324
Nebraska Heat, 357, 718, 742
Necessary Evil, 129, 397
Nellis Air Force Base, 12
Nelson, Craig T., 746
Nelson, Mike, 87, 101
Netherlands, 622
New England, 19, 252, 387, 594
Newman, Kent, 641, 681
New Mexico, 59, 61, 97, 142, 148, 261, 359–60, 367, 430–31, 433, 443, 458, 534
New Orleans, 293
Newport, Curt, 641, 644
Newton, 7, 295
Newton seconds, 2, 26, 31, 34, 45, 70, 143, 146, 205, 289, 322, 349, 443, 463, 505, 557, 559–60
Newton's third law of motion, 10
New York, 83, 159, 182, 210, 236, 380, 440, 461, 491, 509, 516–17, 577, 598–99
New York Times, 503, 577, 634
NFPA 1122, 23, 26, 73, 86–87, 89, 105, 192, 194–95, 198, 235, 248, 287, 289, 292, 348
NFPA 1125, 134
NFPA 1127, 285, 289–93, 301, 348, 363, 403, 408, 446, 450
Niche, Tav, 149
Nightingale Mountains, 201
Nightmare, 145
Nike Hercules, 203, 401, 441

Nike Smoke, 112, 202, 379, 464, 510, 515
Nike Tomahawk, 251
1941 Historical Aircraft Group, 509
1942 *Stearman*, 559
nitrous oxide, 339–40, 342, 642
Nixon, 200
Noble, Richard, 173
North American *XB-70*, 79
North Carolina, 390, 507, 562
North Coast Rocketry, 96, 107, 137, 146, 303, 310, 315, 338
Nossman, Paul, 570
NOTAM, 36, 527
Nowaczyk, Gene, 379, 639–42
N-Sane, 390

O

Oberg, Dean, 210
Oberlin, Ohio, 36
O'Brien, John, 119, 124, 144
Obsession, 388
O'Connor, J. P., 86
Octoberfest, 149, 393
October Sky, 378, 380, 441, 593, 622
odd roc competition, 609, 622
Of a Fire on the Moon, 632
Ogino, Charlie, 622–23
Ogino, Guy, 623
Ohio Nuclear, 98
Old Fire Face, 251
Oldham, Wedge, 397, 401, 441, 565, 568, 635
Olsen, 379, 515
orange book, 499
Orangeburg, 29, 284, 323–25, 327, 345–46, 376, 387, 389–90, 392, 467, 488, 509, 512, 523, 588, 607
Orbital Transport, 51, 53
Oregon, 1, 172, 253
Orell, Rolf, 531
Oren, Rolf, 355
Organized Crime Control Act of 1970, 404
Orient Express, 514
Orr, Graham, 610

Orr, Jay, 213
Orsack, Matt, 675
O'Sullivan, Jerry, 389, 511, 514–16
Otta, Matt, 32
Oullette, Mark, 389
OuR Project, 323, 329–31, 410, 588, 603, 620, 627, 676
Outback Thunda, 623
outer space, 4, 59, 82, 172, 178, 216, 355, 372, 398, 457, 467, 531, 625, 640, 650, 674, 676–77
Overmoe, Brad, 367
Owens, Darren, 379, 434
Ozark, 644

P

Pacific Rocket Society, 210
Pad 39a, 487
Panhandle of Texas Rocketry Society (POTROCS), 434, 436–37, 439, 445, 557, 559, 729
Paralyzer, 510
Park County, Colorado, 121
Parker, Jim, 557, 569
Parker, Warren, 406
Part 101, 577, 693
Patriot, 251, 256, 262, 368, 510, 570, 601
PBAN, 238
PC Magazine, 542
Peacekeeper, 94
Pearce, Scott, 40, 51
Pearson, Chris, 2, 30, 34–35, 39–40, 46, 49, 52, 56, 63, 70, 81, 89, 93, 96, 99, 103, 107, 115, 123, 125, 127, 132, 176, 186–87, 215, 338, 520, 595, 624, 642, 679, 681, 696–97
Pearson, Florence, 30
Pearson, Thomas, 30
Penelope Pig, 251
Pennsylvania, 16, 24, 30, 46, 77, 83, 85, 119, 121, 132, 176, 303, 490
PEPCON, 138–39
Perfectflite, 93, 310
Performer, 367

Perpetual Altitude Trophy, 625–26
Perry, Rick, 377
Pershing, 37, 67
phenolic, 3, 10, 157, 232, 238, 297, 357, 460
Phillips, Ron, 397
Phobos, 460
Phoenix, 318, 352, 510, 674
Phoenix missile, 318
Phoenix XL, 560
Piester, Lee and Betty, 8, 114
Pigs in Space, 251
Pinhead, 387
Pinky and the Brain, 368
Piper, 639–41
Piper, Chuck, 13, 138
Pittsburgh, 78, 80–83, 85, 87–88, 91, 122, 126, 281, 518, 521
Pitzeruse, Rich, 515, 517
Plaster Blaster, 610
Platt, Michael, 291, 405, 412
Point 39 Productions, 388, 680
Polaris, 6, 94
Polish Rojo, 465
Popular Mechanics, 18, 616
Popular Science Magazine, 46, 99, 173, 179, 201, 461, 616
Portales, 59–60
Porthos, 396, 442, 567, 601
potassium nitrate, 6, 83, 305, 351, 433
Potter, New York, 598
"Power Freaks Arise," 3
President's Award, 514
Price, Gary, 121, 241, 276
Primera, 470–74
Primera Technology, 470, 725
Primm Valley Resort and Casino, 563
Princeton University, 237
Prior, Phil, 329–30, 356, 620
Prodyne, 92, 154, 157, 185
Progressive Propulsion, 317
Project 463, 318–21, 505
Project Odyssey, 607
Prometheus, 641

propellant-actuated device (PADs), 6, 404, 411, 413–14, 486–87, 492, 496–97, 504, 545, 583
Propulsion Dynamics, 157
Propulsion Systems, 349
Proteus 6, 641
Proteus 6.5, 641
Proteus 7, 644
Public Missiles Systems, 184, 357, 460–61, 616
Purple Haze, 356
Pursley, John, 234
Purvis, Jeff, 456
Pyramid Lake, 200, 674

Q

Q-Reaction, 641
Qu8k, 648–50, 670–71
Quest, 291
Questionable Mental Health, 638, 663

R

Race Head Service, 274
Rack Rocket, 519
Raden, Lou, 524
radial burn rate, 573
Radio Electronics, 303
Rahkonen, John, 157
Raimondi, David, 674
Ramada Inn, 196, 401, 531
Rancor, 390
range safety officer (RSOs), 119, 123, 129, 183, 259–60, 399, 530, 569, 613
Raumberger, Scott, 371
Reaction Labs, 67, 149, 154
Ready to Rumble, 387
Reagan, Ronald, 15, 33, 108, 490
Ream, Bradley, 157
Redd, Randall, 127, 157, 185, 251
Red Devil, 445, 560, 570
Redemptive Power, 391
Redeye, 358
Red Glare, 2, 562, 593, 681

Redline, 399, 401, 514
Red Rhino, 603
Redstone, 306, 382–85, 440, 567
Redwing, Minnesota, 372
Reed, Allen J., 85
Reese, David, 394, 624, 681
Regulus, 557
Reloadable, 246, 249
Reloadable Motor System (RMS), 239, 327, 747
Rempel, Tim, 526
Repcheck, Randy, 451
Return to Sender, 357
Rhode Island, 19
Richland Township, 77
Rickwald, Ron, 559
Right Stuff, The, 173, 696, 707, 754
RIT Inn and Conference Center, 514
Ritz Hotel, 559
RMS/Hybrid, 342
Roar at the Shore, 41
Robby's Rockets, 266
Roberson, Steve, 325, 681
Roberts, Mary, 288
Roberts, Richard W., 489
Robinson, Dar, 375
Robinson, Paul, 316, 329, 387, 410, 445, 603–4, 620
Robinson, Robert, 266
Rocket Boys, The, 377, 593, 723
Rocket Challenge, 464–66, 523
Rocketflite Silver Streak motors, 262
Rocket Mail, 521
Rocketman, 357, 366, 373–74, 377, 515, 650
Rocket Manual for Amateurs, 83, 285
Rocket Men, 693, 699
Rocket Newsletter, 625–26, 679
rocket propulsion lab, 13
Rocketry of California (ROC), 563–65, 611, 681
Rocketry Planet, 680
Rockets, x, 484, 542–43, 558–59, 562–63, 567, 592, 598, 631, 680
ROC Lake, 526–27, 529, 531–32

ROCstock, 2, 621, 681
Rogers, Chuck E., 126, 143, 145, 152–53, 158–60, 165, 168, 176, 180, 185–87, 192, 204, 208, 210, 271, 278, 288, 291, 300, 312, 331, 337, 393, 405, 450, 482, 626, 677, 679, 681, 717, 748
Rogers, Cynthia, 159
Rogers, Jim, 159
Rogers, Mark, 586–88
Rollins, Shannon, 515
Roosevelt, Franklin D., 97
Roosevelt, Theodore, 676
Rosenfield, Gary, 4–5, 8, 10, 14, 16, 29, 32, 38, 48, 53, 55, 63, 66, 72, 81, 94, 110, 112, 134, 146, 163, 169, 173, 187, 191, 229, 235, 239–40, 243, 248, 263, 273, 279, 287, 291, 333, 337, 342, 349, 366–67, 379, 393, 401, 407, 411, 438, 442, 506, 536–37, 575
Rosenfield, Melodi, 32, 104
Rosenfield, William, 4–5
Rosson, Jim, 329, 356, 365, 410, 603, 620
Roszell, Mark, 387
Rouse, Tom, 632, 674
Rowe, Ed, 560, 570
Russell, Ben, 596
Russell, Chas, 112
Russell, James, 607
Russo, Damian, 328, 358, 369
Russo, John, 510

S

SAAB Missile, 371
Sackett, Chuck, 155, 253, 318, 505
Sacramento, 29, 94
safe distance Table, 118, 290, 292–93
Safety and Motor-Testing Committee, 126
Safety Rocket, 624
Salter, Susanna M., 258
Salt Lake City, 128, 275, 366, 588
Saturn IB, 307, 319

Saturn V, 15, 37, 367, 370, 394, 432, 508, 570, 587, 589–93, 595–98, 601–2, 680
Save Rocketry Now, 459, 461–62, 514
Savoie, Denise, 526
Sawyze, Allen, 256, 735
Scarlet to White, 512
Scarpine, Jim, 387
Schaeffer, Dave, 440
Schatzer, David S., 403
Schechter, Andy, 387
scholarship, 490, 535, 605
Scholl, Curtis, 284
Schoner, Bob, 386
Schultz, Deborah, 29, 749, 755
Schultz, Ron, 53, 97, 110, 112, 125, 132, 140, 142, 145, 148, 184, 276, 319, 367, 680
Schumacher, Fred, 507, 593
Schumer, Charles E., 502–3
Science Channel, 607–9, 622
Scientific American, 18
Scion, 639
Scorpion, 176
Scrapheap Challenge, 453
Selmi, Bruno, 177
Senate Bill 724, 501
Service Master, 275
Seventh Element, 400
Seven-Toe-Dawg, 441
Shaffer, Richard, 20
Shamu the Killer V-2, 380
Shannon, Steve, 532
Shatner, William, 41
Shenosky, Larry, 235
Shepard, Alan B., 31, 306, 382, 384
Shepard-Churchley, Laura, 384
Shock, Ken, 355
Sicker, John, 319
Sidewinder, 12
Sienkiewicz, Caz, 155
Sienkiewicz, Martha, 191, 213, 252, 749
Simpson, Charles, 367
SK3J, 252–53
Skidmark, 387, 566, 570

Sky Angle, 458
Sky in My Eye, 514
Sky Maven, 570
Sleeter, David, 433, 616
Slobodian Avenger, 51
Small Balls, 390–91
Smith, Denny, 641
Smith, Kevin, 339–40
Smith, Larry, 324–25, 717
Smoke Creek Desert (or Smoke Creek), 1–2, 13–14, 17, 24–25, 30–32, 35, 46, 71, 115, 167, 171–72, 174, 177, 625, 642–43, 674, 739, 741, 746
Smokey Sam, 113, 124, 144, 154, 181, 202, 253
Smokin' Rockets, 510
SNOAR, 29–30, 34–35, 96, 99, 107, 116
SNOAR News, 31, 37, 57, 151
Sobczak, Randall R., 334
"Somewhere over the Rainbow," 463
sonic boom, 252, 330
Sonotube, 256, 319, 369, 383
SORTA, 624
Soucy, Guy, 145, 514
South Carolina, 148, 215, 304, 323, 325, 376, 386, 467, 588
Soviet Union, x, 34, 59, 82, 361
Space Launch Act of 1984, 363
Space Ordinance Systems, 337
space program, x, 5, 15, 50, 59–60, 78, 108, 161, 274, 336–37, 445, 468, 519, 531, 585, 614, 642
Space Race, x, 5, 236, 266, 361, 587
Space Shot 2000, 469–70
space shuttle, 50
space shuttle *Atlantis*, 642
space shuttle *Challenger*, 108
space shuttle *Columbia*, 15, 33, 326, 432, 511, 694
space shuttle *Endeavor*, 487
Sparks, Ken, 594
Spartacus, 559
Special Devices Incorporated, 337

Spieth Road, 34–35, 66, 107, 697
Spinal Tap, 560, 570
Spirit of America, 316
Sport Rocketry, 353, 543, 561–62, 680
Sports Illustrated, 281
Springfield, Ohio, 585
Sputnik, 17–18, 46, 59, 82, 236, 296, 361, 500, 525
Stafford, Walt, 445, 560, 570
Standard Missile, 559
Standish, Chris, 440
Stanford University, 633
Starburst, 617
Starchaser, 379
Starfinder, 213
Stark, Jamie, 586
Starshot, 204
Star Trek II, 33, 41–42
State Highway 447, 171, 200
Stateline Hotel, 355
Stateline/Silversmith Hotel & Casino, 366
statute of limitations, 487–88, 494
Steele, Matt, 38, 338
Stephens, Ian, 529, 531
Stephenson, Neal, 365
Stewart, Alene, 430
Stewart, Mary Elizabeth, 430
St. Francis College, 615
Stine, G. Harry, 17, 56, 61, 105, 134, 142, 169, 235, 243, 286, 302, 359
Stone, Al, 386
Strange Message from Another Star, A, 158
Strato Cobra, 510
Stratospheric Dreams, 319, 321–22, 329
Streak, 336
strontium, 603
Stroud, Dan, 441, 445, 465
Sullivan Lake, 525–26
summary judgment, 496–97, 499, 503, 509, 524, 544–46, 556, 571, 579–80, 582
Summerfest, 41, 239, 277, 374, 393
Sumo, 442–43, 601
Super Blue Propellant, 390

super grain, 474
Super Mario, 440
Super Sod, 324–26, 386
Sweden, 30, 314, 355, 368, 589
Sweet T, 567
Swenson, Tom, 283

T

T-6061, 627
Talon, 607
Tarantula, 317
Tatel, David S., 552
Tavares, Chris, 131
Taylor, Jeff, 356, 370, 505, 512, 591, 641, 644
Taylor, John, 387
Team Numb, 638, 663
Team Rage, 567
Team WAC, 315, 358, 369
television, 2, 5, 31–32, 108, 160, 330, 350, 373, 377, 383, 430, 452–56, 461–66, 500, 519, 604, 607–9, 622
Tentative Interim Amendment, 243, 408
tenth anniversary LDRS, 187, 215, 250, 393
Terex, 593
Texas, 18, 137, 401, 429–30, 432–33, 435–36, 439, 441–43, 462, 495, 557, 563
theodolite, 302
Thermalite, 297, 315
thermite, 506, 568, 595
thermoplastic, 342
Thiokol, 6–7, 157
Thirsk, Robert, 531
Thompson, Sonny, 345, 353, 390, 434, 628
Thor, 566–67, 628
Thor-X, 566
ThreeCarbYen, 674
3/48 Rule, 130–31, 182, 192, 346
Thrust 2, 173
Thrust SSC, 173

thrust stand, 168–69, 352, 476
Thuering, Juerg, 368, 395
Thunderbolt 2B, 317
Thunderflame, 350, 353–55, 370, 390–91, 434, 562, 628
Tide, 466
Tiger Team, 192, 235, 278
Time Magazine, 280
Tindell, Edward, 117, 286
Titan, 213, 361
titanium, 317, 387, 603
TooCarbYen, 673
Top Gunn, 125, 132–33, 149, 262
Torrey Farms, 598–600
Tortora, Gary, 622
Total Obsession, 358
Transolve, 303, 310
Triano, David, 637
Triple Threat, 641, 645
Tripoli Advisory Panel (TAP), 298–301, 446, 450
Tripoli Alberta, 527
Tripoli Anaheim, 393
Tripoli Atlanta, 324
Tripoli Class 3 Committee, 451
Tripoli Federation, 46, 82, 282, 334, 518–19
Tripoli Houston, 136–37, 158, 559
Tripoli Los Angeles, 393
Tripoli Middle Tennessee, 346–47
Tripoli Minnesota, 570
Tripoli Nebraska, 382, 387
Tripoli North Carolina, 390
Tripoli Pittsburgh, 148, 509, 521
Tripoli Report, 271–72, 279, 292–93, 406–9, 505, 522
Tripoli Research, 346–48, 354
Tripoli Research Interim Launch Rules, 347
Tripoli Rocketry Association, Inc., 551, 571, 729
Tripoli Rocket Society, 85, 87–88, 90, 92, 104–5
Tripoli Safety Code of 1988, 141
Tripoli San Diego, 610–11

Tripoli Science Association, 82, 85
Tripoli Southern Ontario, 509
Tripoli Tampa, 340
Tripolitan, 82, 84–85, 88, 108, 111, 120, 131, 136, 152, 177, 194, 208, 239–40, 269–73, 276, 278–82, 297, 448, 538, 679
Tripoli Utah, 365
Tripoli Western New York, 598
Trojanowski, Kevin, 439, 622
Truckee River, 200
Truly Recyclable Motors, 240, 242, 410
Tulia, Texas, 435
Tyndall Air Force Base, 556
Tyson, Mike, 570, 580–81

U

Udder Madness, 457, 466, 623
Ultimate Endeavor, 441
Ultimate Max, 132–33, 184
Union Carbide Corporation, 159
United Kingdom Rocketry Association, 637
United States Air Force, 4, 11, 55, 78, 201, 296, 321, 432, 449, 677
United States Air Force Academy, 61, 337
United States Air Force Research Laboratory, 556
United States Constitution, 405, 621
United States Court of Appeals, 548
University of Central Florida, 204
University of Georgia, 615
University of New Mexico, 61
Upscale Rocketry Team, 566
Upton, Art, 557, 681
Urinsco, Ron, 317, 375
Uroda, Frank, 184, 458–59, 461, 514, 537, 679
U. S. News & World Report, 542
U. S. Rockets, 16, 46, 67, 92, 146, 184, 212
Utah, 6, 19, 121, 126, 215, 251, 275–76, 364, 438, 529

Utley, Robert "Bob," 355, 514, 542, 559, 591, 605, 681

V

V-2, 80, 97, 367, 380–81, 395, 397–99, 455
Vanguard, 59, 628
van Norman, John, 604
Vaughn, Mike, 370
Veet-1, 310
Vega, 615
Velledor, Christian, 646
VideoRoc, 145
Vietnam War, 97, 205, 237, 304, 449
Viking, 359, 458
Virginia Polytechnic Institute, 615
Visijet, 338
von Braun, Werner, 20, 295, 484
von Delius, Curt, 674
Vosecek, Ken, 241, 257, 334
Vostok, 319
Vulcanite, 394

W

WAC Corporal, 155, 315–16, 623
Wadsworth, Will, 516
Wagner, Bill, 440
Wagner, Mike, 34, 116
Wagner farm, 35, 42, 66, 107, 114, 165
Wallace, Fred, 390
Walston, 626, 644
Walton, Reginald Bennett, 490
Waltzing Matilda, 623
Ward, Mike, 155, 184, 253, 318–21
Warlock, 121, 156
Warp Factor, 124–25
Warp Factor II, 144
Wasatch Rocketry Ascender, 185
Washington, 1, 413, 440, 490, 502, 532
Washington, D.C., 70–71, 74, 142, 406, 412, 485, 487, 489–91, 499, 548, 571, 577, 579
Waters, Rick, 440
Way High, 626

Wayside, Texas, 557, 563
weather channel, 28
Webb, Charles, 355
Weber, Mark, 39, 86, 332
Weese, Brian, 367, 370
Weikamp, Tom, 87
West, Scott, 529
Westinghouse, 84, 304
West Wendover, 368–69
What's Up Hobbies, 564
Whitakers, 390–91, 507, 562
White, Edward, 108
White Lightning, 146, 239, 315, 386, 442
White Sands Missile Range, 61
Whitney, Steve, 394
Whittaker Corporation, 12
Whittle, Sarah Fallon, 518
Wichita Eagle, 385
Wickman, John, 349, 501
Wilkerson, Jim, 638
Wilkey, Duane, 379
Wilkins, David, 623
Williams, Chris, 607
Williams, Michael, 596
Wills, Rick, 266
Wilsey, Tim, 592, 730–31
Wilson, Brad, 512
Wilson, Quentin, 379, 381, 441
Wimmer, Andrew, 646
Winnemucca Lake, 200
Wisconsin, 160, 283–84
Woerner, Andy, 394, 397, 455, 564–65
Wolfe, Tom, 31
Wonderful World of Disney, 295

Wood, Bill, 169, 241, 292, 337, 339
Wood, Lloyd, 510
Wood, Woody, 637
Woodrum, Lyle, 387
Workbench 2.0, 675
Worthen, Mike, 532
Wright, Brad, 641
Wright, Darren, 457, 505–6, 568, 635
Wright Brothers, 343
Wynn, Brent, 328
Wysack, Dave, 368

X

X-15, 78–79, 159
X-30, 165, 440
X-ray, 519
X-Wing Fighter, 356

Y

Yeager, Chuck, 78
Youghiogheny River, 84
Young, Neil, 315
Yucca Mountain, 577, 584

Z

Z-4, 140
Zareki, Richard, 144
Zelienople, Pennsylvania, 132–33, 140, 148
Zeus, Ruler of the Sky, 441, 445
zookeeper, 603
Zupan, Dave, 441
Zupnyk, Larry, 358, 369

Printed in the United States
By Bookmasters